COPPER

Its Trade, Manufacture, Use, and Environmental Status

Günter Joseph

Edited by
Konrad J.A. Kundig

**The Materials
Information Society**

International Copper Association, Ltd.

First printing, June 1999

Library of Congress Cataloging-in-Publication Data

Joseph, Günter
Kundig, J.A. Konrad
Copper
Its Trade, Manufacture, Use, and Environmental Status
Includes bibliographical references and index.
I. Title
TA169.5.T56   1998   620′.00452-dc21   98-29456

ISBN: 0-87170-656-3
SAN: 204-7586
ASM International®
Materials Park, OH 44073-0002

Printed in the United States of America

ASM International Technical Books Committee (1998-1999)

Contents

Preface

This book was commissioned by the International Copper Association, Ltd. (ICA) to provide a comprehensive overview of technical and commercial topics that affect copper's place as a useful engineering material. Copper is unique among modern metals in that it has been in use for more than 8000 years, yet it is still in the forefront of technology. Copper, more than any other metal except perhaps gold, is intimately related to global economic conditions; this fact is expressed by the importance of the metal in human endeavor.

To understand copper as a modern material of commerce and industrial importance, it is necessary to grasp the complexity that attends all aspects of bringing the metal to the ultimate user: its native properties, the technologies involved in the manufacture of its products, and the ways in which this manufacture influences its ability to efficiently utilize its properties. It is also important to know and to understand copper's role in the human and natural environment. With the exception of geology, mining, and extractive metallurgy, which are amply covered elsewhere (see for example, A.K. Biswas and W.G. Davenport, *Extractive Metallurgy of Copper*, 3rd ed., Pergamon Press, Oxford, Tarrytown, NY 1994), this book attempts to satisfy these needs.

Although the book contains considerable numerical data, it is intended more as an encyclopedia than a handbook because the data presented were selected as being representative. Material was taken from the scientific literature and from industrial sources, including publications by industry-sponsored copper information and development centers located in various countries (The copper centers' addresses are listed in Appendix I).

The book begins with an overview on the commerce of copper because trade in the metal is so closely bound to its viability as an engineering material. Following chapters deal with the properties of copper and copper alloys because these properties make it such a useful material. There are also chapters on the fabrication of copper products, the products themselves, and the applications, which provide utility in industry and everyday life. Finally, it is acknowledged that copper, as one of the so-called "heavy metals," is subject to scrutiny on environmental and human health grounds. A chapter describing contemporary knowledge and understanding of these subjects is, therefore, included.

This book was assembled by Professor Günter Joseph at the Institute of Investigation and Testing of Materials, University of Chile, Santiago. Preparation of the book was conducted under the oversight of an international editorial committee made up of industrial and academic authorities in the various subjects. The content was edited by Dr. Konrad J.A. Kundig, a metallurgical consultant and former editor of *The Journal of Metals*. The International Copper Association, Ltd. wishes to express its appreciation to all who have been associated with the production of this important contribution to the technical literature.

ICA is the leading international organization for the promotion of copper. As such, it is responsible for the development and dissemination of information about copper and its uses. Additional information can be found on *The Copper Page,* http://www.copper.org.

William H. Dresher, Ph.D.
Vice President, Technology
International Copper Association, Ltd.
New York, NY

Trade

History of the Use of Copper

Prehistoric

The discovery that metals could be obtained by smelting has been ranked in importance with the discovery of fire as one of the milestones in man's history. In particular, it was the development of copper metallurgy that contributed so greatly to the development of early civilization and culture. Mankind first started to use copper for domestic implements during the Chalcolithic period (8000 B.C. to 4000 B.C.) — the transitional period from the Stone Age to the Metal Age (Ref 1). Some of the first evidence of man's use of copper have been dated to the 9th millennium B.C. in the discoveries of copper trinkets from Iran. Figure 1.1 shows the middle eastern sites of Timna, Kition, Catal Huyuk, and Cayonu where archaeological evidence of early copper smelting has been found (Ref 2).

Copper exists in nature in both native and combined states. All early copper artifacts were formed from native copper. Archeological metallurgists, working on the basis of arsenic and nickel contents of early artifacts, have verified that, in fact, some of these artifacts had been fabricated through melting procedures. However, extensive use of copper had to wait until the recovery of the metal from its ores.

Malachite was probably the first metallic ore smelted on an important scale. To smelt copper from malachite requires a temperature of at least 1083 °C (the melting point of pure copper) and a reducing (oxygen-poor) atmosphere. At one time, archaeologists suggested that neolithic man could have accidentally satisfied these metallurgical conditions when he used malachite stones to surround his camp fires. This suggestion has now been discarded as tests conducted by the Institute of Archaeometallurgical Studies in London revealed that naturally aspirated wood or charcoal fires can only attain temperatures between about 600 and 700 °C (1112 and 1292 °F). This is hot enough to smelt lead, which has a melting point of 327 °C (621 °F), but not hot enough to reduce and melt copper (Ref 2).

It is more likely that the required smelting conditions were first created in pottery furnaces. Pottery and metal smelting appear in Neolithic period findings in about the same time period. The oldest known pottery works, supplying settlements such as Catal Huyuk in Anatolia (now Turkey), date from 9000 to 10,000 years ago. It appears to have become known by then that when clay is fired at about 450 °C (842 °F), it sinters and becomes irreversibly hard and waterproof (Ref 2).

Archaeological discoveries suggest that pots were first heat treated in open fires. Later, pots were stacked on top of the fuel, and after ignition, the assembled pile was covered with earth. This was the genesis of the so-called "beehive" furnace, which is still used today to make charcoal and distill coke. Man then discovered that efficiency could be improved by building a permanent cover and a built-in flue, which led to the development of the pottery kiln around the beginning of the 6th millennium B.C. With its thick walls and flue-assisted natural draft, this furnace could maintain temperatures well in excess of 1000 °C (1832 °F) for hours. At about this same time, potters who supplied ceramic objects found at Catal Huyuk had learned to produce red or black pots from the same clay. Because the latter require a reducing atmosphere during burning, these artisans must have had some degree of pragmatic knowledge about pigment transformations. In any event, it appears likely that the art of pottery making inadvertently made available the two parameters also needed for successful copper melting: a sufficiently high temperature and a regulated air supply (Ref 2).

Introduction of metallic ores into these furnaces has also been linked with pottery-making skills. From very early times, fine crushed metallic ores were used as pigments for decoration of ceramics. In the process of firing, these ores could have become reduced to metal. For example, even present-day ceramics from Gaipur, India, which are decorated with blue copper pigments and clear lead oxide glaze, sometimes show dark metallic patches. These patches consist of reduced lead from lead oxide contained in the glaze. It is possible that at higher temperatures, under over-firing conditions, copper may have occasionally been reduced as well. Some greenish over-fired pottery from Ur in Mesopotamia, dating from 4000 to 3500 B.C., show that such conditions may have occurred, although no metallic copper is reported to have appeared in those artifacts. Nevertheless, pottery-type ore smelting furnaces found by archaeologists in the 1960s support this theory of the evolution of copper metallurgy (Ref 2).

The earliest known method of copper smelting in pottery kilns was pieced together based on bowl-shaped depressions in the earth of the Timna Valley, close to the Sinai port of Eilat. Clay sides of these depressions were discolored by heat. Mixed with sand that filled and surrounded these depressions were bits of greenish copper slag, which contained small "prills" or blobs of metallic copper. Nearby sandstone blocks could be fitted together to form a furnace with a total depth of 80 cm and a width of 45 cm. Charcoal fuel, made from desert acacia trees and carbon dated at 3500 B.C., was also found at the site. An iron oxide flux was probably gathered from nearby cliffs. Tests on

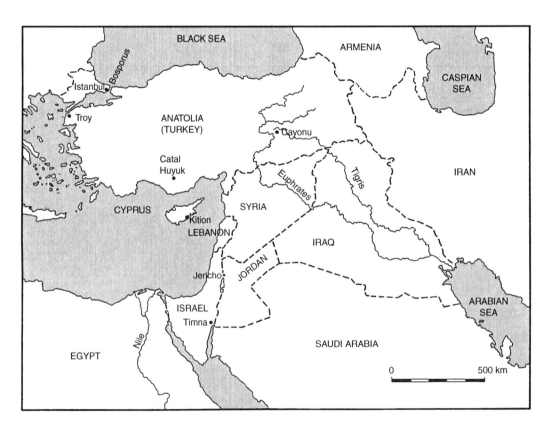

Fig. 1.1 Middle Eastern sites where archaeological evidence of early copper smelting has been found

this slag showed that furnace temperatures reached 1180 to 1350 °C (2156 to 2462 °F). These temperatures would have required some form of forced draft, possibly pumped into the furnace with a goatskin bag and a clay tube (Ref 2).

Modern attempts to reproduce the ancient metallurgical process suggest that smelts lasted several hours. Copper ore, flux, and charcoal had to be added from time to time until a sufficient volume of slag had accumulated in the bottom of the furnace, whereupon the entire charge was probably left to cool overnight. The next day the solidified mass of slag and copper was lifted out and crushed with stone hammers to extract the prills of metal. These were remelted in a crucible, a procedure that allowed impurities to be skimmed off. The remaining metal could then be cast as an ingot of fairly pure copper (Ref 2).

Antiquity

Transition from the Neolithic to the Chalcolithic periods occurred in more than one geographic area, as archaeo-metallurgical evidence suggests that copper smelting was discovered independently in many different parts of the world. For example, excavations at Rudna Glavia in Yugoslavia revealed that a large underground mine was in operation there before 4000 B.C.

One of the most ancient people in Western history, the Sumerians, probably obtained their first supplies of copper from the mountainous country surrounding Lake Van in Armenia. Egyptians probably drew their first supplies of copper as native metal and from the abundant malachite stones found in the hills near the Red Sea in the Eastern Desert. The Egyptian mines lay almost on the natural trade route to the Red Sea.

Around 2800 B.C., traders in Cyprus were receiving copper objects from Egypt and similar articles bearing cuneiform inscriptions, probably from Sumeria. At about the same time, Cyprus developed its own copper mines. These became renowned throughout the Eastern Mediterranean, and the island's name became the root of the Latin name, *cuprum,* and ultimately the English name, *copper,* for the red metal.

Some ores contain both copper and tin. When smelted, these ores yield an alloy of these two metals, which is known as bronze. Bronze is significantly harder and stronger than pure copper, and in utilitarian terms at the time, considerably more valuable. The discovery of a copper-tin alloy and its uses led to the Bronze Age.

The Bronze Age began in Europe around 1500 B.C. In China, it reached its apogee at about 1525 B.C., during the Chang dynasty. Whole series of magnificently ornamented, useful, and ceremonial bronze vessels, exist from that time. Figure 1.1 shows a map illustrating how trade evolved in the Mediterranean during the Bronze Age. In the area of the map shown, tin ore coexisting with copper appears in Turkey and Siam in Asia and in Wales and Spain, in Europe. Phoenicians probably brought bronze ingots from Europe to Egypt. The cake-shaped ingots were a few centimeters thick and were cast with a rounded profile shaped to rest on the backs of the men who had to carry them. They can be seen on Egyptian frescos and Persian reliefs (Ref 2, 3).

Casting was one of the earliest metal manufacturing methods. Clay molds were probably employed at first, although a few wooden molds have also been found. Open molds, which can only produce articles that are flat on one face, ultimately evolved into closed molds similar to those still used today.

To cast copper successfully, one must observe special precautions, particularly with regard to the evolution of gases, which can create an unsound casting. Molds must be constructed with special openings known as sprues, gates, and risers to pour the metal, permit the escape of dirt and gases, and cause the metal to solidify properly. Ancient coppersmiths were well aware of the difficulties involved and became very adept at overcoming them. The majority of surviving relics of early copper work are in cast form. However, the process of beating thin copper sheets against a form, with or without the addition of special ornamentation or engraving, was also used by ancient coppersmiths to make a large variety of artistic objects.

Bronze is easier to cast than pure copper. Once the Egyptians had learned to alloy copper with tin, and frequently also with a little lead to improve the metal's solidification characteristics, casting became a much more viable process. Products began to include such diverse items as axes, bowls, tools of many kinds, weapons, celts (a chisel-like tool), figurines, large vases, and sacred vessels.

Egyptians are also commonly credited with inventing the lost wax method of casting metal, although the method was known in China, as well. The lost wax process provided solid castings upon which a great refinement of ornament or detail could be worked. Hollow castings were also made. These required some kind of core.

Ancient Greeks also used bronze to a considerable degree. In Greece, hundreds of types of products were unearthed ranging from exquisite little figures used as the supports or handles of mirrors and caskets to large products such as statues and armor plates. Somewhat later in history, Romans were the first to apply copper and bronze in building construction. Their architects held stone structures together with copper or bronze ties and clamps. The finest surviving architectural work of this type is the Pantheon in Rome, built by Hadrian in A.D. 115 to 126. The Pantheon is an immense circular temple, 43.2 m (143 ft) in both diameter and height, which is surmounted by a brickwork dome. The dome has a central open hole, 9 m (29.8 ft) across at its apex, which provides illumination to the interior. The interior of the dome reveals a cofferwork originally embellished with bronze ornaments. It is said to have been covered with copper plates. The dome also had an outside covering of copper or bronze tiles. Of all that metal work, only the central ring and bronze door remain (Ref 1).

Romans were the first to use brass, an alloy of copper and zinc, on any significant scale, although Greeks were already well acquainted with the metal in Aristotle's time (330 B.C.). Greeks knew it as "oreichalcos," a brilliant white copper, which was made by mixing tin and copper with a special earth called "calmia," or calamine. Calamine was an impure zinc carbonate, which was rich in silica and found on the shores of the Black Sea. To make brass, ground calamine ore and copper were heated in a crucible. The heat applied was sufficient to reduce zinc to the metallic state but not high enough to melt copper. However, zinc vapor permeated the copper and formed brass, which then melted.

Romans used brass for personal ornaments and for decorative metalwork. Alloys used contained from 11 to 28% zinc, and the value of different grades of brass for different purposes was clearly known.

Greeks used only a few copper coins, but Romans had a large variety of copper money. The earliest Roman currency consisted of copper or bronze bricks, which were cast in stone molds and stamped with the figure of an ox. One example was the Roman As, a copper alloy coin also used as a unit of weight of about 5.4 kgf (12 lb). Coinage as we know evolved somewhat later, when the identifying impression was struck on plain cast discs. The first struck coins had a two-headed Janus in one side and the prow of a ship on the other. In imperial Roman times, coins generally bore an image of the head of the reigning Caesar.

The Middles Ages and Renaissance

Europe. During the early Middle Ages, much of the early use of copper served military purposes. For example, experience gained through handling of comparatively large bronze castings must have proven to be quite valuable when cannons were introduced. Bronze cannons were used by German armies in Italy, at the siege of Cividale in 1331. Bronze cannons used by Edward III at Cambrai in France and Crecy may have led to the establishment of a metallurgical industry in England soon afterward. The first record of the manufacture of brass guns in England was in 1385, when three such cannons are said to have been made by the Sheriff of Cumberland (Ref 1).

Medieval uses of copper were certainly not limited to ordnance. Early artistic applications included bronze bells and the well-known baptistery doors at the cathedral of Florence. Copper and bronze also formed the basis for decorative enameled ware, including amphorae, jugs, plates, and other functional as well as artistic items. Among the best known examples of this art form were those produced in Limoges, France, during the 15th and 16th centuries.

The Orient. In the Middle Ages, copper and bronze craft work also flourished on a grand scale throughout the Orient. Central and South Indian temples contain many fine bronzes, including large and small Buddhas. Some of the immense Buddhas and bells that can be found in India, China, and Japan must have caused artisans many headaches; but once successfully cast, they proved to be durable, having survived to this day (Ref 1).

The Temple of Ananda, at Tirumalai, India, is unique in the sense that it foreshadows modern trends. This temple is entirely sheathed in copper sheets containing elaborate hand-wrought ornamentation. It presents an interesting parallel to the use of copper wall sheathing on a number of today's buildings.

In Japan, the most ancient copper products are copper bells, known as "dokatu," which have been unearthed in many places. Because similar products have not been discovered in China or Korea, dokatu is believed to be an original product of the oldest copper industry in Japan. On the other hand, copper products such as swords, utensils and mirrors were imported to Japan from China in ancient days. Therefore, it is assumed that in early times a major portion of copper raw materials was imported, while a small amount of native copper produced in Japan was added (Ref 1).

In A.D. 708, a large deposit of native copper was discovered in Japan. Forty years later, a large (450 tonnes) copper statue of Buddha was constructed at Nara, then the capital of the country. Considering the size of this statue, it is reasonable to assume that sulfide smelting had been developed by that time. In the 17th century, a matte converting process called "mabuki" was established in Japan. This process made it possible for the Japanese metallurgists to treat low-grade copper sulfide ores.

The Americas. Aztecs, Toltecs, Zapotecs, and Mayas of México and Central America and Moché, Nazca, Chibcha, Quimbaya, Chimú, Chanzay, Tiahuanaco, and Incas of Central America and Perú all apparently possessed a fairly advanced knowledge of metalworking. The first manifestation of Moché metalwork was dated A.D. 200. In México, Toltecs were expert metallurgists as early as the 12th century A.D. Aztecs became the dominant power around the beginning of the 14th century A.D. and worked copper, tin, and gold, as well as an alloy of copper and gold known in Colombia as "tumbaga" (Ref 1–3).

Techniques of smelting, casting, beating, soldering and gilding were understood, and most of the output of metalworkers, who were organized in separate guilds, appears to have been in the form of ornaments. Some utilitarian objects have survived. Those in copper include fishhooks, needles, pincers, mirror frames, small picks, chisels, and axes. Ornamental clapperless copper bells have been found in sites all over México and Central America. These were made in the form of a hollow shell and used a pebble inserted through a vertical slit to make a noise.

Mesoamericans possessed great skill as masons, and it is fairly probable that they used copper tubes with water and abrasives to drill holes in blocks of stone. Research has shown that Peruvian Incas used stone, copper and bronze tools, the latter of which were reserved for more difficult tasks. Analysis has shown that the bronze used consisted of 94% copper and 6% tin. Considering the abundant supplies of both metals in this part of the world, local development of this alloy is hardly surprising (Ref 1).

North American Indians also used copper for tools, weapons, ornaments, and amulets. The metal used was probably native copper, which was and is still abundant around the shores of Lake Superior. A number of copper artifacts were discovered, mainly in burial mounds, during the last decades of the 20th century.

The Industrial Revolution

The Industrial Revolution brought about a tremendous change in the production of copper and its alloys, beginning with a demand for more and better raw material. Mine production rose as steam-driven pumps were applied to remove water from diggings. Smelter throughput increased as well, largely due to faster removal of impurities from ore.

Demand was driven by technical developments, most important of which concerned the use of electricity. In 1729 Stephen Gray used brass wire for the first known attempt to transmit an electric current. In 1747, Sir William Watson succeeded in transmitting a current 735 m (2410 ft) across Westminster Bridge using the Thames River as the return wire. This, among other experiments, proved that metals were the best conductors of "electric fluid" as it was called, and that of these, copper, even in its relatively impure state, was superior to all others except silver. It is most likely for this reason that Benjamin Franklin used copper for lightning conductors, an application in which it has been virtually unchallenged. In 1811, copper wire was first employed to protect the mast of a ship from lightning.

Alessandro Volta assembled the first electric battery in 1799. His invention comprised discs of copper and zinc placed one upon the other with a layer of wet cloth between each pair. The development of large, powerful batteries led directly to the invention of the electric telegraph and, in turn, to today's communications industry. It also created the first large industrial demand for copper wire.

Another significant increase in demand for copper was brought about by the development of the stamping press (and coining press shortly thereafter), which helped speed up the production of simple brass articles. Metal buttons, for example, suddenly became much more readily available, as did many of the brass furniture fittings used by the master cabinetmakers of the period.

Brass has long been the first material of choice in the construction of measuring instruments for use at sea or in any moist, salt-laden atmosphere. This is due not only to the corrosion resistance of brass but also to its good machinability, ease of engraving, and, perhaps most important of all, the fact that it is nonmagnetic. It is because of the latter property that marine compasses are mounted upon brass binnacles, equipped with brass compass bowls and strung in brass gimbals.

From ancient times until well into the 19th century, wooden ship hulls were often clad with copper sheathing to protect against burrowing shipworms and thereby prevent the embarrassing circumstance of having a ship disintegrate and sink in calm seas. An additional and equally important benefit of copper sheathing was that it prevented growth of algae and barnacles, which created considerable hydrodynamic drag. As a result of their smooth copper hull sheathing, the British ships under Admiral Nelson were significantly—and decisively—faster than unsheathed vessels of the French and Spanish flotillas that Nelson defeated at Trafalgar in 1805. Demand for copper hull sheathing virtually disappeared as steel hulls replaced wood; however, there has been a renewed interest in the process during the last quarter of the 20th century. See Chapter 5, Applications.

Electrical engineering in the modern industrial sense followed from Faraday's discovery of electromagnetic induction in 1831. Invention of the electric dynamo by Werner von Siemens in 1866 led to a tremendous increase in demand for copper to supply electric power. This became particularly evident during the second half of the 19th century after Edison's invention of the electric light bulb in 1878 and his construction of the first electrical power generating plant in 1882. Installation of land and submarine telegraph cables and use of electric traction motors for trains were among many important consequences of rapid progress in electrical engineering during the second half of the 19th century. All of these developments naturally made growing demands on the supply of copper (Ref 4).

The Modern Era

The Growth in Consumption. Figure 1.2 graphically describes the enormous growth in world copper demand (actual and projected) between 1850 and 2000. As indicated, this growth was initiated by a series of advances in copper-using technologies. Whereas total demand at the end of the 19th century was about 500,000 mT, 1992 forecasts of world copper consumption were 14,630,000 mT for the year 2000 and 17,900,000 mT for 2005 (Ref 5, 6).

Toward the end of the 20th century, consumption of primary (for example, mined) copper products had climbed to a yearly average (1986–1990 figures) of 8.9 million mT. The 1994 world total consumption reached 13.4 million mT (Ref 7–9).

The increase in the rate of growth in copper consumption, and hence in production of copper in the 19th and 20th centuries, took place in four distinct stages. From about 1875 until the beginning of World War I, copper consumption increased at the rate of approximately 5.5% per year. Between the two world wars, consumption was erratic due to effects of World War I and global economic depression. Following World War II, growth in copper consumption stabilized again at an increasing rate of approximately 4.5% per year. Immediately following the succession of energy crises in 1973 and 1978, the growth rate of copper dropped to about 1% per year. That slump notwithstanding, world copper consumption grew at an average annual rate exceeding 3% for the period between World War II and 1988. For the period 1996 to 2005, copper consumption has been estimated to grow at an average rate in excess of 3%. The largest increases are anticipated to occur in construction, transportation, and the electrical and electronics industries. It is also expected that the largest growth in demand will occur in the Asian market (Ref 9, 10).

Energy and materials conservation concerns of the 1970s and 1980s probably contributed indirectly to the decline in the rate of growth in copper consumption during and since that time. Modernization and rationalization in the form of downsized automobiles, miniaturized electronics, and generally improving efficiency in electrical products have also had a moderating effect on the rate of copper consumption. See Chapter 5, "Applications." Figure 1.3 depicts world and U.S. copper demand between 1950 and 1994. Two sharp dips in 1975 and 1981 coincided with petroleum price crises and the recession of the early 1980s. Demand decreased during the early 1980s. However, world demand recovered by 1985 with U.S. demand not recovering until the 1990s (Ref 6, 11, 55).

The United States produced about one-half of the 500,000 mT of copper consumed worldwide at the beginning of the 20th century. The number of countries supplying the primary copper consumed by what has now become a 20-fold larger market grew to 46 by 1994. In 1994, Chile supplied approximately 24% of the world total mine production and the United States supplied about 19% (Ref 7, 8).

Per Capita Consumption. The 1990 apparent per capita consumption of refined copper for a number of countries is shown in Table 1.1. Note that these data refer to consumption of refined copper; for example, copper as a raw material for manufacturing, and they do not necessarily reflect per capita end-use consumption. To calculate the latter requires an accounting for the amount of copper contained in manufactured goods, such as automobiles and electronic equipment, exported from the manufacturing country and imported into the consuming country.

World Reserves

Copper deposits have been identified on all inhabited continents, as shown in Fig. 1.4. The largest and commercially most important deposits are those in the mountainous spine of South America and western North America, principally in the

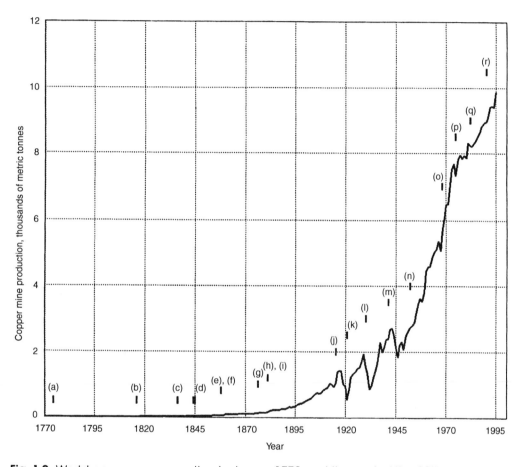

Fig. 1.2 World copper consumption between 1770 and the end of the 20th century correlated with historical events. Consumption between 1850 and 1900 was estimated to have been 200,000 tonnes per year (tpy). Consumption between 1991 and 2000 was projected to average 13,000,000 tpy, and demand for the year 2000 was anticipated to be 14,630,000 tonnes. (a) First electric telegraph demonstrated by Georges Le Sage, 1774. (b) Eight-mile-long electric telegraph demonstrated in England by Francis Ronalds, 1816. (c) Samuel F.B. Morse demonstrates electric telegraph, 1836; first Morse-type telegraph line installed between Washington and Baltimore, 1844. (d) First cable under the English channel by Jacob and Walkins Brett, 1845. (e) First installation in England of an electric generator for lighting a building, 1858. (f) First transatlantic telegraph cable, in operation only from August 5 through August 28, 1858. (g) Alexander Graham Bell invents and demonstrates the telephone, 1876. (h) First prototype electric railroad built by Siemens brothers, 1881. (i) Thomas A. Edison installs first public generating station in New York, 1881. (j) World War I, 1914–1918. (k) Invention of mineral concentration through flotation, 1921. (l) The Great Depression. (m) World War II, 1939–1945. (n) Korean War, 1951-1953. (o) CIPEC founded, 1968; replaced by GIEC in 1991. (p) Petroleum crisis, 1975. (q) Economic recession, 1982. (r) Economic recession 1990. Source: Ref 7

United States and Chile but also including Canada, México, and Perú. Large and important deposits are also located in south-central Africa, principally in Zaire and Zambia. Together, North and South America and Africa contain more than one-half of the known copper deposits in the world; however, Europe (primarily Russia and the Confederation of Independent States, or C.I.S.) and the Pacific Rim (chiefly Australia and Papua New Guinea) also contain important copper mineralization. A database maintained by the U.S. Bureau of Mines recorded nearly 300 copper deposits in the market economy countries in 1983. It should be understood that this number continually changes due to such factors as ore depletion, fluctuating copper prices, and, in some cases, civil unrest (Ref 12, 13).

An additional, enormous, and virtually untapped supply of copper has been identified in the deep-sea nodules found scattered in several regions of the ocean floor. These curious structures, approximately 2 to 4 cm (approximately 1 to 2 in.) in diameter, contain high concentrations of several important metals, including about 1% copper. Commercial interest in deep-sea nodules peaked during the 1970s but has since subsided for a variety of economic and political reasons. Most important among these were the prohibitively high recovery costs (in relation to copper prices) and the issue of ownership of deep-sea mineral rights, which was hotly contested by landlocked less-developed countries, particularly those with copper resources of their own.

The U.S. Bureau of Mines estimates land-based world copper resources at 1.6 billion mT, plus 0.8 billion mT contained in deep-sea nodules. In 1983, world copper reserves were placed at 350 million mT, and the reserve base was at 510 million mT. Note that as defined by the Bureau, resources include the total of all copper mineralization; reserves are the fraction that is economically minable" or "thought to be economically recoverable with existing technology at operating or developing mines," while the reserve base consists of the "measured plus indicated (demonstrated) resource from which reserves are estimated" (Ref 12).

Distribution of world reserves and reserve bases by country and region are listed in Table 1.2. Primarily because of the huge Chilean deposits,

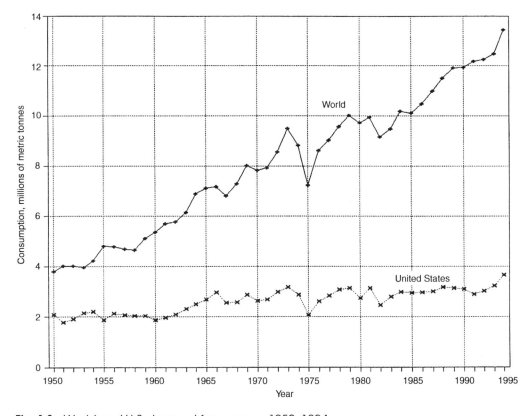

Fig. 1.3 World and U.S. demand for copper, 1950–1994

South America holds the largest shares of both reserves (31.1% of world total) and reserve base (18.5%). North and Central America follow, with 22.7% and 24.9% of world reserves and reserve base, respectively. Among individual countries, Chile has the largest fraction of the total reserve base at 19%, the United States ranks second with 18%; the C.I.S. and Zambia each have 7%; while

Canada, Perú, and Zaire each account for 6% (Ref 6, 11, 12, 14).

Although there have been numerous periods in which copper was in short supply (including a notably severe shortage during portions of the 1980s), copper clearly cannot be thought of as a scarce material or one that is in danger of deple-

Table 1.1 Apparent per capita consumption of refined copper, 1990

Country	Consumption, 10^6 tonnes	Population, millions	Consumption per capita, kg
North America			**8.6**
Canada	184.0	22.0	8.4
United States	2,160.0	250.0	8.6
Central/South America			**1.3**
Argentina	26.3	32.0	0.8
Brazil	145.9	148.0	1.0
Chile	45.2	13.0	3.5
México	133.5	60.0	2.2
Peru	13.2	22.0	70.6
Western Europe			**8.8**
Austria	22.8	8.0	2.9
Belgium	399.0	9.9	40.3
France	417.0	58.0	7.2
Germany(a)	897.0	62.0	14.4
Greece	52.0	11.0	4.7
Italy	478.0	58.0	8.2
Portugal	25.0	12.0	2.1
Spain	146.1	42.0	3.5
Sweden	117.2	9.0	13.0
United Kingdom	317.0	58.0	5.5
Eastern Europe			**4.1**
Bulgaria	56.0	10.0	5.6
C.I.S.(b)	1,250.0	290.9	4.3
Czechoslovakia(b)	98.0	15.7	6.2
Hungary	22.1	12.0	1.8
Romania	24.7	28.0	0.9
Yugoslavia	113.2	25.0	4.5
Middle East			**1.4**
Iran	39.6	52.0	0.8
Turkey	114.0	56.0	2.0
Asia			**1.2**
China (PRC)	475.0	1,120.0	0.5
India	135.0	840.0	0.2
Indonesia	2.0	178.0	0.0
Japan	1,577.0	124.0	12.7
Republic of Korea	324.0	45.0	7.2
Taiwan	312.6	21.0	14.9
Pacific Rim			**5.2**
Australia	112.0	18.0	6.2
New Zealand	2.0	4.0	0.5
Africa			**0.8**
Egypt	6.0	53.0	0.1
R.S.A.	67.5	36.0	1.8
World	**10,308.7**	**3,803.5**	**2.7**

(a) Western States. (b) Before partition. Source: World Bureau of Metal Statistics World Almanac

Table 1.2 World copper reserves and reserve bases in 1983 and 1991

Region	Reserve(a) 1983	Reserve(a) 1991	Reserve base(b) 1983	Reserve base(b) 1991
North and Central America	**92**	**70**	**151**	**146**
Canada	17	11	23	23
México	17	14	23	20
Panama	...	...	...	10
Puerto Rico	...	...	...	2
United States	57	45	90	90
Other	1		15	1
South America	**95**	**96**	**140**	**106**
Argentina	...	...	...	7
Brazil	...	1	...	11
Chile	79	88	97	140
Perú	12	7	32	25
Other	3	...	12	3
Europe	**50**	**70**	**70**	**103**
Finland	...	1	...	1
Norway	...	1	...	1
Poland	...	20	...	36
Portugal	...	3	...	3
Spain	...	1	...	1
Sweden	...	1	...	1
C.I.S.	...	37	...	54
Yugoslavia	...	4	...	4
Other	...	2	...	2
Africa	**60**	**26**	**71**	**70**
Namibia	...	1	...	3
R.S.A.	...	2	...	2
Zaire	26	10	30	30
Zambia	30	12	34	34
Other	4	1	7	1
Asia	**26**	**34**	**37**	**52**
China	...	3	...	8
India	...	3	...	4
Indonesia	...	11	...	17
Iran	...	3	...	5
Japan	...	1	...	1
Mongolia	...	3	...	3
Philippines	12	7	18	11
Other	14	3	19	3
Middle East	...	**1**	...	**2**
Turkey	...	1	...	2
Pacific Rim	**15**	**11**	**35**	**28**
Australia	8	7	16	21
Papua New Guinea	6	4	15	7
Other	1	...	4	...
World total	**338**	**308**	**504**	**587**

(a) Millions of tonnes contained copper. (b) Millions of tonnes contained copper and includes reserves. Sources: World Bureau of Metal Statistics, U.S. Bureau of Mines

Fig. 1.4 Areas of major copper mining districts and reserves

tion. Even excluding scrap, which satisfies approximately 40% of the copper demand in the world, world copper reserves are more than adequate to provide for mankind's copper needs for the foreseeable future.

It is also worth noting that a sizable portion of all copper ever mined is still in use, although it may have been recycled, re-refined, and/or alloyed several times. This fact alone attests to the durability of copper. It also underscores the commercial and environmental importance of the high ability of the metal to be recycled. This subject is discussed in more detail in Chapter 6, "Copper in the Environment."

World Production

Copper production is reported in several ways, depending on the stage of production being considered. Because ore and concentrate grades vary to some extent, mine production statistics are traditionally cited in terms of the recoverable copper content of the products. Smelter production is occasionally reported as the quantity of matte (an unrefined mixture of copper and other sulfides) produced, but it is more often quoted in terms of the output of primary blister and anode copper (from ores and concentrates) and secondary blister (produced from scrap), respectively. Refinery production is reported in terms of fire refined and/or electrolytically refined copper. Historically, refinery feedstocks included blister, anodes, or other primary materials, together with scrap and secondary blister. The accounting system has changed somewhat to accommodate feedstocks and products encountered in solution mining, heap leaching, and solvent extraction/electrowinning (SX/EW) processes. Production statistics cited in this publication are given in terms of copper content unless otherwise specified.

Primary Copper

Primary copper and copper scrap are raw materials consumed in fabrication of semifinished products, which are variously known as semimanufactures, semifabricates, or simply "semis." Semis, in turn, are used in the production of a variety of copper and copper alloy commodities needed for the manufacture of industrial and consumer products.

Total world mine production increased from 2.52 million mT in 1950 to 7.66 million mT in 1974. This year (1974) was the peak year of production before the steady growth of production

and consumption of copper was halted briefly by effects of the first sharp oil price increase. Thus, for that nearly quarter century, production of copper increased at a compound annual rate of 4.7% (compounded average growth rate). The rate of production returned to a steady increase beginning in 1976, however, at a lower rate. After the first oil crisis, which resulted in a slight decline of copper production and consumption, production began a steady rise at an average annual rate of 1.4%, reaching approximately 9.50 million mT in 1994. World smelter production grew from 2.59 million mT in 1950 to 8.02 million mT in 1974 then to 9.90 million mT in 1994.

Refinery production grew from 3.15 to 8.90 million mT between 1950 and 1974 and to 11.38 million mT in 1994 (Ref 8, 15).

Western Europe. Although Western Europe contains several large smelting and refining centers and is one of the leading copper consumers, it is relatively insignificant among the copper mining regions of the world. Production of copper ores and concentrates totaled 292,700 mT in 1975. By 1994 it has risen to 551,000 mT. This represents 5.8% of the world's total production. On the other hand, Western Europe imported more than 27% of the copper in the world in 1994, mainly in the form of refined copper. Western European smelter production was 946,100 mT in 1994. However, refinery production for that year was 1,609,500 mT. The difference between smelter and refinery production is attributable to large quantities of recycled high grade scrap. In addition, 983,000 mT of unrefined copper scrap was used in alloy production in 1994 (Ref 7, 8, 16).

Austria. Production of concentrates from Austrian copper mines, which never was very high, dropped to only 2000 mT in 1975. It ceased altogether in 1976. Despite the loss of native raw materials, Austria retains a modest smelter and refinery capacity. Production is based mainly on scrap obtained locally as well as from southern Germany and northern Italy; it is supplemented by small imports of anode and blister. Production of secondary refined copper increased from a mere 17,000 mT in 1975 to 21,000 mT in 1994, whereas smelter production increased from 11,000 mT in 1975 to 26,100 mT in 1979. By 1994, it had risen to 51,600 mT. Smelter capacity was reported to have been 60,000 tpy in 1989, the year in which existing refining facilities were replaced by a new 50,000 tpy plant. Refined copper production almost doubled from 26,900 mT in 1975 to 50,900 mT in 1994 (Ref 7, 16–19).

Belgium. Belgium enjoys a large international trade in a wide variety of copper-containing materials. In spite of having no mine production, this country is also a significant producer of blister and refined copper and both copper and copper alloy semis. Smelter production in 1994 was 103,200 mT, and refinery production was 304,800 mT. In addition, Belgium re-refines more than 100,000 tpy of scrap into secondary refined copper. In 1990, the Benelux countries also used 35,000 mT of scrap directly in the manufacture of new copper products (Ref 7, 16, 18).

Finland. Finland's mines produced 38,800 mT of copper in 1975. Production rose to 46,800 mT in 1978 but decreased relatively steadily thereafter, dropping to only 9500 mT in 1994. Concurrent increases in the import of concentrates led to a rise in smelter and refinery production during that period. Smelter production rose from 46,200 mT in 1975 to 98,100 mT in 1994, while refinery production rose from 35,800 tpy in 1975 to 69,200 mT in 1994. Finland has also become an important exporter of copper-related (and other) technology. High import levels are likely to be maintained in view of Finnish involvement in mining and industrial projects abroad (Ref 7, 18).

Germany. German production statistics reflect the former political division which, in the present frame of reference, ended in 1991. While Western Germany has had essentially no mined production of copper in modern times, its smelter and refinery production has been significant. Germany as a whole has steadily maintained a smelter production of 250,000 tpy during the 1980s; it reached 295,800 mT in 1994. Germany's 1994 refinery production of 591,900 mT accounted for slightly more than one-half of her domestic needs (Ref 7, 16, 18).

Norway. Production of copper concentrates fell from almost 30,000 tpy (contained copper) in the late 1970s to around 7700 mT in 1994. Low copper prices are generally blamed for the decline. The 1994 smelter and refinery production levels were balanced at 39,300 mT, this being more than double the quantity of concentrates produced domestically. Because the amount of concentrates exported is roughly the same as the amount of copper produced, smelter production must rely to a large extent on scrap. Some blister and anode copper are imported for refining (Ref 8, 17, 19).

Portugal. Portugal was a relatively minor producer of copper concentrates until the start-up of operations at Neves Corvo in 1989, after which the country became the largest producer in Western Europe. In the initial year of operation for the mine, Portugal produced 103,700 mT. Production reached 159,700 mT in 1990 and fell to 130,300 mT in 1994. Nonetheless, this production has contributed significantly to the increase in the European copper trade (Ref 7, 18).

Spain. Mine production declined from 51,600 to 4900 mT between 1975 and 1994. Nonetheless, refined copper production rose relatively steadily from 56,700 to 189,200 mT during the same period. Smelter production has risen from 105,000 mT in 1975 to 172,800 mT in 1994 (Ref 7, 19).

Sweden. Mine production more than doubled between 1977 and 1985, reaching 91,800 mT, but then began to decline markedly in 1989. In 1994, mine production had dropped to 77,600 mT. Smelter production increased steadily and almost doubled between 1977 and 1988, when it reached a peak of 116,000 mT. By 1994, smelter production had dropped to 98,400 mT. Production of primary refined copper stood at 102,500 mT in 1994. Sweden both imports and exports refined copper; the 60,000 mT imported in 1991 were a little less than twice the quantity exported (Ref 7, 16).

Yugoslavia. Copper mine production fell from 292,700 mT in 1975 (when the country was still unified) to 76,000 mT in 1994, a reflection of the crippling conflict that marked the division of the country into several independent states. Smelter production was 50,400 mT and refinery production 69,900 mT in 1994 (Ref 16, 19).

Other Western European Countries. France, Italy, and the United Kingdom played a relatively small role in the production of refined copper, having collectively produced 168,300 mT in 1994. At 121,600 mT in 1991, the United Kingdom was the largest of these minor refined copper producers. However, by 1994, U.K. production had dropped to 46,700 mT, making Italy, at 79,700 mT, the largest. United Kingdom mine production only amounted to a few hundred mT in 1990 and 1991 and was nil in 1994. France produced just 300 mT, in 1991, and production was nil by 1994. The last year of mine production in Italy, 1985, saw a total output of only 100 mT (Ref 7, 18).

Eastern Europe and the C.I.S. The Eastern European states and the Confederation of Independent States (C.I.S.) are major producers of copper. In 1990, the former Soviet Union ranked third worldwide in copper mine production and second in refined copper output. However, throughout the cold war period, 1945 to 1989, the Eastern Bloc carried on a system of trade quite separate from the West. Also, after the formation of the C.I.S., the region underwent sharp political transformations and an acute economic crisis. For these reasons, reliable statistical information regarding the region's industrial output over a period of almost five decades has been difficult to obtain in the West. The decline in the late 1980s is attributed to the introduction of free market economics together with chronic pollution problems and energy shortages.

Among these countries, the C.I.S. members are the largest copper producers. Other Eastern European mine production is relatively small, and concentrates are, and have traditionally been, exported to the C.I.S.

Through 1990, Western countries reported a small volume of trade in concentrates and blister copper with the former Eastern Bloc countries, quantities amounting to only a few tens of thousands of mT per year. Net exports to Western countries, mainly from Russia and Poland, were greater than 100,000 tpy. Overall, this region was a small net exporter of copper in all forms. This condition changed substantially in January of 1993, when Russia reportedly sold 100,000 mT to the West (Ref 20).

Albania. The Albanian copper industry includes mining, smelting, and refining, as well as the manufacture of semis and finished products. Mine production was only about 1000 tpy in the early 1960s, but it rose to 11,500 mT by 1980 and was estimated to have been 17,000 mT in 1988. In 1994, however, it was only 8000 mT. In that year, smelter production was the same as mine production. However, refinery production was only 1000 mT, estimated to be about the same as mine production (Ref 21).

Bulgaria. Bulgaria is the third largest producer of copper in Eastern Europe. Mine production increased slowly from 55,000 mT in 1975 to 68,000 mT in 1983. Thereafter, annual production oscillated rather widely for several years and eventually fell to 32,400 mT in 1990. It had risen to 60,400 mT in 1994. Smelter production remained stable at around 60,000 mT from 1975 to 1983. It then reached 70,000 mT the following two years, returned to its historic 50,000 to 60,000 mT levels until 1989 and then fell to half that value in 1990 to 1991. By 1993, it had climbed again to reach 71,900 mT, a level that it held in 1994. Refinery production remained near 55,000 tpy from 1975 through 1989, after which it fell abruptly to 24,100 mT in 1990 and 12,600 mT in 1991. In 1993 and 1994, refinery production was 26,300 mT. Bulgaria traded some copper with Western countries during the Cold War, but the largest reported imports consisted of only 6500 mT of refined copper in 1988 and the largest exports comprised 7100 mT of concentrates and 700 mT of refined copper in 1986 (Ref 7, 18, 22).

Confederation of Independent States (C.I.S.). In 1994, the mined copper output of the C.I.S.was 884,900 mT making it the third largest producer, behind Chile and the United States. Production peaked in 1985 to 1986 with a mine production of 1,599,200 mT and refinery output of 2,038,600 mT in 1986. In 1994, smelter production was 900,000 mT, and refinery production was 1,319,900 mT. Smelter was slightly higher than those for mine production, probably reflecting imports of concentrates from former Eastern (COMECON) trading partners. The significantly higher refinery production indicates imports of blister copper from the west.

Within the C.I.S., in 1989, Armenia claimed reserves of 322 million mT of ore with an average grade of 0.36% copper. Production was reported to have risen steadily through the 1970s and early 1980s. Statistics made available in 1992 placed yearly production at about 63,500 mT during that period. Concentrates were smelted and refined in Japan. Kazakhstan, another member of the C.I.S. had the highest abundance of copper mineralization among the states of the former Soviet Union, and it accounted for more than 30% of the region's copper output. As is the case for most of the C.I.S. countries, specific production data are not available in the West (Ref 5, 7).

Czech and Slovak Republics. Czechoslovakian concentrate production stood at about 10,000 tpy (copper content) from 1975 through 1981. It remained around half that value until 1988, after which it gradually slipped to 2400 tpy. In 1994, it was 3000 mT. Smelter and refinery production remained fairly constant at roughly 9000 and 25,000 mT, respectively, through the 1980s. However, by 1994, smelter production had reached 22,400, and refinery production had decreased to 23,200 mT. The republic(s) imported copper because refined production amounted to only about one-third of the volume of consumption (Ref 18).

Mongolia. Mine production began in 1978 at a modest 4 tpy but rapidly climbed to more than 100,000 tpy by the early 1990s. As of 1994, Mongolia had no smelting or refining facilities, and most of the mine production was shipped to the C.I.S. and Eastern Europe. Mine production in 1994 was 105,100 mT (Ref 18).

Poland. Poland has come to relative prominence as a copper producer only in the second half of the 20th century. In 1950, mine production was only 400 mT. With the discovery of a major ore body in 1957 at Luban, in southern Silesia, it has grown fairly rapidly since then. Between 1977 and 1991, it remained between about 300,000 and 400,000 mT per year. A peak of 441,000 mT was reached in 1988, but production subsided to 320,300 mT in 1991 after which it has steadily increased, reaching 376,800 mT in 1994. Smelter production had remained somewhat lower, around or slightly above 360,000 tpy, although by 1993 production had risen to 411,200 mT where it remained during 1994. Refined copper production ranged from 306,600 to 400,600 mT between 1977 and 1990. In 1994, Polish refined copper production was 405,200 mT. Exports of concentrates fell from 49,000 mT (copper contained) in 1984 to only 5000 mT in 1989. In 1994, exports of refined copper were 267,600 mT (Ref 7, 18, 22).

Romania. Romania has a small but fairly stable mine production. Beginning in 1966 with a modest 8700 mT, production has since ranged between approximately 30,000 and 40,000 tpy. Smelter production during this period remained around 40,000 tpy, although it fell to less than 30,000 tpy in the early 1990s. Refinery output was 45,000 mT in 1975, but it rose to as much as 65,000 tpy in the early 1980s. Production has fallen steadily since then, with a production in 1994 of 28,000 mT. Smelter production was 26,200 mT, and refinery production was 16,600 mT in that year. In the late 1970s, as much as 20,000 tpy of refined copper was imported to Romania from Western countries. These imports ceased in 1982 to be resumed in 1994 when 8000 mT were imported. Domestic refinery production began at 10,000 mT in 1966. It grew to 54,000 mT in the mid 1970s but fell off to only 16,600 mT in 1994 (Ref 15, 17).

North America. In 1994, with a mine production of 2,713,400 mT, almost a third of the copper mined worldwide was produced in North America. The smelter production of 2,551,800 mT in 1994 was several hundred thousand tonnes a year less than the production of concentrates because Canada, México, and the United States export copper in this form. On the other hand, the production of refined copper in the region is larger than smelter production owing to the import of concentrates. Refined copper production was 2,952,800 mT in 1994 (Ref 7, 19, 22).

Canada. Since 1970, Canadian mine production has ranged between 690,000 and about 802,000 tpy in 1986. In 1994, it was 626,300 mT. During the 1980s, more than 300,000 tpy of copper concentrates were exported from Canada, while smelter production ranged between 400,000 and 550,000 tpy. In 1994, smelter production was 560,000 mT, and 218,000 mT of copper as con-

centrates was exported in that year. Imports of concentrates increased during the 1980s, reaching 76,200 mT in 1985 before falling to 56,100 mT in 1989. However, by 1994, imports of concentrates had increased to 218,000 mT. Production of refined copper ranged between 500,000 and 561,000 tpy between 1984 and 1993, with 549,900 mT produced in 1994. About 64% of this was exported (Ref 9, 22).

México. Mine production in México rose from 78,200 mT in 1975 to 291,000 mT in 1994. Smelter production in 1994 was 559,900 mT. In 1989, exports of concentrates fell as low as 55,200 mT, but recovered to 137,800 mT in 1990. Production of primary refined copper peaked at 143,900 mT in 1989 and then fell to 129,000 mT in 1990 to rise to 183,000 mT in 1994. A further 22,000 and 31,000 mT of refined copper were produced by re-refining secondary material in 1989 and 1990, respectively. The country consumed 137,000 mT in 1994, a year in which exports of blister and anode copper climbed to 93,900 mT (Ref 7, 8, 19).

The United States of America. Between 1975 and 1994, mine production rose from 1,282,000 to 1,795,900 mT; in the latter year, smelter production stood at 1,715,000 mT, and the production of refined copper was 2,219,900 mT. The 1970s and 1980s were an extremely difficult period for copper mines in the United States due to low world prices and high labor and energy costs. In 1970, mine production was greater than 1,560,000 mT, and although this total was exceeded in 1972 and again in 1973, it was not until 1990 that production returned to this level. During this time, some U.S. copper mines were forced to close and many others undertook enormous cost reductions to remain in production. Also, whereas the United States produced more copper in ores and concentrates than any other country in 1970, by 1989 its level of output had been overtaken by production in Chile, which had more than doubled since 1970. As of 1994, Chile's mine production exceeded that of the United States by about 423,900 mT, although U.S. smelter production surpassed Chile's by 254,600 mT (Ref 7, 8, 15, 22).

South America. Some of the largest and most productive copper mines in the world are located in South America. Chile is the leading producer in the region, followed by Perú. In 1994, together, Chile and Perú accounted for about 2,579,700 mT of mined copper, or about 30% of world production. Small quantities are produced in other South American countries; of these, Brazil, with a mine production of 37,000 mT in 1994, was the next largest producer (Ref 7, 19, 22).

In 1994, South American smelter production amounted to only about 74% of mine production, the difference being largely accounted for by exports of concentrates from Chile and, to a lesser extent, Perú.

Similarly, during 1994, production of refined copper in South America was about 250,000 mT less than the production of unrefined blister and anode copper. This was mainly because Chile and Perú also exported substantial quantities of blister and anode. The largest net exports from South America were made in the form of refined copper (Ref 7).

Argentina. Argentina is a very small copper producer whose output ranges in the hundreds of mT per year, mostly in the form of byproducts from other mining operations (Ref 5, 7, 22).

Brazil. Copper mine production in Brazil first received attention in 1959 when some 2200 mT were produced. Production fell thereafter and, in fact, ceased entirely at the end of the 1970s. Production restarted in the early 1980s, eventually reaching several tens of thousands of mT per year over the following decade. In 1994, mine output was 37,000 mT of copper. In 1982, copper smelting restarted as well and grew to nearly 160,000 mT in 1992. Imports of concentrates accounted for roughly two-thirds of the total refinery production of 150,000 mT in 1994. Refinery production was 143,200 mT, and consumption of refined copper was 141,800 mT in 1994. Because both smelter production and refinery production exceeded the refined consumption, it is evident that both blister and refined copper were exported (Ref 5, 8, 19, 22).

Chile. Chile became the largest copper producer in the world throughout the mid to late 19th century after installation of its first reverberatory furnace in 1842 and its first blast furnace in 1857. The country was displaced by the United States as the world leading producer in 1882 but regained its leading position exactly 100 years later. Mine production in Chile surpassed the one million tonne mark in 1976. The rise in production thereafter was impressively rapid, and by 1994, Chile accounted for nearly one-quarter of the copper production in the world. Most mine production is smelted or electrorefined domestically; however, with the return of foreign-owned mines, Chile has also become a major exporter of copper concentrates, shipping more than 759,000 mT in 1994. Most exports are destined for Western Europe and the Far East. Japan, the Republic of Korea, and Taiwan together account for about one-third of the

total shipped; less than 20,000 tpy are exported to the United States (Ref 7, 15, 19).

In recent years, Chile's smelter production has increased at a somewhat slower rate than mine production, as new, largely foreign-owned, mining operations came on stream. Refinery production also increased steadily in keeping with smelter capacity increases. Almost all the refined copper produced in Chile is exported. Most of this goes to Europe, with lesser quantities destined for the Far East and the United States. Chile is also the leading source of refined copper imported into other South American countries, mainly Argentina and Brazil. In 1994, Chile's mine production was 2,219,800 mT; smelter production was 1,460,300 mT, and refined production was 1,277,300 mT (Ref 5).

Perú. Peruvian mine production grew steadily between 1950 and 1977. It has remained greater than 340,000 tpy since then, peaking in 1987 at 406,000 mT. In 1994, mine production was 359,900 mT. Smelter production has generally kept pace with mine expansion, as has the production of refined copper. By the early 1990s, it had grown to more than 240,000 tpy. In 1994, smelter production was 315,000 mT, and refined production was 253,000 mT. Exports account for over 90% of the total production (Ref 7, 8, 19).

Africa. The African countries of Zaire and Zambia accounted for approximately 16% of world production in the early 1970s. However, this had fallen to less than 5% in 1994. In 1994, African countries accounted for about 7.2% of the world production of copper concentrates, 6.4% of world smelter production, and 4.9% of world refinery production. A large portion of the concentrates produced in Africa are smelted domestically, but substantial quantities of unrefined blister and anode copper are exported to refineries elsewhere. Most of the refined copper produced on the continent was formerly exported to more industrialized countries, but increasing amounts are now committed to the development of domestic infrastructures (Ref 5, 7, 18, 22).

Botswana. Copper mining in Botswana began in 1974 and peaked at 24,400 mT in 1988. Since 1980, production generally proceeded at a rate of about 20,000 tpy. Concentrates are smelted in Botswana, and a copper-nickel-cobalt matte is exported for refining. Botswana does not produce refined copper (Ref 18).

Morocco. Mine production began at a modest 800 mT in 1952. By the early 1990s, it had settled near 15,000 mT, off by about 50% from an earlier maximum. There is no copper smelter in Morocco, and all copper concentrates are exported (Ref 7, 15, 18, 22).

Namibia. Production of concentrates rose from an initial 35,000 mT (copper contained) in 1975 to more than 50,000 tpy in the mid 1980s but then declined to earlier levels in the 1990s. In 1994, mine production was 28,394 mT. Smelter production has generally paralleled mine production with 29,300 mT smelted in 1994. As of 1994, there were no refineries in the country, and all products were exported (Ref 7, 18, 22).

Republic of South Africa. Mine production slowly grew until 1984 when it reached a peak of 212,000 mT. Since then, it has slowly decreased to a production of 183,900 mT in 1994. Only small quantities of concentrates were imported during this period. Smelter production of copper has traditionally stood at more than 75% of mine output, with 165,900 mT smelted in 1994. The balance of mine production was accounted for in the form of exported concentrates. Some blister copper is also exported. Refinery production in 1994 was 131,500 mT with about 50,000 mT of refined copper exported (Ref 7, 18, 19).

Zaire (now the Congo). This country has been one of the world's leading producers in the world, but political and economic turmoil have severely affected production. Mine production was nearly 500,000 mT in 1975, but it has fallen substantially since 1985 with only 40,000 mT mined in 1994. Most domestic concentrates are smelted to blister or anode or are obtained by electrowinning from leachate. Some concentrates are exported, and smelter production has consequently been slightly lower than mine production with 38,400 mT smelted in 1994. However, exports of concentrates have declined sharply. A precipitous decline in exports took place in 1989, which is attributed to the September 1988 collapse at the Komoto mine. Exports of Zairian blister and anode copper also fell as local refinery production became starved for lack of feedstock. Refinery production decreased from 25% of the tonnage of unrefined copper in the late 1970s to about one-half that in 1989. In 1994, refinery production was only 26,000 mT. About 85% of the refined copper produced in Zaire is exported (Ref 7, 18, 23).

Zambia. Zambia has historically been one of the major copper producers in the world, but here as well, production of concentrates has fallen by more than 40% between 1975 and 1994, with 698,000 mT produced in 1974 and only 384,000 mT produced in 1994. Smelter production closely followed the production of copper in concentrates with 380,600 mT produced in 1994. For a brief period, small quantities of Zairian concentrates were toll smelted in Zambia. Of the 369,500 mT of refined copper produced in 1994, about 90% of this was exported (Ref 5, 7).

Zimbabwe. While rich in minerals, Zimbabwe has been a relatively minor copper producer. A brief period of activity took place in the 1970s, but the production of copper concentrates has declined gradually since about 1980. Mine production in 1994 was only 9800 mT. Smelter production includes production from imported concentrates and was 19,200 mT in 1994. This was nearly matched by refinery output, with 24,000 mT produced in 1994 (Ref 18, 22).

Asia. For the purposes of statistics, the countries of Asia, the Middle East, and the South Pacific are grouped together. In this group, mine production increased steadily between 1975 and the early 1990s, nearly doubling during that period. Analysts have attributed this partly to the 5.5% per annum growth in the Asian consumer market during the 1980s. The more than 1,130,800 mT produced in 1994 represented nearly 12% of that year's world mine production. In 1994, Asia accounted for 21.2% of the world smelter production and 21.4% of refined copper production (Ref 7, 8, 22, 24).

China. Mine production doubled between the late 1970s and the early 1990s, reaching a respectable 345,700 mT in 1994. Smelter and refinery production have increased similarly because of the import of both concentrates and blister. In 1994, smelter production was 443,700 mT, and refinery production was 684,100 mT. The Chinese copper industry is expanding rapidly to keep up with the expanding requirements of the country's newly developing industrial economy (Ref 7, 8, 9).

India. Mine production in India has grown steadily since 1950, but remains near 50,000 tpy with 50,710 mT produced in 1994. Mine, smelter, and refinery production are nearly balanced, with 1994 smelter production at 50,300 mT, and refinery production at 48,800 mT (Ref 8, 18).

Indonesia. Copper mining in Indonesia began in 1971. By 1975, production had grown more than tenfold. It has continued to increase steadily since then, reaching a respectable 333,845 mT in 1994. The country had no leaching or smelting operations as of 1994, and all mine output was exported in the form of concentrates, a major fraction of which were shipped to Japan (Ref 5, 7, 8, 22).

Iran. In the last quarter of the 20th century, Iranian mine production rose 20-fold. Most of the increase occurred in the mid 1980s, during and after a period of major industrialization. Mine and smelter production followed suit; negligible before 1980. Major mine production was initiated in 1982 with production of 43,000 mT. While this production was sustained during the Iran-Iraq war, it rose to 120,000 mT in 1994. Smelter and refinery production were both 88,400 mT in 1994 (Ref 5, 8, 18, 22).

Israel. The Timna mine, said by some to be the site of King Solomon's mines, was operated until 1983 by a subsidiary of Israel Chemicals, Ltd. Copper output at that time was sold for chemical uses. Proven reserves are reported to be 21 million mT of ore, of which 7.5 million mT are reported to contain 1.5% copper. Mine production ceased in 1974 (Ref 5).

Republic of Korea. The Republic of Korea had a small mine production of about 2000 tpy until 1983. However, by 1987, this had dropped to nil. On the other hand, the Republic of Korea's smelter and refinery production is quite large, at 159,700 and 221,300 mT, respectively, in 1994 (Ref 5).

Japan. Japan has been traditionally a copper-producing country. However, its 50,700 mT of mine production in 1982 has dwindled to less than 6000 mT in 1994. Nevertheless, both smelter and refinery production have approached or exceeded 1 million mT per year since the 1970s. Japan is the largest producer of unrefined and refined copper of Asia. The copper industry is based almost exclusively on imports of raw material. In 1994, Japanese smelter production was 1,122,000 mT, and its refinery production was 1,119,200 mT. Japan is both an importer and exporter of refined copper products (Ref 7, 8).

Turkey. Production of concentrates, as well as smelter output, has remained at several tens of thousands of mT per year during the last quarter of the 20th century. In 1994, mine production was 39,800 mT, and smelter production was 36,000 mT. During this period, however, refinery production tripled to a high of 106,000 mT in 1990. In 1994, it was 104,000 mT (Ref 15, 19).

Malaysia. Copper mining began in Malaysia with the opening of the Mamut open pit mine in 1975. Mamut is a porphyry deposit with reserves estimated at 77 million mT, averaging 0.61% Cu. Mine production in 1994 was 27,700 mT. There are no domestic smelting or refining facilities, and concentrates are exported, mainly to Japan (Ref 5, 18).

Myanmar. Mine production was negligible until 1983, but it has since risen to and remains at several thousand tonnes per year, with 4800 mT produced in 1994 (Ref 18).

Oman. Copper mining in Oman began in 1983 when the fully integrated mine, smelter, and refinery at Sohar came on stream. There are two underground mines, Lasail and Bayda, in operation. A third deposit, Aarja, had not yet been exploited by 1989. Production has fallen from a high of 18,100 mT in 1988 to 4300 mT in 1994. Because the country imports concentrates, smelter and refinery production have always exceeded mine production, albeit both were erratic during the 1980s and early 1990s. Smelter production was 21,600 mT in 1988. It dropped to 15,100 mT in 1990 and 1991 and rose to 31,200 mT in 1994. Refinery production was equally irratic with a 1994 production of 24,200 mT (Ref 18, 19, 22).

Philippines. From a high of more than 300,000 mT in 1981, mine production has continually decreased to just about 102,000 mT in 1994. Until 1982, concentrates were exported and smelted in other countries. In 1983, government-owned smelting and refining facilities were opened, and production at these plants roughly tripled by the early 1990s, reaching a peak of 212,400 mT in 1993. In 1994, smelter production was 168,000 mT. Refined copper production also reached a peak in 1993 at 166,000 mT. In 1994, refinery production was 152,900 mT. As a result of both falling mine production and the rise in local smelting and refining, exports of concentrates have fallen steadily since the 1970s. Increased exports of refined copper have paralleled the decline in concentrate shipments (Ref 7, 8, 18, 19).

Pacific Rim. In 1994, this region, consisting of Papua New Guinea and Australia, represented approximately 6% of world mine production. Mine production rose rapidly in the third quarter of the 20th century, a period of major industrialization. Except for political difficulties with the Bougainville mine, production has been more or less stable.

Australia. Mine production has remained comparatively steady, especially when compared with the large fluctuations that have occurred in other countries. It has risen from the 250,000 tpy level sustained during the 1970s and 1980s to 430,000 mT in 1994. Smelter production at the beginning of the 1990s was exceeded by refinery output because of the production of a modest amount of secondary refined copper from scrap. In 1994, smelter and refinery production were 325,000 and 335,900 mT, respectively. The export of concentrates, once significant, has ceased (Ref 7, 8, 22).

Papua New Guinea. In 1994, mine production was 203,000 mT. Mine production began in 1972 with the opening of the Bougainville mine. Between 1987 and 1990, production at the once prosperous mine fell significantly because of a combination of political and technical problems, but it appeared to have recovered by 1991. However, in 1992, the mine was closed for political reasons. The Ok Tedi mine accounts for all current production. All copper concentrates produced in Papua New Guinea are exported (Ref 8, 16, 18).

Secondary Copper

Secondary copper is the product of remelted copper scrap. The ability to recycle copper, as well as that of copper alloys, is extremely high because most copper is used in the form of metal or alloys whose intrinsic value is sufficiently high to encourage recovery and recycling. This subject will be addressed in more detail in Chapter 6, which discusses the environmental aspects of copper.

Copper-bearing scrap takes several forms, referred to in metallurgical works as "runaround" or "home scrap." Scrap that is generated during manufacture of products and quickly returned to its suppliers is called "industrial," "prompt," or "new scrap," while scrap salvaged from old buildings, transportation systems, etc., is designated "old" scrap. At the end of 1992, it was estimated that the world had accumulated a reservoir of potential. For example, future old scrap totaled 194 million mT, of which the United States held more than 70 million mT (copper content) (Ref 8).

Accounting for all of the scrap existing at any given time is difficult, because copper scrap is recycled by adding it to the melt in conventional smelting and refining processes. There are few metallurgical plants that do not process recycled scrap to some degree. This scrap may or may not be counted in copper production statistics. In an industrialized country such as the United States, about 72% of the total scrap volume is generated locally, for example, in the plant; the other 28% is returned from end-use applications. In the United States, about one-third of all the scrap reprocessed returns to the market as refined copper, and two-thirds is reused as unrefined or alloyed copper.

In the United States, scrap is classified according to its purity. The Institute for Scrap Recycling recognizes 53 classes of copper and copper alloy scrap. Scrap containing more than 99% Cu is designated No. 1 copper; it can be directly remelted and used without further processing. So-called No. 2 copper contains a minimum of 94.5% Cu and usually must be re-refined if it is to be used as copper metal. Other common scrap grades include leaded yellow brass; yellow and low brass; car-

tridge brass; automobile radiators; red brass, and low-grade ashes and residues. Alloy grades are consumed by ingot makers who supply melting stock to foundries. Widely used alloys such as free-cutting brass are recycled directly in the remanufacture of the same composition. This is regarded as a major advantage because it significantly reduces net metal costs to manufacturers. Scrap is also used for the production of copper chemicals, but quantitative consumption data are difficult to obtain (Ref 25).

Scrap is an important alternative source of copper when the market for newly mined copper is tight. As might be expected, scrap plays an important role in influencing the market price of copper. Generally speaking, more scrap is available when the market for copper rises than when it is depressed. However, the significant economic fact about copper scrap is that its availability is relatively insensitive to market price; therefore, the availability of scrap does little to stabilize copper prices. In fact, scrap prices have tended to be more volatile than refined copper prices (Ref 26).

Records of scrap trades are regularly reported by the World Bureau of Metal Statistics (Ref 7, 16).

Copper Commodities in International Trade

The preceding section dealt with the production of copper products emanating from mines, mills, smelters, and refineries. This section considers the international trade in those products, as well as the trade in semimanufactures, or semis.

International trade in copper during the second half of the 20th century took place in two separate circles, the Western capitalist or free-market economies and the Eastern socialist or centrally planned economies. While trade within the two circles was active throughout the period, there was nearly no trade between the two economic blocs. Trade statistics for the Western economies are well documented, presumably accurate and widely available. Similar data for the C.I.S., Eastern Europe, and China were slowly becoming available at the time the present publication was prepared; however, such data were still considered to be unreliable (Ref 6, 26, 27, 28, 29).

Copper enters into international trade mainly in the form of concentrates, blister, anodes, refined copper ingots and cathodes, and copper semis. The volume of international trade in copper castings, copper powder, and copper compounds is small by comparison, and statistics for such products are unavailable or are incomplete.

Trade in concentrates is carried out mainly by mines that do not have an associated smelter, or mines that produce more than their smelters can process, so the surplus is sold. Other mines produce concentrates strictly for sale on world markets or for processing in foreign smelters. Concentrates also become available from time to time when the smelters that generally treat them are closed down for one reason or another.

Blister copper may enter international trade when there is insufficient refinery capacity in the country in which it is produced, or as the result of contractual arrangements for the shipment of blister from the smelter for refining elsewhere. Blister copper is about 98% Cu; however, it also contains the noble metals carried over from the ore.

There is a substantial international trade in refined copper, because manufacturers of copper and copper alloy semis often obtain their supplies from refineries in other countries. It may be helpful to note that smelter and refinery products are sometimes classified as "unwrought" products, in comparison with semis, which are "wrought." Copper-base castings are, of course, also unwrought, but they are generally finished or nearly finished products and are not classified in the same manner as the semis considered here. The exception is continuously or semicontinously cast products, such as wire rods, which are true semis.

Concentrates

Most of the international trade in concentrates involves highly industrialized countries lacking adequate copper ore reserves. For example, while Japan was one of the earliest countries to mine, smelt, and refine copper, the present ore deposits in the country are insufficient to satisfy its needs. The basis for the 1990s trade in concentrates, therefore, originated largely during the 1970s, after the Japanese built smelters and refineries to integrate their copper industry to the largest extent possible. In 1994, Japan imported 1,230,200 mT of copper concentrates, accounting for 39% of the net world concentrate market. Germany was the next largest importer, followed by Portugal, the Republic of Korea, Belgium, and Finland. Asia imported 56% and Europe imported 38% of the net global market in copper concentrates (Ref 7).

Leading exporters of copper concentrates in 1994 were Chile, with 56% of net exports, followed by Indonesia, with 19% of net exports, and Papua New Guinea, with 11% of net exports. Aus-

tralia, Mongolia, the United States, and Canada were the other major exporters. As noted earlier, exports from Papua New Guinea gradually increased since 1984, but closure of the Bougainville mine had a severely adverse effect on this trade. Concentrate exports from the United States began increasing in 1984 and were 262,400 mT in 1994. Shipments were largely to Canada (54%), Japan (33%), and the Philippines (8%). Exports from Indonesia had increased by $4\frac{1}{2}$ times during this period (Ref 7).

It should be noted that the United States is both an importer and exporter of concentrates. While exporting 262,400 mT in 1994, it imported 80,900 mT; largely from Chile and Canada.

Worldwide trade in concentrates, amounted to approximately 1.2, 1.5, and 2.0 million tonnes in 1976, 1988, and 1994, respectively, a 3.1% compound annual increase over the period. Between 1988 and 1994, such trade increased at the rate of 5.5% (Ref 7, 22).

Blister and Anode Copper

Blister copper is produced at smelters in converter furnaces. Converting represents the first metallic stage of refining. This product is full of gas pores, hence its name. It contains about 98% Cu, and all of the precious metals contained in the ore. These are removed in subsequent refining steps.

Blister copper is traded in the form of cakes, a type of ingot, and anodes. International trade in blister copper is smaller than trade in concentrates because refineries are more inclined to be located near smelters than smelters are to mines. Mine output concentrates, therefore, tend to be more mobile on the world market than blister produced at smelters. Anode copper is a fire-refined copper grade made from blister. It is cast and traded in the form of large flat anodes suitable for electrolytic refining, the next purification stage.

In 1994, international world trade in blister and anode copper amounted to 573,100 mT. Chile, México, and Perú accounted for 73% of all recorded exports of unrefined copper in 1994. Significant exports are also recorded from Spain, Finland, Namibia, and South Africa.

Until 1990 Zaire was the most important exporter of blister copper, shipping between 235,000 to 255,000 tpy between 1984 and 1988. This quantity, which accounted for more than one-half of the total production of blister and anode copper, declined significantly after 1989 because, among other factors, of a collapse at the Kamoto mine.

Zaire's 1991 blister exports amounted to only 117,800 mT. Between 1987 and 1990, Chile ranked second in blister exports, shipping between 150,000 and 250,000 tpy. In 1994, Chile's exports were 198,100 mT. In spite of this increase in exports, Chile's exports are much smaller as a proportion of domestic blister production than those of other major blister producers. Perú was the third largest exporter of unrefined copper during that period; exports from that country reached 126,700 mT in 1985. In 1991, Perú became the second largest blister exporter in the world, with 101,000 mT shipped. However, by 1993, exports of blister and anode copper by México had exceeded exports by Perú with each country shipping 120,300 and 98,000 mT respectively in 1994 (Ref 7, 23, 28).

Principal blister importers are located largely in Western Europe, where, for example, Belgium imported 214,000 mT in 1984 and 147,600 mT in 1994. Germany and Portugal are also major European importers. Total imports into Western Europe, however, fell from 477,900 mT in 1984 to only 282,900 mT in 1994. The United States is a significant importer, with 120,200 mT of blister and anode copper imported in 1994 (Ref 7, 19).

Refined Copper

The term "refined copper" now refers to metal with a copper content of at least 99.99%, although some special grades are as pure as 99.9999%. Refined copper is the product of either electrolytic refining of anode copper or electrowinning copper from hydrometallurgical or leach solutions. The purest commercial refinery shapes are electrorefined or electrowon copper cathodes and, with the exception of minor differences in oxygen content, certain types of continuously cast rod, such as those obtained by the dip forming and Outokumpu up-cast processes. Since the mid-1980s, conventional continuously cast wire rod has completely replaced traditional wirebars as the raw material from which electrical wire is produced. As a result, refined cathode and wire rod copper are now by far the most important forms used in international trade (Ref 30).

Refined copper is the most important copper commodity in terms of volume traded. In 1994, about 4 million mT of refined copper were traded internationally. This represented about 34% of the total world consumption of both primary and secondary refined copper.

Since 1978, the largest source of refined copper in the world has been Chile, which, at least until 1989, exported nearly all of the roughly 1 million mT of refined copper it produced annually, with 1994 exports of 1,193,000 mT. Zambia had been the next largest exporter, at somewhat less than 0.5 million mT per year; however, it was replaced by Canada in 1994. Zambia's shipments have been declining since 1984 along with a decline in refinery output. Zambia's 1994 exports were 351,500 mT, whereas Canadian exports were 388,600 mT. Belgium and Germany are also significant exporters of refined copper with 151,700 and 149,100 mT, respectively, in 1994 (Ref 7, 19).

Western European countries, which were large importers of concentrates and unrefined copper at the beginning of the 1990s, also imported more than 1.5 million tpy of refined copper during that period. Their total refined export volume was about 500,000 tpy, much of it in reciprocal trade among one another. Germany was the leading European importer of refined copper followed by France, Italy, and the United Kingdom. In 1994, these refined copper imports were 522,700, 462,700, 394,300, and 259,000 mT, respectively. Japan, which ranked first among the importers of the world of unrefined and refined copper in 1994, imported 354,800 mT of refined copper, or 28% of its total copper imports. The United States was the second largest importer of refined copper in 1994 with 466,800 mT. However, in the same year, the United States exported 157,600 mT of refined copper, making it a net importer of 306,700 mT (Ref 8, 20).

Recent import leaders among the less industrialized, non-OECD (Organization for Economic Cooperation and Development) countries include the Republic of Korea with 1994 imports at 263,400 mT and Brazil with 64,100 mT (Ref 7, 19).

Semis

Total world production of copper-base semis in 1994 was almost 12 million mT, of which 7.3 million mT were pure copper and 4.4 million mT were copper alloys (Ref 7, 16, 18, 19, 27–29).

World export trade in copper semis had increased at about 6.3% per annum between 1986 and 1989, reaching a volume of nearly 2,300,000 mT, valued at U.S. $10 billion. The largest fraction, about 43% of the total, was traded in the form of copper and copper alloy wire. Other significant market fractions included plate, sheet, and strip (26%); bars and sections (17%), and tubes (14%). After 1989, the rate of growth in trade volume decreased to 0.3% per annum, reflecting the global economic recession of the time (Ref 24).

The major exporter in 1994 was Germany, followed by Belgium, France, Japan, and Italy. Together, these countries accounted for 58% of the total semis exports. Western Europe was the largest copper semis exporter in the world having accounted for 49% of the total export activity. Japan and the United States were the next largest exporters, at 8.7 and 6.1% of the total, respectively. Export growth rates increased rapidly for the United States, Taiwan, Portugal, and Spain.

Western Europe also imported the largest share of copper and copper alloys semis, followed by the United States, Canada, and Taiwan. At the end of the 1980s, the Republic of Korea, Spain, Hong Kong, and Japan had the highest growth in import rates for semis. The bulk of these imports originated in the United States, Taiwan, Portugal, and Spain. Between 1986 and 1991, corresponding export growth rates for these countries increased from 13 to 25% per year (Ref 7, 28).

World Consumption of Copper and Copper Products

In addressing this subject it is necessary to clarify two important points—nomenclature of the product forms and the quality of the statistics that are available.

Copper consumption is normally reported by product form, therefore, it may be helpful to review a few terms and definitions. Thus, "refined copper" includes "new" copper derived from mine production, which is called "primary copper," plus that which is re-refined from recycled scrap, "secondary copper." The two basic types of refined copper are designated "electrolytic" and "fire refined" coppers, depending on the refining process used. Also, although not acknowledged in technical standards, minor distinctions can be made between electrowon and electrorefined copper cathodes. The former are derived from solvent leach/solvent extraction (SL/SX) copper recovery; the latter from electrolytic refining of pyrometallurgically obtained anodes. There are a number of subclassifications of electrolytic and fire refined copper, but these mainly describe minor compositional nuances.

Both types of refined copper are brought to the world market in the form of what are known as refinery shapes. The most common refinery shape from the mid-1980s on, has been the cathode, which replaced wirebars as the standard unit of trade. The high-grade or Grade A cathode is now the standard. Other important shapes include continuously or semicontinuously cast billets and wire rods, ingots or cakes, slabs, and powder.

Primary consumers of refined copper are semifabricated copper commodities or semis manufacturers. These industries comprise wire rod mills, brass mills, ingot makers, foundries, powder plants, and a few additional industries whose individual consumption is so small that they may be conveniently grouped as "others." Semis manufacturers process refinery shapes into alloyed and unalloyed wire, strip, rod, and other semifinished copper and copper alloy products. These products are in turn consumed by the manufacturers of a countless variety of finished products for end-use applications.

A more detailed description of copper products is given in Chapter 3 "Products"; however, it can be noted at this point that wire is by far the most common copper product. In 1990, wire accounted for 51% of all copper consumed in the Western world. Copper tube ranked next in importance at 12% of consumption, followed by alloy rods, bars, and sections, 11%; copper sheet and strip, 8%; alloy sheet and strip, 7%; casting alloys, 6%; copper rods, bars, and sections, 2%; alloy tube, 2%; alloy wire, 1%; and chemicals and powder, estimated to constitute less than 1%, but may be somewhat higher (Ref 31).

Accurate statistics for the production and consumption of copper and copper alloy semis are not uniformly available on a country-to-country, basis other than in gross tonnage. Further, product definitions differ from country-to-country, and competitive factors impede the public reporting of production and consumption numbers of individual product forms.

Copper semis are sold either to an end-use customer; for example, a manufacturer, or to a distributor. While a company knows the use of its product when it is sold to a manufacturer, it does not know who is ultimately using the product when it is sold to a distributor. Consequently, data representing the use of copper in consumption sectors must be an estimate based on what knowledge is available.

Copper consumption is often reported in terms of uptake of refined copper by semis fabricators. According to this convention, total copper consumed by all semis fabricators in a particular country are reported as the "apparent" copper consumption of a country, because neither origin of the refined copper nor ultimate destination of the semi fabricator's products can be explicitly identified. Similarly, the net consumption (domestic production plus imports minus exports) of a country of copper semis in the manufacture of end products reflects only its apparent consumption of copper goods, because neither the origin of the semis nor the ultimate consumer of the end products are known.

The nature of trade is such that countries might concurrently export and import copper, perhaps even in the same product form. This is one reason why it is difficult to differentiate between real and apparent copper consumption. The true consumption picture is also clouded by the quantity of metal held in stocks by semis fabricators and end users as a hedge against, or speculation with, fluctuating copper prices.

Finally, the degree to which consumption of refined copper is supplemented by the direct reuse of copper scrap without re-refining must be understood. Re-refined and directly used scrap together add more than 40% to the supply of copper available from primary sources. Estimates place the ratio of refined scrap to direct use scrap at 80% in Europe, 47% in the United States, and 18% in Japan. From 1982 to 1994, an average of 44% of all copper consumed in the United States was derived from new and purchased old scrap; the corresponding figure for European economic community was 41%.

Global Copper Consumption Patterns

Table 1.3 shows the average annual apparent copper consumption in the world for the first nine decades of the 20th century. Copper consumption has grown from 700,000 mT between 1901 and 1910 to more than 11 million mT per year by 1994. The bulk of this consumption was, of course, directly attributable to the growing use of electricity in an increasingly industrialized world economy. Estimates by the U.S. Bureau of Mines state that world refined copper consumption will rise from 11.1 to 14.6 million mT, an average rate of about 3.1% per year, between 1990 and 2000 (Ref 5, 6).

For the period between 1970 and 1974, the average apparent consumption of copper semis in Japan, the United States, and Western Europe represented 92.5% of the total Western world. The next two decades saw the emergence of semis manufacturing facilities in developing countries, as well as falling exports from developed to developing countries and changes in U.S. home demand for both semis and finished goods. Together, these factors led to a realignment in the distribution of copper semis consumption. Whereas Japan, the United States, and Western Europe together could still account for about 92% of the apparent semis consumption in the world in 1989, Japan's share had risen from about 18% in 1970 to 1974 to 21% at the end of the 1980s. Western Europe maintained roughly 40% of world semis consumption, while U.S. consumption fell from about 34 to 31% during the same period. In 1991, the so-called market economy countries accounted for 80% of the world's copper consumption, virtually all of it in the Americas, Europe, and Asia, as shown in Fig 1.5.

There was a remarkable increase in the consumption of refined copper by Malaysia, Thailand, Indonesia, and Singapore in the last quarter of the 20th century. Collectively, these Southeast Asian countries increased their consumption by more than a factor of 100. Between 1974 and 1991, consumption increased nearly 15-fold in the Republic of Korea and nearly 20-fold in Taiwan, clearly reflecting the movement of copper-consuming manufacturing operations to this region of the world.

Western Europe maintained about 27 to 29% of the apparent refined copper consumption in the world during the second half of the 1980s to 1994. Consumption in the United States during this period was 22 to 23%, and Japanese consumption was about 12%. Non-Japanese Asian consumption rose from 5.4 to 22% of the world's total during the same period, largely on the strength of increases in the Republic of Korea and Taiwan. Eastern Europe's share of the world's refined copper consumption fell from 24% in 1985 to 12% in 1994 (Ref 7, 15, 16, 27–29).

The following paragraphs briefly describe copper consumption in selected countries and regions of the world. Trade (import/export) data are included to place copper consumption figures in perspective wherever appropriate. More detailed discussions regarding the consumption of specific

Table 1.3 Growth in the consumption of refined copper, 1901 to 1994

Period	Average consumption, tonnes per year	Annual average growth, %
1901 to 1910	689	6.4
1911 to 1920	1,110	4.7
1921 to 1930	1,360	2.1
1931 to 1940	1,655	2.0
1941 to 1950	2,380	3.7
1951 to 1960	3,774	4.6
1961 to 1970	6,145	4.6
1971 to 1980	8,609	2.5
1981 to 1990	10,037	2.2
1991 to 1994	11,095	1.9

products, and consumption in various market sectors, is contained in a later section of this chapter.

Western Europe. Western Europe absorbed more than 28% of all the refined copper produced worldwide in 1994. The region also imported more than 21% of the worlds' copper concentrates, which were smelted, refined, and consumed in that year. In addition, recycling, re-refining, and direct use of scrap for local consumption all take place to a significant extent. Most semis produced in Western Europe are consumed within the region, although a small net export trade in copper and copper alloy semis docs cxist. The major fraction of copper exports is in the form of finished products (Ref 7, 24).

Between 1986 and 1989, Western Europe's increase in industrial production spurred a huge demand for copper semis, which, in many cases, were imported at rates three to four times the level of domestic production. In 1994, the West European market consumed 5.1 million mT of copper semis. This accounted for more than 43% of the world's copper semis production (Ref 7, 24).

Austria. In 1994, apparent consumption of refined copper in Austria was 23,262 mT. However, its net consumption (production plus imports minus exports) of copper semis was 137,741 mT. Statistics on the semi product forms consumed domestically are not available. However, Austria's considerable industrial base includes a small semis production sector. In 1994, these semis industries consumed 54,148 mT of refined copper. Most of the consumption went for the production of wire, which accounted for 30,574 mT, and tube, which consumed 16,827 mT.

France. Between 1950 and 1974, refined copper consumption in France rose at an average annual rate of 5.4%, reaching some 414,000 tpy at the end of that period. Consumption flattened thereafter and did not change appreciably until the late 1980s, when annual consumption suddenly approached 460,000 tpy. By 1994, consumption of copper semis had grown to 584,714 mT, although growth rates in the early 1990s were only about 2.4%, probably reflecting the general economic conditions of the time (Ref 7, 24).

Germany. In the Western states of Germany, the consumption of refined copper increased by between 5 and 6% per year between 1950 and 1985, the period corresponding to the country's so-called "Wirtschaftswunder," or economic miracle. Consumption, which was little more than 180,000 mT in 1950, grew to 896,000 mT in 1989, an average rate of some 10.1%. In the former German Democratic Republic, consumption also grew, but at a rate a little more than one-half that in the West. The apparent refined copper consumption in unified Germany surpassed 1 million mT in 1990 and was 999,500 mT in 1994. Net consumption of copper semis in 1994 was 1,295,814 mT (Ref 7, 15, 29).

Germany was the world leader in semis consumption in 1989 and 1990 when consumption exceeded 1 million mT. In those years, German apparent semis consumption corresponded to 24.9 and 27.9%, respectively, of the European totals (Ref 32, 33).

As might be expected, the commercially most important product in the period of major economic growth (1950–1985) in Germany was building wire, followed by plumbing and heating tube, for which Germany is Europe's largest consumer. Aside from these uses, other major applications in Germany for copper during the late 1980s in-

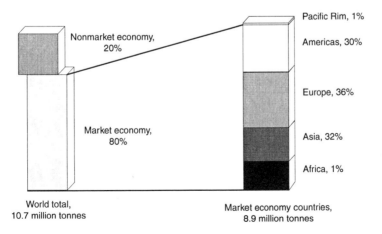

World total,
10.7 million tonnes

Nonmarket economy, 20%

Market economy, 80%

Pacific Rim, 1%

Americas, 30%

Europe, 36%

Asia, 32%

Africa, 1%

Market economy countries,
8.9 million tonnes

Fig. 1.5 World copper consumption by region, 1991

cluded industrial and consumer electronic products, automotive accessories, machinery, and appliances. In 1990, the electrical industry took up 43.3% of the semis; the building, machinery, transportation, and consumer goods manufacturers used 50.3%. Castings made up 5.6% of the market and chemicals and other products, 1% (Ref 33).

German trade reports in the early 1990s spoke of competition from aluminum for building wire and optical fibers for communications, which, until that time, had not been estimated to capture more than 5% of the German market. Aluminum and optical fibers were, in fact, thought of as factors that would expand the existing market for copper. The 1970s trend to replace copper tubing with plastics in Western Germany, and with plastics and galvanized steel in the Eastern states, had been completely overcome by the early 1990s. Instead, copper tubes were preferred and at times even used as replacements to improve existing buildings. The use of copper sheet for roofing and other architectural applications was also favored. The increased use of copper for electric motors, controls, and wiring harnesses in automotive vehicles were supposed by some to compensate at least in part for the replacement of copper and brass in radiators by aluminum, but copper usage remained strong in chemical and industrial equipment, coinage, and other applications. All of these trends tended to maintain a high demand for copper semimanufactures (Ref 33).

Greece, Portugal, and Spain. These countries joined the European Economic Community (EEC) in 1985. In the three years prior to that event, the countries' combined consumption of copper semis had decreased by nearly 3%. Consumption rose between 1985 and 1990, a fact attributed to growing investment in the countries' industrial and construction sectors following their entry into the EEC. Greece ran counter to this trend, however, posting decreases in copper exports during the early 1980s, while increasing imports of all types of copper and copper alloys by fully 105% between 1982 and 1984, before entry in the EEC. Portugal and Spain also increased their imports of copper semimanufactures, but at significantly smaller rates (25.4% per annum for Spain; 16.4% per annum for Portugal). Despite the use of plastic for plumbing tube in some low-income housing, there has been a general increase in the use of copper in building construction in these countries, amounting to 81% in Greece, 31% in Portugal, and 42% in Spain.

In 1994, consumption of refined copper was 33,600, 10,400, and 178,000 mT, respectively, for

Greece, Portugal, and Spain. Net semis consumption was 47,633 and 293,085 mT, respectively, for Portugal and Spain. Data for semis consumption for Greece are not available. As elsewhere, building wire is the major product in these countries, followed by copper tube, heat exchangers, and a variety of others in both building and nonbuilding market sectors (Ref 7, 21, 34, 35).

Italy. Refined copper consumption in Italy grew at an average annual rate of 5.7% between 1950 and 1974, then fell to near 1% per annum for a decade, after which, beginning in 1984, it resurged at the rate of 6.8% per annum. During the latter period, however, the import of copper semis increased at a somewhat higher rate. Telecommunications and electronic equipment, automotive components, electrical connectors, measuring instruments, and tubes and wires for building construction and renovation accounted for most of the increased demand. In 1994, refined copper consumption was 480,000 mT; net copper semis consumption was 1,051,744 mT (Ref 7, 21, 24).

The Netherlands. The pattern of copper consumption in Holland can be inferred from data for 1988, when the country consumed 20,000 mT of refined copper and imported 103,021 mT of copper and copper alloy semis. Among the imported semis, wire accounted for about one-half of the total, followed by alloy semis of various kinds (approximately 35%), and miscellaneous copper products. By 1991, semis imports had risen to more than 110,000 mT. In 1994, refined consumption was down to 18,630, mT whereas net semis consumption was an estimated 146,000 mT (Ref 7, 19, 24).

Japan. Following its reconstruction after World War II, Japan has consistently been second only to the United States in the production and consumption of refined copper and copper semis. In 1994, refined consumption was 1,374,873 mT, and net consumption of semis was 1,923,793 mT. Wire constituted 52.6% of consumption; however, tube constituted only 10.3% (Ref 7, 21).

Scandinavia mines on the order of 100,000 mT of copper annually, making the region a moderate-sized primary producer. However, the region is more important as a semis producer, having consistently turned out more than 300,000 mT in recent years. Apparent consumption of refined copper has grown steadily, especially in Norway, which saw consumption double between the late 1970s and 1989. Net consumption of copper semis for the region in 1994 was 222,375 mT. Imports were 199,261 mT, and exports were 125,423 mT (Ref 7, 21).

Switzerland produces no primary copper. It does have a few copper melting facilities for local and imported scrap and some semis manufacturers, but it is predominantly a semis importer. In 1994, the country consumed 5220 mT of refined copper. It imported 97,663 mT of copper semis, produced an additional 84,074 mT, and exported 27,095 mT for a net domestic consumption of 154,642 mT of copper semis in 1994 (Ref 7, 21).

United Kingdom. The consumption of refined copper in the United Kingdom has grown by about 40% since 1950; however, periods of severe economic depression and shrinkage in the domestic semimanufactured products industries have caused consumption to fall far below even mid-century levels. However, by 1990, there had been a significant increase in consumption over earlier years. In 1994, the United Kingdom had a net consumption of 465,330 mT of refined copper. Net consumption of copper semis was 480,591 mT in that year. Major increases were seen in products for telecommunications and electronics, automotive components, electrical goods, instrumentation, and building construction, mainly tube and wire (Ref 7, 24).

It has been said that the copper industries in France, the United Kingdom, and Italy have been slower to change and innovate than those of other Western European countries, and that this has affected—or has at least been reflected in—their growth of copper consumption. On average, these countries exported 28% of their semis production, mainly to European, Middle East, and African countries, and imported 20% of their semis from other European countries and the United States.

Eastern Europe and the C.I.S. Little reliable data are available on copper consumption in Eastern Europe in years prior to 1991. Before the dissolution of the Soviet Union in 1991 and the institution of "glasnost," whatever consumption statistics that did exist were centrally processed in Moscow and were treated as state secrets. It has been estimated, however, as the former Eastern Bloc accounted for about 20% of the world's copper consumption in those years. By 1994, however, this region of the world consumed only 6.5% of the copper in the world. The following comments should be viewed with the uncertainties of the Soviet era in mind (Ref 6).

Albania increased its consumption of refined copper from 6400 tpy for the period between 1977 to 1981 to 16,000 mT in 1989. In 1994, however, it consumed only 1300 mT. It cannot be stated that the country is self sufficient in copper because

there have been no recorded imports or exports of consequence (Ref 18, 21).

Armenia. No data on copper consumption are reported.

Bulgaria was estimated to have consumed 56,000 mT of copper in 1989; however, by 1994, this was down to 16,900 mT. Small trade volumes have been recorded between Bulgaria and Western countries, the largest of which consisted of 6500 mT of refined copper imported in 1988 (Ref 18).

C.I.S. In 1991, C.I.S. copper consumption was 750,000 mT, 40% lower than that of 1981. By 1994, this had fallen further to 559,700 mT, or less that half of that of Soviet years. In the early 1990s, the C.I.S. began responding to free market competition by, among other acts, closing or replacing inefficient production facilities. As a result, some analysts predicted as much as a fourfold decrease in raw materials consumption for some products. Little information is available regarding copper consumption in the Ural region, as production and consumption were included among the annual statistics of the former Soviet Union. The bulk of refined copper production was exported to the industrialized countries of the USSR; however, there is an important semimanufactures plant located in Revdensk. No data are available for copper consumption in Uzbekistan and Western Siberia in years prior to 1994 (Ref 18).

Czech and Slovak Republics. The former Czechoslovakia is estimated to have consumed about 97,000 mT of refined copper in 1986, a figure that had been slowly developing during the previous several years. Small quantities of refined copper were imported. All of the copper and copper alloys produced were consumed domestically. However, by 1994 an 86% decrease in consumption, to 13,900 mT, had occurred (Ref 18, 21).

Hungary. Despite its rapid restructuring after the dissolution of the Eastern Bloc, Hungary remains a relatively small copper consumer, utilizing 34,000 and 15,900 mT in 1990 and 1994, respectively. Consumption is expected to rise in the near future, however (Ref 18).

Kazakhstan. There are no data available on domestic copper consumption prior to 1994. At least until the early 1990s, the country lacked a sufficient industrial base to absorb local copper production (Ref 18).

Poland. One of the largest Eastern European copper consumers over the last quarter of the 20th century, Poland has posted refined copper consumption levels ranging between 140,000 tpy in 1974 and 241,000 tpy in 1989. As a consequence of the restructuring of the former Soviet economies and by then the worldwide economic recession, consumption levels decreased somewhat thereafter. In 1994, consumption was down to 137,600 mT. A positive trend was anticipated, however. Despite the fact that the exports of semis dropped significantly, Poland's exports of refined copper to Western countries appeared quite strong (Ref 7, 21, 36).

Romania. There are four producers of semi-manufactures in Romania, whose output include wire and cable, copper and brass rolled products, rods, sections, and tubes. The refined copper consumption in 1994, at 25,700 mT, was down sharply from that of a decade earlier when it had been as high as 64,000 mT (Ref 7, 21, 36).

North America. Over 25% of the copper mined worldwide is consumed in North America, making the continent both one of the largest producers and largest consumers of the metal.

Canada. Canada consumes about 200,000 mT of new copper per year. In addition, 60,000 to 70,000 tpy are produced by re-refining scrap, or are consumed directly as scrap in the manufacture of copper products. In 1994, consumption of refined copper in Canada was 199,400 mT. Its net consumption of copper semis in that year was approximately 25,000 mT (Ref 7, 9, 21).

United States of America. The United States is the leading copper consumer in the world. The country's total consumption was 3,652,900 mT in 1994, of which slightly more than two-thirds, 2,678,000 mT, was refined copper and the balance scrap. This represents an average annual growth rate of about 2.2% per annum for the period between 1984 and 1994 and a virtually flat average growth of only 1.3% per annum between 1974 and 1994. Steeper growth did occur, however, during the housing and construction boom between 1982

and 1986, during which consumption rose at an average annual rate of 6%. In 1994, net consumption of copper semis was 3,378,277 mT (Ref 7, 8, 21).

The supply and consumption of copper in the United States is shown in a diagram in Fig 1.6. The chart tracks the flow of both primary and secondary copper through its consumption by wire rod mills, brass mills, foundries, powder plants, and other industries, and ultimately to its consumption within the five major industrial market sectors: building construction, electrical and electronic products, industrial machinery and equipment, transportation equipment, and consumer and general products. Approximately 78% of the newly mined and refined copper goes to wire rod mills. Brass mills consume about 20%, and all other manufacturing industries combined consume less than 2% (Ref 8).

During the early to mid 1980s, U.S. demand for copper was largely driven by the strong performance of the residential and nonresidential construction markets, which by 1992 accounted for more than 42% of U.S. copper consumption. Building wire experienced a 72% increase in consumption during the 1980s, followed by plumbing tube, 27%, and power cable, 19%. The increased wire consumption could be attributed to greater utilization of electrical appliances and electronic systems in offices and houses, while increased consumption of copper tube resulted from larger houses and flats equipped with larger kitchens and more bathrooms. Table 1.4 shows the distribution of copper products in average American homes and apartments and the amounts of copper used in passenger cars (Ref 8, 38).

South America. Only 302,400 mT of refined copper were consumed in Latin America in 1994.

Table 1.4 Copper and copper alloy usage in U.S. dwellings and automobiles

Dwelling applications	Single-family house, 2100 ft^2		Multi-family unit, 100 ft^2	
	kg/unit	lb/unit	kg/unit	lb/unit
Building wire	88.3	195	56.6	125
Plumbing tube, fittings, valves	68.4	151	37.1	82
Plumbers' brass goods	10.9	24	9.1	20
Built-in appliances, including heating, ventilating, and air conditioning	21.3	47	17.2	38
Builders' hardware	5.4	12	2.7	6
Other (wiring devices, telephone system, heating system, water heater, etc.)	4.5	10	3.2	7
Total	**198.8**	**439**	**125.9**	**278**

Automobile applications	Weight, kg/car	Weight, lb/car
Wiring	15.9	35
Heat exchangers, alloys, etc.	7.2	16
Total	**23.1**	**51**

Source: Copper Development Association Inc.

This underscores the position of the region as the leading copper producer, but a relatively small per capita consumer of the metal.

Argentina. A sharp decline in copper consumption toward the end of the 20th century can be attributed to the combined influence of global recession and internal political/economic difficulties. In 1994, consumption of refined copper was 55,100 mT.

Brazil was the leading refined copper consumer in South America during the last decades of the 20th century. After 1950, consumption grew briskly to somewhat more than one-quarter million mT by 1986. Continued rapid expansion was forecast at the time, but consumption gradually decreased thereafter. It recovered to just above 170,000 mT of refined copper in 1991, but dropped back to 141,700 mT in 1994. Net consumption of semis in 1994 was 138,975 mT. Consumption is mainly relegated to electrical conductors (55% of total consumption), building construction (9%), the transport industry (9%), general mechanical manufacturing (7.7%), and other applications (14.5%) (Ref 39).

Chile. Annual domestic consumption of refined copper in Chile had remained near 50,000 mT for the late 1980s and early 1990s; however, in 1992, it began to increase such that in 1994 it was 86,300 mT (Ref 7, 29).

In contrast to its enormous mine and smelter production, of more than 1,000,000 mT, the output from Chilean semis manufacturing facilities is relatively meager. However, in addition to its small domestic semis industry, Chilean companies have formed joint ventures for continuous casting in Germany and France and a partnership in the semis manufacturing industry in China. Semis produced locally, apart from continuously cast electrolytic tough pitch copper rod and oxygen-free copper, include semicontinuously cast billets, wire, plate, sheet, strip, foil, shapes, and seamless and welded tubes. Finished copper products manufactured in Chile include cables, architectural hardware, electric motors and connectors, gas, steam and water valves and fittings, and domestic appliances. These products are consumed domestically, but are also exported to other Latin American countries, the United States, and Europe (Ref 7, 39).

México. Until the end of the 1970s, the Mexican copper industry concentrated on smelting and refining with the aim of becoming self-sufficient in refinery products. However, in 1980, Mexican refined consumption rose sharply, an event some analysts suggested was related with the oil price increase of 1979. In 1987, Mexican refined copper consumption peaked in 1993 at almost 137,000 mT, and remained near that level through 1994 (Ref 40).

Perú. Like several of the major copper-producing nations, the population and industrial and social infrastructures in Perú are far too small to absorb the quantity of copper the country produces. Therefore, like Chile, Zaire, and other large raw material suppliers, Perú produces and exports far more copper than it consumes. Perú does, however, have a modest wire and cable industry, and this is largely responsible for the refined copper consumption. In 1994, Perú consumed 19,200 mT.

Africa. Because it is so closely linked with the use of electricity, copper consumption is a reliable barometer of a region's degree of industrialization. Thus, while Africa is among the leading copper producers in the world, it has been, on the whole, only a minor consumer. Per capita copper consumption on a scale approaching that of Europe or North America can only be seen in the relatively more industrialized districts of the Republic of South Africa and to a lesser extent, in Zimbabwe.

Republic of South Africa (R.S.A.). In 1977, the R.S.A. exported nearly twice as much refined copper as it consumed domestically. By 1989, for exemple, after the imposition of trade restrictions by the United Nations, the balance had changed to the extent that exports exceeded domestic consumption by only about 10%. Ironically, the trade restrictions created a modest 20,000 tpy market expansion for copper within the Republic. In 1994, refined copper consumption was 82,700 mT— nearly twice the level of exports (Ref 18).

Zambia. Most of the modest consumption of refined copper in Zambia can presumably be accounted for by a small wire rod mill at Luanshaya. However, consumption has been steadily increasing, from 6000 mT in 1985 to 8000 mT in 1990 and to 18,000 mT in 1994 (Ref 18).

Zimbabwe typically consumes 10,000 to 15,000 mT of refined copper per year, largely for the manufacture of wire and cable. In 1994, its consumption was 12,000 mT. Not unlike the R.S.A., Zimbabwe was obliged by the rigors of economic sanctions, imposed during the 1970s, to develop her own copper-intensive industrial infrastructure. Sanctioned copper-containing consumer products such as transformers, switchgear, and appliances, and automotive motors were simply manufactured domestically, using semis produced largely within the country.

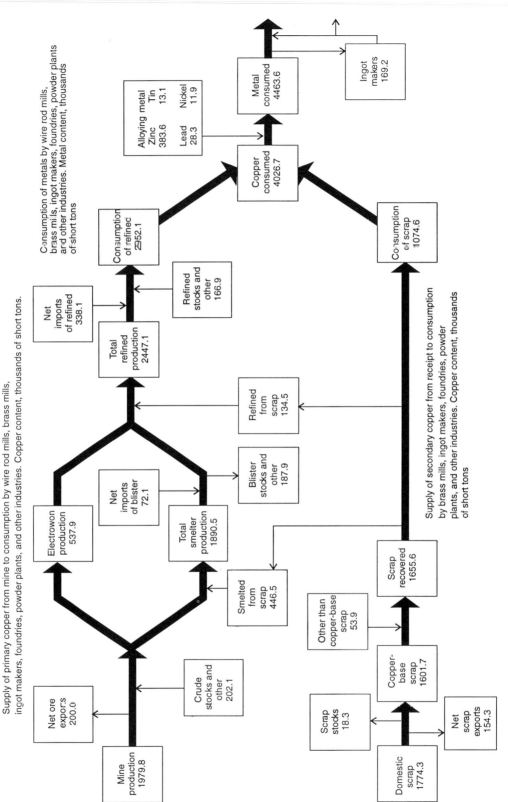

Supply of primary copper from mine to consumption by wire rod mills, brass mills, ingot makers, foundries, powder plants, and other industries. Copper content, thousands of short tons.

Consumption of metals by wire rod mills, brass mills, ingot makers, foundries, powder plants and other industries. Metal content, thousands of short tons

Supply of secondary copper from receipt to consumption by brass mills, ingot makers, foundries, powder plants, and other industries. Copper content, thousands of short tons

Fig. 1.6 Copper supply and consumption in the United States, 1994

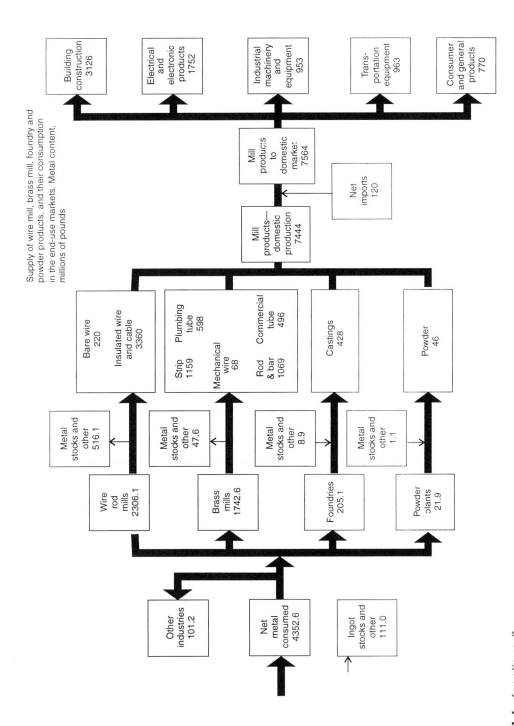

Fig. 1.6 (continued)

Asia. As in the discussion on production, the Middle Eastern countries of Iran and Turkey are included in the consumption statistics for Asia. During the 1980s, Asia was the fastest growing refined copper consuming market in the world, with a 5.2% per annum average growth rate for the decade. Short-term rates spurted as high as 8.5% between 1986 and 1989. The rise in consumption has continued to grow since then, beginning at 2,500,000 mT in 1989, reaching 3,100,000 mT in 1990, and then 3,900,000 mT in 1994. These figures are about 10% higher when recycled scrap is included. Japan was the major consumer, followed by China, Taiwan, and the Republic of Korea (Ref 24).

Consumption also grew rapidly in smaller Asian nations during the 1980s, principally in Malaysia, Hong Kong, Indonesia, Singapore, and Thailand. Most of these countries do not produce refined copper and must rely entirely on imports. The similarity ends there, however, because these countries are at very different levels of economic development, which is reflected in their patterns of copper consumption.

At the beginning of the 1990s, imports into the Asian market accounted for between 18 and 20% of the world semis trade. About 87 to 90% of this tonnage could be accounted for as intra-Asian imports, while the remaining 10 to 13% were exports from Europe and the United States.

Peoples' Republic of China. The Peoples' Republic of China has been steadily increasing its consumption of refined copper. Consumption of 400,000 mT in 1982 more than doubled by 1992, reaching a peak of 943,300 mT in 1993. In 1994, however, consumption had fallen to 745,700 mT. In the same year, net consumption of copper semis was 995,524 mT. Most of this consumption was for wire and cable. The country has one tube mill whose 50,000 tpy capacity had not yet been reached by 1994.

Hong Kong has become the largest copper semis importing market in Asia. In recent years, the market has experienced rapid growth and high volatility. In 1989, for example, semis imports rose 22.7%, reaching about 1,300,000 mT. Semis consumers include countless small manufacturers of electrical products, accessories for the garment industry, and decorative goods. Manufacturing units are mainly family-owned businesses. The highest-value semis imported include copper foils; copper tubes, pipes, and fittings; copper powder and flakes; and brass tube, pipe, and fittings. Major products in terms of import volumes include copper wire; copper-clad laminates; brass plates,

sheet, and strip; and brass rod, bar, and sections. In 1994 Hong Kong consumed only 15,000 mT of refined copper. In 1994, Hong Kong imported 227,701 mT of copper semis for internal consumption (Ref 7, 24).

Brass sheet, strip, and rod are used mainly in Hong Kong's large and diverse consumer products manufacturing industry. Garment accessories, such as buttons, buckles, and fasteners, constitute the largest end use. The construction sector accounted for about 16% of sheet and strip usage in 1989, mainly in the form of builder's hardware, locks, keys, and hinges. Water tube and roofing sheets, major products in other parts of the world, find comparatively little use in Hong Kong.

Hong Kong is also a major Asian trading hub for copper. During the second half of the 1980s, Hong Kong firms exported approximately 16% of all imported products. About 4% of imported semis were exported to other Asian countries. Refinery shapes such as cathode copper are generally transshipped, usually to the Peoples' Republic of China.

Iran. Consumption of refined copper in 1994 was 42,000 mT, down from a peak of 61,000 mT in 1992. During the 1980s, the Iran-Iraq War did significant damage to the heavy industries; however, massive military production maintained and even increased the level of copper consumption throughout most of the conflict (Ref 7, 29).

Japan is among the leading producers, manufacturers, and traders in copper products in the world, with activity that extends from unrefined metal to finished products. In 1994, for example, Japanese imports of concentrates, blister, anodes, and refined copper exceeded 1,300,000 mT. Japan's apparent consumption of refined copper exceeded 1,375,000 mT in 1994, down from a high of 1,613,000 mT in 1991. Japan's net consumption of copper semis was 1,923,793 mT in 1994 (Ref 7, 29).

As might be expected, Japan is a very large consumer of copper. However, as other Asian countries have increased their consumption, Japan's fraction has been diminishing. In 1989, for example, Japan consumed more than 75% of the refined copper consumed in the region. By 1994, however, this fraction had dropped to only 35%. About one-half of the refined copper consumed went for the manufacture of wire and cable.

Between 1983 and 1989, copper semis imports rose at an average rate of about 44% per year. The extraordinary import growth rate during the 1980s has been attributed to Japan's high production rates, the relatively depressed value of the yen at the time, and a set of government measures designed to induce domestic consumption and increase imports. In the early 1990s, total Japanese semis imports slipped significantly as the worldwide economic recession and a strengthening yen brought the steep growth to a standstill. However, by 1994, it was beginning to return and imports of 35,155 mT that year exceeded those of 1990 (Ref 7, 19, 24).

Also since the 1980s, the growing demand for Japanese products, the recovery of the yen, and the overall high rate of technological innovation have accelerated the relocation of several lines of production and/or plants to other countries. By 1991, there were approximately 54 Japanese-owned brass mills and wire and cable manufacturers operating in Asia, outside of Japan. A growing number of copper-intensive, Japanese-owned, Japanese-licensed, or joint-ventured industries have arisen as the formerly all-Japanese automobiles, motorcycles, spare parts, computers, appliances, industrial equipment, and other products increasingly began to be produced throughout the region.

Republic of Korea (R.O.K.). Although the country has no copper mines, the R.O.K. has, since 1970, managed to build up a modern copper semis manufacturing industry. This industry has a total installed capacity of more than 400,000 tpy. About 10% of the semis production was exported to major Asian markets. The remainder was used to satisfy demand in Korea's growing computer, motor, electronic, and metal industries. Leading copper-intensive product categories include building wire and telecommunications and power cables. These products, all based on wire rod, consume approximately 60 to 65% of the semis production. In 1994, Korea consumed 478,700 mT of refined copper.

The R.O.K. represents a very large market for rolled products, about half of which was, in past years, accounted for by the vehicle radiator market. In 1990, however, Korea's two vehicle manufacturers began converting their radiator production to aluminum. About one-third of flat-rolled product production was in the form of coinage strip, most of which was exported. Electrical and electronic goods account for about one-fifth of the apparent consumption of alloy sheet and strip. Most of this material was brass strip used in items such as lighting products, plugs, and switches (Ref 41).

In 1990, the R.O.K. was the second largest market in the world for copper tube, primarily for air conditioning and refrigeration (ACR) tube and plumbing tube for under-floor heating systems. The tube market underwent considerable growth during the late 1980s and early 1990s, capturing significant fractions of the Hong Kong and Japanese trade. It is interesting to note that Korea exported 6000 mT of tube in 1988 at an average price of U.S. $4120/tonne (U.S. $1.85/lb), while importing 2200 mT at U.S. $6900/tonne (U.S. $3.14/lb), a price difference of about 67%. This difference was partly due to the higher quality of imported tube. It, however, could also be attributed to the fact that exporters of copper semis obtain a tariff discount when they produce semis from imported copper, which is usually taxed at 20%.

Taiwan. Taiwan underwent rapid industrial growth during the last decades of the 20th century. This was clearly reflected in its consumption of copper, which grew at an average rate of 15.5% per annum, compared with 1.8% per annum worldwide. By the end of the 1980s, Taiwan had become the third largest copper consumer in Asia. In 1994, Taiwan consumed 547,000 mT of refined copper, all of which was imported. Imports of semis reached a record level of just more than 315,300 mT in 1989.

By the end of the 1980s, growth in Taiwan in apparent copper consumption had increased to an astonishing 26% per annum. About one-half of the growth represented copper used in the production of wire and cable, which increased at an average of 12% per year and reached 221,000 mT in 1989 before falling off slightly in 1990. The remainder of Taiwanese semis production involved extruded and rolled products and tube.

Exports of semis increased by 3.6 times during the late 1980s, reaching 92,300 mT in 1989. Most of these exports were in the form of bars, rods, wire, and strip. Major imported product forms included bare brass strip, copper-clad plates, tube and pipe, and copper foil less than 0.07 mm in thickness. Import tariffs on copper semis were cut to between 10 and 15% in 1988 for most-favored nations and between 15 and 20% for other countries. This reduction has benefited both domestic consumers and exporters (Ref 24).

More than 35% of the rolled copper products consumed were used in electrical and electronic goods. Strip and sheet for coinage and ordnance applications accounted for about 20%; screw machine products, about 13%; and builder's hardware, 9%.

The apparent consumption of copper tube in Taiwan grew by about 300% between 1982 and 1987, mainly because of rapid growth in the production of air-conditioning units for both domestic and export markets. Gas heating systems for export markets were another important application. Copper plumbing tube is used in large nonresidential buildings, especially for hot water distribution; however, plastic and galvanized iron pipe dominate residential systems (Ref 41).

Turkey. Refined copper consumed in Turkey is produced mainly from domestic concentrates and imported blister. Substantial quantities of blister have been imported in recent years, but imports of refined copper were generally only slightly higher than exports. Consumption of refined copper was 94,800 mT in 1994, down significantly from previous years (Ref 18).

Indonesia. Like other countries in the region, Indonesia enjoyed a rapid rise in investment for manufacturing during the 1980s. During 1994, apparent consumption of refined copper in Indonesia was 47,700 mT. The country has several wire mills, with an output that is used largely for domestic infrastructure development. Most of the refined copper consumed as wire and cable is in the form of power cable (60%) and telecommunications cable.

Malaysia. In 1994, apparent consumption of refined copper in Malaysia was 108,400 mT. Malaysia has three wire rod mills with a combined capacity of 72,000 tpy. Their output feeds five major domestic wire mills, plus a number of smaller mills. All brass mill products consumed in Malaysia are imported. In terms of quantities, these are divided about equally among sheet and strip, rod and bar, and tube. Most of the tube imported is used for the manufacture of air-conditioning units. It has been projected that Malaysia will produce one-half of the expected Southeast Asian output of air-conditioning units by the mid 1990s.

Philippines. The Republic of the Philippines is the only Southeast Asian country, that exports a significant amount of refined copper; however, a domestic market for copper does exist in the form of new manufacturing industries and a developing infrastructure. In 1994, the apparent consumption in the Philippines of refined copper was 12,000 mT. In addition to cathodes, the country produces about 16,000 mT of wire and cable annually, some of which is exported and the remainder used domestically. Imports include about 5000 tpy of brass mill products, mostly in the form of alloy sheet and strip, plus 1000 tpy of tube.

Pacific Rim. A significant fraction of copper consumption in this region occurs in the relatively industrialized nations of Australia and New Zealand, although changes in the distribution of consumption can be expected as infrastructures become developed elsewhere in the Pacific Rim.

Australia. Australia is by far the largest refined copper consumer among the Pacific Rim countries; its yearly uptake is similar to that recorded by countries such as India, Sweden, Yugoslavia, and Mexico. Australian copper consumption peaked in 1981 and had not recovered the same levels by the early 1990s. Consumption in 1994 was 141,000 mT. To put this figure in perspective, it should be pointed out that consumption of refined copper in Australia was only 33,300 mT in 1950, and the significant increases in the consumption of both refined copper and semis during the second half of the 20th century were the result of considerable industrial development during that period (Ref 7, 15, 27, 34, 42).

By 1981, Australia had developed a semis manufacturing industry, which produced copper rod, profiles, wire, plate, sheet, strip, and tubes, some of which were exported. However, about 10% of domestic consumption continued to be imported. The principal suppliers were Japan, the United Kingdom, and New Zealand (Ref 43).

Singapore. Singapore has been a modest importer of brass mill semis and wire and cable. The country consumed no refined copper until 1989, when Sumitomo Electric Company commissioned a 14,500 tpy rod mill. In 1994, the apparent consumption in Singapore of refined copper was 20,000 mT (Ref 21).

Thailand. In the early 1990s, Thailand was rapidly becoming the principal manufacturing center in Southeast Asia. During the previous decade, wire and brass mills were erected to supply improving infrastructure and provide feedstock to the growing industrial base. Domestic consumption of copper wire and cable in Thailand reached approximately 60,000 mT annually in the late 1980s, while local consumption of brass mill products (tube, rod and bar, sheet and strip) was about 11,000 tpy. At that time, most of the latter demand was satisfied by imports. Since 1989, however, domestic requirements for such products as air-conditioning tube, coin blanks, and radiator strip were met by local manufacturers. By 1994, domestic apparent consumption of refined copper had grown to approximately 92,000 mT (Ref 21).

Consumption by Market Sectors and End-Use Applications

The commercially most important information from the standpoint of producers and semis is the distribution of copper consumption among recognized market sectors. Data on the consumption of specific product forms is also important, because it often signals trends in the way the metal is utilized. There are, however, several methods to partition end-use copper consumption, and because this can lead to confusion, it is important that the differences among them are understood (Ref 8, 24, 44).

For example, the Copper Development Association Inc. (CDA), using data supplied by the American Bureau of Metal Statistics (ABMS), the U.S. Bureau of Mines and a variety of other commercial and government sources, tracks the macroscale domestic consumption of both refined and scrap copper as uptake by the semi manufacturing industries. These industries, in CDA's broad frame of reference, include wire rod mills, brass mills, ingot makers (which supply foundries with prealloyed melt stock), foundries (which, in addition to ingot makers' products, also consume copper directly, mostly in the form of scrap but also as cathode or refinery ingots), powder plants, and others. Table 1.5 lists the consumption of copper among U.S. semis manufacturers beginning in 1972 (Ref 8).

Consumption can also be interpreted in terms of the supply of specific manufactured product forms:

Wire rod mill products
- Bare wire
- Communication wire and cable
- Building wire
- Magnet wire
- Power cable
- Apparatus wire and cordage
- Automotive wire and cable (except magnet wire)
- Other insulated wire and cable

Brass mill products
- Strip, sheet, plate, and foil
- Mechanical wire
- Rod and bar
- Plumbing tube and pipe
- Commercial tube and pipe

Foundry products
- Castings, exclusive of ingots

Table 1.5 Consumption of copper by U.S. semis manufacturing industries, 1972 to 1994

Consumer of copper	Copper content, thousands of short tons					
	1972	1977	1982	1987	1992	1994
Wire rod mills						
Refined	1526.3	1511.2	1359.0	1757.0	1846.4	2270.8
Scrap	16.3	15.0	12.6	16.6	21.2	24.1
Total	1542.6	1526.2	1371.6	1773.6	1867.6	2294.9
Brass mills(a)						
Refined	667.2	628.6	433.4	538.6	503.2	626.1
Scrap	562.1	516.9	435.6	589.3	641.0	731.8
Total	1229.3	1145.5	869.0	1127.9	1144.2	1357.9
Ingot makers(b)						
Refined	6.0	6.6	4.8	5.8	5.8	5.6
Scrap	250.5	195.0	173.2	201.6	193.9	183.8
Total	256.5	201.6	178.0	207.4	199.5	191.3
Foundries(c)						
Refined	17.9	22.5	15.1	16.3	15.2	19.0
Scrap	121.7	101.2	72.1	80.3	74.3	65.9
Total	139.6	123.7	87.2	96.6	89.5	84.9
Powder plants(a)						
Refined	9.5	11.3	7.5	8.5	7.9	10.0
Scrap	16.0	18.3	10.1	12.8	12.0	13.5
Total	25.5	29.6	17.6	21.3	19.9	23.5
Other industries(a)						
Refined	10.7	13.0	14.6	18.1	16.7	20.8
Scrap	76.4	88.1	66.1	66.6	61.5	55.1
Total	87.1	101.1	80.7	84.7	78.2	75.9
Total copper consumed	**3280.6**	**3127.7**	**2604.1**	**3311.5**	**3398.9**	**4028.4**

(a) Direct consumption only; not including consumption of copper in ingots from ingot makers. (b) Ingot makers consume refined copper, scrap copper, and alloying metal, and ship to foundries, brass mills, powder plants, and other industries. (c) Chemical, steel, aluminum, and other industries. Sources: U.S. Department of the Interior, Bureau of Mines, American Bureau of Metal Statistics

Powder metallurgy (P/M) products

● A large variety of (usually) finished, pressed, and sintered products

Table 1.6 lists the consumption of copper among these products from 1972 to 1994 in the United States. The importance of wire mill products, particularly building wire, is evident (Ref 8, 37, 45).

Another way of looking at copper consumption is by markets. Here are the five major end-use sectors:

● Building construction
● Electrical and electronic products
● Industrial machinery and equipment
● Transportation equipment
● General and consumer products, including ordnance items

Table 1.7 shows the percentage distribution of copper among these market sectors for the three major copper-consuming regions of the world in 1994. Differences in distribution reflect different patterns of copper usage in these regions. In Western Europe and Japan, for example, with the exception of the United Kingdom, copper is not used as extensively for plumbing systems as in the United States. In Japan, for instance, plastic-lined steel tube is used extensively for potable water systems whereas copper is reserved for hot water systems. The distribution of products from U.S. semi manufacturers in 1994 among these market sectors is shown in Table 1.8 (Ref 8, 31).

A somewhat different approach is taken by the U.S. Bureau of Mines, among others. The distribution categories used in these systems are:

● Electrical
● Construction
● Machinery
● Transportation
● Ordnance
● Other

However, it is important to recognize that these categories are based on uses, rather than market sectors. The U.S. Bureau of Mines system in ef-

Table 1.6 Supply of copper and copper alloy semis and their consumption by U.S. end-use markets 1992

Product form	Metal content, millions of pounds					
	1972	1977	1982	1987	1992	1994
Bare wire	270	230	179	161	179	220
Communication wire and cable	766	836	751	674	487	464
Building wire	580	658	701	1181	1078	1140
Magnet wire	496	516	396	461	540	670
Power cable	273	241	247	268	287	301
Apparatus wire and cordage	169	196	157	231	172	233
Automotive wire and cable (except magnet)	80	108	105	217	225	258
Other insulated wire and cable	79	72	94	94	80	86
Total insulated wire and cable	2443	2627	2451	3126	3049	3393
Total wire mill products(a)	2713	2857	2630	3287	3228	3613
Strip, sheet, plate, and foil	1110	1004	754	915	934	1159
Mechanical wire	128	94	61	67	69	68
Rod and bar	910	900	620	913	963	1069
Plumbing tube and pipe	463	439	369	593	565	598
Commercial tube and pipe	432	340	257	362	398	496
Total brass mill products	3043	2777	2061	2850	2929	3390
Total foundry products	784	642	444	477	375	428
Total power products	54	64	34	39	35	46
Total domestic products	6594	6340	5169	6653	6567	7477
Net imports of mill products	229	202	224	455	5	51
Mill products to domestic market(b)	6823	6542	5393	7108	6572	7528
Building construction	2218	2123	1857	3047	2638	2997
Electrical and electronic products	1749	1694	1489	1633	1672	1897
Industrial machinery and equipment	1164	1044	849	942	831	901
Transportation equipment	807	943	568	817	764	957
Consumer and general products	885	738	630	669	635	776

(a) Copper content. (b) Markets include: (1) Building construction: building wiring; plumbing and heating, air conditioning and commercial refrigeration, builders' hardware, and architectural. (2) Electrical and electronic products: power utilities, telecommunications, business electronics, lighting, and wiring devices. (3) Industrial machinery and equipment: plant equipment, industrial valves and fittings, nonelectrical instruments, off-highway vehicles, and heat exchangers. (4) Transportation equipment: automobile, truck and bus; railroad; marine, and aircraft, and aerospace. (5) Consumer and general products: appliances, cord sets, military and commercial ordnance, consumer electronics, fasteners and closures, coinage, utensils and cutlery, and miscellaneous. Sources: U.S. Department of Commerce, Bureau of the Census; IPC Technology/Marketing Research council; Metal Powder Producers Assn., Copper Development Association Inc.

fect acknowledges that electrical and electronic uses dominate copper consumption, and it therefore, assigns all electrical and electronic copper usage to the "electrical" category, irrespective of the end-product or market sector in which that use occurs. Thus, automotive wiring harnesses and house wiring, for example, are listed under "electrical" rather than "transportation" and "construction," respectively. Table 1.9 illustrates this for the United States by showing the consumption of copper in the top ten usage categories.

Consumption can also be defined in terms of other physical, mechanical, chemical, or even aesthetic properties of copper. Table 1.10 lists the consumption of copper as defined by its useful properties in the United States.

Patterns in Worldwide Copper Product Consumption

It is possible to combine consumption data in such a way as to provide a picture of copper consumption by product and geographic region. Such combined data can be interpreted in terms of market-sector usage in the regions described. However, to develop such an analysis requires considerable effort in the interpretation of reported statistics to avoid overlapping data and the differences in the nomenclature used by different countries in reporting data. The following interpretation was

Table 1.7 Copper consumption by market sector and region

Market sector	U.S.A., %	Japan, %	Western Europe, %
Building construction	41	21	24
Electrical and electronic products	23	44	52
Industrial and machinery equipment	14	16	10
Transport equipment	13	11	4
General consumer products, including ordnance items	10	7	10

done for 1990 data by Market Analysis Company (MARCO) of Birmingham, England, for the International Copper Association, Ltd., and is presented here as an example of the pattern of consumption at that time.

In 1990, Japan, Western Europe, and the United States accounted for 79% of the apparent copper consumption in the world. It is instructive to examine the distribution of this consumption among major semifabricated product forms.

Wire and Cable. Wire and cable have almost always been the largest and most important copper product forms. In 1990, copper wire and cable consumption in the Western world was 4,259,000 mT, representing 51% of all copper consumed. This was distributed by:

Region	Consumption, thousands of mT	Percent of world total
Western Europe	1,559	18.7
Japan	1,163	13.9
United States	1,537	18.4
Total	**4,259**	**51.0**

The main uses of copper wire include current-carrying bare wire and insulated conductors (including power cables); telecommunications cables; wiring cables; magnet (winding) wires; and a variety of miscellaneous wires and cables. In 1990, building wire constituted 28% of all wire and cable consumed; magnet wire, 18%; telecommunications wire, 16%; power cable, 16%; automotive wire, 7%; appliance wire, 7%; bare wire, 6%, and other wire categories, 2%.

Tube. Copper tube is the second most important product form among copper semis. Three quarters of all copper tube is used for general heating, for water service, and for the manufacture of fittings. The remainder is principally applied as industrial tube in refrigeration and air-conditioning equipment. In 1990 tube consumption totaled 967,000 mT, or almost 12% of all copper

Table 1.8 Supply of copper and copper alloy semis to U.S. markets, 1994

Market	Metal content, millions of pounds			
	Wire mills	Brass mills	Foundry and powder mills	Total
Building construction	1301	1496	168	2965
Electrical and electronic products	1200	694	38	1942
Industrial machinery and equipment	429	382	199	1010
Transportation equipment	405	404	36	845
General consumer equipment products	267	414	32	714
Domestic total	**3602**	**3390**	**473**	**7476**
Net imports of mill products	(196)	252	(5)	51
Total	**3406**	**3642**	**468**	**7527**

Source: Copper Development Association Inc., 1995

consumed in the Western World. The breakdown was:

Region	Consumption, thousands of mT	Percent of world total
Western Europe	435	5.2
Japan	176	2.1
United States	356	4.3
Total	**967**	**11.6**

Sheet and Strip. Copper sheet and strip are used mainly in the electrical industry for electronic components; in the construction industry for architectural applications; and in the transportation industry as components of heat exchangers, primarily automobile radiators. In 1990, copper sheet and strip consumption amounted to 634,000 mT, or 8% of the copper consumption, in the Western world. The distribution was:

Region	Consumption, thousands of mT	Percent of world total
Western Europe	303	3.6
Japan	188	2.3
United States	143	1.7
Total	**634**	**7.6**

Rod. Copper rod is mainly used in electrical applications, as bus bar. In 1990, 207,000 mT, or 2% of Western copper consumption, went to this application. By region, the distribution was:

Region	Consumption, thousands of mT	Percent of world total
Western Europe	104	1.2
Japan	37	1.5
United States	66	0.8
Total	**207**	**3.5**

Alloy Wire. Of the products classified as copper alloy wire, typically 83% are brass, 10% are bronze, and the remainder are nickel-silver and other alloys. These materials are most commonly used in fasteners and fittings. In 1990, only 1% of the copper consumption in the Western world, 84,000 mT, was in the form of alloy wire. By region, the distribution was:

Region	Consumption, thousands of mT	Percent of world total
Western Europe	44	0.5
Japan	23	0.3
United States	18	0.2
Total	**85**	**1.0**

Alloy Tube. Of products classified as copper alloy tube, about 60% are brass, 30% are copper-nickels, and the remainder are aluminumbrass, and other alloys. Copper alloy tube is mainly used in marine condensers and heat exchangers and seawater piping systems. In 1990, 129,000 mT of alloy tube (2% of total consumption) were used in the Western world. Consumption distribution was:

Region	Consumption, thousands of mT	Percent of world total
Western Europe	78	0.9
Japan	18	0.2
United States	33	0.4
Total	**129**	**1.5**

Alloy Sheet and Strip. Of all products classified as alloy sheet, 70% are brass. Of the remainder, 18% are copper-nickel, 8% are phosphor-bronze, 3% are nickel-silver, and 1% are composed of other alloys. Most products are in the form of light gage sheet and strip; heavy plate represents only about 4% of consumption. Among the many uses for flat-rolled products are wiring connectors and accessories, telecommunications equipment, low-voltage switch gear components, decorative panels, automotive radiators and heaters, clocks and instruments, and marine heat exchangers. In 1990, 612,000 mT of alloy sheet and strip was consumed in the Western world, representing approximately 7% of all copper consumed in the region. Distribution was:

Region	Consumption, thousands of mT	Percent of world total
Western Europe	247	3.0
Japan	168	2.0
United States	198	2.4
Total	**613**	**7.3**

Alloy Rod. In terms of tonnage consumed, rod is the most important copper alloy product. Free-cutting brass, the predominant grade, typically accounts for 97% of all such rod. Phosphor-bronze, nickel-silver, and other alloys make up the remaining 3%. Alloy rod is chiefly used for screw machine products, small forgings, and cold-headed parts. Alloy rod accounted for 11%, or 909,000 mT, of Western copper consumption in 1990. In 1990, the consumption of alloy rod in the three major regions of the Western world was:

Region	Consumption, thousands of mT	Percent of world total
Western Europe	519	6.2
Japan	190	2.3
United States	199	2.4
Total	**908**	**10.9**

Casting Alloys. More than 100 standardized copper alloys are used in the manufacture of castings. Among the more important applications for the countless variety of cast copper products are plumbing fixtures, industrial valves and fittings, corrosion-resistant pumps, ship propellers, and statuary and decorative items. Alloys used are mainly made from scrap, although virgin metals are occasionally added to control composition precisely. In 1990, 483,000 mT of copper were consumed in the form of casting alloys in the Western world, representing approximately 6% of total copper consumption in this region. The consumption of copper in casting alloys in 1990 was:

Region	Consumption, thousands of mT	Percent of world total
Western Europe	222	2.7
Japan	81	1.0
United States	180	2.2
Total	**483**	**5.9**

Chemicals and Powder. Less than 1% of all copper consumed is in the form of chemicals, predominantly copper sulfate, and copper powder. Chemicals are widely used in agriculture as bactericides and fungicides, in animal feed, and as wood preservatives. Copper powder is used in the fabrication of bearings, electrical components, and a large variety of other small parts using the P/M process. A total of 69,000 mT of copper powder and chemicals were consumed in 1990, representing approximately 0.8% of the copper consumption in the Western world. The breakdown was:

Region	Consumption, thousands of mT	Percent of world total
Western Europe	42	0.5
Japan	4	0
United States	23	0.3
Total	**69**	**0.8**

Production Costs and Market Price

Production Costs

The cost of producing refined copper varies from country to country and from mine to mine. Cost at any individual mine also varies over time as new technological advances are adopted and as new government-imposed regulations are implemented. It also varies by the ore grade being mined at any individual mine at any one point in time. Table 1.11 lists estimates of world copper production costs on a country-by-country basis for 1988. Because copper is a commodity that is internationally traded, its price is established by the markets. Thus, copper companies compete with one another on the basis of their production costs.

Production cost depends on a number of variables, the most important of which are ore grade, the amount of overburden to be removed, and energy and labor rates. Mining costs are strongly influenced by mining strategy, which depends on the type, grade, and disposition of the ore. Environmental impact/reclamation considerations have become an additional and significant cost factor. Mining and milling costs typically represent about two-thirds of the total cost of producing refined copper by conventional mining, milling, smelting, and refining. Costs are often partially offset by byproduct credits. Molybdenum is the most common byproduct recovered from copper operations; others include such diverse materials as potash (South Africa), cobalt (Zaire and Zambia), and precious metals (Ref 11, 46, 47).

Sulfuric acid is a significant byproduct of copper smelting because the normal porphyry copper ore contains about 2 lb of sulfur for each pound of copper. However, most of the sulfuric acid produced is used internally for the acid leaching of oxidized ore minerals. In some instances, where shipping costs are not a major factor, it is sold for

Table 1.9 Top ten copper uses in United States, 1994

Use	Percent of total
Plumbing and heating	16
Buiding wire	14
Power utilities	8
Air conditioning and commercial refrigeration	8
In-plant equipment	7
Telecommunications	7
Automotive electrical	7
Electronics	5
Industrial valves and fittings	4
Automotive, nonelectrical	4
All others	20
Total	**100**

Source: Copper Development Association Inc., 1995

Table 1.10 Consumption of copper in the United States by its property

Property	Percent of total
Electrical conductivity	55
Corrosion resistance	25
Heat transfer	11
Structural capability	8
Aesthetics	2

Source: Copper Development Association Inc., 1995

other purposes. The magnitude of U.S. byproduct sulfuric acid production from copper sulfide minerals can be judged from data given in Table 1.12. The use of this acid for the leaching of waste dumps and other already mined material enables the recovery of copper from these sources for significantly less cost than that recovered by smelting. Traditional mining and extraction techniques are becoming increasingly supplemented by low cost, sometimes bacterial-assisted, leaching and SX/EW processing. Thus, while sulfuric acid itself does not represent a significant byproduct credit, the low-cost copper recovered by its use does offset the average cost of copper recovery at the mine site in which it is used.

In some copper-producing countries, currency exchange rates could also be counted among factors influencing production costs, because such rates have a strong influence on the cost of equipment and supplies, among other things. In the United States, estimated aggregate real-time operating costs were reduced from U.S. $0.79/lb in 1981 to U.S. $0.545/lb by 1986. Chile, which had already been the lowest cost producer, further reduced its costs from U.S. $0.446/lb in 1981 to U.S. $0.299/lb in 1986. However, by 1990, de-

creasing ore grades, operating problems, and higher labor rates caused Chilean costs to rise to U.S. $0.52, an increase of 74%. U.S. production costs rose only 10%, to U.S. $0.60/lb, over the same half decade (Ref 15, 46).

Market Price

Primary Copper. The price of primary copper is extremely sensitive, not only to normal supply and demand forces, but also to world events. No mathematical model has as yet been able to predict either short- or intermediate-term prices. This is in contrast to long-term trends in copper consumption, which are correlated well with fundamental economic indices of industrial production (Ref 49).

Market price sensitivity of copper also makes it difficult for producers to accommodate changes in consumer demand in a timely manner. Demand spikes can, therefore, cause temporary shortages, which cause prices to rise. This spurs capital investment to increase metal production. But higher copper prices also permit competing alternative materials to enter the market, and this ultimately lowers the demand for copper. By the time the increased supply of copper finds its way to market, demand may have already fallen, causing cop-

Table 1.11   Estimated 1988 production costs at producing copper mines

Region	Number of mines	Mine operating cost	Mill operating cost(a)	Smelter refinery cost(b)	Byproduct credit (less)	Net operating cost	Taxes(c)	Cash costs	Recovery of capital(d)	Total production cost(c)(e)
Average annual production costs(f)										
Australia	4	$0.33	$0.14	$0.16	$0.23	$0.39	$0.0	$0.39	$0.07	$0.47
Canada(g)	18	47	30	52	92	36	0	37	18	54
Chile	7	15	16	09	05	34	0	34	05	39
Peru	5	21	30	33	07	76	2	79	14	92
Philippines	7	28	37	24	49	39	4	43	10	53
United States	18	18	28	17	10	52	1	53	07	61
Zaire	4	27	14	29	25	45	1	46	03	48
Zambia	9	32	26	25	08	76	8	84	06	89
Other	40	35	36	28	42	57	1	59	15	74
Total(e) or average	**112**	**26**	**25**	**24**	**27**	**47**	**1**	**48**	**09**	**57**
Life of the mine production costs(h)										
Australia	4	$0.47	$0.22	$0.19	$0.44	$0.44	$0.00	$0.44	$0.07	$0.51
Canada(g)	18	61	31	68	1.23	37	0	37	18	55
Chile	7	18	21	09	05	42	0	43	05	48
Peru	5	19	30	36	22	63	1	64	15	80
Philippines	7	24	31	24	28	52	3	55	10	65
United States	18	16	26	17	09	50	1	50	07	57
Zaire	4	29	15	25	19	51	1	52	03	54
Zambia	9	35	26	25	08	77	8	85	06	91
Other	40	28	34	28	34	56	2	58	16	74
Total(e) or average	**112**	**25**	**26**	**21**	**23**	**50**	**1**	**51**	**09**	**60**

(a) Includes copper recovery leaching. (b) Includes cost of transportation and cost of byproduct and coproduct smelting. (c) Taxes and production costs are at zero percent rate of return and do not include state or federal revenue based taxes. (d) Average over life of the mine capital cost. (e) Data may not add to totals shown because of independent rounding. (f) Based on annual production rates and ore grades for 1988. (g) Includes Inco Ltd's and Falconbridfe Ltd's Sudbury nickel-copper operations. (h) Based on life of the mine production rates and ore gardes. Does not necessarily reflect 1988 operating grade and production. Source: U.S. Bureau of Mines, Minerals Availability System cost analysis

per prices to fall again. Events such as wars and industrial recessions add additional instability to the copper market and contribute to its cyclic character (Ref 48, 49, 50).

Base prices of several grades of refined copper are established by buying and selling on the London Metal Exchange (LME), the Commodity Exchange of New York (COMEX), and the Chicago-based Mid-America Commodity Exchange (MACE). The main function of these commodity exchanges is to separate the manufacturing and speculating activities of the participants. The exchanges also provide the means by which natural market fluctuations can be mitigated to the benefit of both producers and consumers.

Until 1981, the LME's principal copper contracts were based on wirebars. Changing technology caused cathodes to replace wirebars as the fundamental unit of trade after that year. At the same time, a new contract was introduced for high-grade (Grade A) cathodes required for production of continuously cast wire rod. This contract eventually assumed such a high degree of importance that it replaced the standard cathode contract at the end of 1988. The COMEX adopted the high-grade contract as its standard beginning in 1990.

The COMEX and LME provide contracts for the purchase and sale of copper for both immediate, "spot," and future, "forward," delivery. Spot copper can be purchased directly from exchanges, which maintain warehouses for this purpose. This option enables suppliers to fulfill their contractual obligations when they are unable to do so from their own production. Levels of these warehouse stocks exchanges are published weekly, as undue changes can affect market sentiment, and, therefore, prices.

Forward sales permit both producers and consumers to hedge against the risk of fluctuating prices. For example, producers may sell a portion of their output for future delivery to fix their prices and to facilitate production planning. Consumers may buy a portion of their needs for future delivery to fix their costs.

Metal exchange contracts for copper are also used by investors and speculators who never physically receive delivery of the metal. Speculators' trades help finance transactions required by producers and consumers and thus help stabilize the market.

In recent years, private sales of concentrates and refined copper based on prices negotiated directly between producers and buyers have become increasingly popular. List prices posted by U.S. producers are generally based on the first-position COMEX price. List prices do not, of course, necessarily reflect discounts or premiums negotiated between buyers and sellers (Ref 18).

In any event, sales prices are related to exchange prices by means of formulae agreed upon in advance. It is the negotiation of these formulae that establishes the basis for competition among copper producers. The purchaser's objective is to reduce the average price paid each month by careful selection of quantities priced on particular days, when prices are low. Thus, terms offered by sellers may include such considerations as the proportion of monthly deliveries that may be priced on a single day (perhaps 12 to 15% of the monthly quota), or in any week (25 to 50%), and the extent to which back pricing may be used. The purchaser pays for the freedom to price higher proportions of the monthly total on the price prevalent in a single day or week in the form of a premium over the LME price. The premium may also be negotiable.

Annual negotiations for copper trades generally begin in October. Major exporters of refined copper are Chile, Zaire, and Zambia, whose copper production facilities are at least in part controlled by state-owned corporations. These powerful bodies are major competitors in annual price negotiations.

Table 1.13 shows both the LME copper prices and the U.S. producer prices from 1935 to 1980. Producer pricing gradually gave way to exchange-based pricing during the 1970s and 1980s. Since then, however, direct price negotiations between producers and consumers have once again gained in importance. Figure 1.7 tracks the U.S. producer price, in both actual and in constant 1994 dollars, from 1934 through 1994 (Ref 22). Since 1974, U.S. producer prices have consistently tended

Table 1.12 U.S. production of byproduct sulfuric acid versus copper production 100% basis

Year	Acid production(a), thousands of tonnes	Copper smelter production, thousands of tonnes	Ratio of acid to copper
1984	2251	1305	1.72
1985	2230	1313	1.69
1986	2307	1310	1.68
1987	2543	1377	1.85
1988	2893	1515	1.38
1989	3075	1631	1.89
1990	3381	1613	2.05
1991	3819	1639	1.84

(a) Excludes acid made from pyrite concentrates. Source: Commodity Research Unit Ltd., Copper Development Association Inc.

downward in constant dollar value—a reflection on the level of competition within the industry.

Copper Concentrates. The market for concentrates is much smaller than the market for refined copper, but it is an important part of the overall trade in copper. While most mines smelt their concentrates at mine-associated smelters, concentrates are also brought to market by mines that produce concentrates strictly for sale. A number of such mines were financed by owners of smelters in countries such as Germany and Japan, which

consume large quantities of copper but are deficient in copper ores.

Copper concentrates are generally sold under longer term contracts than refined metal. Such contracts are obviously advantageous to mines in that they provide operating stability, cash flow, and the basis for development loans; smelters benefit from the guarantee of a steady supply of feed stocks. Concentrates supplied under long-term contracts are, however, generally balanced by short-term contracts or by spot sales. In addition to price, quantities, and delivery schedules, concentrate contracts contain provisions for smelting and refining charges. Fluctuations in market price for copper are taken into account; these benefit mines when prices rise, but harm them most when prices fall (Ref 18).

As with refined copper, pricing formulae are based on the LME cash price, and are generally modified by provisions for back-pricing and for the proportion of deliveries, which may be priced on a single day. Payment terms are generally nearly net 30 days, with a balance due upon final assay.

Copper Scrap. The quantity of copper available for recycling increases both due to increased consumption (primary scrap) and to the number of installations coming to the end of their useful life (old scrap). Scrap arises mostly in the industrialized countries, in which it may be reused, or from which it may be exported to the international scrap market.

Industrialized countries have well-established networks for collection and recycling of scrap. These networks, comprised of countless commercial enterprises, form an important facet of the market for a valuable and recyclable metal such as copper. Availability of scrap varies with the price users are willing to pay, which is, in turn, related to the price and availability of unrefined copper. When the copper price is low, scrap may be held in inventory; when prices rise, scrap becomes more plentiful. Scrap trade, therefore, acts as a flywheel to moderate, however slightly, changes in the price of copper.

Currency Exchange Rate. Foreign exchange or currency relationships also affect the price of copper. Because copper is usually sold internationally in terms of U.S. dollars or British pounds, its real cost depends on the relationship of these currencies to each other and to the currency of the producing country. Currencies of countries that furnish equipment and supplies to mines, mills, and refineries are also important economic factors. Together, these currency rates affect the cost to the

Table 1.13 Comparision of historic and constant prices of copper

Year	LME price, pounds/tonne	U.S. producer price in actual dollars, cents/lb	U.S. producer price in constant 1994 dollars, cents/lb
1935	31.867	8.649	93.561
1936	38.441	9.474	101.011
1937	54.465	13.167	135.510
1938	40.707	10.000	105.106
1939	42.689	10.965	116.907
1940	62.000	10.662	119.576
1941	62.000	11.296	118.933
1942	62.000	11.797	107.059
1943	62.000	11.775	100.870
1944	62.000	11.775	99.151
1945	62.000	11.775	96.947
1946	77.170	13.820	105.032
1947	130.540	20.958	139.281
1948	134.000	22.038	135.520
1949	133.039	19.202	119.569
1950	178.782	21.235	130.582
1951	220.362	24.200	137.940
1952	259.275	24.200	135.337
1953	256.275	28.798	159.845
1954	249.349	29.694	163.593
1955	352.276	37.491	207.320
1956	329.081	41.818	227.847
1957	219.642	29.576	155.984
1958	197.846	25.764	132.119
1959	237.833	31.182	158.803
1960	246.004	32.053	160.482
1961	229.792	29.921	148.304
1962	234.100	30.600	150.163
1963	234.775	30.600	148.200
1964	352.879	31.960	152.789
1965	469.875	35.017	164.747
1966	554.471	36.170	165.444
1967	417.338	38.226	169.614
1968	523.975	41.847	178.211
1969	621.254	47.534	191.949
1970	587.902	57.700	220.390
1971	444.430	51.443	188.207
1972	427.960	50.617	179.460
1973	726.820	58.852	196.438
1974	877.000	76.649	230.413
1975	556.810	63.535	175.016
1976	782.400	68.824	179.257
1977	750.250	65.808	160.936
1978	710.500	65.510	148.905
1979	936.198	92.334	188.483
1980	941.334	101.416	182.401

Note: Copper prices are yearly average.

producer and the price paid by the consumer (Ref 12, 44).

International Trade Controls

Free international trade in copper as well as its market price is governed primarily by the state of the global economy, which establishes the level of demand. Trade and price are also influenced by factors such as the cost of energy and transport and a number of historic political and economic forces. These forces are a legacy of the trade imbalances that arose roughly between the Industrial Revolution and the mid 20th century, when production and trade in copper was tightly controlled, particularly in Africa, the Caribbean, and the Pacific Rim, as well as in India and Canada.

In post-colonial times, economic domination replaced political hegemony. For example, in developing countries of the southern hemisphere, particularly Chile, copper mining was developed with capital and knowledge supplied by American companies. This led to a condition whereby, as late as the early 1950s, more than 75% of world copper production was in the hands of no more than eight North American and Western European mining companies.

An additional element of control arose after the end of the Second World War. Eastern Europe and some small Asian countries fell under the absolute domination of either the Soviet Union or the Peo-

ples' Republic of China. Copper trade within the European portion of this political alliance was controlled from Moscow, whereas what control existed or remained in the Asian communist bloc was dominated by China. There was little trade in copper between the so-called centrally controlled and free-market economies until the collapse of communism in the early 1990s (Ref 18, 22).

The situation has changed dramatically since colonial and post-colonial times, and by the last decades of the 20th century—for the Western World—more than 40% of the copper supply in the world was controlled by eight producing countries, which had allied themselves in an organization known as the Conseil Intergouvernamental des Pays Exportateurs de Cuivre (CIPEC). This organization had attempted to control copper trade and prices, but its efforts were unsuccessful and it ceased to function in 1992. The International Copper Study Group (ICSG) was then devised by the governments of the major copper producing countries in an attempt to make the elements of copper supply and demand more transparent. A number of parallel trade conventions were also under study. Some of these involved socioeconomic issues and attempts to provide support to poorer nations, including those of the former Eastern Bloc (Ref 51).

Because copper consumption is so strongly tied to the use of electricity, per capita copper consumption tends to be highest in highly industrialized countries. Among the industrialized countries,

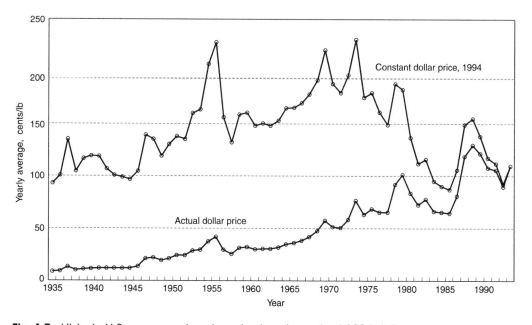

Fig. 1.7 Historic U.S. copper prices in actual and constant 1994 dollars

only Canada, the United States, Russia, and the Republic of South Africa are geologically self sufficient in copper. Others, principally Western Europe and Japan, smelt, refine, and recycle imported copper and concentrates to such an extent that they can satisfy internal needs and export surplus semimanufactures and finished goods. It is in the interests of such countries to produce and/or recycle indigenous copper and, when necessary, to purchase the lowest cost raw materials (concentrates, blister, scrap, etc.) on the international market. This self interest is reflected in, for example, the nature of the import tariffs for 1987 listed in Table 1.14 (Ref 52).

The traditional source of such low value-added commodities are less developed countries, many of which are former European colonies located in the Southern Hemisphere and Asia. A number of less developed countries, principally Chile, Perú, México, Zaire, Zambia, and Papua New Guinea, have extensive copper ore reserves, and are equipped with smelting and refining facilities with capacities that far exceed the domestic requirements of these countries. It is in the interest of these generally capital-poor nations to raise the value of their copper products and thereby maximize inflow of hard currency needed for industrialization and social welfare. This strategy favors the export of refined copper, and, where possible, semimanufactures, in direct competition with established producers and fabricators in already industrialized countries.

Many, but by no means all, copper production facilities in developing countries are at least partially state owned. To generate foreign exchange, promote development, avoid labor unrest, or pursue social programs, such facilities have at times been subsidized by their governments. Producers in developed countries claim such subsidization has contributed to the oversupply of copper and

has led to inequitable competition. It should be noted that virtually all such countries have provided, and continue to provide, some form of subsidy to their copper-related industries. This point has prompted strong disagreements among European, Japanese, and American copper producers.

The situation has been compounded to some extent by the action of international financial institutions, such as the World Bank, the International Monetary Fund, and the Inter-American Development Bank, among others. It is the goal of these institutions to raise living standards while promoting international trade in a stable monetary system. Copper producers in industrialized countries, for example, countries that underwrite these funds, complain that when such international aid is used to fund copper mining, smelting, and refining projects in less-developed countries, it reduces production costs in those countries to unrealistic levels, and this, in effect, leads to a restriction of free competitive trade. Recipient countries argue that such funding organs provide the only means to underwrite expensive mining ventures other than those that might once again lead to foreign ownership.

A number of international conventions have sought to address this complex situation. Their general goal has been to promote Third World industrial development by favoring poorer nations with reductions in, or elimination of, copper import duties. Early postwar efforts included the Yaoundé Conventions of 1963 and 1969. In the 1970s and early 1980s, copper producing countries participated in numerous trade conferences organized under the auspices of the United Nations Council on Trade and Development (UNCTAD). Among other things, these meetings led in 1984 to the formation of a working group within the General Agreement on Tariffs and Trade (GATT), whose objective was and remains to study and aid in the resolution of problems related to trade in nonferrous metals. That year also saw statutory renewal, in the United States, of the Generalized System of Preferences (GSP). The GSP favors imports to industrialized countries from developing countries, with certain customs liberalizations, including duties suppression. Japan, European Union countries, and the United States place quantitative limits on those imports based on dollar values and/or import fractions for the country in question (Ref 1, 44, 53).

It was also through this mechanism that the Africa, Caribe, Pacific (ACP) countries, which include Zaire, Zambia, and Papua New Guinea, were accorded preferential treatment by the EEC

Table 1.14 Copper import tariffs in developed countries, 1987

Imported product	Percentage of value		
	U.S.A.	Japan	EEC
Product 1(a)	1.1 (0:4.6)	Free (0:0)	Free (0:0)
Product 2(b)	3.3 (0:6.0)	4.3 (0:5.7)	5.2 (0:12)
Product 3(c)	4.9 (1:7.3)	6.8 (5.8:7.2)	5.0 (0.5:6.5)
Product 4(d)	4.2 (0:11.2)	6.6 (4.9:10)	5.9 (3.5:6.5)
Product 5(e)	5.5 (0:10.0)	6.0 (4.3:10.0)	3.8 (0.0:5.6)
Product 6(f)	5.4 (0:10.0)	6.5 (4.2:7.2)	4.6 (0.0:7.0)

(a) Ores and concentrates. Unrefined copper, blister, scrap, and residues. (b) Compounds and principal alloys. (c) Bars, plates, sheet, foil, wire, powders, and others. (d) Tubes and conductors, cables, gratings, and net. (e) Valves and similar, tools, and household appliances. (f) Insulated electrical conductors

through what have become known as the Lomé Conventions, the first two of which were signed in 1975 and 1979. The Second Lomé Convention came into force in 1980. It established duty-free access to EEC countries for 99% of all goods, including copper products, produced in ACP countries. The Fourth Lomé Convention was signed on December 15, 1989; it will succeed the Third Convention, which expired in March 1990, and will remain in effect for ten years.

The principal stated goal of the four Lomé conventions was to establish a system of cooperation to promote and expedite economic, cultural, and social development of the ACP states. These conventions also favor Algeria, Morocco, Tunisia, Egypt, Jordan, Syria, Turkey, Israel, Lebanon, Cyprus, and Malta. These conventions eliminate tonnage restrictions on exports of semis or finished copper goods from ACP countries to the EEC. Further, EEC customs duties and "other charges having similar effect" are not to be applied to goods originating in ACP countries. These conventions, therefore, clearly promote the development of integrated copper production facilities in the beneficiary countries at the expense of existing European industries, as well as that of exporting nations not favored in the treaty (Ref 1, 29, 55).

Thus, at least to the extent that Lomé and other conventions are actually effective in international trade, copper has become a little less of a technical commodity and, at least to a small extent, an instrument of social and political action. The ability of such a system to ensure stable and realistic prices, which do ultimately benefit all nations, remains to be demonstrated.

Efforts by copper producers themselves toward ensuring a healthy and growing trade in their commodity should be mentioned. In recent years, traditional uses for copper have been challenged by aluminum, polymers, and, most recently, optical fibers. Success of these challenges is at least partly reflected in the diminishing annual growth rate in copper consumption. The copper industry has responded by creating technical and market development organizations throughout the world. The first such organization was the Deutsches Kupfer Institut, founded in Germany in 1927. The Copper Development Association of the United Kingdom followed in 1933. In later years, several others appeared, the largest of which was the Copper Development Association Inc., which serves the U.S. market. In 1988, copper producers founded the International Copper Association, Ltd. (ICA), using as an administrative and technical base the International Copper Research Association (INCRA), whose functions the new umbrella organization absorbed. The International Copper Research Association was founded in 1959 for the purpose of developing new uses for copper and enhancing existing uses through the development and application of technology. The present organization, ICA, coordinates the individual efforts of various national and regional copper centers and establishes global market development strategies. In 1994, funding for such activities amounted to approximately U.S. $45,000,000.

REFERENCES

1. B.W. Smith, *Sixty Centuries of Copper*, CDA Publication (No. 69), London, 1965
2. R. Raymond, *Out of the Fiery Furnace, The Impact of Metals on the History of Mankind*, MacMillan Company of Australia Pty, Ltd, South Melbourne, Australia, 1984
3. J. Pijoam, *History of the World (Historia del mundo)*, Vol 6, Salvat Editores S.A., Spain, 1979
4. *Knaurs Lexicon, A–Z (Knaurs Lexikon A–Z)*, Droemersche Verlagsanstalt, Germany, Aug 1956
5. Roskill Reports on Metals & Minerals, *The Economics of Copper 1992*, 6th ed., Roskill Information Services, Ltd., London, June 1992
6. J. Jolly, *Annual Report: Copper 1991*, U.S. Department of the Interior, U.S. Bureau of Mines, Washington, D.C., March 1993
7. *World Metal Statistics,* Vol 46 (No. 2), Feb 1993
8. *Annual Data 1993, Copper/Brass/Bronze, Copper Supply and Consumption 1972–1992*, Copper Development Association, 1993
9. G. Bokovay, *Copper, 1995 Canadian Minerals Yearbook*, Natural Resources Canada, 1996
10. J. Leibbrandt, *Challenges for the Copper Industry*, Minerals Industry International, March 1992, p 28–31
11. J.L.W. Jolly and D. Edelstein, Copper, *Minerals Yearbook 1988*, U.S. Government Printing Office, Washington, D.C., 1988, p 309–364
12. W.C. Butterman, Copper, *Mineral Commodity Profiles 1983*, U.S. Bureau of Mines, U.S. Department of the Interior, Washington, D.C., 1983
13. *World Mining Map*, Metallgesellschaft AG, Frankfurt, Germany, 1987
14. Copper, *Mineral Facts and Problems,* Bureau of Mines Bulletin 675, U.S. Bureau of Mines, U.S. Department of the Interior, U.S. Government Printing Office, Washington, D.C., 1985 p 197–221
15. *World Copper Statistics Since 1950*, World Bureau of Metal Statistics, Ware, Herts, 1956
16. *World Metal Statistics August 1992*, Vol 45 (No. 8), London, Aug 1992
17. Roskill Reports on Metals & Minerals, *The Economics of Copper 1988*, 4th ed., Roskill Information Services Ltd., London, 1988
18. Roskill Reports on Metals & Minerals, *The Economics of Copper 1990*, 5th ed., Roskill Information Services, Ltd., London, 1990

19. *World Metal Statistics,* Vol 44, World Bureau of Metal Statistics, London, 1991
20. C. Gazitúa, Copper Again (Otra Vez el Cobre), *Mercury (El Mercurio),* newspaper, (Santiago, Chile), 17 April 1993
21. *Nonferrous Metal Data, 1994,* American Bureau of Metal Statisics, 1995
22. *World Metal Statistics Year Book 1992,* World Bureau of Metal Statistics, London, May 1992
23. Production Problems Continue, *CRU Copper Studies,* Vol 19 (No. 3), 1991
24. S. Alanoca, *The World Wrought Copper Market: Situation and Export Perspectives,* International Trade Center UNCTAD/GATT (Geneva), 1990
25. J.L.W. Jolly and D.L. Edelstein, *Copper Annual Report 1990,* U.S. Department of Interior, U.S. Bureau of Mines, U.S. Government Printing Office, Washington, D.C., April 1992, p 9
26. T.F. Bower, private communication, Jan 1994
27. *World Metal Statistics August 1985,* Vol 38 (No. 8), Aug 1985
28. *World Metal Statistics August 1990,* Vol 43 (No. 8), Aug 1990
29. *World Metal Statistics August 1988,* Vol 41 (No. 8), Aug 1988
30. S. Yagioka, *Copper Consumption Trends in Japan,* Minerals Industry International, March 1992, p 26–27
31. A. Hall, *Market Analysis Research Company,* Birmingham, England, 1991
32. S. Alanoca, *Unexploited Market Opportunities for Copper Semimanufactures (Oportunidades de mercado inexplotadas para las semimanufacturas de cobre),* International Commerce Forum, April to June 1991, p 14–19, 34
33. *Developing Trends in the Use of Nonferrous Metals (Entwicklungstendenzen in der Verwendung von NE-Metallen),* Metallgesellschaft AG, Frankfurt, Germany, 1992
34. W. Bauer, *Metal Statistics 1977–1987 (Metallstatistik 1977–1987),* 75th ed., Metallgesellschaft AG, Frankfurt, Germany, 1988
35. *Copper and Copper Products,* International Trade Center UNCTAD/GATT, Geneva, 1990
36. *Developing Trends in Nonferrous Metal Commerce in the East (Entwicklungstendenzen der NE-Metallwirtschaft im Osten),* Metallgesellschaft AG, Frankfurt, Germany, 1991
37. *CRU Copper Studies,* Vol 19 (No. 10), 1992
38. CDA Market Data, *Usage of Copper and Copper Alloys in the USA,* Copper Development Association Inc., Greenwich, 1991
39. *Copper and Copper Products (El Cobre y los Productos del Cobre),* Centro de Comercio Internacional UNCTAD/GATT, Geneva, 1990
40. Copper Demand in North America: Prospects for Growth, Opportunities for Suppliers, *The Structure of Supply and Demand,* Vol 1, Commodities Research Unit, Ltd., London, 1989
41. L. Gustafsson, Markets for Copper, Paper 93, presented at Copper '90 Symp. (Vesterås, Sweden), 1990, p 7–13
42. I. Castles, *Yearbook of Australia 1991,* (No. 74), Australian Bureau of Statistics, 1991
43. Minister of Exterior Relations, Chile, Republica de Chile, Ministerio de Relaciones Exteriores, *Market Profile, Copper Semimanufactures (Perfil de mercado: Productos Semi-Manufacturados de Cobre),* PROCHILE, Santiago, Chile, 1995
44. D. Empey, Base Metal End Use Markets, *Base Metal Outlook,* McCarthy Securities, Ltd., Sept 1990
45. *Copper: Technology and Competitiveness,* OTA-E-367, U.S. Government Printing Office, Washington D.C., Sept 1988
46. Cost: Mining and Milling Cost Reductions, *CRU Copper Studies,* Vol 15 (No. 12), 1988, p 1–3
47. G. Muñoz, *Environmental Management Systems, A Technical Seminar, Codelco-Chile: An Environmental System in a Developing Country,* The International Council on Metals and the Environment (ICME), Santiago, Chile, 30 May to 4 June 1993
48. *Analysis for the Copper Industry and Market (Análisis de la Industria y del Mercado del Cobre),* Comisión Chilena del Cobre, Chile, 1991
49. Metals Analysis & Outlooks, *Five Year Outlook 1991–1995,* The International Review of Metal Market Trends, 1990
50. W.H. Dresher, The Influence of Technology on the Copper Market, *Copper '87 Symposium* (Santiago, Chile), Vol 1, Nov 1987, p 15–29
51. *Annual Report, 1992 (Memoria anual 1992),* Comisión Chilena del Cobre, Santiago, Chile, 1993
52. I. Valenzuela, *Overview of the Copper Metalworking Industry in Chile (Panorama de la Industria Elaboradora de Cobre en Chile),* Ed. Universitaria, Santiago, Chile, 1990
53. Cost Reduction at CODELCO, *CRU Copper Studies,* Vol 16 (No. 4), 1987, p 1–7
54. *History of the Commercialization of Copper Semimanufactures (Antecedentes de Comercialización de Semimanufacturas de Cobre),* Comisión Chilena del Cobre, Santiago, Chile, 1991
55. Metallgesellschaft, AG, Frankfurt, Germany, 1995

2

Metallurgy and Properties

Atomic Structure

Copper, with atomic number 29 and atomic weight 63.54, occupies the first position of subgroup IB in the periodic chart of the elements (Fig. 2.1). Subgroup IB also includes silver and gold, and in fact, copper shares many characteristics with these other noble metals as a result of its atomic and electron structure.

Electron Structure

The copper atom is composed of a positively charged nucleus containing 29 protons and 34 to 36 neutrons surrounded by 29 electrons. The electrons are arranged in a structure described by the notation $1s^2 2s^2 2p^6 3s^2 3p^6 3d^{10} 4s^1$, which implies that the 1s and 2s energy states contain two electrons each, and the 3s state contains two electrons; the 2p states contain six electrons and so forth. The structure is essentially that of an argon atom core plus the filled 3d state and the one 4s electron; it is sometimes written as $Ar3d^{10}4s^1$. The single "outer shell" 4s electron is responsible for many of copper's important physical properties, including its high electrical conductivity, its chemical stability, and its reddish color. In metal-

lic crystalline copper, as with other metals, the 4s electron does not remain associated with any particular atom but becomes part of the electron cloud that pervades the crystal lattice (Ref 1).

The ionization potential of the 4s electron, 7.724 eV, is relatively low, and the "cuprous" ion, Cu^+, is easily formed. The ionization potential of the 3d state is only slightly higher, and copper, therefore, also displays a "cupric" valence state, Cu^{2+} (Ref 1–3).

Brillouin Zones. Quantum theory limits the allowable energy of electrons to specific values, or "quanta." According to quantum theory, electrons can be described as wave functions of wavelength $\lambda=h/mv$, where h is Planck's constant and mv is the momentum of the electron. A free electron, therefore, has kinetic energy $E=\frac{1}{2}mv^2$, which by substitution is $h^2/2m\lambda^2$ or $h^2k^2/2m$, where k is the wave number, $1/\lambda$. Free, or valence electrons occupy states of increasing energy, for example, of increasing k. Within a metal crystal, however, there exist certain critical wavelengths, and corresponding critical k values, for which electron energies satisfy the Bragg condition for reflection, $\lambda=2d\sin\Theta$, and electrons having such energies are in effect excluded from the crystal. These forbidden energies are depicted as the gaps $b_1 - a_1$ and

Orbit

K

K-L

K-L-M

-L-M-N

-M-N-O

-N-O-P

-O-P-Q

Non-metals

Metals

Key to chart

Atomic number	50	+2	← Oxidation states
Symbol	Sn	+4	
Atomic weight	118.69		← Electron configuration
	-18-18-4		

Transition elements

Fig. 2.1 Arrangement of element groups in the periodic table

-N-O-P
-O-P-Q

*Lanthanides	58 +3 Ce +4	59 +3 Pr	60 +3 Nd	61 +3 Pm	62 +2 +3 Sm	63 +2 +3 Eu	64 +3 Gd	65 +3 Tb	66 +3 Dy	67 +3 Ho	68 +3 Er	69 +3 Tm	70 +2 +3 Yb	71 +3 Lu
	140.12 -20-8-2	140.9077 -21-8-2	144.24 -22-8-2	147 -23-8-2	150.4 -24-8-2	151.96 -25-8-2	157.25 -25-9-2	158.925 -27-8-2	162.50 -28-8-2	164.9304 -29-8-2	167.26 -30-8-2	168.9342 -31-8-2	173.04 -32-8-2	174.967 -32-9-2

**Actinides	90 +4 Th	91 +5 +4 Pa	92 +3 +4 +5 +6 U	93 +3 +4 +5 +6 Np	94 +3 +4 +5 +6 Pu	95 +3 +4 +5 +6 Am	96 +3 Cm	97 +3 +4 Bk	98 +3 Cf	99 +3 Es	100 +3 Fm	101 +2 +3 Md	102 +2 +3 No	103 +3 Lr
	232.038 -18-10-2	231.0359 -20-9-2	238.029 21-9-2	237.0482 -22-9-2	239.052 -24-8-2	(243) -25-8-2	(247) -25-9-2	(247) -27-8-2	(251) -28-8-2	(254) -29-8-2	(257) -30-8-2	(258) -31-8-2	(259) -32-8-2	(260) -32-9-2

Numbers in parentheses are mass numbers of most stable isotope of that element.

Fig. 2.1 (continued)

$b_2 - a_2$ in the plot of energy versus k shown in Fig. 2.2. When extended to the three-dimensional structure of the face-centered cubic (fcc) crystal structure exhibited by copper, the excluded energy states take the form of the polyhedron shown in Fig. 2.3. This is the first Brillouin zone for the fcc crystal structure (Ref 3).

According to Pauli's exclusion principle, only two electron states are available within a Brillouin zone. In a univalent metal like copper, the single valence electron occupies one-half of the zone, and its position may be visualized by a model such as that depicted in Fig. 2.4, where the ruled area is the spherical Fermi surface bounding the occupied energy states. Under simplified electrical and thermal conduction theory, the space, for example, energy between the Fermi surface and the Brillouin polyhedron represents the allowable free electron displacement. Because it is related to lattice dimensions, the Brillouin zone shrinks under the influence of applied compressive stress.

The most important consequence of the electronic structure of copper is that copper behaves like a monovalent alkali metal except that its behavior is modified by the existence of the filled d-shell states at an energy about 2 eV below the unfilled 4s valence state. For example, the filled d states cause the compressibility of copper to be much smaller than that of other monovalent metals, such as the alkali metals. Since the electrical resistivity of copper at room temperature and above is dominated by the scattering of electrons by thermal vibrations, the low amplitude of these vibrations resulting from the metal's low compressibility gives rise to the observed high electrical conductivity (Ref 4).

Nuclear Properties

The copper nucleus has a diameter on the order of 10^{-15} m. The nominal radius of the electron cloud surrounding an atom can be taken as the "diameter" of the atom; it is nominally five orders of magnitude larger than that of the nucleus. The radius, which depends in part on the number of nearest atomic neighbors, can be determined from the dimensions of the unit cell. Thus, for crystalline copper in the fcc structure, the distance of closest approach (Burgers vector) is 2.556×10^{-10} m, and the so-called Goldschmidt atomic radius for twelve-fold coordination is 1.28×10^{-10} m (Ref 5).

Isotopes. The international atomic weight of copper is 63.546, although the value 63.54 is adequately precise for common calculations and is more commonly used. Copper has two stable isotopes, Cu^{63} (occurring 69.09% in nature) and Cu^{65} (30.91%). The unstable isotopes of copper include Cu^{58}, Cu^{59}, Cu^{60}, Cu^{61}, Cu^{62}, Cu^{64}, Cu^{66}, Cu^{67} and Cu^{68}; these have half lives of 7.9 min, 81.6 s, 23.4 min, 3.33 h, 9.8 h, 12.88 h, 5.07 min, 58.6 h and 32.0 s, respectively (Ref 6–8).

Nuclear Absorption Cross Section. The absorption cross section of copper, σ, with respect to low energy proton bombardment is 10^{-27} barns (1 barn = 1 cm^2). For high energy (134 ± 4 MeV) protons, $\sigma = 725 \pm 68$ barns. The absorption cross section for fast neutrons is 10^{-24} barns, while for thermal neutrons, values of 3.63 and 3.8 barns have been reported (Ref 2, 8).

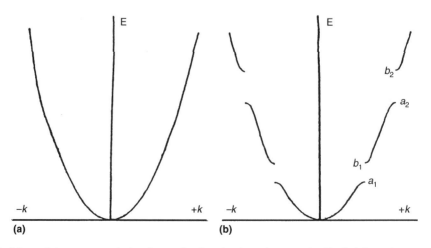

Fig. 2.2 Allowable energy states for a single electron in a periodic field, such as a crystal lattice as a function of the wave number, k. Source: Ref 37

Physical Properties

Mass Constants

Compressibility. Compressibility is a measure of the ease with which atoms of the metal can be brought closer together. Since the process is reversible, a low compressibility at zero pressure indicates strong bonding. Like other subgroup IB metals, copper has high compressibility. When the temperature of a crystal is raised, the amplitude of the atomic vibrations is increased and the crystal expands. Generally, the stronger the bonding forces, the smaller the coefficient of expansion. Figure 2.5 shows coefficient of expansion of copper relative to other elements. The coefficients follow the same general trend as compressibilities; however, the two terms cannot be compared directly since corrections are required to allow for the effect of temperature on the coefficient of expansion.

Density. The density of copper varies from 8.90 to 8.95 g/cm³, depending on the metal's thermomechanical history (Table 2.1). Density initially decreases with increasing degrees of cold work owing to the generation of vacancies and dislocations. It then increases to a value higher than that of recrystallized copper. This has been explained by the presence of persistent subgrain boundaries. The accepted value for the density of copper is 8.94 g/cm³ at 298 K (77 °F). This density is slightly different from that used in electrical standards, as described below (Ref 9).

Electrical Properties

Conductivity and Resistivity. In 1913, the standard conductivity of pure annealed copper was fixed by the International Electrotechnical Commission (IEC) as that of an annealed copper wire 1 m long, weighing 1 g and having a density of 8.89 g/cm³. The wire exhibited a resistance of exactly 0.15328 Ω. This value was assigned a volume conductivity of 100% of the International Annealed Copper Standard, written 100% IACS. It corresponds to a volume resistivity of 0.017241 Ω-mm² /m (Ref 10).

Conductivity and resistivity depend strongly on purity. Purity levels and processing techniques have improved considerably since the IACS was established, and a more precise value for density has also been measured. However, the standard value continues to be used in engineering practice. Currently, the highest measured room-temperature (20 °C, 68 °F) volume conductivity for very pure copper is about 103.6% IACS; the corresponding volume resistivity is 0.016642 Ω-mm²/m (Ref 10–12).

Current practically attainable conductivity values for electrolytic tough pitch and deoxidized low residual phosphorus coppers stand at about 101% IACS. Oxygen-free (OF) copper is certified under ASTM B 170 to meet this as a minimum value. Conductivity data for copper and Unified Numbering System (UNS)-listed copper alloys and for pure copper are shown in Tables 2.2 and 2.3 (Ref 11–19).

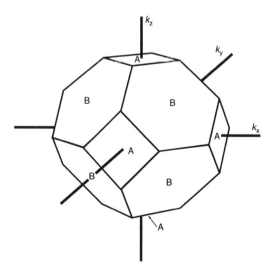

Fig. 2.3 Polyhedron representing the first Brillouin zone for a face centered cubic metal such as copper. Source: Ref 37

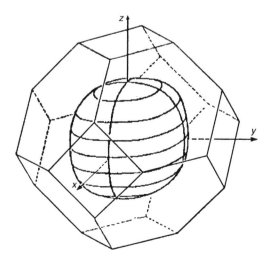

Fig. 2.4 Spatial relationship between the Fermi surface and the first Brillouin zone in copper. Source: Ref 37

Matthiessen's Rule. This rule states that the product of a metal's specific electrical resistance and its temperature coefficient of resistance is equal to a constant. In effect, the rule relates the effect of thermal vibrations, foreign atoms in solid solution (as alloying additions or impurities), and plastic deformation on electrical and thermal conductivity. All of these factors lower the electrical and thermal conductivities of copper. Ordinarily, the electrical conductivity of copper decreases with increasing temperature. The effects produced by small amounts of alloying elements in solid solution and by cold deformation are independent of temperature and are additive to the effect of thermal vibrations. The increase in resistivity brought about by small amounts of secondary elements in solid solution are shown in Fig. 2.6 and Table 2.4 (Ref 5, 13).

Temperature and Pressure Coefficients of Resistance. Thermally induced lattice vibrations reduce electrical conductivity by shortening the mean free path for electron movement and by distorting the crystal structure, for example, by changing the shape of the Brillouin zone. Thermally induced stresses can also give rise to the formation of crystallographic twins, which can be interpreted as another form of lattice distortion. The rate of rise in electrical resistivity with temperature follows Matthiessen's rule and is independent of alloy content and the state of deformation if both of these factors are small. For pure copper, the temperature coefficient of resistance is 0.068 n$\Omega \cdot$ m/K at 20 °C (68 °F). The pressure coefficient of resistance is −0.228 a$\Omega \cdot$ m/Pa (0.013 f$\Omega \cdot$ in./psi) for pressures from 100 kPa to 9.8 GPa (1.45 psi to 1420 ksi) (Ref 12, 20).

Thermal Properties

The thermal properties of pure copper at 1 atm pressure are:

- Melting point: t_m = 1084.88 °C (1984.78 °F) (k2 – 14), 1083 ± 1 °C (Ref 13)
- Heat of fusion: ΔH_f = 134 J/g (57.7 Btu/lb) (Ref 5); 205 J/g (88 Btu/lb) (Ref 2); 204.9 J/g (88.07 Btu/lb) (Ref 21); 206.8 J/g (88.89 Btu/lb) (Ref 22), and 12,804 J/mol (201.49 J/g, 86.61 Btu/lb) (Ref 12)
- Heat of vaporization: ΔH_v = 3630 J/g (1463 Btu/lb) (Ref 5)
- Heat of sublimation at 1299 K: ΔH_s = 3730 J/g (1606 Btu/lb) (Ref 5)
- Boiling point: t_b = 2595 °C (4703 °F)(6911) (k2 – 1); 2567 °C (4652 °F) (Ref 2); 2565 °C (4649 °F) (Ref 5)

Table 2.1   Densities of copper at 20 °C

State	ρ, g/cm³
Single crystal	8.95285
Melted and solidified in vacuum	8.94153
Commercial soft-drawn wire, annealed at 970 °C (1778 °F) in vacuum for 12 h	8.92426
Sample reduced 67.90% in cross section by drawing	8.90526
Diameter reduced 96.81%	8.91187
Sample annealed at 880 °C (1616 °F) in vacuum for 12 h	8.93003
Sample annealed at 1035 °C (1895 °F) in vacuum for 12 h	8.92763

Source: Ref 13

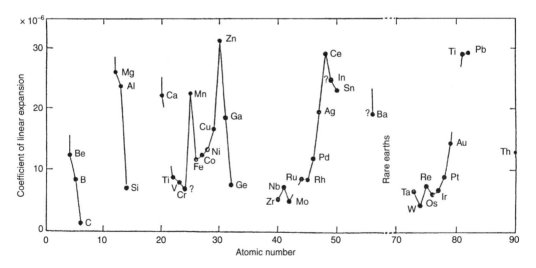

Fig. 2.5 Coefficient of thermal expansion of copper relative to other metals. Source: Ref 37

- Vapor pressure: $p_v = 6.9214 - 17546/T(\text{K}) - 0.0001659T(\text{K})$ atm (Ref 13); 1.3×10^{-1} atm 1141 °C (2086 °F) (Ref 2)
- Thermal conductivity: 398 W/m · K at 27 °C (Ref 2); (226 Btu/ft^2/ft/h/°F at 68 °F) (Ref 15)
- Coefficient of thermal expansion at 20 °C (68 °F): 16.7 μm/m/°C (9.4 μin./in./°F) (Ref 2, 15)

- Specific heat: $C_p = 0.092$ cal/g/°C at 20 °C, 0.092 Btu/lb/°F at 68 °F (383 J/kg · K) (Ref 15); 384.6 J/kg · K (Ref 2), otherwise: $C_{p, 0-1-t°} = 0.009619 + 6.139 \times 10^{-6}t + 1.928 \times 10^{-9}t^2$ (cal/g/°C). $C_{p, 1038-t°} = 0.11 + 45.6t^{-1}$ (cal/g/°C) For higher temperatures and solidification and cold worked structures, C_p can be calculated by the formulae: $C_p = 0.092 +$

Table 2.2 Physical properties of selected wrought copper alloys

Alloy	Description or previous tradename	Form of specimen measured	Tensile strength, MPa	Yield strength, 0.5% extension (0.2% extension), MPa	Elongation in 50 mm, %	Shear strength, MPa	Fatigue strength, 100 million cycles, MPa	Hardness Rockwell B/F
C10200	Oxygen-free	Rod	221–375	69–345	10–55	152–186	117	47–60/40–94
C11000	Electrolytic tough pitch	Rod	221–379	69–345	16–55	152–200	117	45–60/40–87
C12200	Phosphorous deoxidized	Tube	221–379	69–76	8–45	152–200	76–131	35–60/40–95
C14500	Tellurium-bearing	Rod	221–296	76–338	10–50	152–200	NA	36–54/40–43
C15720	...	Rod	441–552	400–531	16–25	NA	NA	66–74/NA
C17200	Beryllium copper	Rod	469–1379	(172–1227)	3–48	NA	NA	3–95/NA
C18200	Chromium copper	Rod	310–593	97–531	5–40	NA	NA	65–83/NA
C19500	...	Plate	552–669	(448–655)	2–15	NA	179–200	NA
C26000	Cartridge brass, 70%	Rod	331–483	110–359	30–65	234–290	NA	60–80/65
C28000	Muntz metal, 60%	Rod	359–496	138–345	25–52	269–310	NA	78/78–80
C36000	Free-cutting brass	Rod	338–400	124–359	25–53	207–262	138	75–80/68
C37700	Forging brass	Rod	359	138	45	NA	NA	NA/78
C50500	Phosphor bronze, 1.25% E	Plate	276–517	97–345	4–48	NA	221	64–79/68
C51000	Phosphor bronze, 5% A	Rod	483–517	400–448	25	NA	NA	78/NA
C54400	Phosphor bronze, B-2	Rod	469–517	393–434	15–20	NA	NA	80–83/NA
C61300	Aluminum bronze	Rod	552–586	331–400	35–40	276–331	NA	88–91/NA
C61800	Aluminum bronze	Rod	552–586	269–293	23–28	296–324	179–193	88–89/NA
C63000	Aluminum bronze	Rod	689–814	414–517	15–20	427–483	248–262	96–98/NA
C65500	High silicon bronze A	Rod	400–745	152–414	13–60	296–427	NA	60–95/NA
C70600	Copper-nickel 10%	Tube	303–414	110–393	10–42	NA	NA	15–72/65–100
C71500	Copper-nickel 30%	Tube	372–414	172	45	NA	NA	35–45/77–80

Note: These are ranges of properties over a limited span of tempers. Manufacturer's specifications should be consulted before choosing a suitable alloy. Source: Ref 17

Table 2.3 Physical properties of selected cast copper alloys

Alloy	Description or previous tradename	Tensile strength, MPa	Yield strength, 0.5% extension (0.2% extension), MPa	Elongation in 50 mm, %	Compressive strength 0.1 in. set, MPa	Creep strength, 0.1% in 10,000 h, %	Fatigue strength 10 million cycles, MPa	Hardness Brinell (lb load)
C80100	Copper	172	62	NA	NA	NA	NA	44 (300)
C81500	High copper	351	276	17	NA	NA	NA	105(300)
C82500	High copper	414–1035	(310–1035)	1–35	NA	NA	NA	NA
C83600	Leaded red brass	205–255	95–117	20–30	262	48–86	NA	60(300)
C86300	Leaded HS yellow brass	270–821	414–462	12–18	689	3–390	172	223(300)
C87500	Silicon brass	414–462	165–207	12–21	517	10–193	152	154(300)
C90200	Tin bronze	262	110	30	NA	NA	172	70(500)
C90500	Tin bronze	275–310	124–170	20–25	276	NA	90	75(500)
C90700	Tin bronze	240–303	117–152	10–20	NA	NA	172	65–80(500)
C92200	Leaded tin bronze	234–276	110–138	22–30	262	43–110	76	65(500)
C93200	High-leaded tin bronze	205–241	95–124	15–20	317	NA	110	20(500)
C94700	Nickel-tin bronze	310–517	138–414	5–35	NA	NA	97	85–180(500)
C95200	Aluminum bronze	448–552	170–186	20–35	483	54–145	152	110(300)
C95500	Nickel-aluminum bronze	620–827	275–469	5–12	827–1034	17–72	214–262	190–230(300)
C95700	Mn aluminum bronze	620–655	275	20–26	1034	29–66	228	180(300)
C96200	Copper-10 nickel	310	170–241	20	255	NA	90	NA
C96400	Copper-30 nickel	415–469	220–265	10–28	NA	NA	124	140(300)
C97600	Nickel silver	276–310	117–255	10–20	393	153–224	107	80(500)
C99700	Manganese bronze	379	172	25	NA	NA	NA	110(300)

Note: These are ranges of properties over a limited span of tempers. Manufacturer's specifications should be consulted before choosing a suitable alloy. Source: Ref 17

0.000025t cal/g/°C for solid state between 0 and 1083 °C. For molten copper, $C_p = 0.112$ cal/g/°C (Ref 13).

Thermal Conductivity. The thermal conductivity of pure copper ranges from 2780 W/m · K (1607 Btu/ft²/ft/h/°F) near 0 K to 177 W/m · K (102 Btu/ft²/ft/h/°F) at 2273 K (3632 °F), with a maximum of 19,600 W/m · K (11,300 Btu/ft²/ft/h/°F) at about 10 K. Alloying copper reduces its thermal conductivity. In practical terms, however, alloying markedly improves other characteristics such as mechanical properties and corrosion resistance, and the value of these effects override losses in thermal conductivity for many applications. The changes in thermal conductivity induced by alloying are of the same relative magnitude as those for electrical conductivity (Ref 2, 22).

Surface area and film coefficients frequently affect total heat flow more than the inherent thermal conductivity of the bulk metal. This is shown in Fig. 2.7 in which heat transfer coefficients are compared for samples of six alloys of the same size and wall thickness. Inhibited admiralty metals C44300 or C44400 are among the most commonly used materials for heat exchangers because of their high heat transfer coefficients. Copper-nickel alloys such as C71500 are also used for this purpose; these alloys have lower conductivity than copper-zinc alloys, but their higher corrosion resistance and their especially high erosion-corrosion resistance makes them the more cost-effective choices for long-term service (Ref 9).

Optical Properties

Color. The distinctive reddish color of copper corresponds to the sharp onset in frequency of strong interband optical transitions between filled states at the upper part of the 3d band and empty conduction-band 4s-like states. Copper alloys range in color from the reddish pink of pure copper to golden yellow and silvery white depending on alloy composition. The color of brasses ranges from reddish to pale yellow depending on their zinc content. Thus, low zinc casting alloys are categorized as "red" or "semired" brasses, while high-zinc brasses are called yellow brasses. Beryl-

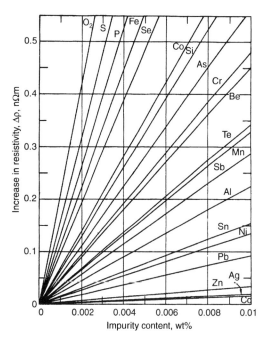

Fig. 2.6 The change in electrical resistivity of high-purity copper with the addition of various elements in solid solution. Source: Ref 5

Table 2.4 Increase in resistivity of copper due to small solute additions

Solute	Room temperature solubility, wt %	Δρ/at.%, μΩ-cm	Observed range of Δρ/at.%, μΩ-cm
Ag	0.1	0.6	0.1–0.6
Al	9.4	0.95	0.8–1.1
As	6.5	6.7	6.6–6.8
Au	100.0	0.55	0.5–0.6
B	0.06	1.4	1.4–2.0
Be	0.2	0.65	0.6–0.7
Ca	< 0.01	(0.3)(a)	...
Cd	< 0.5	0.3	0.21–0.34
Co	0.2	6.9	6.0–7.0
Cr	< 0.03	4.0	3.8–4.2
Fe	0.1	8.5	8.5–8.6
Ga	20.0	1.4	1.3–1.5
Ge	11.0	3.7	3.6–3.75
Hg	?	(1.0)(a)	...
In	3.0	1.1	1.0–1.2
Ir	1.5	(6.1)(a)	...
Li	< 0.01	(0.7)(a)	...
Mg	1.0	(0.8)(a)	...
Mn	24.0	2.9	2.8–3.0
Ni	100.0	1.1	1.1–1.15
O	~0.0002	5.3	4.8–5.8
P	0.5	6.7	6.7–6.8
Pb	0.02	3.3	3.0–4.0
Pd	40.0	0.95	0.9–1.0
Pt	100.0	2.0	1.9–2.1
Rh	20.0	(4.4)(a)	...
S	~0.0003	9.2	8.7–9.7
Sb	2.0	5.5	5.4–5.6
Se	~0.0004	10.5	0.2–10.8
Sn	1.2	3.1	3.0–3.2
Te	~0.0005	8.4	2.8–3.5
Ti	0.4	(16.0)(a)	...
U	~0.1	(10.0)(a)	...
W	?	3.8	...
Zn	30.0	1.3	0.3

(a) Estimated. Source: Ref 13

lium coppers change from red to gold with increasing beryllium content, while copper-nickels are pink at low nickel concentrations and distinctly silvery at higher nickel contents. The wide range of available colors make copper alloys useful for decorative purposes (Ref 23, 24).

Reflectivity and Emissivity. The reflectivity of copper decreases continuously with wavelength from the infrared through ultraviolet regions. For incandescent light, the spectral reflection coefficient of copper is 0.63. Tables 2.5 and 2.6 list indices of refraction, extinction coefficients, and reflectance values (measured and calculated) for several types of copper surfaces.

Among copper's other published optical properties are:

- Nominal spectral emittance for polished copper at $\lambda = 655$ nm and 800 °C (1470 °F): 0.15
- Coefficient of absorption (absorptivity) for solar radiation: 0.25
- Emissivity of heavily oxidized copper surface: ~0.8

Magnetic Properties

Susceptibility and Permeability. Magnetic susceptibility (χ) is defined as the ratio of induced magnetization to applied magnetic field, for example, $\chi=M/H$, where H is the applied magnetic field, and M is the magnetization. Each material

has a characteristic magnetic susceptibility, although the property varies somewhat with thermomechanical history. Copper is diamagnetic because its magnetic susceptibility is negative. Data from several sources are listed in Table 2.7. The magnetic susceptibility of several copper-nickel alloys is given in Table 2.8. Note that the alloys' magnetic behavior changes from diamagnetic to paramagnetic between 0.94 and 5.0% Ni (Ref 26, 28).

Some copper alloys and intermetallic compounds, such as Cu_8Sm, Cu-Ni-Fe and Cu-Ni-Co alloys, exhibit ferromagnetic behavior. The so-called Heussler alloys based on Cu-Mn-Al are also ferromagnetic. These materials have a maximum permeability of 100 to 1000, a saturation induction of 6 T, a saturation hysteresis of 300 to 450 erg/cm³, and a coercive force ranging from 79.6 to 15,900 A/m. In this case, ferromagnetism arises not from copper but from a superlattice structure in which manganese atoms occupy appropriately spaced positions (Ref 29).

Cunife is an alloy of 80% Cu, 20% Fe, and 20% Ni. As heat treated (quenched from 1000 °C, 1830 °F), the alloy contains ferromagnetic iron- and nickel-rich clusters in a copper-rich matrix. The clusters apparently arise through spinodal decomposition. By cold working and aging, the alloy develops a crystallographic (and magnetic) texture (Ref 30).

In commercial coppers and copper alloys, magnetic susceptibility is determined largely by the iron contained as an impurity. In OF copper, the small amount of iron present in solid solution has only a small effect on magnetic properties. In elec-

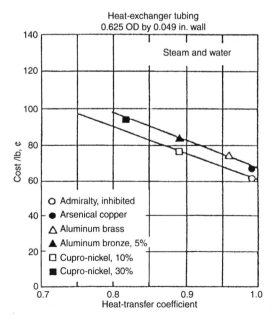

Fig. 2.7 Heat transfer coefficients for selected copper alloys. Source: Ref 9

Table 2.5 Optical properties of copper: normal-incidence reflectance of freshly evaporated mirror coating copper

Wavelength, μm	Reflectance, %	Wavelength, μm	Reflectance, %
0.220	40.1	0.750	97.9
0.240	39.0	0.800	98.1
0.260	35.5	0.850	98.3
0.280	33.0	0.900	98.4
0.300	33.6	0.950	98.4
0.315	35.5	1.0	98.5
0.320	36.3	1.5	98.5
0.340	38.5	2.0	98.6
0.360	41.5	3.0	98.6
0.380	41.5	4.0	98.7
0.400	47.5	5.0	98.7
0.450	55.2	6.0	98.7
0.500	60.0	7.0	98.7
0.550	66.9	8.0	98.8
0.600	93.3	9.0	98.8
0.650	96.6	10.0	98.8
0.700	97.5	15.0	99.0

Source: Ref 22

tronic-tough-pitch (ETP) copper containing about 200 ppm oxygen, the iron is present as Fe_3O_4, which has a much greater effect on the magnetic properties than iron in solid solution. Iron in alloys, likewise, has little effect when present in solid solution, such as after high-temperature solution annealing followed by rapid cooling or quenching. But it creates a detectable positive susceptibility when precipitated in the form of iron-rich phases, as shown in Table 2.9.

Magnetic permeability (μ) is defined as the magnetic opacity of a material in a given magnetic field. It is expressed as $\mu = B/H$, where B is the flux density and H is the applied magnetic field or magnetizing force. It is the instantaneous slope of the magnetization. For pure copper, μ is very low. This is very useful because low permeability in the presence of magnetic fields avoids energy losses in applications such as electric motors and generators.

Table 2.6   Optical properties of various copper surfaces

Wavelength, μm	Index of refraction	Extinction coefficient	Reflectance, calculated
Bulk copper			
0.3650	1.0719	2.0710	0.5004
0.4050	1.0769	2.2890	0.5491
0.4360	1.0707	2.4610	0.5860
0.5000	1.0308	2.7843	0.6528
0.5500	0.7911	2.7177	0.7013
0.5780	0.3250	2.8923	0.8716
0.6000	0.1491	3.2867	0.9508
0.6500	0.1074	3.9104	0.9740
0.7500	0.1034	4.8847	0.9835
1.000	0.1471	6.9334	0.9881
Single crystal copper			
0.4400	1.1070	2.5565	0.5965
0.4600	1.0942	2.6320	0.6131
0.4800	1.0618	2.7124	0.6341
0.5000	1.0836	2.7684	0.6390
0.5200	1.0438	2.7784	0.6490
0.5400	0.9324	2.7348	0.6674
0.5600	0.6470	2.7200	0.7440
0.5800	0.2805	2.9764	0.8931
0.6000	0.1360	3.3464	0.9565
0.6200	0.1040	3.6525	0.9714
0.6400	0.0972	4.0692	0.9798
0.6600	0.897	4.0692	0.9798
Evaporated copper			
0.1025	1.05	0.70	0.098
0.1113	0.95	0.73	0.115
0.1215	0.95	0.78	0.137
0.1306	0.96	0.83	0.148
0.1392	1.00	0.91	0.165
0.1500	1.02	1.02	0.192
0.1603	0.98	1.04	0.2219
0.1700	0.94	1.12	0.254
0.1800	0.90	1.21	0.296
0.1900	0.88	1.36	0.335
0.2000	0.94	1.51	0.378
0.500	0.88	2.42	0.625
0.600	0.186	2.980	0.928
0.700	0.150	4.049	0.966
0.800	0.170	4.840	0.973
0.900	0.190	5.569	0.977
1.000	0.197	6.272	0.981
1.35	0.45	7.81	0.971
1.69	0.58	9.96	0.977
2.28	0.82	13.0	0.981
3.00	1.22	17.1	0.984
3.4	1.53	20.3	0.985
3.97	1.94	23.1	0.986
4.87	2.86	28.9	0.987
5 0	2.92	27.45	0.985
5.8	3.71	34.6	0.988
7.00	5.25	40.7	0.988
7.3	5.79	43.2	0.988
8.35	7.28	49.2	0.988
9.6	9.76	57.2	0.988
10.25	11.0	60.6	0.988
10.8	12.6	64.3	0.988
12.25	15.5	71.9	0.989

Source: Ref 22

Table 2.7   Magnetic susceptibility of pure copper

Temperature, °C	$\chi \times 10^6$, cgs
−267	−0.0836(c)
−259 to −253	−1.22(b)
−250	−0.0856(c)
−242	−0.0863(c)
−233	−0.0879(c)
−211	−0.0873(c)
−174	−0.0869(c)
−171	−0.0867(c)
−65	−0.0863(c)
−26	−0.0859(c)
18	−1.08(b)
19	−0.0855(c)
20	−0.086(a)
300	−0.085(a)
600	−0.082(a)
900	−0.079(a)
1080 (solid)	−0.077(a)
1080	−0.97(b)
1090	−0.68(b)
1090 (liquid)	−0.054(a)

(a) Ref 24. (b) Ref 21. (c) Ref 26. Source: Ref 29

Table 2.8   Magnetic susceptibilities of copper-nickel alloys at 22 °C (72 °F)

% Ni	$\chi \times 10^6$, cgs
0	−0.080
0.94	−0.062
5.0	−0.042
10.0	0.185
11.0	0.221
20.3	0.55
30.0	1.28
35.6	1.99
38.0	2.84
50.2	11.8

cgs units, emu/g; data precise to ± 0.004. Source: Ref 26, 30

Hall Effect. The Hall effect is the electromagnetically induced electrical potential oriented 90 degrees to the direction of the magnetic field. The Hall voltage for pure copper is -5.24×10^{-4} V at 0.30 to 0.8116 T_m where T_m is melting point, and the Hall coefficient is -5.5 mV $\cdot$ m/T.

Influence of Heat Treatment. Heat treatment has no influence on the magnetic properties of pure copper; however, in alloys such as copper-rich Cu-Ni-Fe (Cunife) and Cu-Ni-Co, heat treatment aids the formation of ferromagnetic phases. Ferromagnetic phases may also arise from precipitation hardening, spinodal decomposition, and order-disorder transformations. Alloys used in electronic instruments, mine sweepers, and similar applications may be quenched from solution annealing temperatures to retain magnetic phases in solid solution and thereby avoid their magnetic effects.

Acoustic Properties

Brittle fracture occurs when the velocity of a propagating crack exceeds the velocity of sound in a material, for example, when fracture energy is dissipated faster than (elastic) acoustic energy. For pure annealed copper, the velocity of sound is 4759 m/s for longitudinal bulk waves, 3813 m/s for transverse waves, 2325 m/s for shear waves, and 2171 m/s for Rayleigh waves. These values are very high. As a result, brittle fracture and fatigue failure occurs infrequently in the pure annealed metal. The possibility of brittle failure and fatigue should, however, be taken into account for hard tempers and for hard copper alloys (Ref 26).

Liquid State Properties

Dynamic Viscosity

Viscosity is defined as the internal friction of a fluid that causes it to resist flowing past a solid surface or other layers of the same fluid. Table 2.10 shows the change in viscosity in pure liquid copper at temperatures at and slightly above the melting point.

Electrical Resistivity and Thermal Conductivity

Table 2.11 lists thermal conductivity and electrical resistivity data for liquid-state oxygen-free and phosphorus-deoxidized low residual phosphorus coppers at temperatures up to 1400 °C (2550 °F) (Ref 31).

Surface Tension

Every interface has a tension associated with unequal forces between atoms close to the surface that forms the phase boundary. In the case of molten metal, this interfacial energy is important to nucleation processes and to phenomena such as slag separation. Table 2.12 contains surface ten-

Table 2.9 Magnetic susceptibilities of selected copper alloys

Nominal composition, % Cu-Sn-Pb-Zn	Casting alloys, $\chi \times 10^6$			
	0.01% Fe		0.15% Fe	
	As cast	After solution-quench treatment	As cast	After solution-quench treatment
85-5-5-5 (C83600)	0.1	0.1	0.9	0.1
81-3-7-9 (C84400)	0.1	0.14	1.3	0.4
67-1-3-29 (C85400)	...	...	0.9	0.2
87-8-1-4 (C92300)	...	...	4.5	0.4
86-6-1-1.5-4.5 (C92200)	...	...	6.3	0.8
88-8-4 (C90300)	...	...	3.4	1.1
71-1-3-25 (C85200)	0.1	0.1	16	6.0
80.5-19Sn-0.5P (C91300)	0.1	-0.1	67	8.5
	Wrought alloys, $\chi \times 10^6$			
	0.01% Fe		0.15% Fe	
	As rolled	After solution-quench treatment	As rolled	After solution-quench treatment
92Cu-8Al	-0.2	-0.1	0.3	0.1
62Cu-19Zn-18Ni-1Pb (C76300)	+0.4	+0.5	0.8	0.7
60-3-37	0.1	0.0	2.6	308
Electrolytic Cu	-0.1	-0.1	6.1	0.1
95Cu-5Sn+P (C51000)	0.0	0.0	17	43
62Cu-0.5Sn-32.5Zn (C46400)	0.1	-0.1	42	1.7
67Cu-33Zn (C27000)	0.1	-0.1	58	1.0

In cgs units, emu/g. Source: Ref 26

sion data for molten copper of several levels of purity.

Gas Solubility

The solubility of gases in liquid copper depends on temperature and pressure, but it is quite high. In contrast, the solubility of hydrogen and oxygen in solid copper is very small, and the metal, therefore, rejects considerable amount of these (and other) gases upon solidification. Figures 2.8(a–c) give oxygen and hydrogen solubility data for pure copper.

Oxygen is retained in copper after the converting stage, but it is removed, along with other impurities, in subsequent electrolytic refining operations. Cathodes produced by electrolytic refining must be remelted to make useful shapes. If suitable precautions are not taken, oxygen can be reintroduced at this point; however, copper can be maintained in the "oxygen-free" (OF) condition by remelting and casting under reducing or neutral atmospheres.

Both oxygen and hydrogen interfere with conductivity; however, small and controlled amounts of oxygen are actually beneficial to conductivity in that they combine with and remove from solution impurities such as iron, which are far more detrimental. Tough pitch coppers, which can contain between 0.02 and 0.05% oxygen and which do not undergo the expensive electrolytic refining

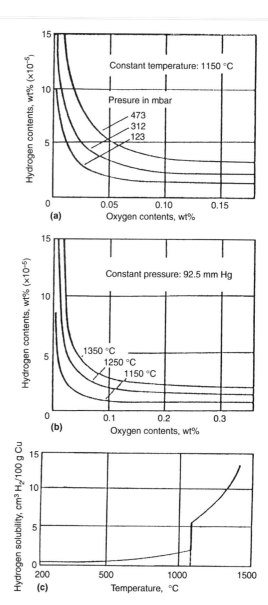

Fig. 2.8 Solubility of oxygen and hydrogen in copper. Source: Ref 11

Table 2.10 Temperature dependence of viscosity for pure copper

Temperature		Viscosity,
°C	°F	mPa · s
1085	1985	3.36
1100	2012	3.33
1150	2102	3.22
1200	2192	3.12

Source: Ref 2

Table 2.11 Thermal conductivity (λ) and electric resistivity (ρ) for molten pure copper and molten phosphorous deoxidized copper

Temperature, K	OFHC copper		C12200 (DHP copper)	
	λ,W/m · K	ρ, 10^{-8} Ωm	λ,W/m · K	ρ, 10^{-8} Ωm
1373	163.8	21.8	139.0	22.45
1423	165.2	22.2	141.6	22.9
1473	166.5	22.55	144.2	23.3
1523	167.8	22.9	146.9	23.75
1573	139.8	23.3	149.5	24.2
1623	170.5	23.65	152.2	...
1673	171.8(a)	24.1(a)	154.8(a)	25.1(a)

(a) Extrapolated values. Source: Ref 32

process, are, therefore, the most economical grades for common electrical conductors. Oxygen content must be carefully controlled, however, because excessive oxygen forms detrimental quantities of Cu_2O, which decreases both hot and cold workability. Also, in molten copper, oxygen can react with dissolved hydrogen to form water vapor, which evolves as voids during solidification. The voids cause severe cracking during hot rolling and produce a variety of defects on the surface of wire rods. Equilibrium oxygen solubility can be interpolated from the copper-oxygen phase diagram shown in Fig. 2.9. (Equilibrium diagrams are discussed in a following section.) (Ref 11, 32).

Hydrogen can be introduced into copper by (1) the presence of excessive hydrogen in the fuel of the melting furnace, (2) an excessively reducing flame in furnaces and launders, as in rod casting machines, or (3) by entrapped electrolyte in the copper cathodes. The latter is believed to be the overriding factor.

Solid State Properties

Copper exhibits the fcc structure at all temperatures below the melting point (structure symbol: $A1$; space group: O_h, $Fm3m$, $cF4$). Reported values for the lattice parameter include $a = 0.3615090 \pm 0.000004$ nm (Weast), $a = 0.36147$ nm at 293 K (20 °C, or 68 °F)(Simon et al.), and a

$= 0.36073$ nm at 290 K (17.8 °C, or 64 °F)(Hume-Rothery et al.). Based on the value reported by Weast, the distance of nearest approach, or Burgers vector, in the [110] direction at 293 K is 0.255625 nm (Ref 2, 5, 22, 23).

Slip Systems. In fcc metals such as copper, the {111} octahedral planes and the ⟨110⟩ directions contain closest atomic packing and, therefore, constitute the most active slip systems. There are four independent (111) slip planes in the fcc unit cell. Each (111) plane contains three [110] directions; therefore, the copper lattice has 12 possible slip systems. Slip occurs when the shear stress in a slip direction in the slip plane reaches a critical value, τ_{cr}. Table 2.13 lists room temperature values of τ_{cr} for copper single crystals of two degrees of purity (Ref 34).

Twinning. Face-centered metals such as copper undergo twinning; for example, they deform by forming mirror-image orientations, across {111} planes in the [112] direction. Twinning is a less important deformation mechanism in fcc metals than it is in body-centered cubic (bcc) or hexagonal close packed (hcp) metals, which have less favorable slip systems; however, copper does form annealing twins to accommodate deformation arising from recrystallization following cold work. Thus, the presence of annealing twins in the microstructure is evidence that the metal has been mechanically deformed prior to annealing.

Table 2.12 Liquid surface tension for coppers of several degrees of purity

Copper purity, %	Atmosphere	Temperature °C	°F	Surface tension, N/m
99.95	Vacuum	1083	1981	1.300
99.93	N_2	1100	2012	1.341
99.93	N_2	1150	2102	1.338
99.93	N_2	1200	2192	1.335
99.997	He/H_2	1083	1981	1.355
99.997	Ar	1083	1981	1.358
99.997	Vacuum	1083	1981	1.352

Source: Ref 21

Table 2.13 Room-temperature slip systems and critical (threshold) resolved shear stress for copper single crystals: slip on (111) plane in [110] direction

Purity, %	σ_c, g/mm^2
99.9999	65
99.98	94

Source: Ref 21

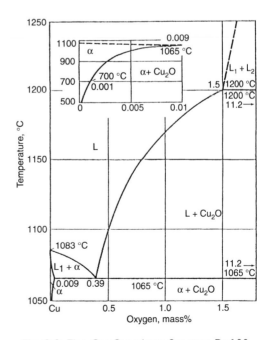

Fig. 2.9 The Cu-O system. Source: Ref 33

Stacking Faults. The stacking fault energy of copper, that is, the energy associated with imperfectly positioned adjacent lattice planes, is approximately 40 erg/cm^2. Stacking fault energy is related to such deformation-related properties as flow stress and strain hardening. It has been shown by x-ray diffraction that the stacking fault energy of brass decreases with zinc content, and this is in agreement with the observation that alpha brass forms a greater number of mechanical twins than copper.

Copper Alloys

Solid Solutions. Perhaps the largest commercial tonnage of copper alloys are those that exist as solid solutions. A solid solution results when atoms of different elements can occupy equivalent sites randomly on a crystal lattice. Such elements form what are known as "substitutional solid solutions." This behavior is most likely to occur (but will not always do so) when atoms of the elements involved differ in size by less than 14 to 15%. Where the mismatch is less than 8%, complete mutual solubility is possible. (It should be understood that atomic size depends, among other factors, on the immediate environment in which the atom finds itself. In simple terms, it is convenient to assume the size of atoms in a crystalline metallic state to be constant, and it is upon this assumption that the 15% rule is based.) Solid solubility is also favored when constituent metals have the same number of valence electrons, although this is not necessarily limiting, and when they exhibit the same crystal structure (Ref 35–37).

Copper and nickel, whose atomic sizes differ by only 2.5% and which both exhibit the fcc structure, form a continuous series of solid solutions over their entire binary composition range. This is illustrated in the Cu-Ni equilibrium diagram, Fig. 2.10. Copper-gold alloys, represented in Fig. 2.11, as well as other bivalent transition metals such as palladium and platinum, also exhibit complete solid solubility with copper at high temperatures; however, intermediate alloys with these metals dissociate into one or more ordered phases at lower temperatures. During ordering, Cu and Au atoms, for example, abandon their random distribution to take up preferred positions on the crystal lattice. The atomic arrangement in this and analogous cases, is such that nearest-neighbor atoms are of different species. Such atomic arrangements are known as superlattices.

Size similarity alone does not guarantee complete solubility, as can be seen in the copper-silver system (Fig. 2.12). Among copper alloys, cases of complete solid solubility are relatively rare, but partial or limited solubility is quite common. Some of the commercially important examples of alloys with limited solubility are the systems copper-zinc (brasses), copper-tin (tin or phosphor bronzes), copper-aluminum (aluminum bronzes), and copper-silicon (silicon bronzes).

Interstitial solid solutions result when the size difference between atoms is sufficiently large such that the smaller atoms can fit into interstices in the space lattice of the larger atoms. Such solutions occur, for example, when relatively tiny atoms such as those of hydrogen, boron, oxygen, nitrogen or carbon dissolve in a metal lattice. While interstitial solid solutions are very important in steels and a few other systems, they have little practical use in commercial copper alloys. The presence of interstitial elements are, in fact, often deleterious to useful properties. For example, oxygen and hydrogen readily form interstitial solutions with copper and, except for OF grades, a small amount (up to several tens of ppm) of oxygen is contained in many copper products. When present together in sufficient quantity, these atoms can combine at elevated temperatures to form water molecules, which coalesce as voids, causing severe embrittlement. Carbon and boron are nearly insoluble in solid copper.

Precipitation Hardening. Limited solid solubility is taken advantage of in several copper systems, most notably Cu-Be, Cu-Cr, and Cu-Zr. These systems contain alloys that can be precipitation hardened. The phenomenon, which is also called precipitation strengthening and age-harden-

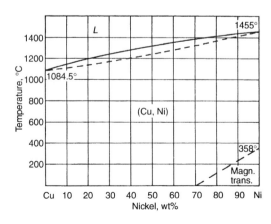

Fig. 2.10 The Cu-Ni system

ing, is possible because the limit of solid solubility contracts with decreasing temperature, a condition known as retrograde solubility.

An example is illustrated in the Cu-Be equilibrium diagram shown in Fig. 2.13. Commercially important copper-beryllium alloys nominally con-

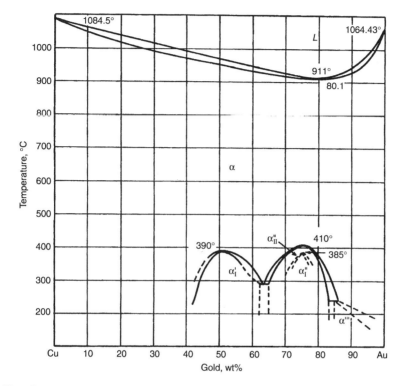

Fig. 2.11 The Cu-Au system

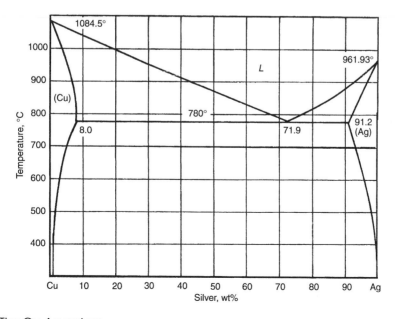

Fig. 2.12 The Cu-Ag system

tain less than about 2.5 wt% Be (0.15 at.% Be) (plus other elements that needn't be discussed here). They are heat treated to remarkably high strengths in two steps. The first step involves solution annealing at temperatures above those defined by the retrograde solubility line, between 790 and 925 °C (1450 and 1700 °F) depending on composition, followed by quenching in water. The rapid drop in temperature inhibits formation of the beryllium-rich γ phase which, according to the phase diagram, is stable below the solubility line. Next, alloys are reheated to between roughly 300 and 500 °C (575 and 930 °F). This provides sufficient thermal energy to cause γ phase precipitation from microscopic regions known as Guinier-Preston zones, where local Be supersaturation is at its highest. The resulting fine precipitates are coherent with (bound to) the matrix copper. This coherence gives rise to strain fields surrounding each particle. The strain fields induce stresses that inhibit the movement of dislocations, thereby strengthening the alloy.

Other commercially important alloy systems in which precipitation strengthening is exploited include copper-chromium (Fig. 2.14), copper-zirconium (Fig. 2.15), and copper-chromium-zirconium alloys. Alloys containing chromium, iron, or iron and phosphorus also benefit from the precipitation

of second phases, but these materials are not normally heat treated in the same sense of the word.

Precipitation leaves the surrounding matrix nearly devoid of the precipitated element, thereby raising the matrix's electrical conductivity. Dilute precipitation-hardened alloys have, therefore, found extensive use in electrical and electronic applications ranging from resistance welding electrodes (Cu-Cr, Cu-Cr-Zr) to electrical/electronic connectors and leadframes (Cu-Zr, Cu-Be). The so-called "high conductivity" beryllium coppers were specifically designed for heavy-duty electrical/electronic connector applications.

Order-Disorder Transformations. The crystal structure of alloy phases can be ordered or disordered. Ordered phases display what is known as translational symmetry of their Bravais (space) lattice. That is, the various constituents' atoms occupy specific sites in the unit cell of the alloy's crystal structure, and that pattern is repeated over long distances, producing symmetry. This type of atomic arrangement, which is readily detectable through the generation of extra lines in x-ray diffraction patterns, is known as "long-range order." When a crystal orders, it is likely that the ordering starts at a large number of positions throughout the crystal. At each position, it is pure chance

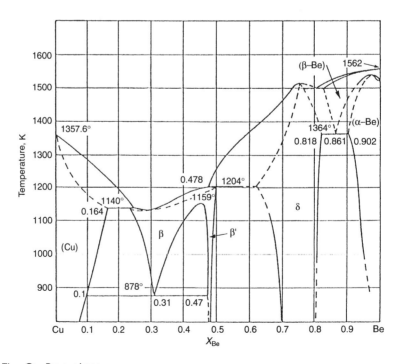

Fig. 2.13 The Cu-Be system

which will be chosen as α sites and β sites, and hence there can be neighboring regions in which there has been an opposite choice. Such regions are called anti-phase domains, and the boundary between the two regions is called antiphase domain boundary (Ref 38, 39).

Beta brass (brass with relatively high zinc content, for example, approaching 40% Zn) exhibits an ordered phase in which the two components (copper and zinc) have equal concentrations. The phase's structure is analogous to that of cesium chloride (CsCl), Fig. 2.16. This structure is equivalent to a simple cubic lattice with Cu atoms at $(0,0,0)$ and Zn at $(a/2)$ $(1,1,1)$.

In the disordered beta phase of brass, which is stable at elevated temperatures, zinc and copper atoms are placed randomly on the space lattice. For example, proceeding along a ⟨111⟩ direction in ordered β-brass, one would encounter copper and zinc atoms in the sequence: Cu Zn Cu Zn Cu Zn Cu Zn..., whereas in the disordered phase, a typical sequence might be: Cu Zn Zn Zn Cu Zn Cu Cu, etc. That notwithstanding, a certain degree of "short-range order" is retained at temperatures above the range where the long-range ordered structure is stable. Short-range order is explained by the persistence of chemical affinity between the atomic species involved above the temperature range where differences in atomic size govern the crystal structure.

Intermetallic Compounds. Copper forms many intermetallic compounds with other metals.

The compounds may be simple binary phases or they may have highly complex structures composed of dozens of several different atoms. Bonding may be ionic or covalent, depending on the electronegativities of the constituents.

The British metal physicist Sir William Hume-Rothery proposed on the basis of atomic theory that the formation of structures now known as "electron compounds" should be favored at electron:atom ratios (in the unit cell) of 3/2, 21/13, and 7/4. The structures are not compounds in the ionic sense and are better described as intermediate phases. The 3/2 compounds adopt the bcc structures, complex cubic (β-manganese) structures, and occasionally hcp structures. The 21/13 compounds exhibit complex cubic structures (γ-brass) containing 52 atoms per unit cell, while the structure of 7/4 compounds is hcp (ε-brass). Electron compounds may be disordered or ordered, and they may display composition ranges on either or both sides of the idealized composition represented by the stated electron:atom ratios. Examples of electron compounds in copper-base systems include $CuZn$, Cu_3Al and Cu_3Sn (3/2, bcc), Cu_3Ga (3/2, hcp); Cu_5Zn_8, $Cu_{31}Zn_8$ and Cu_9Al_4 (21/13, complex cubic); and $CuZn_3$, Cu_3Ga, and Cu_3Ge (7/4, hcp). Electron compounds are of interest in a variety of alloy systems, particularly where their precipitation produces strengthening (Ref 38).

Another class of intermetallic compounds known as "Laves phases" have the general formula

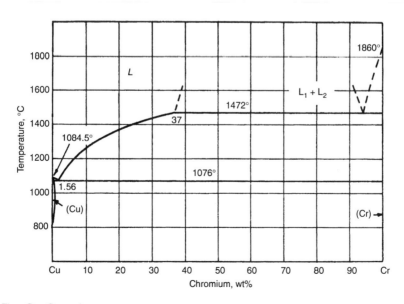

Fig. 2.14 The Cu-Cr system

AB$_2$. Here, too, structures are complex, either of the cubic or hexagonal systems. Copper-bearing examples of the former include MgCu$_2$ and CuNi$_2$.

Diffusionless (Martensitic) Transformations. A few copper alloys, when quenched from elevated temperatures, form a martensitic phase that is structurally analogous to the martensite seen in heat-treated steels. The phase forms by a shear transformation that proceeds at acoustic velocities, for example, far too fast for diffusion to take place.

Copper-aluminum alloys (aluminum bronzes such as C95300, C95400, C95500) containing more than about 8.5 to 9.25% Al, depending on overall composition, can be heat treated. In practice, however, the alloys are not always heat-treated since acceptable mechanical properties can often be realized without the additional cost that heat treatment entails. (It should be understood that many aluminum bronzes are not simply copper-aluminum binary alloys, but are actually mul-ticomponent and multiphase materials, which may contain iron, manganese, nickel, or silicon. These additional elements are not essential to the martensitic transformation but may influence it.) When heat-treated, the alloys are typically solution-annealed at temperatures in the range 788 to 927 °C (1450 to 1700 °F), then quenched in water or oil. The high-temperature anneal creates a uniform structure containing mostly the beta phase, and it is this phase that transforms martensitically when quenched.

Other commercially important alloy systems in which martensitic transformations are known include Cu-Zn-Al and Cu-Al-Ni. In these alloys the reversible nature of the shear transformation is responsible for the shape memory effect. See Chapter 3, "Products."

Nonequilibrium Solid Phases

Metallic Glasses. Cooling rates during solidification on the order of 10^5 to 10^6 °C/s suppress crystallization, and the metal or alloy freezes as an

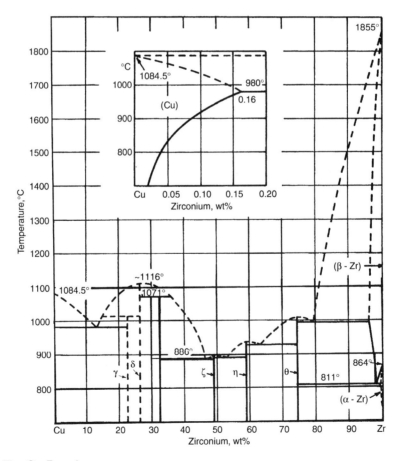

Fig. 2.15 The Cu-Zr system

amorphous structure known as a metallic glass. Copper-base metallic glasses can be obtained from the pure metal or from alloys, such as $Cu_{50}Ti_{50}$, $Cu_{67}Ti_{33}$, Cu-5 at.% Ag, Ag-15 at.% Cu, $Cu_{20}Ta_{30}$, $Cu_{70}Ta_{30}$, and $Cu_{60}Zr_{40}$. The metallic glass $Pd_{77.5}Cu_6Si_{16.5}$ exhibits a hardness of 500 HV, a yield strength of 1570 MPa (227 ksi) and a Young's modulus of 88.2 GPa (24.2×10^6 psi) (Ref 41).

Quasi-Crystals. Copper-containing materials are capable of forming quasi-crystalline phases. These are structures with very high defect densities, whose domains extend for less than a few tens of nanometers. Quasi-crystals exhibit so-called icosahedral symmetry. One such occurrence can be found in the Al-Li-Cu ternary system, where a thermodynamically stable quasi-crystalline phase is formed by peritectic reaction. The composition has been given as $Al_{5.1}Li_3Cu$ (Ref 42).

Nanocrystals. Nanocrystals are polycrystalline materials with grain sizes smaller than 50 nm. They can be produced by mechanical alloying and a variety of other techniques. Nanocrystalline materials have been produced from high purity copper, $Cu-Cu_2O$ and Cu-5 at.% Zr. Nanocrystalline materials offer the possibility of developing superplasticity at moderate and elevated temperatures and of developing high strength and hardness as a result of their ultrafine grain size (Ref 43–45).

Thermodynamics of Copper Alloys

The thermodynamics of copper alloys has considerable technological and commercial significance, particularly in the understanding of alloy systems, structure-property relationships, and corrosion phenomena. Only a brief synopsis of this extensive subject can be presented; however, excellent treatments are available in publications of the International Copper Association (ICA, formerly the International Copper Research Association, INCRA) and ASM International, as well as numerous standard texts (Ref 46–48).

Equilibrium Diagrams

Equilibrium diagrams, also called phase diagrams, are a graphic representation of the stability ranges for phases in metal and alloy systems. Stability regions (phase fields) are depicted as functions of composition, temperature, pressure, or, in the case electrochemical systems, pH and Eh. A phase may simply entail a physical state—gaseous, liquid, solid—or, as in the case of metals and other solids, a homogeneous and crystallographically distinct structure. Phase diagrams are very useful tools to understand alloy compositions and heat treatments.

Copper Alloy Systems. Of the several types of phase diagrams that can be constructed for metallic systems, the most common are those that define regions of phase stability in terms of composition and temperature. These are most often constructed for two-element systems. A selection of such binary, copper-base phase diagrams are shown in Fig. 2.9 through 2.15 and 2.17 through 2.21. Ternary (three-element) diagrams, being three-dimensional, are usually represented either as isothermal or isocompositional sections. Four and higher component diagrams cannot be visualized, and phase stability in such cases can only be represented in terms of fixed temperatures and compositions.

For compounds that at least nominally obey Hume-Rothery's electron:atom ratio relationships, certain general observations have been made (Ref 49):

- The first stable phase to appear in binary Cu-X alloys after the (X) solubility limit is exceeded has an electron:atom ratio of 3/2. As stated earlier, 3/2 phases tend to exhibit either the bcc β structure (if disordered, and β′ if ordered); the hcp structure, usually called ζ or ζ′, or the β-manganese structure, μ.

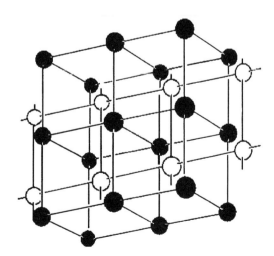

Fig. 2.16 CsCl type superlattice displayed by ordered beta brass. Spheres represent Cu and Zn atoms, which occupy equivalent lattice sites. Source: Ref 38

- Increasing valency of the solute favors formation of the ζ and μ structures at the expense of β or β′ structures; while elevated temperature, which accelerates diffusion, favors the disordered β structure at the expense of β′ or ζ.
- Increased solute atom size favors the β structure while narrowing the composition range of 3/2 compounds. It also moves the compound's stability range to lower electron concentrations.
- Large differences in electronegativity favor the formation of valency compounds such as Cu₃P. Such compounds differ from true intermetallic compounds in the nature of their bonding. Valency compounds also characteristically display long-range order.
- Still higher solute concentrations promote the formation of 21/13 electron compounds, which have the complex cubic γ-brass structure (also called the β-manganese structure, with 52 atoms/unit cell). These compounds are characterized by having 87 or 88 electrons per unit cell. Increasing the size of the solute atoms shifts the electron concentration to lower numbers, but no γ phases form beyond a size difference of 20%. Increasing the electrochemical difference between solute (Cu) and

solvent shifts the electron concentration to higher numbers.

A few of commercially useful alloy systems in which second phases play a role deserve mention. The most important of these is the copper-zinc system (Fig. 2.17). Here, a solid solution field extends to 32.5% Zn at 903 °C (1657 °F) and to approximately 39% Zn at lower temperatures. (Compositions are in weight percent unless otherwise specified.) This region comprises the alpha brasses, whose structure and forming characteristics are similar to those of pure copper. Increasing zinc content beyond the limit of the α-phase field leads to the formation of bcc β-phase (CuZn), a 3/2 compound. (The designation CuZn should not be taken to suggest β is a line compound of invariant composition. Like most intermetallic compounds, β displays a range of solid solubility for its constituents. The composition represented by CuZn falls within that solubility field.) Other intermetallic phases in the system include γ, a 21/13 compound with a complex cubic structure, nominally Cu₅Zn₈, and the 7/4 compound ε, nominally CuZn₃, which has a hcp structure (Ref 49).

The copper-tin system, Fig. 2.18 is the basis for the common tin bronzes, which are also called phosphor bronzes. Tin forms a range of alpha

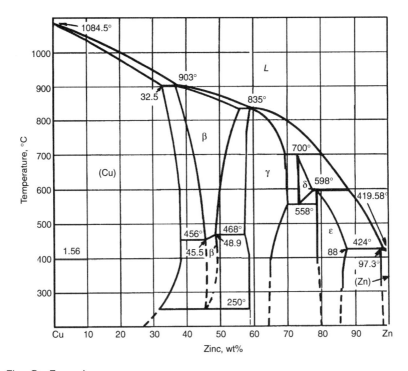

Fig. 2.17 The Cu-Zn system

solid solutions with copper extending to 15.8% Sn at temperatures between 520 and 586 °C (968 to 1087 °F). One of the important practical differences between the copper-rich ends of the Cu-Sn and Cu-Zn systems is that in the former, the temperature range between liquidus and solidus is quite broad. The extended freezing range causes tin bronzes to pass through a semisolid or mushy stage during solidification. Castings molds must be designed to take this weak structure into account. The extended liquidus-solidus gap is also responsible for the dendritic segregation, or coring, seen in tin-bronze castings (Ref 49, 50).

The Cu-Sn system contains several electron compounds. The β phase is a 3/2 compound found at compositions surrounding Cu_5Sn. It has the expected bcc structure. The 21/13 ratio is represented by the complex cubic δ phase (nominally $Cu_{31}Sn_8$), and the 7/4 ratio is satisfied by the orthorhombic ε phase at approximately 38% Sn (Cu_3Sn). The β phase, as well as the ordered γ phase (which is not an electron compound) are stable only at elevated temperatures and cannot be retained by quenching. Because of coring, the last metal to freeze in nominally all α compositions may undergo the peritectic reaction to form β,

which upon cooling dissociates to γ and eventually the brittle δ phase, which can be retained at room temperature. However, the low-temperature dissociation of α to α + ε rarely occurs due to the sluggishness of the ε-forming reaction in this temperature range; therefore, for practical purposes, the diagram can be thought of as containing a vertical line extending downward from the 15.8% Sn limit of the α solid solution field (Ref 50).

The copper-aluminum system (Fig. 2.19), contains numerous intermetallic phases and, several of which obey conventional electron:atom relationships. The copper-rich end of the system is marked by a solid solution field that extends to 9.5% Al at 565 °C (1050 °F), and alloys with less than about 8.5% Al comprise the single-phase alpha-aluminum bronzes. Because of segregation and nonequilibrium solidification and cooling, cast alloys containing between 8.5 and 9.5% Al contain substantial concentrations of β, which is stable between about 8.5 and 15% Al at elevated temperatures. The phase contains the composition Cu_3Al, a 3/2 electron compound. The β phase has the bcc structure. Upon cooling, it dissociates eutectoidally into α + γ₂. The χ phase includes the composition Cu_9Al_4, a 21/13 compound, which is

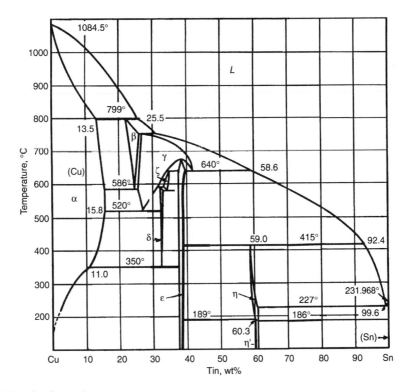

Fig. 2.18 The Cu-Sn system

stable above 963 °C (1765 °F). The complex cubic structure that is characteristic of 21/13 compounds is displayed by the γ_2 phase, one of the χ phase's ultimate decomposition products. The γ_2 phase contains somewhat more aluminum than would be required for a 21/13 electron compound (Ref 28, 50).

Eutectoidal decomposition of the β phase can be avoided by rapid cooling, in which case an ordered phase, designated β_1 forms. The β_1 phase transforms martensitically at lower temperatures to yield the hcp structures β' and γ', the particular phase depending on aluminum concentration.

Nonequilibrium Conditions

It is important to understand that the phase diagrams described above represent conditions existing under thermodynamic equilibrium. Such conditions may require very long times for sluggish or kinetically inhibited reactions and are not always attainable in practical terms. When this is the case, it is more useful to create nonequilibrium diagrams that describe, for example, phase changes that occur over a range of times and temperatures. The so-called time-temperature-transformation (TTT) diagrams and continuous cooling diagrams (CCT) that describe such conditions have found only limited use in copper alloy systems.

Segregation

For many practical thermal processes, surface- and interface-related phenomena can be more important than bulk physical properties. Several examples of this phenomenon are related to chemical heterogeneity brought about by segregation. Segregation occurs when the first metal of a multiconstituent alloy to freeze differs in composition from that which freezes later on. In castings, for example, solute elements are enriched at the chilled surface, which freezes first. Hot working and annealing enable the segregated structures to homogenize through the process of solid-state diffusion.

Influence of Segregation on Properties. Segregation also occurs within grains and dendrites, where it can lead, among other things, to intergranular embrittlement and corrosion. Electrical and thermal resistivity changes can also be detected. Segregation can influence phase changes

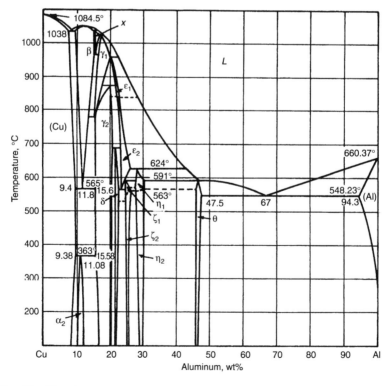

Fig. 2.19 The Cu-Al system

that nucleate at grain boundaries, and thereby affect the alloy's structure and properties. Segregation can also alter the surface diffusion that occurs at grain boundaries and thereby influence the rate of sintering during powder metallurgical processing.

Effects of Alloying Elements and Impurities

Alloying Elements versus Impurities

Whether a secondary element is considered an impurity or a true alloy constituent depends on its origin and on the nature (beneficial, detrimental, benign) of its effect(s) on properties. In general, alloying elements are species that are intentionally added to obtain certain desired properties. Impurities also produce specific effects, but they differ from true alloying additions in the unintentional nature of their occurrence. Impurities originate in the ore from which the metal is refined and/or are picked up accidentally during processing.

In commercially pure copper, impurities such as phosphorus, tin, selenium, tellurium, and arsenic are detrimental to economically important properties such as electrical conductivity and recrystalli-

zation temperature. When used in controlled amounts, however, the same elements provide specific benefits. Phosphorus, a common deoxidizer, improves weldability; tin provides solid solution strengthening and improves corrosion resistance; tellurium improves machinability, while arsenic (along with phosphorus, tin and antimony) inhibits dezincification corrosion in alpha brasses.

Effect on Properties

The most important effects of secondary elements on the properties of copper can be categorized by their influence on:

- Thermal and electrical conductivities
- Mechanical properties, including processing
- Chemical behavior, principally with regard to corrosion resistance
- Magnetic properties
- Processing and use

Specific effects are described in the following section.

Electrical Conductivity/Resistivity. Metals such as copper conduct electricity by virtue of the movement of electrons through the crystal lattice. Conduction is impaired (for example, resistivity

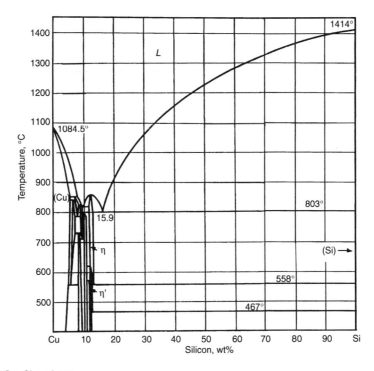

Fig. 2.20 The Cu-Si system

rises) by thermally induced atomic vibrations and by perturbations (defects or solute atoms) in the lattice, both of which scatter the mobile electrons. At temperatures approaching absolute zero, the effect of thermal vibrations on conductivity diminishes and the effect of impurities (solutes) dominates. The effect of solutes diminishes at elevated temperatures, although it can remain technically significant. Matthiessen's rule, mentioned earlier, is a statement of this phenomenon (Ref 11).

All secondary elements found in electrical coppers reduce conductivity and are, therefore, seen as impurities. Silver, which is present in some copper ores and can be carried through the refining process, is tolerated or ignored because the element exerts a negligible effect on conductivity in the concentrations at which it is normally present. The effect of other impurities ranges from slight to severe, as shown in Fig. 2.22 to 2.24. Figures 2.22 and 2.23 distinguish between impurities that oxidize readily in molten copper and are rejected from the lattice upon solidification, and those which are not.

Figure 2.24 illustrates the rule that changes in resistivity are proportional to solute concentration when the latter are small. Impurities also exert a much stronger effect when they are in solid solution than when they exist partly or entirely in a second phase. The effects of multiple impurities are cumulative in a single-phase alloys (Ref 51–57).

Thermal Properties. Solute atoms reduce thermal conductivity by perturbing the distribution and interaction of phonons in the lattice. In general, the reduction in thermal conductivity is similar to the effect on electrical conductivity. Solutes also raise the metal's specific heat and, therefore, its enthalpy and free energy. As solute content increases, the drive to minimize free energy can lead to the formation of a new phase, which generally has different thermal properties than the matrix metal (Ref 13, 58).

Magnetic Properties. While pure copper is diamagnetic, a number of copper alloys can develop paramagnetic or even ferromagnetic behavior through the precipitation of magnetic phases. Iron is the most common source of ferromagnetism in copper and copper alloys. The susceptibility of copper becomes positive (for example, paramagnetic) when as little as 0.1 wt% Fe is present, even when the copper is annealed and quenched to retain as much iron in solution as possible. Other examples of magnetic phenomena include copper-cobalt alloys, which generate magnetic structures through spinodal decomposition; Cu-Ni-Co phases, which are produced through age-hardening re-

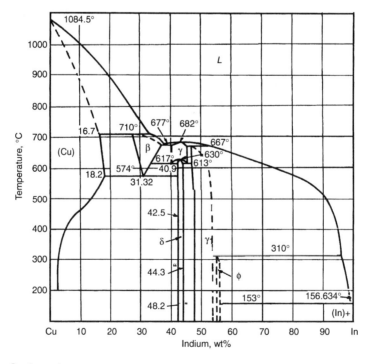

Fig. 2.21 The Cu-In system

actions; Heussler phases such as $MnCu_2Al$ (so-called σ phases, which are ferromagnetic by virtue of the spacing of manganese atoms in the crystal lattice), and Hume-Rothery electron compounds (Ref 11, 59–61).

Mechanical Properties. Strength, ductility and toughness are strongly affected by alloying. Impurities can also exert significant effects on mechanical properties, and such effects may or may not be detrimental. It is the job of the alloy designer to balance these effects to obtain the optimum combination of required properties.

The mechanisms by which solute elements affect mechanical properties include:

- Solid solution hardening
- Formation of hard phases
- Precipitation hardening
- Grain refinement
- Order-disorder transformations
- Shear (martensitic) transformation
- Dispersion strengthening

Solid solution hardening is by far the most common strengthening mechanism in commercial copper alloys. It is the dominant source of strength in alpha brasses; high copper alloys (exclusive of age-hardening systems); tin (phosphor) bronzes; copper-nickels and nickel-silvers, and alpha-aluminum bronzes. Mechanical strength generally increases, and ductility decreases, with increasing solute content.

Hard secondary phases are primarily responsible for the high mechanical properties of yellow brasses, high-strength yellow brasses, which are also called manganese bronzes, and silicon bronzes and brasses. The principal hardening constituent in these alloys is the bcc δ phase, although other structures also contribute. Strengthening by secondary phases is very effective, but it can entail the risk of reduced corrosion resistance in certain alloys.

Precipitation hardening, discussed earlier, is employed in high copper alloys such as chromium-, zirconium- and beryllium-coppers. Precipitation of hard phases from supersaturated solid solutions is also used to improve properties in copper-nickels and aluminum bronzes containing iron, nickel, chromium, or cobalt.

Strength increases with decreasing grain size. This is particularly useful since unlike most other strengthening mechanisms, raising strength by reducing the grain size does not strongly affect ductility. Castings especially benefit from small grain sizes, and small amounts of iron, a known grain refiner, are frequently included in the composition

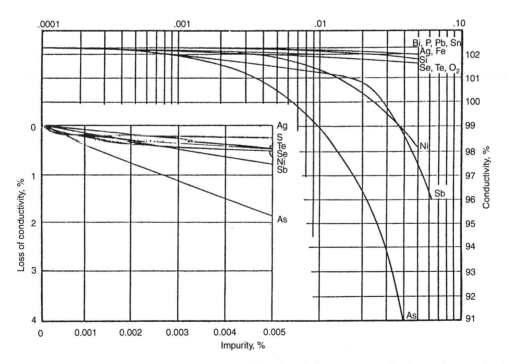

Fig. 2.22 Decrease in the conductivity of tough-pitch copper with impurity content. Source: Ref 51–54

of copper-base casting alloys. Nickel has a similar effect.

Order-disorder reactions and spinodal decompositions abound in copper-base systems, and a few of these are utilized commercially to improve properties. Spinodal decomposition in copper-nickel-tin alloys such as C71900, C96800, and C96900 produce strengths comparable to those attainable in beryllium coppers. Ordering and spinodal decomposition are discussed in the section of this chapter dealing with phase transformation.

Martensitic transformations occur in certain aluminum bronzes, which can be quenched and tempered much like heat-treatable steels. Martensitic transformations are also responsible for the shape memory effect (SME) displayed by Cu-Zn-Al and Cu-Al-Ni alloys.

Dispersion strengthening (DS) involves the incorporation of insoluble secondary structures in a copper-base matrix. The dispersants are not necessarily metallic phases but may be oxides, carbides, or other compounds. DS materials are called alloys, but they are more accurately characterized as composites because the secondary structure is insoluble in the matrix. The best-known commercial examples of copper-base DS materials are the aluminum oxide-bearing alloys C15715 to C15760. These materials are principally used for heavy-duty (RWMA Class III) resistance welding electrodes. Cu-W powder metallurgical compacts have been used for GMAW welding electrodes and other wear-resistant electrical products. Cu-Nb composites, which combine high thermal conductivity and good elevated-temperature mechanical properties, are used for liquid-cooled rocket motor cases. Composites based on Cu+(Zr or Ti)B_2 offer similar combinations of mechanical and thermal properties.

Influence of Solutes on Processing and Fabrication

Solute elements can modify the fabrication characteristics of copper and its alloys, even when they are present at impurity-level concentrations. For example, copper is most commonly deoxidized with phosphorus, which has little detrimental effect in concentrations below about 0.04% but does produce a slight increase in strength and a corresponding decrease in ductility. Manganese, magnesium, and boron are also used as deoxidizers, but they may be detrimental if used improperly. More than 0.5% manganese, for example, causes harmful inclusions in castings (Ref 50).

Annealing. Mechanical working is the most common, most effective and commercially most

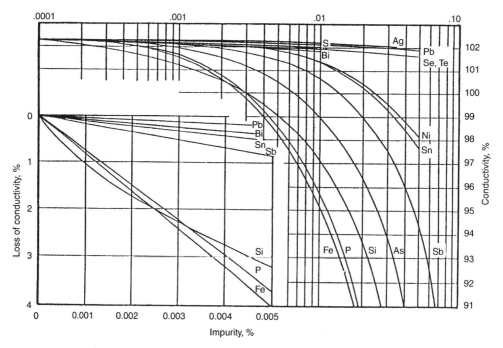

Fig. 2.23 Decrease in conductivity of oxygen-free (OF) copper of with impurity content. Source: Ref 51–54

important process for raising strength in copper metals. Wire and tube products are cold drawn; sheet and strip are cold rolled and formed into useful products. Such processes are classified as cold working because they are performed at temperatures below which recovery and recrystallization take place. Cold working renders the material stronger, harder, and less ductile than it was before deformation. In copper alloys, the degree of cold working is commonly referred to as the metal's temper. Because hard tempers may not be satisfactory for all applications, cold-worked material is often partially or fully annealed before use.

Recovery and recrystallization relieve residual stresses remaining after working. Energy supplied by heating (annealing) accelerates the processes. The metal first undergoes recovery, during which the more mobile dislocations produced during deformation redistribute and to some extent annihilate each other, creating a cell or subgrain structure. Internal stresses are partially relaxed, and the deformed structures begin to arrange themselves into a state of lower internal energy. No phase change accompanies these reactions (Ref 62).

Recrystallization, the nucleation of new grains, can occur simultaneously with recovery or begin after recovery has taken place. Final grain size depends on the amount of previous cold work and,

to a lesser extent, on the annealing temperature. The recrystallization temperature itself depends on the amount of cold work. It is particularly sensitive to even small concentrations of dissolved impurities, which modify the activation energy and kinetics of the grain-boundary migration phenomena that constitute nucleation. The activation energy for recrystallization of copper has been reported as 29,900 kcal/mol (oxygen-free high-conductivity copper cold rolled 99.7%); 22,400 kcal/mol (spectrographically pure copper cold rolled 98%) and 28,800 kcal/mol. Recrystallization is discussed further under "Phase Transformations" (Ref 62–64).

Recrystallization can lead to the development of preferential crystal orientations known as textures. Rolling produces textures oriented such that [110] planes and $\langle 112 \rangle$ directions tend to align parallel to the rolling direction. This is the so-called brass texture. Drawing produces a double fiber structure with both $\langle 111 \rangle$ and $\langle 100 \rangle$ directions parallel to the wire axis. Compression produces a [110] texture, while a $\langle 111 \rangle$ texture is produced by torsional deformation (Ref 46, 65).

Impurities elevate both the recrystallization temperature and the threshold temperature at which recrystallization stabilizes after a fixed time period. The effects increase with increasing vol-

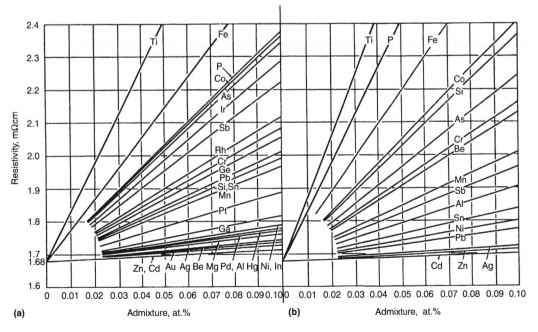

Fig. 2.24 Increase of electrical resistivity of (a) tough pitch copper and (b) oxygen-free copper with admixtures of various elements. Source: Ref 56

ume fraction of impurities. For comparable degrees of prior cold work, annealing at a given temperature takes longer when impurities are present. The presence of impurities, therefore, raises processing costs.

On the other hand, the annealing resistance of, for example, silver-bearing coppers is beneficial to soldered products such as electrical springs and automobile radiators. Phosphorus is another element capable of raising copper's softening temperature when it is present in sufficient concentration. Phosphorus-deoxidized, high residual phosphorus copper, C12200, is therefore used for plumbing tube, which is normally soldered for assembly. Other elements known to retard softening in deoxidized coppers include cadmium, iron, cobalt, and zirconium.

Oxygen, present as Cu_2O in tough pitch and undeoxidized coppers, restricts grain growth at high annealing temperatures. Annealed grain sizes in oxygen-bearing coppers are, therefore, more uniform than those found in high-purity copper.

Workability. Copper itself is very easily hot or cold worked, with reductions in thickness of more than 99% possible before annealing is required. Workability is usually impaired by the presence of impurities and by alloying. The effects are particularly strong when impurities segregate to grain boundaries, which can lead to embrittlement, and when alloying gives rise to the formation of a brittle phase such as β in the copper-zinc system.

Soluble impurities such as silver, gold, nickel, and arsenic have no effect on hot workability. Because these elements do not form stable oxides in copper, they are benign in both oxygen-bearing and deoxidized or OF copper. Cadmium and antimony, which are also soluble in copper but do form stable oxides, are also innocuous, in this case because their concentration is normally too low to cause concern. Oxides of iron, tin, zinc, phosphorus, silicon, and aluminum are not deleterious because they are rarely entrained during solidification. Selenium and tellurium form brittle sulfides and oxides, and the presence of these compounds in copper seriously degrades both hot and cold workability.

Zinc improves cold formability, which reaches optimum levels at the familiar 70Cu-30Zn composition of cartridge brass, C26000. The presence of δ phase in higher zinc brasses impedes cold formability, but hot workability (forgeability) is markedly improved. Forging brass, C37700, is one example of a material that takes advantage of this effect. The alloy's hot forgeability is high enough to overcome the detrimental effect of the small amount of lead added to improve machinability.

Bismuth and lead are nearly insoluble in solid copper, and their melting points are within or below the hot-working range for copper and copper alloys. The elements seriously impair hot workability by causing a form of embrittlement known as hot shortness. Cold workability is also reduced, particularly by bismuth, which readily segregates to grain boundaries.

Weldability. Oxygen-bearing coppers become embrittled when exposed to hydrogen or hydrocarbon gases at elevated temperatures. Hydrogen diffuses into the metal, where it combines with oxygen to form water. Water vapor collected in naturally occurring voids can develop pressures sufficient to deform or rupture the matrix metal. This phenomenon causes cracking when oxygen-bearing coppers are welded with oxyacetylene flames. Other welding conditions that favor the entry of hydrogen into the metal include high humidity and, for arc processes, use of cellulosic flux-covered electrodes.

Sensitivity to impurities varies with alloy composition. Bismuth, sulfur, and phosphorus are particularly detrimental to copper-nickels. Lead content should not exceed 0.01% in these alloys. Bismuth, selenium, antimony, and boron are also harmful. No more than 0.01% P is permissible in aluminum bronzes that are to be welded, because phosphorus produces hot shortness in these alloys. As little as 0.01% Mg also impairs ductility in aluminum bronzes.

Castability. Because they do not have to withstand mechanical working stresses, cast copper metals generally tolerate higher impurity levels than their wrought counterparts. For example, lead is severely detrimental to hot workability, and high-leaded wrought brasses seldom contain more than 3.5 to 4% of the element. Cast leaded brasses normally contain as much as 6 to 8% Pb, and lead content may be as high as 25% in certain bearing bronzes and leaded coppers.

Lead improves the fluidity of casting alloys, making it easier to reproduce fine detail. Lead also seals the microscopic pores formed when certain alloys solidify; this sealing action renders such alloys pressure tight. On the other hand, even small amounts of lead are severely detrimental to strength and ductility in cast high-strength yellow brasses. Phosphorus also improves fluidity, but allowable concentrations are severely restricted if aluminum is also present (Ref 50).

Machinability. Lead, tellurium, sulfur, and bismuth are added to copper metals to cause machining turnings to break up into small fragments, making the alloys free cutting. Lead also lubricates the cutting surface. Lead contents in free-cutting alloys are usually limited to about 3.5% since the reduced ductility brought about by higher lead contents makes such operations such as thread rolling or knurling difficult. Similar restrictions apply to other free-cutting additives.

Solidification

This section describes the solidification process and, based on that information, discusses a few of the practical factors that influence the structure of cast copper alloy products. For readers seeking further information, a comprehensive review of the solidification behavior of copper alloys is available in a monograph published by the ICA. Solidification principles are also discussed in numerous metallurgical texts, a few of which are listed in the references to this chapter (Ref 66–68).

All metals begin to solidify by the nucleation of microscopic crystals. As solidification proceeds, the crystallites grow and ultimately coalesce into the grains seen in the frozen structure. The crystals initially formed may grow as cellular structures or as tree-like bodies known as dendrites. The mode of solidification depends on a number of factors, the most important of which is composition, whether it is pure copper or an alloy. Other significant factors include the partition of solute between solid and liquid phases as solidification progresses; the width of the liquidus-solidus interval; temperature gradients at and near the liquid-solid interface; the degree of supercooling; turbulence and convection currents, and the occurrence of monotectic, peritectic, or eutectic reactions during solidification.

Factors that influence the structure and quality of the cast product include the rate and direction of solidification; fluidity of the melt; the degree of restraint offered by the mold (which, in conjunction with the structure and mechanical properties of the solidifying metal, determines the likelihood of cracking or hot tearing); the severity of solidification shrinkage (a function of alloy composition); cleanliness, such as, the presence of dirt, sand particles, dross, etc., and the concentration of dissolved gases. All metals, including copper, can retain more gas (primarily hydrogen and oxygen) in solution when molten than they can after solidification, as illustrated in Fig. 2.9. When the metal freezes, the excess gas is either rejected from the casting, retained in macroscopic voids, or precipitated as internal porosity.

Crystal Formation

Pure Copper. As temperature decreases below the melting point, where T_m is 1083 °C (1981 °F), metal crystals begin to nucleate at the mold wall. The process is one form of "heterogeneous" nucleation in that it involves an agent (in this case the mold wall) other than the pure metal. Heterogeneous nucleation can also occur on entrained particulates and on fragments of dendrites. "Homogeneous" nucleation, which involves no external agent or surface, can theoretically begin anywhere in an isothermal melt (Ref 66).

Homogeneous nucleation has little practical significance, but in principle, the phenomenon can be used to estimate the maximum degree of supercooling attainable. Supercooling is a depression of the freezing point; a phenomenon common to many pure metals.

Consider the free energy of formation, ΔG, of a solid nucleus as the sum of the difference in chemical (Gibbs) free energy between liquid and solid and the increase in energy attending the formation of new surface area. The term chemical free energy, is a function of temperature and, being volumetric, varies with the cube of the radius of the nucleus. It is positive above the melting point, exactly zero at the melting point, and negative below. The term interfacial energy is a weak function of temperature and varies with the square of the nucleus' radius. It is assumed to be positive at all temperatures near the melting point. Given a spherical nucleus particle:

$$\Delta G = 4\pi r^2 \sigma - \frac{4}{3}\pi r^3 \Delta G_p$$

where r is the radius of the solid nucleus, σ is the interfacial (liquid:solid) surface energy per unit volume, and ΔG_p is the change in chemical free energy per unit volume.

At temperatures below the melting point, the first term is positive, the second is negative (denoting thermodynamic spontaneity), and expression for the sum of the terms exhibits a maximum value at a particular "critical radius," r^*. Nuclei with radii larger than r^* are stable, whereas those with radii smaller than r^* tend to redissolve in the melt. It can be shown that for a temperature, T, below the melting point, the critical radius size is:

$$r^* = \frac{2\sigma}{\Delta G_p} = \frac{2\sigma T_m}{\Delta H_f \Delta T}$$

where T_m is the melting point of pure copper, 1356 K, ΔH_f is the heat of fusion of pure copper, 3280 cal/mol, and ΔT is the $T_m - T$, the degree of supercooling.

The solid:liquid interfacial energy of pure copper is 177 erg/cm² therefore, the maximum supercooling possible is 236 °C. This degree of supercooling is rarely attained, however, due to the inevitable presence of impurities and, with the exception of levitation-melted samples, a mold wall. In practical cases, however, stable nucleation often begins at temperatures that are significantly below the melting point. As grains eventually nucleate and grow at the mold wall, release of the latent heat of solidification causes temperature to rise back to the melting point until all metal has frozen.

As temperature continues to fall, the crystals grow until they touch, forming a shell of equiaxed grains at the wall of the casting. The speed at which new crystals nucleate and grow depends on cooling rate; higher rates promote the formation of more grains, hence smaller grain sizes. Growth continues up the temperature gradient, for example, toward the hotter liquid at the center of the casting. Because homogeneous nucleation within the liquid copper ahead of the solidification front is energetically difficult, no stable new grains form there during the initial stages of solidification.

Growth is faster in some crystal directions than others, therefore, crystals that are oriented such that a favorable growth direction points at or toward the hottest metal (and away from interfering neighbors) continue to grow, and others are crowded out. In copper, which is face-centered cubic (fcc), preferential growth takes place in the cube edge directions, crystallographically designated $\langle 100 \rangle$. The result, in pure copper and some high copper alloys, is a radial array of columnar grains directed from the equiaxed zone near the casting's surface toward its thermal center. Pure copper ingots and continuous-cast wire rod exhibit this structure (Fig. 2.25).

A second region of equiaxed grains may be found near the center of the casting. These grains nucleate heterogeneously, either on fragments of crystals formed at the mold wall and swept into the melt by convective currents or on impurity particles, which tend to concentrate there, driven ahead of the solidifying metal (Ref 66).

If insufficient liquid metal is available at the center of the casting, for example, if the region is "frozen off," an irregularly shaped void, or "shrinkage cavity," results. When such a cavity

breaks the surface of an ingot, it is known as pipe. Metal near the cavity is normally quite sound. Cavity formation is characteristic of the solidification of pure copper (and other pure metals) and alloys with narrow freezing ranges.

Copper Alloys. The solidification behavior of impure copper and dilute copper alloys depends on the amount of solute present, the manner in which the solute partitions between the liquid and solid phases, and the degree of undercooling attending solidification. Dilute alloys solidify in a manner similar to pure copper if solubility of the alloying element in copper is high. Alloys containing elements with low solubility in copper show little tendency to form columnar grains, and structures are generally equiaxed over most of the casting's cross section. There is less supercooling than in pure copper, and the release of latent heat produces a rather gradual increase in temperature during, solidification. In addition, solute-containing phases may precipitate before, during, or after bulk solidification occurs.

Unlike pure copper, which has a fixed freezing point, the solidification of copper alloys takes place over a range of temperatures bounded by an upper liquidus and a lower solidus temperature. Individual alloying elements produce varying degrees of separation between liquidus and solidus, for example, they produce a variety of freezing ranges, as seen in the phase diagrams shown in Fig. 2.9 through 2.15 and 2.17 through 2.21.

Fig. 2.25 Structure of as-cast pure copper. Source: Outokumpu Copper Partners AB

Tin produces a relatively wide solidification range in copper alloys, and tin bronzes and high-tin alloys go through an aptly named mushy stage as dendrites nucleate and grow throughout the liquid. The extended solidification interval results in compositional segregation, known as coring, between the first and last metal to solidify on the growing dendrites. Coring is usually not detrimental to bulk properties, although it may affect corrosion behavior. On the other hand, mushy solidification can also result in microscopic shrinkage voids as the growing dendrites seal off small nearly-frozen areas from the supply of liquid metal. The process of microporosity formation can also accompany or promote the rejection of dissolved gases. While microporosity is detrimental to ductility, it can also be tolerated if it is not extensive or if it does not occur in highly stressed locations. However, when microporosity forms as a continuous network through the wall of a casting, the resulting product will not retain pressure (Ref 69).

Alloys with narrow freezing ranges are better able to accommodate the solidification and shrinkage stresses imposed by rigid mold walls than wide-freezing range alloys. This is one reason why short-freezing alloys are favored for permanent mold and die casting processes. Examples of short-freezing alloys include yellow brasses, silicon brasses, and aluminum bronzes. Long-freezing alloys include tin bronzes (gunmetals), cast red and semired brasses and bronzes, and some nickel silvers.

Liquid-Liquid and Liquid-Solid Reactions

Solidification can be preceded or accompanied by reactions. Reactions that accompany solidification can occur either at the onset of freezing or at a fixed temperature during the freezing interval, for example, between liquid metal and solid that formed earlier. The reactions described in the following section are based on binary systems; similar reactions also occur in multicomponent alloys.

Systems in which there is little or no solubility between components in the liquid state are characterized by "monotectic" reactions. Above the monotectic temperature but below the so-called consolute temperature, the primary liquid dissociates into two liquid phases:

$$(Liquid) \rightarrow (Liquid)_1 + (Liquid)_2$$

Upon solidification, one of the liquid phases reacts according to the monotectic reaction:

$$(Liquid)_1 \rightarrow (Solid)_1 + (Liquid)_2$$

Monotectics solidify as fibrous composites, which, due to interfacial energy, tend to degenerate into discrete or interconnected globules. The most common example among binary copper systems is Cu-Pb, in which the solidified structure contains islands of nearly pure lead dispersed in a matrix of nearly pure copper. Other binary monotectics include Cu-O (Fig. 2.9), Cu-S, Cu-Se, Cu-Te, and Cu-Th.

The reaction of a liquid solution to form two solid phases at a fixed temperature is known as a "eutectic." It is described, for binary alloys, by the general expression:

$$(Liquid) \rightarrow (Solid)_1 + (Solid)_2$$

Eutectic reactions are common in copper alloys. The copper-aluminum system (Fig. 2.19), contains two:

$$(Cu-Al)_{l_1} \rightarrow \alpha + \beta$$

which occurs at 963 °C (1765 °F) and at approximately 8.5% Al, and

$$(Cu + Al)_{l_2} \rightarrow \theta + \kappa$$

at 548 °C (1018 °F) and 67% Al. Common binary copper alloy systems containing eutectics include Cu-Ag (Fig. 2.12), Cu-Cd, Cu-Cr (Fig. 2.14), Cu-Fe, Cu-O (Fig. 2.9), Cu-Si (Fig. 2.20), and Cu-Zr (Fig. 2.15), among others.

Eutectic structures can be lamellar (sometimes called normal), rod-like, or irregularly dispersed. The reaction $(Cu+Al)_1 \rightarrow \alpha+\beta$ produces a lamellar structure; the structure formed by $(Cu-Si)_1 \rightarrow \eta + \theta$ is irregular.

Reaction between liquid solution and a preexisting solid is known as a "peritectic":

$$(Liquid) + (Solid)_1 \rightarrow (Solid)_2$$

Examples include Cu-Al (Fig. 2.19),

$$(Liquid)_1 + \beta \rightarrow \chi$$

and

$$(Liquid)_2 + \chi \rightarrow \gamma_1$$

and Cu-Zn (Fig. 2.17),

$$(Liquid) + \alpha \rightarrow \beta$$

In the case of brasses containing between 32.5 and 38% Zn, solidification begins with the primary precipitation of α. Upon cooling to 903 °C (1657 °F), this phase reacts at the α-liquid interface to form β. Here, as in most peritectic situations, islands of β initially form at discrete points on the α surface. While still at the peritectic temperature, the primary phase, α, then partially dissolves and solute atoms migrate through the liquid to the nearest β-liquid interface, causing the β island to grow. Once all properitectic α is completely covered with a layer of β, the β phase grows by solid-state diffusion of solute. Since the latter reaction is hindered by relatively slow solid-state kinetics, the reaction may not go to completion unless the alloy is maintained at the peritectic temperature for a long period of time. In practice, the final (nonequilibrium) structure is normally cooled may, therefore, contain residual α grains in a matrix of β phase.

Porosity and Inclusions

Porosity in cast copper (and other) alloys results from the precipitation of dissolved gases and, in long-freezing alloys, interdendritic shrinkage. The two phenomena are related and occasionally occur simultaneously. Porosity can also originate outside the casting as a result of improper feeding and gating practice.

Solidification shrinkage is best avoided by adequate feeding and the use of casting designs that take proper advantage of directional solidification. When it occurs, shrinkage-related porosity is usually found at so-called "hot spots," locations which, because of their relatively large section size, retain heat and remain liquid longest. The objective, therefore, is to minimize hot spots by inducing the casting to solidify in such a way that the last metal to freeze is not isolated from a supply of liquid metal, such as from a riser or a free surface. This is particularly difficult to accomplish in long-freezing alloys such as copper-zinc-tin-lead brasses (or bronzes), in which metal flow is hindered by the inherent mushy state of the remaining liquid. In this case, the size of the resulting pore depends on the volume fraction of shrinkage, liquid viscosity, rate of advance of dendrite tips, and size of the mushy zone. As mentioned earlier, dendritic (mushy) solidification can also promote gas porosity. As freezing proceeds, the supply of liquid metal to isolated regions may be impeded by a metallodynamic pressure drop in narrow interdendritic channels. This can reduce the metallostatic head below the partial pressure

of dissolved gases, and, sufficiently encourage precipitation of bubbles (Ref 69).

Inclusions are classified as primary and secondary. Primary inclusions, the most common type, are further classified as exogenous inclusions, which originate outside the casting in the form of sand and refractory particles. Fluxes and salts, which are suspended in the liquid and become entrapped in the solidified casting, and oxides of constituent elements, which float on the surface of the liquid as dross, become entrained in the pouring stream and remain trapped in the casting due to turbulent flow in the mold cavity. Secondary inclusions are those that form after solidification, as by the diffusion of oxygen into the metal to form oxides of copper and reactive alloy constituents (Ref 69).

Cryogenic Properties

Current interest in the cryogenic properties of copper stems from its use as a thermal and electrical stabilizer in superconducting (SC) conductors. (See Chapters 4 and 5.) Copper's high thermal and electrical conductivities, combined with its excellent formability and reasonable cost, make it the best practical material for SC magnet applications. Corresponding interest in the cryogenic properties of copper-tin alloys is based on a manufacturing process for niobium-tin-based superconductors. Continuous filaments of SC Nb_3Sn are formed "in situ" by encapsulating pure niobium wires in a copper-tin bronze, then heating the composite to diffuse the tin from the bronze and react it with niobium.

Superconducting magnets operate under high mechanical stresses, and SC conductors must resist these stresses to avoid distortion. Pure copper and copper-tin alloys can be strengthened by cold work or combinations of cold work and solid solution strengthening, respectively, but there are practical limits to the degree of strengthening obtainable. Very high strengths can be gained through combinations of cold work and age-hardening, and beryllium coppers, in which this strengthening method is especially efficient, have also been specified for use in high-field magnets. Regardless of the strengthening mechanism, high strength and high conductivity are difficult to achieve simultaneously, and a compromise between strength and conductivity is usually necessary.

Cryogenic properties of copper-base materials of interest to SC magnet designers have been extensively analyzed under a program sponsored jointly by the U.S. National Institute of Standards and Technology (NIST) and the International Copper Association, Ltd. (ICA). The data presented in this section were taken from that study (Ref 5).

Pure Copper

The cryogenic (4 K) properties of OF coppers are tabulated in Tables 2.14 and 2.15. Total impurity contents of the series of OF coppers represented in the tables range from 0.001 to 0.10 wt%. The principal impurity is silver, which has a small but incompletely documented effect on resistivity and mechanical properties within this range.

Yield Strength. The temperature dependence of 0.2% tensile yield strength, σ_y, in annealed OF copper is small, as shown by the following regression equation. The expression is valid over the range 4 K $\leq T \leq$ 300 K. This expression, like others presented in this section, was developed by Simon et al. (Ref 5).

$$\sigma_y = -8.60 - 0.0329T + 292d^{-1/2} + 150I$$
$$S.D. = 9 \text{ MPa}$$

where d is the grain size in μm, I is the impurity content in wt%, and S.D. is the standard deviation. The large apparent impurity coefficient (150) results from the limited amount of available data in the specimens upon which this equation is based. With the exception of one unexplainably high value, measured yield strengths at 4 K range from 29.9 to 38.7 MPa (4.3 to 5.6 ksi). Separate regressions on data taken at 295 K (72 °F) yield a grain size coefficient of 112, which is probably more accurate. See "Mechanical Properties" in a following section.

The tensile yield strength of cold worked OF copper in the temperature range 4 to 300 K (−452 to 81 °F) can be described by the expression:

$$\sigma_y = 124 - 0.241T + 14.1(CW) - 0.166(CW)^2$$
$$S.D. = 32 \text{ MPa}$$

where CW is the degree of cold work as expressed by the percent reduction in thickness. The temperature coefficient for cold worked material is larger than that for annealed copper. Measured tensile yield strengths at 4 K range from 315 MPa (45.7 ksi) after 5% CW to 465 MPa (67.4 ksi) after 50% CW.

Tensile Strength. The ultimate tensile strength (UTS) of annealed OF copper over the range 4 K $\leq T \leq$ 300 K (−452 to 81 °F) is represented by the regression equation:

$$\sigma_u = 419 - 1.19T + 0.00144T^2 + 156d^{-1/2}$$
$$S.D. = 18 \text{ MPa}$$

Temperature dependence is nearly linear, although the small quadratic term produces a better fit. There is no observed effect of impurity content on UTS in the stated temperature range; however, better determinations of impurity content may show that purity is indeed influential. Temperature dependence is considerably larger than that for yield stress (YS) in annealed material. Measured values of UTS at 4 K range from 393 to 460 MPa (57 to 66.7 ksi), with the observed variation possibly resulting from differences in grain size.

The UTS of cold worked OF copper in the range 4 K $\leq T \leq$ 300 K (−452 to 81 °F) can be described by the multivariant expression:

$$\sigma_u = 412 - 0.664T + 2.73(CW) - 0.00695(CW)^2$$
$$S.D. = 32 \text{ MPa}$$

Results calculated from this expression correlate well with those from an equation developed separately for room-temperature data. Measured values of UTS at 4 K range from 410 to 514 MPa (59.5 to 78.5 ksi) for degrees of cold work ranging from 0 to 60%. There is no apparent saturation in the effect of up to 60% cold work on UTS at temperatures below 76 K (−323 °F); however, some saturation is observed at 295 K (72 °F).

Table 2.14 Yield strength dependence of OF copper on grain size and purity at 4 K (−452 °F)

Yield strength, MPa	Grain size $^{-1/2}$, 10^3 m$^{-1/2}$	Impurity, wt%
30.5	0.112	0.05
38.6	0.154	0.01
35.9	0.154	0.01
31.4	0.154	0.01
29.9	0.154	0.01
36.9	0.154	0.01
32.2	0.154	0.01
35.6	0.154	0.01
38.7	0.137	0.0068
100.0	0.302	0.0053

Source: Ref 5

Table 2.15 Tensile strength dependence of OF copper on grain size at 4 K (−452 °F)

Tensile strength , MPa	Grain size$^{-1/2}$, 10^3 m$^{-1/2}$
410	0.112
460	0.302
393	0.137

Source: Ref 5

Ductility. Over the temperature range 4 K ≤ *T* ≤ 300 K (–452 to 72 °F), tensile elongation (%El) of annealed and of cold worked OF copper is described by the expression:

%El = 58.4 – 0.0553*T* – 0.516CW   S.D. = 12%

Specimens had been cold worked up to 96%, but data was taken from both round and flat samples, and the normally higher elongation values in the former were not taken into account in the derivation. Whereas annealed samples exhibited elongations ranging from 29.6 to 101% at 4 K, samples reduced 96% had little (1 to 6.3% El) residual ductility.

There is a small but consistent variation in tensile elongation with silver content at room temperature; data taken over the range 4 K ≤ *T* ≤ 300 K (–452 to 81 °F) indicate that this variation extends to the cryogenic range; however, these data were not quantified.

There is no significant dependence of reduction in area (%RA) upon temperature in the range 4 K ≤ *T* ≤ 300 K (–452 to 81 °F). For example, even specimens cold worked 21, 37, and 60% exhibited reductions in area of 80.6, 71.7, and 75%, respectively when tested at 4 K.

Stress-Strain Curves. Examples of engineering stress-strain curves for annealed C10400 OF copper plate tested at 4, 76, and 295 K (–452, –323, and 72 °F) are shown in Fig. 2.26. Serrated yielding, indicated in the figure, is observed at the lowest test temperature. The phenomenon persists in samples cold worked to ½ hard and hard tempers, Fig. 2.27 and 2.28.

Impact Properties. Metals such as copper with fcc crystal structures characteristically do not become embrittled at low temperatures, and it is not surprising that the impact strength of pure copper should be rather high at cryogenic temperatures. Charpy V-notch impact energies for annealed and cold worked OF electronic copper (C10100) measured at 76 K (–323 °F) range from 168 to 182 J (124 to 134 ft-lb), with values for cold worked material being slightly lower than those for annealed stock. Due to adiabatic heating during testing, data for tests below this temperature are not considered reliable (Ref 70).

Fatigue Properties. The fatigue life of annealed OF copper at cryogenic temperatures appears to be more related to UTS than to YS. UTS increases with decreasing temperature while YS does not. The observed maximum stress varies somewhat with the rate of loading, which accounts for lack of complete agreement in the data reported by various investigators. Nevertheless, the maximum fatigue stress to cause failure at 4 K at

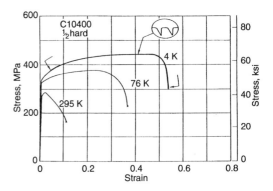

Fig. 2.27 Engineering stress-strain curve for ½ hard C10400 OF copper at 4, 76, and 295 K. Source: Ref 5

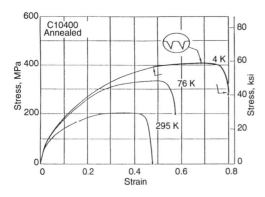

Fig. 2.26 Engineering stress-strain curves for annealed C10400 OF copper at 4, 76, and 295 K. Source: Ref 5

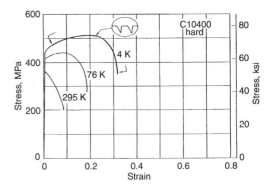

Fig. 2.28 Engineering stress-strain curve for hard C10400 OF copper at 4, 76, and 295 K. Source: Ref 5

large N values falls in the relatively narrow range of values near 240 MPa (34.8 ksi) at $N = 2 \times 10^6$, 268 MPa (38.9 ksi) at $N = 3 \times 10^6$, and 207 MPa (30 ksi) at $N = 3.7 \times 10^5$ cycles. Reported low-cycle fatigue strengths are predictably higher: 304 MPa (44 ksi) at $N = 52{,}000$, 320 MPa (46.4 ksi) at $N = 16{,}000$, and 263 MPa (38.1 ksi) at 27,000 cycles, respectively. A compilation of available fatigue data for annealed OF copper over a series of temperatures between 4 and 298 K (–452 and 77 °F) are shown in Fig. 2.29. Specimens of C11000 deoxidized copper tested in vacuum at 77 K (–321 °F) exhibit somewhat higher axial fatigue strengths than similar samples tested in liquid nitrogen.

There is little effect of temperature on fatigue strength in axially loaded, strain-controlled specimens of C11000 copper (Fig. 2.30), although a clear temperature dependence in flexurally loaded specimens has been reported. On the other hand, several observations on fatigue crack growth rates (da/dN) in annealed and cold worked C11000 copper indicate that rates are lower at 77 K (–321 °F) than at room temperature (Ref 5).

Creep Properties. The temperature dependence of creep properties in the cryogenic range has not been studied systematically; however, an indication of such a dependence can be seen by comparing Fig. 2.31 and 2.32.

Elastic Moduli. Regression analysis of measurements on dynamic elastic moduli for annealed OF coppers at temperatures between 4 and 300 K (–452 and 81 °F) showed that data could be fit to an expression (for polycrystalline material) in the form:

$$E(\text{GPa}) = 137 - 1.27 \times 10^{-4} T^2 \quad \text{S.D.} = 2.5\ \text{GPa}$$

The measured value at 4 K, 138 GPa (20,000 ksi), is very near that predicted by the expression. The calculated value at room temperature, 126 GPa (18,200 ksi), is in reasonable agreement with standard handbook data (117 GPa, or 17,000 ksi) for commercial OF and deoxidized coppers.

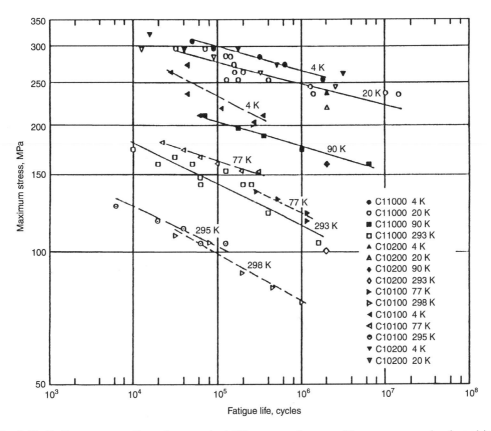

Fig. 2.29 Fatigue properties of annealed OF copper (some with oxygen contents as high as 0.03%) between cryogenic and room temperatures. All R-ratios = –1. Source: Ref 5

Elastic modulus decreases between about 5 and 15% in 99.999% pure copper after cold work at 77 K (–321 °F); little recovery is observed at temperatures below about 120 to 140 K (–243 to –207 °F). The effect of cold work on E is reduced by about 30% in dilute copper-gold alloys containing 99.97% Cu.

Over the range 4 K $\leq T \leq$ 300 K (–452 °F to 81 °F), the influence of temperature on the shear modulus of annealed polycrystalline OF copper is described very closely by the expression:

$$G(\text{GPa}) = 51.2 - 4.63 \times 10^{-5}T^2 \qquad \text{S.D.} = 0.5 \text{ GPa}$$

Available data are not plentiful, although it appears that cold work has little effect ($\Delta G/G < 3\%$) on shear modulus in OF copper that had been cold worked up to 37% at room temperature and permitted to recover for several months. Material cold worked and tested at 77 K (–321 °F) exhibits a 12 to 14% decrease in G.

The temperature dependence of the modulus of compressibility or bulk modulus, B, for annealed OF copper between 4 and 300 K is described by the expression:

$$B(\text{GPa}) = 142 - 5.7 \times 10^{-5}T^2 \qquad \text{S.D.} = 2 \text{ GPa}$$

and Poisson's ratio fits the expression:

$$\nu = 0.339 + 7.03 \times 10^{-8}T^2 \qquad \text{S.D.} = 0.002$$

Specific Heat. The specific heat of annealed and cold worked OF copper over the temperature range 4 K $\leq T \leq$ 300 K (–452 to 81 °F) is given by the expression:

$$\log C_p(\text{J/kg} \cdot \text{K}) = 1.31 - 9.454(\log T) + 12.99(\log T)^2$$
$$- 5.501(\log T)^3 + 0.7637(\log T)^4$$
$$\log \text{S.D.} = 0.023$$

For copper, the difference between C_p and C_v is less than 0.05% at temperatures below 90 K (–297

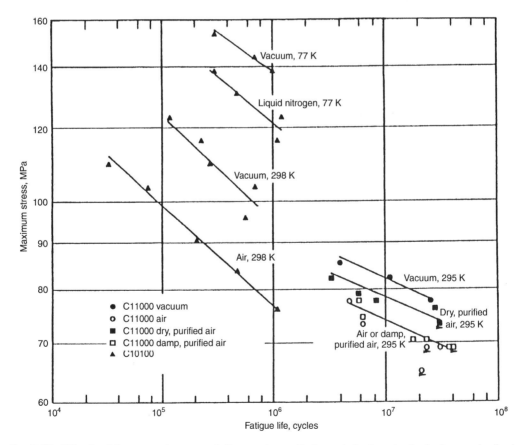

Fig. 2.30 Effect of temperature on fatigue strength in axially loaded, strain-controlled specimens of C11000 copper. Source: Ref 5

°F), but it rises to almost 3% at room temperature. C_p itself rises only about 0.2% as a result of cold work, although the effect is somewhat larger in heavily cold-worked high purity material (Ref 5).

Thermal Conductivity. Thermal conductivity at cryogenic (and higher) temperatures is a complex function of a number of variables, but analysis is simplified by definition of a residual resistivity ratio, RRR, the ratio of resistivity at 273 K (32 °F) to that at 4 K (–321 °F). The resistivity at 4 K is approximately equal to the temperature-independent resistivity, ρ_o, while the resistivity at 273 K is approximately equal to the intrinsic or temperature-dependent resistivity, ρ_i, resulting from thermal vibrations. Thus, $\rho(T) = \rho_o + \rho_i(T)$ according to Matthiessen's rule, and for material of commercial or higher purity, $\rho_i(273\ K) > \rho_o$.

For copper, RRR varies irregularly with purity and cold work, with measured values ranging from several tens to several thousands. As a result, regression equations are most conveniently developed for specific RRR values corresponding to particular samples of copper. The following expression for the effect of temperature on the thermal conductivity of copper has been proposed:

$$\lambda(W/m \cdot K) = (W_o + W_i + W_{io})^{-1}$$

where W_o is β/T,

$$\beta = \frac{\rho_o}{2.443} \times 10^{-8} \frac{V^2}{K^2} \approx \frac{0.634}{RRR} \quad \text{(in SI units)}$$

$$W_i = \frac{P_1 T^{P_2}}{(1 + P_1 P_3 T^{(P_2 + P_4)} \exp\left(-\left(\frac{P_5}{T}\right)^{P_6}\right)) + W_c}$$

and

$$W_{io} = \frac{P_7 W_i W_o}{(W_i + W_o)}$$

The constants are: P_1 is 1.754×10^{-8}; P_2 is 2.763; P_3 is 1102; P_4 is –0.165; P_5 is 70; P_6 is 1.756; P_7 is $0.838/\beta_r^{0.1661}$ where $\beta_r = \beta/0.0003$.

A graphical depiction of the change in λ with temperature for several RRR values is shown in Fig. 2.33.

Thermal Expansion. The thermal expansion coefficient, α, of annealed and cold worked OF copper over the range $4\ K \leq T \leq 300\ K$ (–452 °F to 81 °F) is very closely approximated by the logarithmic regression equation:

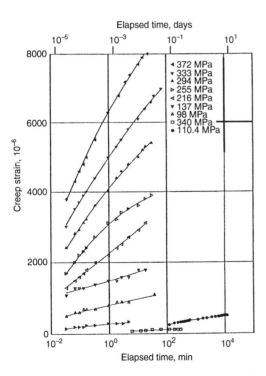

Fig. 2.31 Creep behavior of annealed C10100 OF copper at 77 K (–321°F). Citations pertain to Ref 5

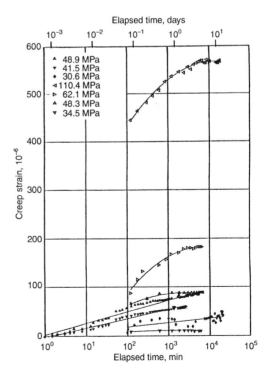

Fig. 2.32 Creep behavior of annealed C10100 OF copper at 4 K (–452 °F). Citations pertain to Ref 5

$$\begin{aligned} \log \alpha = &-11.27 + 37.36 \log (\log T) - 66.59(\log T)^2 \\ &+ 63.49(\log T)^3 - 31.49(\log T)^4 + 7.748(\log T)^5 \\ &- 0.7504(\log T)^6 \qquad \log \text{S.D.} = 0.03 \end{aligned}$$

It appears that α is insensitive to both impurity content and cold work. The temperature dependence of thermal expansion for OF copper is shown in Fig. 2.34.

Mean thermal expansion over the same temperature range is given by the regression equation:

$$\frac{1}{L}\frac{\Delta L}{\Delta T}(10^6/\text{K}) = 11.32 + 3.993 \times 10^{-2}\,T - 7.306$$
$$\times\ 10^5\,T^2 \qquad \text{S.D.} = 0.31 \times 10^{-6}/\text{K}$$

where

$$\Delta L\,/L \cdot \Delta T = [L\,(293\ \text{K}) - L(T)]\,/$$
$$[\,L(293\ \text{K})(293\ \text{K} - T)]$$

for $4\ \text{K} \le T \le 300\ \text{K}$ (-452 to $81\ °\text{F}$) except that at 293 K, it is $(1/L)(\Delta L/\Delta T)$.

A graphical representation of the regression equation is presented in Fig. 2.35. The scatter band in the figure represents two standard deviations about the curve.

Electrical Resistivity. The effects of impurities on resistivity at room temperature, as described in Fig. 2.6, are also valid in the cryogenic range.

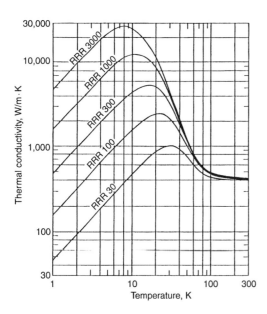

Fig. 2.33 Temperature dependence of thermal conductivity for OF coppers with selected RRR values at cryogenic temperatures. Source: Ref 5

Copper is not a superconductor, although its resistivity at cryogenic temperatures decreases to as little as one thousandth of its room-temperature value. The RRR defined earlier can be used to predict the resistivity at any temperature between 2 and 900 K. Since the ratio depends on a number of factors, it must be determined experimentally for each lot of copper. The ratio for commercially pure copper ranges from 50 to 500.

The temperature dependence of resistivity has been fitted to the regression equation:

$$\rho(\text{n}\Omega \cdot \text{m}) = \rho_o + \rho_i + \rho_{io}$$

where

$$\rho_{io} = P_1 T^{P_2} / (1 + P_1 P_3 T^{(P_2 + P_4)})\exp$$
$$-\ (P5\,/\,T)^{P_6}) + \rho_c$$

and

$$\rho_{io} = P_7\rho_i\rho_o\,/\,(\rho_i + \rho o)$$

The constants are: P_1 is 1.171×10^{-17}; P_2 is 4.49; P_3 is 3.841×10^{10}; P_4 is 1.14; P_5 is 50; P_6 is 6.428; and P_7 is 0.4531.

The value of ρ_o is determined from the RRR, as:

$$\rho_o = \frac{\rho(273\ \text{K})}{\text{RRR}}$$

where $\rho(273\ \text{K})$ is $15.53\ \text{n}\Omega \cdot \text{m}$. A graphical representation of the above regression equation, along with data for coppers of several RRR values, is shown in Fig. 2.36.

Regression equations have also been developed to describe the effect of room-temperature cold work on changes in resistivity, $\Delta\rho$, at 4 and 77 K. At 77K:

$$\Delta\rho\ (\text{n}\Omega \cdot \text{m}) = 9.84 \times 10^3\ (\text{CW}) - 1.04 \times 10^4$$
$$(\text{CW})^2 \quad \text{CW} < 47\%$$

$$\Delta\rho = 0.23\ \text{n}\Omega \cdot \text{m} \qquad \text{CW} > 47\%$$

At 4 K:

$$\Delta\rho = 6.24 \times 10^{-3}\ (\text{CW}) - 5.11 \times 10^{-5}\ (\text{CW})^2$$
$$\text{CW} < 61\%$$

$$\Delta\rho = 0.19\ \text{n}\Omega \cdot \text{m} \qquad \text{CW} > 61\%$$

Graphical representations of the effect of room-temperature cold work on resistivity are shown in Fig. 2.37 and 2.38.

Changes in low-temperature resistivity due to cold work at 77 K ($-321\ °\text{F}$), for example, without recovery, have been fitted to the regression equation:

$$\Delta\rho \ (n\Omega \cdot m) = 2.35 \times 10^{-2} \ (CW) + 1.63 \times 10^{-4}$$
$$(CW)^2 \ S.D. = 0.16 \ n\Omega \cdot m$$

The absence of recovery can be seen in the larger effect of cold work shown in Fig. 2.39 and 2.40.

Magnetic Susceptibility. The temperature dependence of magnetic susceptibility for OF copper have been fitted to the regression equation:

$$\chi \ (10^{-6}) = \frac{3.59}{T} - 9.84 + 6.66 \times 10^{-4} \ T$$

for 1.4 K $\leq T \leq$ 300 (–457.2 to 81 °F). Note that the $1/T$ term arises from the paramagnetic contribution of iron, which was present in some samples. Pure copper is diamagnetic at all temperatures up to the melting point. Thus, the lowest data points shown in Fig. 2.41 represent samples of ultra-low iron concentration. Quantitative effects of iron concentration are discussed in Ref 5.

Beryllium Coppers

Physical and Mechanical Properties. Physical and mechanical properties at cryogenic and higher temperatures are extensively described for several beryllium coppers in Ref 5. In view of the large number of alloys and metallurgical variables analyzed in that work, detailed data selections cannot be presented here. Mechanical properties for annealed, aged, and cold worked and aged alloys tend to increase with decreasing temperature. The degree of increase is similar to that observed for pure copper. Thus, tensile yield strengths in the cryogenic range are slightly higher (10 to 20%) and ultimate tensile strengths are moderately higher (20 to 50%) than at room temperature. As with pure copper, ductility is not strongly affected by temperature in these fcc alloys. Fatigue properties generally increase with decreasing test temperature; inconsistencies in the data may be due to undocumented grain size effects (Fig. 2.42).

Thermal Properties. The temperature dependence of specific heat for beryllium copper C17510 has been fitted to the regression equation:

$$C_p \ (J/kg \cdot K) = 104.5 = 1.883T - 0.002987T^2$$
$$S.D. = 8.4 \ J/kg \cdot K$$

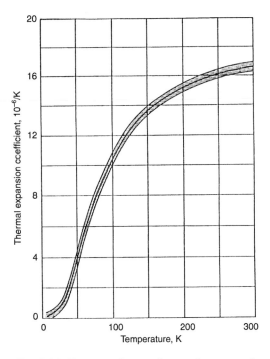

Fig. 2.34 Temperature dependence of thermal expansion coefficient, α, for OF coppers at cryogenic temperatures. Source: Ref 5

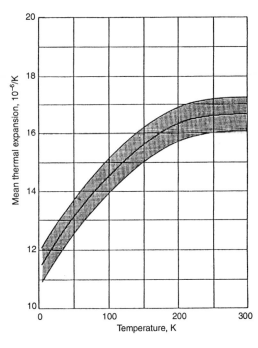

Fig. 2.35 Temperature dependence of mean thermal expansion of OF coppers at cryogenic temperatures. Source: Ref 5

for 83 K $\leq T \leq$ 345 K (–123 °F $\leq T \leq$ 162 °F). Data on the temperature dependence of thermal conductivity for alloy C17200 have been fitted to the equation:

$$\lambda \text{ (W/m} \cdot \text{K)} = 0.93 + 0.492T - 0.000594T^2$$

for 2 K $\leq T \leq$ 300 K (–456 °F $\leq T \leq$ 81 °F), while the regression equation for alloy C17510 is:

$$\lambda \text{ (W/m} \cdot \text{K)} = 64.7 + 0.987T - 0.00138T^2$$

for 77 K $\leq T \leq$ 293 K (–321 °F $\leq T \leq$ 68 °F). The curves are shown in Fig. 2.43.

Temperature dependence of the thermal expansion coefficient, α, for alloy C17200 in various metallurgical conditions is given by the expression:

$$\log \alpha = 25.30 + 95.96(\log T) - 164.5(\log T)^2$$
$$+ 147.6(\log T)^3 - 70.83(\log T)^4 + 17.28(\log T)^5$$
$$- 1.689(\log T)^6$$

for 6 K $\leq T \leq$ 300 K (–449 to 81 °F). The curve for this expression is shown as a $\pm 2\sigma$ scatter band in Fig. 2.44.

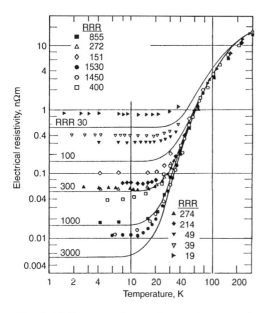

Fig. 2.36 Temperature dependence of electrical resistivity for OF coppers with several RRR values at cryogenic temperatures. Curves plotted from regression equation fit data to ±15%. Source: Ref 5

Mean thermal expansion for alloys C17200 and C17510 fits the second order regression equation:

$$\frac{1}{L}\frac{\Delta L}{\Delta T}(10^{-6}\text{K}^{-1}) = 10.66 + 0.04023T - 7.362$$
$$\times 10^{-5} T^2 \qquad \text{S.D.} = 0.36 \times 10^{-6}\text{K}^{-1}$$

over the temperature range 6 K $\leq T \leq$ 300 K (–449 to 81 °F). The parameter $\Delta L/(L \cdot \Delta T)$ is as previously defined. A graphical representation of the equation is shown in Fig. 2.45.

Electrical Properties. The decrease in electrical resistivity for alloy C17200 in several metallurgical conditions is shown in Fig 2.46. The effect of progressive removal of beryllium from solution is evident in the relative position of the curves. There appears to be little effect of cold work.

Copper-Tin (Phosphor) Bronzes

Yield Strength. The 0.2% tensile yield strength of copper-tin bronzes (phosphor bronzes) is a complex function of composition, grain size, cold work and temperature. A regression equation relating all of these terms over the temperature range 4 K $\leq T \leq$ 297 K (–452 °F $\leq T \leq$ 75 °F) has been developed:

$$\sigma_y = -5.972 + 28.61[\text{Sn}] - 1.581[\text{Sn}]^2$$
$$+ 84.14[\text{Sn}]^{1/2} + 12.02(\text{CW}) - 0.1024(\text{CW})^2$$
$$+ 277.9d^{-1/2} + 88.08[\text{P}] + 0.6416(\text{CW})[\text{Sn}]$$
$$- 0.02421(\text{CW})[\text{Sn}]^2 - 13.01T^{1/3}[\text{Sn}]^{1/2}$$
$$\text{S.D.} = 34.523 \text{ MPa}$$

A study conducted at 4.2 K (–452 °F) measured tensile yield strengths for annealed, phosphorus-free tin bronze wire ranging from 110 MPa (16 ksi) at 1.35 wt% Sn to 241 MPa (35 ksi) at 8.3 wt% Sn. The grain size was 50 μm. A 4.85 wt% Sn bronze, cold worked 85% and having a grain size of 101 μm exhibited a tensile yield strength of 692 MPa (100 ksi). The temperature dependence of yield strength for bronzes of several tin contents is plotted in Fig. 2.47.

Ultimate Tensile Strength. The temperature dependence of UTS for annealed and cold worked phosphorus-free tin bronzes is given by the expression:

$$\sigma_u = 813.9 + 149.2[\text{Sn}]^{1/2} + 3.005(\text{CW})$$
$$+ 0.3748(\text{CW})[\text{Sn}] - 0.01653(\text{CW})[\text{Sn}]^2 - 2.120T$$
$$- 451.8T^{-0.4}[\text{Sn}]^{1/2} \qquad \text{S.D.} = 69.8 \text{ MPa}$$

for 4 K $\leq T \leq$ 297 K (–452 °F $\leq T \leq$ 75 °F).

Very little cryogenic data are available for these materials; however one 4.85% Sn bronze cold worked 60% exhibited an UTS of 805 MPa (117 ksi) at 4 K (–452 °F).

Ductility. Tensile strain to failure values calculated from elongation data have been fitted to the equation:

$$\varepsilon = 0.477 + 0.0677[\text{Sn}] - 0.00303[\text{Sn}]^2$$
$$- 0.0123(\text{CW}) + 0.000152(\text{CW})^2 - 0.474d^{-1/2}$$
$$- 0.00136(\text{CW})[\text{Sn}] + 0.0000796(\text{CW})[\text{Sn}]^2$$
$$+ 0.00637(\text{CW})d^{-1/2} - 0.0000132(\text{CW})T$$
$$\text{S.D.} = 0.066$$

for 4 K $\leq T \leq$ 297 K (–452 to 75 °F).

The equation predicts a decrease in ductility with decreasing grain size; however, this anomaly is based on the expected decrease in ductility with increasing tensile strength.

Available data (67 tests) on reduction in area were fitted to the equation:

$$\text{R.A.}(\%) = 91.0 - 4.84[\text{Sn}] + 0.139[\text{Sn}]^2$$
$$- 1.80(\text{CW}) + 1.21[\text{Sn}]^{1/2}T^{1/3} - 21.5[\text{P}]$$

$$+ 0.558(\text{CW})[\text{Sn}] - 0.0439(\text{CW})[\text{Sn}]^2$$
$$\text{S.D.} = 3.7\%$$

for 4 K $\leq T \leq$ 297 K (–452 °F $\leq T \leq$ 75 °F).

Stress-Strain Curves. The change in engineering stress-strain behavior for spring hard (84% reduction) 4.85% Sn bronze at a series of temperatures from 4 to 295 K (–452 to 72 °F) is illustrated in Fig. 2.48. Evidence of discontinuous yielding can be seen in the 4 K curve.

Impact Properties. Some data on impact properties of C51000 tin bronze (in bar or unspecified forms) extending to cryogenic temperatures suggest a 50% decrease in Charpy V-notch energy as temperature is lowered from 300 to 4 K; however, other data indicate that impact energy may rise or remain unaffected by temperature in upper portions of this range. Charpy V-notch impact energy of 37% cold worked alloy C52400 plate decreases with temperature, from about 70 J at 300 K (52 ft-lb at 81 °F) to about 27 J at 24 K (20 ft-lb at –416 °F).

Elastic Moduli. Limited data on the temperature dependence of Young's modulus for cold worked tin bronzes C51000 and C52100 are plotted in Fig. 2.49. The differing degrees of cold work for the two bronzes may account for the behavior shown in the figure (Ref 5).

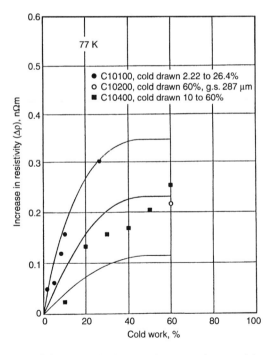

Fig. 2.37 Effect of room-temperature cold work (% reduction) on electrical resistivity for OF copper at 77 K (–321 °F). Source: Ref 5

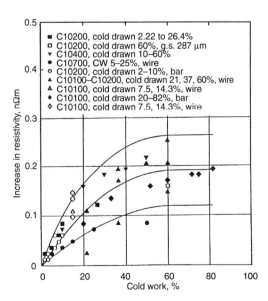

Fig. 2.38 Effect of room-temperature cold work (% reduction) on electrical resistivity for OF copper at 4 K (–452 °F). Source: Ref 5

The shear modulus of copper-tin alloys at 295 K (72 °F) has been fitted to the equation:

$$G(GPa) = 47.34 - 0.8950[Sn] \qquad S.D. = 2.78 \text{ GPa}$$

Limited available data at 77 K (–321 °F) has been fitted to:

$$G(GPa) = 55.32 - 1.025[Sn] \qquad S.D. = 0.86 \text{ GPa}$$

Thermal Properties

Specific Heat. The observed specific heat of a crystalline material is the sum of the lattice and electronic contributions, C_e and C_l, respectively. The electronic contribution is significant only at low temperatures. Thus, in the temperature range between 1.5 and 4 K,

$$C_p(J/kg \cdot K) = \frac{\gamma}{M}T + \frac{1.94 \times 10^6}{M}\left(\frac{T}{\theta_D}\right)^3$$

where γ is the Grüneisen constant, θ_D is the Debye temperature, and M is the molar mass number.

Analysis of available data yields the following regression equations for the relevant quantities:

$$\gamma \, ([Sn]) \, (J/kmol \cdot K^2) = 0.698 + 0.00911[Sn] - 0.00128[Sn]^2 + 6.54 \times [Sn]^3$$

$$\theta_D \, ([Sn]) \, (K) = 344 - 3.11[Sn]$$

$$M \, (g/mol) = 1.187[Sn] + 0.6354 \, (100 - [Sn])$$

Values of specific heat over the temperature range $2 \text{ K} \le T \le 300 \text{ K}$ (–456 to 81 °F) can be obtained from the equation:

$$C_v \, (J/kg \cdot K) = \frac{1}{M} \, [\gamma \, ([Sn]) \, T + D \, (\theta_D / T)]$$

for which values of the Debye function, θ_D/T, can be interpolated from Fig. 2.60 (Ref 5)

Thermal Conductivity. Dependence of thermal conductivity on temperature and tin content in copper-tin bronzes has been fitted to the regression equation:

$$\log \lambda \, (W/m \cdot K) = 0.4145 + 1.563 \log T - 0.22285(\log T)^2 - 0.3234[Sn] + 0.02500[Sn]^2$$
$$S.D. = 0.0796$$

for $4 \text{ K} \le T \le 300 \text{ K}$ (–452 to 81 °F).

Thermal Expansion Coefficient. Limited available data on the variation of thermal expansion coefficient, α, with temperature over the range $10 \text{ K} \le T \le 300 \text{ K}$ (–441 to 81 °F) are plot-

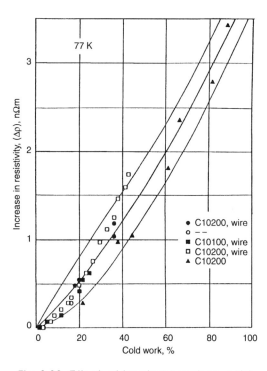

Fig. 2.39 Effect of low-temperature cold work (77 K) on electrical resistivity at 77 K (–321 °F). Source: Ref 5

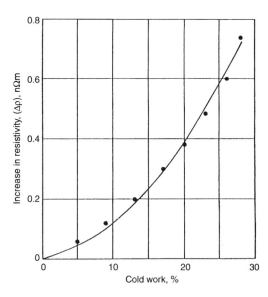

Fig. 2.40 Effect of low-temperature cold work (77 K) on electrical resistivity at 4 K (–452 °F). Source: Ref 5

ted in Fig. 2.51 for cold worked copper-tin alloys C51000, C51900, and C52400 (Ref 5).

Electrical Resistivity. The temperature dependence of electrical resistivity for annealed copper-tin alloys as a function of tin content is given by the regression equation:

$$\rho \ (n\Omega \cdot m) = -4.513 + 17.96[Sn] - 0.4930[Sn]^2 + 0.07202T \quad S.D. = 2.564 \ n\Omega \cdot m$$

Data was taken on specimens in wire form processed from conventional melts. Bronze wire produced by diffusing tin into pure copper wire, the method used to produce niobium-tin superconductors, yields somewhat different results at 4 K (−452 °F), possibly due to nonuniform composition. Dilute, very pure copper-tin alloys exhibit a minimum resistivity at temperatures below about 15 K (−432 °F), a phenomenon that is most noticeable at 0.01 wt% Sn. The effect of phosphorus content on resistivity can be estimated from the equation:

$$\rho \ (n\Omega \cdot m) = 16.52 + 16.24[Sn] - 0.2906[Sn]^2 + 176.2[P] - 101.8[Fe] \quad S.D. = 3.09 \ n\Omega \cdot m$$

Resistivity in cold worked copper-tin alloys is given by the expression:

$$\rho \ (n\Omega \cdot m) = -92.50 + 17.72[Sn] - 0.3198[Sn]^2 + 0.07430T + 1.146(CW) + 0.001793[Sn]T$$
$$S.D. = 2.20 \ n\Omega \cdot m$$

for 4 K ≤ T ≤ 295 K (−452 to 81 °F).

Despite the relatively small value of the standard deviation, data are based on a limited number of samples and a small range of cold work.

Magnetoresistance. Little reliable data on magnetoresistance of fully homogeneous copper-tin alloys is available; however, measurements on alloy C51000 in a 14-T field produced a resistance change at 4 K (−452 °F) of less than 0.01%.

Phase Transformation

A *phase* is defined as a homogeneous portion of a physical system described by the variables temperature, pressure, and composition. Pressure can be disregarded in solid systems. Because metals are ordinarily crystalline, homogeneity implies a fixed crystal structure. Crystal structures in alloy systems are often stable over ranges of temperature and composition; therefore, it is convenient to define a phase as encompassing all compositions that fall within a "phase region."

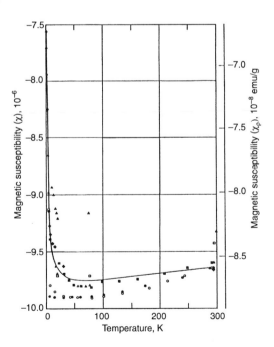

Fig. 2.41 Temperature dependence of magnetic susceptibility for OF copper. Lowest points represent coppers with ultra-low iron contents. Citations pertain to Ref 5

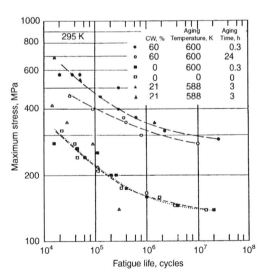

Fig. 2.42 Temperature dependence of fatigue behavior for annealed, annealed and aged, and cold-worked and aged beryllium copper C17200. Source: Ref 5

Broader interpretations of phases in metallic systems have come to include virtually all structures than can be identified metallographically. Examples include highly cold worked structures; metastable structures such as supersaturated solid solutions; structures resulting from order-disorder transformations and spinodal decomposition, and even mechanical mixtures. In such cases, "phase" transformations are sometimes easier to observe than they are to define.

The driving force for all transformations is a reduction of free energy, and structures that exhibit minimum free energy are by definition thermo dynamically stable. The free energy change associated with a transformation is a function of temperature, pressure, and composition, plus energy contributions arising from interfaces and strain. For a transformation to be thermodynamically possible, this free energy change must be negative. In practical terms, however, the observed alloy structure is more often governed by the kinetics of transformation because many thermodynamically possible reactions occur at such slow rates that they can be disregarded.

Kinetics depend on the specific atomic mechanism by which a reaction takes place and on the way that mechanism is affected by such factors as temperature, structure, composition (including impurity content), surface area, and deformation, among others. Most phase transformations in copper alloy systems occur as a result of atomic diffusion, which is a thermally activated process. Important metallurgical processes that rely on diffusion include homogenizing; recovery, recrystallization, and grain growth; precipitation hardening; eutectoid and peritectoid reactions; spinodal decomposition; and crystallographic ordering.

Solid-state reactions can occur in bulk throughout the entire structure or they can take place at discrete locations by a process of nucleation and growth. Homogenizing, recovery, spinodal decomposition, and ordering are bulk transformations. Some metallurgists also consider the formation of highly deformed structures by hot or cold working a bulk transformation. Recrystallization, phase transformation, and precipitation hardening are examples of nucleation and growth transformations. The martensitic transformations that occur in beta brasses, aluminum bronzes, and other copper alloys are also nucleation and growth phenomena, but these differ from the preceding transformations in that they do not involve diffusion. An overview of nucleation and growth transformations in copper alloys is presented in Table 2.16 (Ref 72).

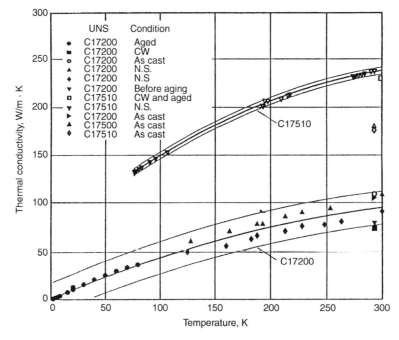

Fig. 2.43 Temperature dependence of thermal conductivity for beryllium coppers C17200 and C17510 in various metallurgical conditions. Citations pertain to Ref 5

Bulk Transformations

Cold Working. With a few exceptions, cold worked materials do not undergo a thermodynamic phase transformation in the course of deformation. Nevertheless, the fact that cold worked materials exhibit significant changes in mechanical and physical properties compared with the annealed state has led some metallurgists to include deformation processes as a form of transformation. For example, the yield strengths of pure copper, 70/30 brass and 95/5 tin (phosphor) increase by a factor of between 4 and 6 as a result of severe cold rolling. Increased strength in these cases results from high dislocation densities, which increase by as much as 10^7 in heavily cold worked material. Property changes brought about by cold working are metastable, and they endure permanently only below the temperature at which deformation took place (Ref 71).

Recovery. Recovery, the relaxation of internal stresses resulting from cold working, is a bulk-type or continuous transformation. It proceeds by the annihilation of dislocation networks and elimination of vacancies through self diffusion. Together, these processes result in the formation of stress-free cells within existing grains. For a further discussion of the recovery process, see "Effect of Impurities on Processing and Fabrication" (Ref 69, 71).

Homogenization. Cored microstructures that arise during the solidification of wide freezing range alloys can transform, during cooling, to phases that differ significantly from those predicted by an equilibrium phase diagram. Such local variations in chemical composition are reduced or eliminated by diffusion during an elevated temperature treatment known as homogenization. Homogenization may be followed by quenching if the intent is to retain the single-phase structure produced at high temperature. Apart from microstructural changes, homogenization may also substantially affect the mechanical and physical properties of the material. Cold or hot working prior to the high-temperature treatment can accelerate homogenization (Ref 69, 71).

Mechanical Alloying. It is possible to create or alter structures by milling mixtures of copper powder and elements which have little or no solubility in copper. The process is appropriately called mechanical alloying, and it can be classified as a bulk transformation because phase changes do occur. X-ray diffraction and differential thermal analysis of mechanical alloys containing from 50 to 75% Cu, for example, show a mixture of two crystalline solid solutions, one rich in tungsten and one rich in copper. An amorphous phase can also be detected. Lattice parameter expansions of up to 0.4% for the tungsten-rich phase and 0.9% for the copper-rich phase are evidence that alloying on an atomic scale has taken place (Ref 73).

Spinodal Decomposition. Spinodal decomposition is the name given to phase transformations by which a single homogeneous phase decomposes spontaneously into two phases of the same or

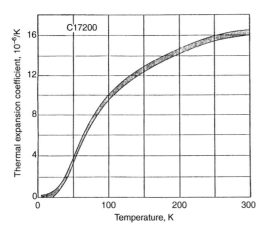

Fig. 2.44 Temperature dependence of thermal expansion coefficient for annealed, annealed and aged, cold-worked, and cold-worked and aged beryllium copper C17200. Source: Ref 5

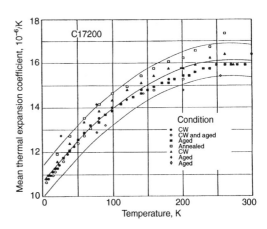

Fig. 2.45 Temperature dependence of mean thermal expansion for beryllium copper C17200 in several metallurgical conditions. Citations pertain to Ref 5

similar crystal structure as the parent phase but having different compositions. Spinodal decomposition arises in systems in which free energy of a solution of the constituent elements varies with composition in the wavy fashion illustrated in Fig. 2.52. The diagram depicts a binary system, but the principle applies as well to multicomponent alloys. Within the range of compositions bounded by c_a and c_a' in the figure, the free energy of the solid solution is higher than that of two separate solid solutions, whose compositions are defined by the points of tangency of a line drawn between minima on the free energy curve. Solutions with compositions such as c_o, therefore, tend to decompose into two separate phases to reduce the system's free energy. Systems described by the right-hand illustration in the figure are unstable, and

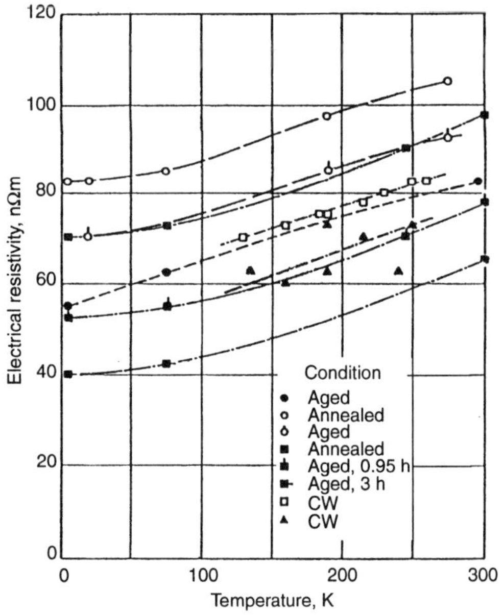

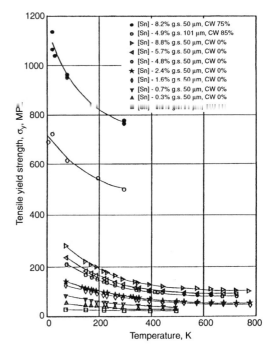

Fig. 2.46 Temperature dependence of electrical resistivity for beryllium copper C17200 in various metallurgical conditions. Source: Ref 5

Fig. 2.47 Temperature dependence of yield strength for tin bronzes. Upper curves represent material cold worked 75% (8.2% Sn, grain size (g.s.) = 50 μm) and 85% (4.9% Sn, g.s. = 101 μm). Source: Ref 5

Table 2.16   Classification of nucleation and growth-type solid-state transformations in copper alloys

Type	Nondiffusional	Diffusional			
Effect of temperature change	Athermal shear	Thermally activated			
Interface type	Glissile (coherent or semicoherent)	Sessile (coherent, semicoherent, incoherent, solid/liquid or solid/vapor)			
Composition of parent and product phases	Same composition	Same composition	Different compositions		
Nature of diffusion processes	No diffusion	Short-range diffusion across interface	Long-range diffusion through lattice		
Interface, diffusion or mixed control	Interface control	Interface control	Mainly interface control	Mainly diffusion control	Mixed control
Specific transformations	Martensite twinning symmetric tilt boundary	Massive ordering recrystallization grain growth	Precipitation dissolution bainite	Precipitation dissolution	Precipitation dissolution eutectoid & peritectoid cellular precipitation

Source: Ref 72

separation into two solid solutions occurs spontaneously and rapidly.

Systems described by the left-hand illustration also decompose spontaneously. However, the free energies of solutions having compositions such as c_0, which lie between the minima and the inflection points in the total free energy curve (the socalled "spinodes"), are higher than that of the original solid solution. Decomposition of the original solution, therefore, takes place much more slowly. Also, it is important to note that the difference in composition of the two phases formed must be large enough to account for the interfacial free energy associated with the two-phase structure. Interfacial energy constitutes a positive free energy term, and it, therefore, decreases the driving force for the reaction (Ref 74).

The free energy and equilibrium phase diagrams, respectively, for a system capable of spinodal decomposition are shown in Fig. 2.53 and 2.54. The phase diagram contains a solid-state miscibility gap in the alpha phase below a critical temperature T_c. In systems such as Cu-Be that form coherent Guinier-Preston (G-P) zones, the miscibility gap is metastable. In such cases, free energy is elevated by the reversible elastic work needed to overcome the coherency strain between the two lattices formed when the phases separate. In fact, a "coherent" phase diagram is always metastable unless the two phases formed by decomposition have identical lattice parameters. It is, however, a real phase diagram involving a reversible, metastable equilibrium that is subject only to the constraint that the lattice remains continuous. Redissolution (reversion) of the spinodal phases occurs when the elastic energy exceeds the magnitude of the (unstressed) free energy of precipitation, so that the system can reduce its total free energy by homogenizing (Ref 58, 69, 74).

Spinodal decompositions occur as precursors to many precipitation and ordering reactions, but they are best known in systems where they are particularly instrumental in the development of useful properties. Examples of the latter include Cu-Ni alloys such as C96800 (nominally Cu-10% Ni-0.20% Mn-0.20%Nb) and Cu-Ni-Sn alloys such as Cu-10% Ni-6% Sn, Cu-15% Ni-8% Sn-0.2% Mn (C96900), and 82% Cu-8% Sn-10% Ni-0.2% Nb. The permanent magnet alloys found in the Cu-Ni-Fe an Cu-Ni-Co systems also involve spinodal decompositions (Ref 12, 69).

Cold Precipitation Hardening. This phenomenon, which is also called room-temperature aging and homogeneous phase demixing, is indicated by an initial rise in hardness with time followed by a fairly constant horizontal plateau. There is no optically detectable change in microstructure; however, Laue diffraction patterns characteristically exhibit diffuse streaks. These are explained by the formation of submicroscopic (~150 Å in diameter), planar elements composed of solute atoms. In Cu-Be alloys, for example, homogeneous nucleation of G-P zones is followed by precipitation of coherent γ' and eventually optically resolvable γ'' particles (Ref 12, 75).

Ordering Reactions. The chemical affinity between certain elements leads to a tendency for unlike atoms to form nearest-neighbor pairs. Thermodynamically, the enthalpy of mixing for such alloys is not zero due to a difference in energy

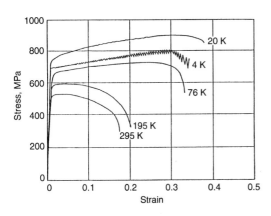

Fig. 2.48 Engineering stress-strain curves for spring-hard 4.85% tin bronze bar at several temperatures. Source: Ref 5

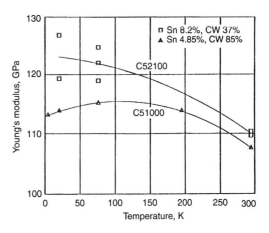

Fig. 2.49 Temperature dependence of Young's modulus for tin bronzes C51000 (85% CW) and C52100 (37% CW). Citations refer to Ref 5

between A–B and A–A or B–B bonds. When the phenomenon applies for a small number of atoms and over a few atomic distances, it is called "clustering" or "short-range order" (Fig. 2.55b). When the phenomenon extends over sizeable regions of the crystal lattice, it is known as long-range order (Fig. 2.55a). Long-range order leads to the formation of so-called superlattice structures, which are in effect two crystal lattices superimposed on one another. Examples of superlattice types found in copper alloys are shown in Fig. 2.56 (Ref 72).

The final products of spinodal decomposition in Cu-Ti and Cu-Ni-Sn alloys are ordered phases that arise in solute-enriched regions of the structure. The schematic diagrams of Fig. 2.57 describe qualitatively the complex processes that can take place in such alloys. For example, alloys represented by the curve in Fig. 2.57(a) are stable over the entire composition range. An alloy of composition C_0 can lower its free energy by ordering, but because it now becomes unstable, the ordered phase can decompose spinodally. This type of process is described as a conditional spinodal. An alloy of lower solute content C_0', even after ordering, will not decompose spinodally but will precipitate the ordered phase only through a nucleation process (Ref 69).

In Fig. 2.57(b), the disordered curve also shows an instability such that a composition γ_0' initially,

decomposes in the normal spinodal fashion. In this case, the solute-rich cluster orders when the composition of the solute-rich regions has reached a value where the ordered phase has a lower free energy than the disordered phase. The phenomena exemplified by alloys of compositions C_0' and γ_0' alloy illustrate the difference between the Ni-Al and Cu-Ti systems.

In the Cu-Au system, the ordered α' phase, Cu₃Au, first appears on cooling to 396 °C (745 °F). Extrapolation of free energy curves suggests there is an instability temperature T_{instab}, below which continuous ordering can be expected. Between T_c and T_{instab}, ordering occurs by a nucleation and growth mechanism. Thus, the Cu₃Au phase appears by continuous ordering at 390 °C (734 °F) without change in composition. Similar observations have been reported for the α'' phase, CuAu.

Additional descriptions of ordering in copper alloys is presented in the section on solid-state properties.

Transformation by Nucleation and Growth

Thermodynamics. Formation of a new phase or structure in an existing matrix is driven by a re-

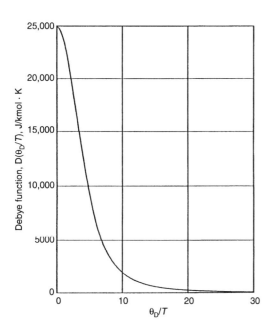

Fig. 2.50 Debye function, θ_D/T, for copper-tin alloys. Source: Ref 5

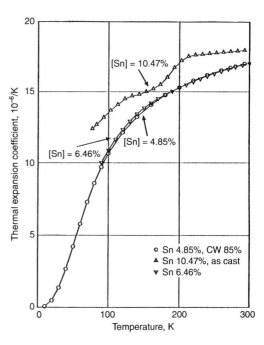

Fig. 2.51 Thermal expansion coefficient of several copper-tin alloys over the temperature range 10 K ≤ T ≤ 300 K (–441 °F ≤ T ≤ 81 °F). Source: Ref 5

duction in free energy. That is, the free energy of the transformed phase or structure must be lower than that of the pre-existing matrix. In addition, free energy changes associated with nucleation and growth reactions must account for the energy associated with newly created interfaces, and they must include a term for the elastic strain energy due to coherency between the precipitate and its parent phase.

Nuclei of the new phase or structure form and revert continuously as a result of diffusion and random atomic motion; however, only nuclei of a critical radius, r^*, will be stable and able to grow. Smaller nuclei either remain immobile (in the case of recrystallization) or revert to solution (if present as precipitates).

Recrystallization. Recrystallization involves the transformation of a cold worked structure into new strain-free grains. Nuclei of the new grains form at high-energy sites such as grain boundaries and dislocation pile-ups on crystallographic slip planes. In this case, nucleation is irreversible because strain-free nuclei cannot revert to their former state. Nuclei, therefore, incubate in a submicroscopic state until sufficient strain-free structure has developed around them (by diffusion) to enable growth. Because diffusion is thermally activated, recrystallization is accelerated by elevated temperatures.

What is generally referred to as a recrystallization temperature is the lower end of a range of temperatures at which highly cold worked material will recrystallize in a fixed period of time, usually one hour. Recrystallization temperature, time, and the degree of prior cold work are interrelated. The complex nature of the relationship is shown graphically for electrolytic tough pitch copper (C11000) in Fig. 2.58 (Ref 13, 71).

Eutectoidal Decomposition. Eutectoidal decomposition involves the transformation of a homogeneous solid phase, usually a solid solution, into two new phases, one with low solute content and one with high solute content. It is described in general terms by the expression:

$$\gamma \rightarrow \alpha + \beta$$

The $(\alpha + \beta)$ dispersion can take several forms. Lamellar or plate-like eutectoids are sometimes called pearlite because of their iridescent appearance; spherical and rod-like eutectoidal structures are also known. A number of eutectoid reactions occur in copper-base systems. In copper-aluminum alloys containing between approximately 9.5 and 15% Al, the β phase decomposes at 565 °C (1049 °F) to form α and γ_2. The high temperature χ phase (approximately 16% Al) dissociates eutectoidally at 963 °C (1765 °F) to form β and γ_1 (Fig. 2.19).

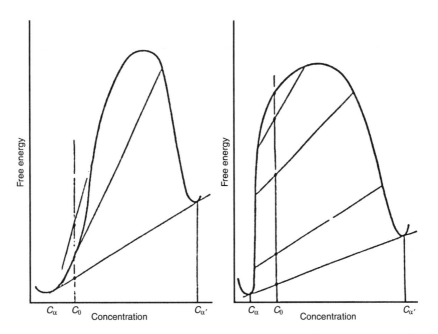

Fig. 2.52 Free energy changes during spinodal decomposition for a metastable (left) and unstable (right) phase of composition, C_0. Source: Ref 3

The β phase in copper-beryllium alloys decomposes at 575 °C (1067 °F) to form a eutectoidal dispersion of α and γ (Fig. 2.13). This eutectoid reaction does not, however, play a role in the hardening treatments applied to beryllium coppers.

Eutectoid reactions can be complex, involving intermediate phases, as in the case of Cu-5.2%Si. As can be seen in Fig. 2.20, the reaction for an alloy of this composition slowly cooled from elevated temperatures is (Ref 76):

$$\kappa \to \kappa + \gamma \text{ (pro-eutectoid)} \to \kappa + \alpha + \gamma \to \alpha + \gamma$$

Similarly, hypoeutectoid copper-aluminum alloys (for example, with approximately 4 to 11% Al) containing significant quantities of β decompose according to:

$$\beta \to \beta + \alpha \text{ (pro-eutectoid)} \to \beta + \alpha + \gamma_2 \to \alpha + \gamma_2$$

and at 12.2% Al,

$$\beta \to \beta + \beta_1 \to \beta + \beta_1 + \alpha \text{ (pro-eutectoid)} \to \beta + \beta_1 + \alpha + \text{pearlite } (\alpha + \gamma_2) \to \alpha + \text{pearlite } (\alpha + \gamma_2)$$

In this case β_1 is an ordered phase that forms at about 530 °C (Ref 76).

Precipitation Hardening. Precipitation is one of the most common and commercially most important diffusional phase transformation processes in copper alloys. It proceeds by the formation, in supersaturated solid solutions, of solute atom clusters that eventually transform to a new phase. Atomic rearrangement takes place by diffusion.

The precipitation process can be visualized with help of a suitable phase diagram, such as that of the copper-zinc system picture in Fig. 2.17. Alloys containing slightly more than 36.9% Zn solidify as the bcc phase designated β. The phase remains stable, upon cooling, to approximately 800 °C

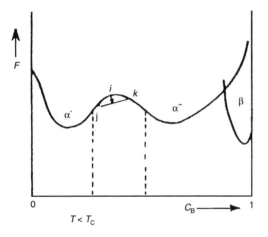

Fig. 2.53 Free energy-composition diagram for a system capable of undergoing spinodal decomposition. Curvature is negative between the dashed lines. Source: Ref 69

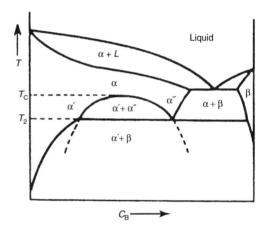

Fig. 2.54 Phase diagram corresponding to Fig. 2.53. Source: Ref 69

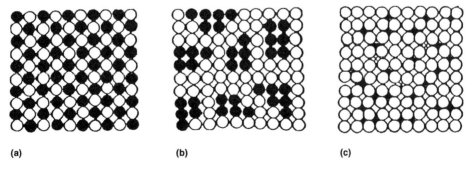

Fig. 2.55 Schematic representation of (a) ordering, (b) clustering, and (c) random interstitials in solid solutions. Source: Ref 72

(1472 °F), the temperature of the α-β solvus line for that composition. Below that temperature, α begins to precipitate. The precipitate takes the form of coarse plate-like particles (the so-called Widmanstätten pattern) aligned on α lattice planes at temperatures near the solvus, as shown in Fig. 2.59. Precipitates that form at lower temperatures will be correspondingly finer. In the copper-zinc alloy, precipitation occurs preferentially near grain boundaries where diffusion is enhanced and coherency strains can be more readily accommodated by the distorted lattice structure (Ref 75).

Holding at elevated temperatures causes the precipitates to coalesce, with larger, less soluble particles growing at the expense of smaller, more soluble ones. This process is sometimes called *Ostwald ripening*. For compositions slightly leaner than 39% Zn, for example, those that pass almost immediately into the α + β phase field

upon cooling, blunt precipitates appear at beta-phase boundaries even at high cooling rates. Such reactions are called "*massive transformations*" (Ref 69, 75, 77).

A second type of precipitation can take place in systems such as copper-beryllium (Fig. 2.13). A Cu-Be alloy containing, for example, 1% Be can be heated and equilibrated at 600 °C (1112 °F) to produce a structure consisting of a uniform solid solution, the fcc phase designated α. When such an alloy is cooled to room temperature, it crosses the solvus line into the two-phase field in which both α and γ are stable. If the alloy is rapidly quenched, the high-temperature solid solution can be retained because there is insufficient thermal energy available at low temperatures to drive the diffusion process needed for transformation. The solid solution remains in a metastable, supersaturated state.

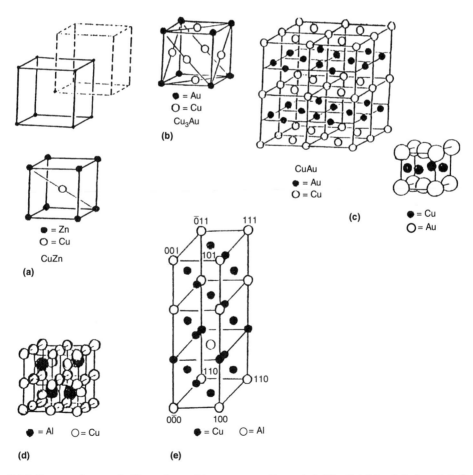

Fig. 2.56 Types of superlattices found in copper alloys: (a) 12_0; (b) 11_2; (c) do_3; (d) CuAl$_3$ with antiphase plane. Source: Ref 72

Reheating to an intermediate temperature below the solvus line enables diffusion to take place, and precipitation follows. As described earlier, the process first involves the formation of solute-enriched regions known as G-P zones, followed by the precipitation of metastable, submicroscopic γ'' and eventually γ'. The γ'' particles are highly coherent, and the lattice strains that accompany their formation cause considerable strengthening. The heat treatment applied to generate this effect is, therefore, known as precipitation-hardening or age-hardening because it takes place as the alloy ages over time. The γ' particles are only partially coherent, and their formation and growth to optically resolvable dimensions brings about a reduction in strength, a phenomenon called overaging.

Precipitates form preferentially at grain boundaries in alloy systems where diffusion proceeds more rapidly through grain boundaries than through the bulk of the lattice. When there is relatively little difference between grain boundary and bulk diffusion rates, precipitation can take place

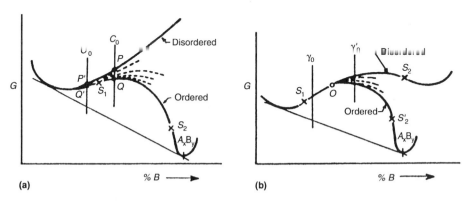

Fig. 2.57 Free-energy composition curves for several modes of spinodal decomposition and continuous ordering. Source: Ref 69

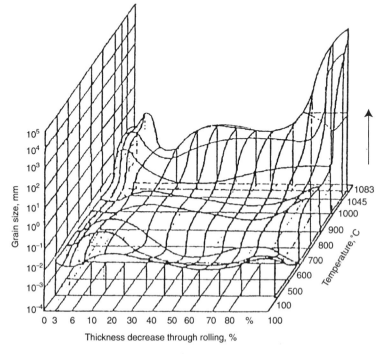

Fig. 2.58 Effect of degree of cold work and annealing temperature on grain size in C11000 copper. Source: Ref 13

within the grains. This is the case in beryllium coppers. General precipitation is more favorable to mechanical properties because grain boundary precipitation can give rise to embrittlement (Ref 69).

In addition to beryllium coppers, precipitation-hardening copper alloy systems include Cu-Cr, Cu-Zr, Cu-Fe, Cu-Co, and Cu-Co-Si, among others. Precipitation in multicomponent systems often involves the formation of complex intermetallic phases containing more than two elements.

Precipitate Dissolution. Upon heating above the solvus line, precipitate particles redissolve in the matrix by a simple diffusion mechanism. This occurs in the beryllium coppers described above and similar age-hardening alloys. The rate of dissolution depends, among other things, on particle size and shape.

Plastic deformation can promote dissolution at temperatures below the solvus if precipitate particles fracture into fragments smaller than the critical size, r^*, required for stability. Enhanced diffusion in the plastically deformed zone at the tip of an advancing fatigue crack has also been suggested as a means by which dissolution can occur (Ref 78).

Related Phenomena. Other phenomena related to precipitation in copper-base systems include a coarsening reaction in the eutectoid structure at Cu-6.5at.%In, which appears when the alloy is annealed at 560 to 580 °C (1040 to 1076 °F), for example, close to the eutectoid temperature. Coarsening occurs when a lamellar $\alpha + \beta$ eutectoid product grows out from a grain boundary and consumes a fine coherent precipitate nucleated in the bulk of an adjacent grain.

Diffusion-induced grain boundary migration is seen in Cu-Ni and other systems. This phenomenon has been observed at relatively low temperatures, where grain boundary diffusion prevails. The result is a discontinuous concentration interval across the boundary. The boundary migrates as diffusion, possibly aided by elastic energy, takes place (Fig. 2.60). In a few systems, boundary migration and recrystallization occur simultaneously (Ref 69, 79).

Diffusionless Transformations. Transformations that occur without long-range atomic movement are classified as diffusionless, shear, or more commonly, martensitic reactions. (Nomenclature used to describe diffusionless transformation reactions is based on steel metallurgy, where such transformations are well known. Thus, high-temperature parent phases are often called austenite and the low-temperature transformation product are labeled martensite.) Martensitic transformations bear certain similarities to crystallographic twinning. As shown in Fig. 2.61, however, they differ in that twinning does not result in a change in the basic lattice type, whereas martensitic reactions do (Ref 69).

When it occurs, martensite usually appears in a eutectoid composition range such as the β range in Cu-Al alloys. Transformation takes place when the high-temperature solid solution phase is rapidly cooled (quenched) to low temperature at a rate that is fast enough to prevent diffusion. Transformation occurs by nucleation and growth, although "growth" rates approach acoustic velocities. The new phase is metastable. The following are a few other characteristics of martensitic transformations (Ref 69, 80):

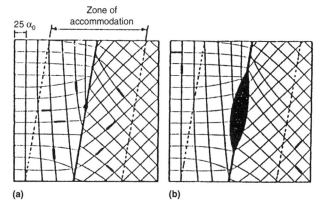

Fig. 2.59 Precipitation near grain boundaries: (a) Widmanstätten precipitates; (b) precipitate coalescence. Source: Ref 75

- Shear takes place on a fixed habit plane and there is an unvarying orientation relationship between the parent and the product crystals. Table 2.17 lists crystallographic relationships in the martensite transformations of copper-zinc, copper-tin, and copper-aluminum alloys (Ref 69, 75).
- The interface between parent and martensitic phases consists of an array of dislocations; it can be coherent or semicoherent. The interface is glissile; it can move at high velocities over a wide temperature range, and its motion can be aided or inhibited by elastic and plastic deformation.
- On cooling, martensite begins to form at a particular threshold or martensite-start temperature, M_s. The quantity of martensite formed depends on the extent of cooling below this temperature. The temperature at which all available material has transformed is called the martensite-finish or M_f temperature. If cooling is stopped in the M_s-M_f interval, remaining untransformed material may stabilize.
- Plastic deformation below the M_s temperature increases the amount of martensite formed. Deformation of the subcooled unstable matrix phase above but close to the M_s causes martensite to nucleate and grow.
- Martensite reverts to the high-temperature phase on heating well above the M_s.

Boundary motion

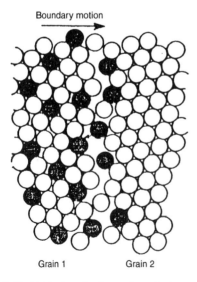

Grain 1   Grain 2

Fig. 2.60 Diffusion-induced grain boundary migration. Diffusion of solute (filled circles) causes grain 1 to grow. Source: Ref 69

In addition to those in aluminum bronzes, martensitic transformations occur in certain copper-tin and copper-zinc alloys. The transformations can be exploited to produce high strengths. The process is less efficient in copper alloys than in quenched steels because ordering and segregation, which occur more frequently in copper-base alloys than in steels, reduce the degree of strengthening attainable. Attempts have been made to delineate martensitic transformations in the type of time-transformation-temperature (TTT) or continuous-cooling-transformation (CCT) diagrams commonly used in heat treatable steels. Figure 2.62 shows a CCT diagram for a eutectoid Cu-11.3%Al alloy (Ref 81).

Shape Memory Effect Alloys. The shape memory effect (SME), also called the martensitic memory effect (MARMEM), describes the behavior of some alloys which, when deformed at a relatively low temperature, can recover their original shape when heated. This is called one-way shape memory. In some cases, alloys revert to their deformed shape upon recooling; such alloys are said to have two-way memory. The SME has been detected in a number of copper-base systems including Cu-Al, Cu-Zn, Cu-Sn, Cu-Al-Zn, Cu-Zn-Sn, Cu-Al-Ni, Cu-Zn-Si, Cu-Au-Zn, and Cu-Zn-Ga. Of these, only the Cu-Al-Ni and Cu-Zn-Al alloys are commercially significant.

The SME derives from a reversible, for example, thermoelastic martensitic reaction. Transformation from the host phase to martensite is accompanied by a change in crystal structure. In Cu-Zn-Al alloys, for example, the host phase is bcc, whereas the martensite is monoclinic. There are as many as 24 crystallographically equivalent and, therefore, potential habit planes and orientations in the host phase but very few in the less symmetrical product phase. Transformation produces an anisotropic volume change, for example, the direction of dimensional changes that take place when martensite forms in a host crystal depends on the particular orientation relationship (of the many available) between the two crystals involved.

The orientation favored by a thermoelastic martensite crystal is that which best accommodates the applied stress. Unfavorably oriented martensite crystals can also reorient themselves by twinning. Both twin boundaries and martensite-matrix boundaries are highly glissile: they move readily under applied stress. Thus, when a martensitic structure is deformed in the temperature range between M_s and M_f, martensite crystals rearrange themselves so as to accommodate the applied stress. The dimensional change resulting from the preferential alignment of orientations is reversible on heating above the martensite transformation temperature range (for example, when the martensite reverts to the parent phase). This reversibility creates the SME. The sequence is illustrated schematically in Fig. 2.63 (Ref 82).

Two-way shape memory can be programmed into some SME alloys by incorporating microstructural stress raisers and by applying appropriate stress and/or thermal cycles. Once the appropriate conditions are achieved, a specimen will spontaneously "bend" when the parent transforms into martensite, and "unbend" to the in-

itial shape during the reverse transformation (Ref 82).

SME alloys have a number of useful properties. For example, when many of the copper-base SME alloys are deformed beyond the single-crystal martensite stage, a new, stress-induced martensite-martensite transformation occurs. This successive mode of martensite deformation enables recoverable strains of more than 17% to be produced. The relative ease of movement of internal martensite-martensite boundaries under relatively small stresses is strongly attenuating, and SME alloys are, therefore, also excellent damping materials (Ref 83–85).

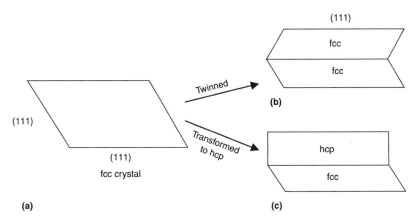

Fig. 2.61 Shape changes of (a) an fcc crystal by (b) twinning or (c) shear. Shear displacement is in the plane of paper

Table 2.17 Crystallographic relationships in copper-base alloy martensitic transformations

System	Composition	Crystal lattice Parent phase	Crystal lattice Martensite	Lattice relationships	Habit plane	Low-temperature equilibrium phases
Cu-Al	11–13.1% Al	β_1-bcc (ordered)	β'- distorted hcp (ordered) [001] direction 2° from normal to (001) plane. Angle between (100) and (010) planes differs from 120° by 1°	$\{011\}$ β_1 4° from $\{00.1\}$ $\beta'\langle111\rangle$ $\beta_1 \| \langle01.0\rangle$ β'	2° from (133)β_1	α-solid solution (fcc) + δ-solid solution (ordered cubic, γ-brass structure)
	12.9–14.7% Al	As above	γ'-hcp (ordered). With increasing Al becomes distorted into an orthorhombic lattice	Two relationships: (110)$\beta_1 \|$ (00.1)γ' [111]$\beta_1 \|$ [01.0]γ' and (011)β_1 4° from (00.1)γ' [111]$\beta_1 \|$[01.0]γ'	3° from (122)β_1	As above
Cu-Sn	25%Sn	β-bcc	β'-structure unknown		(133)β	α-solid solution (fcc) + Cu$_3$Sn (complex fcc)
Cu-Zn	40%Zn	β_1-bcc ordered	β'-structure unknown		(155)β_1 or (166)β_1	α-solid solution (fcc) + β_1-solid solution (ordered bcc)
	40%Zn + 1%Pb + Sn	As above	β' - face centered tetragonal			As above

Source: Ref 76

Metallography of Copper and Copper Alloys

Metallography is one of the most valuable metallurgical tools. It is widely applied to copper alloys for quality control, material evaluation, and alloy development. Common applications of metallographic examination in the production and use of copper alloys include:

- Measurement of grain size in cold worked and annealed products. Grain size has a strong influence on properties and formability in such materials. Qualitative estimates of grain size in sheet and strip products can also be obtained by cup testing.
- Evaluation of the dispersion of second phases, particularly lead and other additives that promote machinability.
- A check on heat-treating conditions. Complex alloys such as aluminum bronzes and high tensile brasses rely on the development of

proper microstructures for mechanical properties and optimum corrosion resistance.
- Failure analysis. Metallographic examination readily reveals stress-corrosion cracking, dezincification, dealuminification, and other common corrosion mechanisms. The technique can also uncover defects in wrought, cast, and welded structures.

Specimen Preparation

Copper and its alloys present no extraordinary problems to the metallographer, and techniques for sectioning and mounting specimens can be found in standard texts on the subject. As a general cautionary rule, however, it should be noted that pure copper and soft single-phase alloys work harden readily, and surface layers can potentially be affected by the mechanical work of grinding and polishing. Grinding and polishing should, therefore, be done carefully or performed electrolytically. Also, unless the effects of prior cold working are of particular interest, areas to be

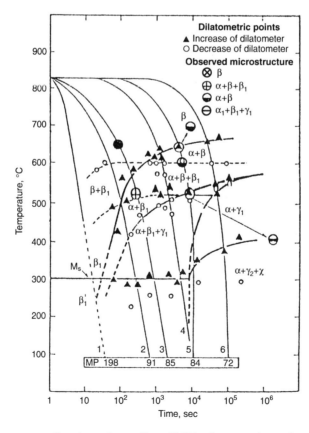

Fig. 2.62 Continuous-cooling transformation (CCT) diagram for a Cu-11.3% Al aluminum bronze. Source: Ref 83

examined should be reasonably far removed from deformed portions of the sample (Ref 86–90).

Macroexamination. Specimens for macroexamination are prepared by sectioning, followed by rough machining to remove grossly cold-worked material. A final cut, taken with a V-shaped tool, completes the process. Machined surfaces can then be ground with 180-grit or finer abrasive, depending on the degree of detail required. Examination of very fine details may call for abrasives as fine as 600 grit (Ref 86).

Table 2.18 lists macroetchants, applicable alloys, and etchant characteristics. Note that several etchants have similar properties and uses; optimum performance is best obtained by experimentation.

Microexamination. Specimens should be wet-ground using progressively finer papers, normally though 600 grit. Finer grits can be used if necessary. Alternatively, initial grinding can be performed dry through 320 grit, followed by wet grinding on 1, 00, 000, and 0000 emery papers. Rough polishing is best accomplished using nylon cloth. Duck canvas, wool broadcloth, and cotton (in that order) are also acceptable. The preferred polishing abrasive is 3 to 9 μm diamond paste. A distilled water suspension of 400-grit or finer alumina (Al_2O_3) can also be used. Finish polishing is performed with distilled water suspensions of 0.3 μm α-Al_2O_3 or 0.05 μm γ-Al_2O_3. Magnesia (MgO), ferric oxide (Fe_2O_3), and colloidal silica (SiO_2) can also be used (Ref 86).

Attack polishing using ferric nitrate ($FeNO_3$), acidified or neutral, or ammonium persulfate [NH_4OH-$(NH_4)_2S_2O_8$] is recommended by several authorities, possibly in light of the soft nature of copper and its alloys, which makes fine scratches difficult to remove. Attack polishing is more safely performed using automatic equipment. Attack polishing conditions are listed in Table 2.19 (Ref 86, 88).

Electrolytic polishing of copper metals offers numerous advantages, and it should be considered first whenever possible. The process can also be applied after mechanical polishing. Electropolishing is not without disadvantages, however, particularly with respect to accelerated attack at edges, voids, and nonmetallic inclusions. Table 2.20 lists several recommended electrolytic polishing solutions. When electrolytic polishing is used as a adjunct to mechanical polishing, etching times are less than 1 min (Ref 86).

Table 2.21 lists chemical etchants used with coppers and copper alloys for microscopic examination. Of the etchants listed, the ammonium hydroxide/hydrogen peroxide/water solution (No. 1) is most widely applicable and most commonly used. The potassium dichromate etch (No. 4) is especially useful for examining welded and brazed structures.

Electrolytic etching can be performed with the etchants listed in Table 2.22. The process is claimed to be useful for revealing cold worked structures in brasses, to give contrast to β phase in brass and to reduce the appearance of coring in copper-nickels. Other uses are listed in the table.

It is sometimes useful to examine unetched or lightly etched specimens. This technique makes voids, oxides, and other metallic and nonmetallic inclusions more readily visible than they are in etched specimens. In some cases, inclusions can be identified. Cuprous oxide (Cu_2O), for example, appears ruby red under polarized light and can thus be differentiated from cupric oxide (CuO) and other inclusions. Polarized light also gives contrast to oxides of arsenic and antimony.

Representative Microstructures

The broad array of copper-base metals offer examples of nearly all microstructural phenomena

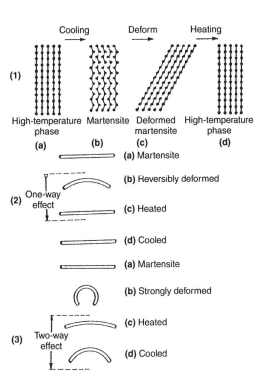

Fig. 2.63 Shape memory effect (SME): (a) lattice changes; (b) one-way SME; (c) two-way SME. Source: Ref 84

discernable by light microscopy. As a group, they are, therefore, valuable objects of study for both students and accomplished metallographers. A complete portfolio of copper-alloy microstructures is beyond the scope of this volume; however, micrographs of several representative structures are contained in Fig. 2.64 to 2.74. The alloys, their metallurgical state, and in some cases the metallographic methods are described in the figure captions (Ref 91).

Mechanical Properties

Useful mechanical properties in copper-base metals and alloys are generated by cold work, composition, thermal or thermomechanical treatment, and combinations thereof. The mechanism by which these processes occur have been discussed in previous sections. Cold work is generally the only means available to strengthen pure or nearly pure copper. It is also the most effective way to strengthen simple single-phase alloys such as alpha brasses, although such alloys are also strengthened by solid-solution effects. Grain size must also be considered because it alone can have a strong influence on mechanical properties.

Composition-based phase transformations are effectively utilized to strengthen such alloys as duplex brasses and high aluminum bronzes, which transform upon cooling from casting or hot working temperatures. In many cases, transformations are manipulated through appropriate heat treatments. The heat treatments are designed either to maximize strengthening, as in precipitation hardening, or to minimize detrimental effects such as embrittlement, decreased corrosion performance, and reduced conductivity.

Copper alloys are customarily classified according to their compositions, but this is mainly done for convenience and easy identification. It would be just as logical to group alloys according to mechanical properties. Thus, pure coppers and ductile single-phase alloys would be classified for

Table 2.18 Etchants for macroscopic examination of coppers and copper alloys

Procedure for use: immerse at room temperature, rinse in warm water, dry, unless otherwise indicated

Composition	Copper or copper alloy	Comments
1. 50 mL HNO_3, 0.5 g $AgNO_3$, 50 mL H_2O	All coppers and copper alloys	Produces a brilliant, deep etch
2. 10 mL HNO_3 and 90 mL H_2O	Coppers and brasses	
3. 50 mL HNO_3 and 50 mL H_2O(a)	Coppers, all brasses, aluminum bronze (b)	Grains: cracks and other defects
4. 30 mL HCl, 10 mL $FeCl_3$, 120 mL H_2O or methanol	Coppers and all brasses	Same as above; reveals grain contrast
5. 20 mL acetic acid, 10 mL 5% CrO_3, 5 mL 10% $FeCl_3$, 100 mL H_2O	All brasses	Produces a brilliant, deep etch
6. 2 g $K_2Cr_2O_7$, 4 mL saturated solution of NaCl, 8 mL H_2SO_4, 100 mL H_2O (3)	Coppers, high copper alloys, phosphor bronze	Grain boundaries, oxide inclusions
7. 40 g CrO_3, 7.5 g NH_4Cl, 50 mL HNO_3, 8 mL H_2SO_4, 100 mL H_2O	Silicon brass, silicon bronze	General macrostructure
8. 45 mL acetic acid and 45 mL HNO_3	Copper	Grain boundary and macroetch by polish attack
9. Saturated $(NH_4)_2S_2O_8$ (ammonium persulfate)	Copper and copper alloys	Use after the acetic acid listed above; increases contrast of brass
10. 40 mL HNO_3, 20 mL acetic acid, 40 mL H_2O	Copper and copper alloys	Macroetch 90-10, 70-30, and lead brass
11. 1 part $FeCl_3$ in H_2O, 1 part 5% CrO_3 in sat. brine, 2 parts 20% acetic acid in H_2O	Brass	Reveals strain pattern
12. 25 g $FeCl_3$ (or 5 or 10 g), 25 mL HCl (or 10 mL), 100 mL H_2O	Copper and copper alloys	Develops grain contrast, requires a good surface
13. 59 mL FeCl3, 2 mL HCl, 96 mL ethanol	Copper and copper alloys	Wash with alcohol or acetone to avoid staining
14. Solution A: 1 part 30% NH_4OH, a part H_2O; solution B: 3% H_2O (25–35 vol. % of solution A); solution C: 1 part HNO_3, 1 part H_2O	Cartridge brass	Reveals flow lines. Use ground surface. Cover specimen with solution A under chemical hood; add solution B to increase etch volume by 25–35%; do not let reaction get too violent; cover, let stand 12 h; brush solution C onto specimen; when flow lines are visible, rinse in water and dry
15. 5 g $Fe(NO_3)_2$, 25 mL HCl, 70 mL H_2O	Copper and copper alloys	Requires a polished surface. Swab sample 3–4 min then immerse 15–60 s
16. Conc. HNO_3	Copper and copper alloys	Rapid etch for rough surfaces; contrast improved by small addition of $AgNO_3$
17. Solution A: 1 g $Hg(NO_3)_2 \cdot 8H_2O$, 100 mL H_2O; solution B: 1 mL HNO_3, 100 mL H_2O	Brass	Verifies stress. Use equal parts A and B; time to failure indicates stress level
18. 1 part 0.880 ammonia, 2 parts aqueous $(NH_4)_2S_2O_8$, 1 part H_2O	β-containing brasses	Reveals grain size

(a) Solution should be agitated during etching to prevent pitting of some alloys. (b) Aluminum bronzes may form smut, which can be removed by brief immersion in concentrated HNO_3. (c) Excellent for grain contrast. (d) Amount of water can be varied as desired. (e) Immerse specimen 15–50 min then swab with fresh solution. Source: Ref 87, 89

their high formability; more highly alloyed compositions would be considered for their generally higher strengths, and heat-treated alloys of any composition would fall into a yet higher strength category. Under such a system, designers seeking materials within a particular strength range could choose from a manageable selection of candidates among the hundreds of standard copper alloys now available.

Unfortunately, copper alloys are rarely considered for their strength alone. More often, it is a combination of mechanical properties and conductivity, corrosion resistance, and even color that dictate the choice of copper alloys over other engineering materials. Ideally, it should be possible to formulate an expert system capable of delineating material selections based on desired ranges for several properties. A designer would then be able

to pick from compositions (and treatments) that offered, say, a given tensile strength and a given electrical conductivity and adequate corrosion resistance in a particular medium. Several such systems are currently in use; among the better known are ASM International's MatDB and the (U.S.) Copper Development Association's CopperSelect. Comparable systems have been published in Europe and the United Kingdom (Ref 92, 93).

Another hindrance to selecting copper alloys on the basis of mechanical properties is that it is impossible to identify a unique set of properties with a specific alloy. Rather, individual alloys can be expected to exhibit a range of properties due to normal variations in composition (including impurity content), imprecise control of heat-treatment parameters and processing-related factors, which are often manifested as an effect of section size.

Table 2.19 Attack polishing solutions and procedures for copper and copper alloys

Composition	Alloy	Comments
1. Aqueous $(NH_4)_2S_2O_8$: 10–15 g/L (for copper), 15–30 g/L (for brass)	Copper, brass with high zinc content	Use with MgO abrasive; insert plastic between cloth and wheel to prevent pitting; use skid-polish technique
2. 2–10% CrO_3 in water	Copper	Use napped cloth and rouge or alumina abrasives
3. 1% $Fe(NO_3)_3$ in water	Copper	Use medium-nap cloth and alumina or colloidal silica abrasives
4. 5 g $Fe(NO_3)_3$, 25 mL HCl, 370 mL H_2O	Copper	Use with γ-alumina; attack-polish for 3 min, wash off wheel, add solution only and polish for 2–5 s
5. 12 g/L aqueous $(NH_4)_2S_2O_8$	Cu-30% Zn brass	Use with MgO abrasive; skid-polish for 20 min with high-nap cloth
6. Aqueous solution of CrO_3 and HCl	Cu-Pb alloys	Concentration and abrasive not given

Source: Ref 89

Table 2.20 Electrolytes and conditions for electrolytic polishing of copper and copper alloys

Composition	Voltage	Current density A/cm^2	A/in.^{2}	Cathode	Duration	Copper or copper alloy
1. 825 mL H_3PO_4 and 175 mL H_2O	1.0–1.6	0.02–0.1	0.13–0.65	Copper	10–40 min	Unalloyed copper
2. 250 mL H_3PO_4, 250 mL ethanol, 50 mL propanol, 500 mL distilled H_2O, and 3 g urea	3–6	0.4–0.8	2.6–5.2	Stainless	50 s	Coppers and copper alloys
3. 700 mL H_3PO_4 and 350 mL H_2O	1.2–2.0	0.06–0.1	0.39–0.64	Copper	15–30 min	Copper; α, β, and α–β brasses; aluminum, silicon, tin, and phosphor bronzes; beryllium, iron, lead, or chromium coppers
4. 580 g $H_4P_2O_7$ and 1000 mL H_2O	1.2–1.9	0.08–0.12	0.05–0.77	Copper	10–15 min	Coppers, brasses
5. 300 mL HNO_3 and 600 mL ethanol	20–70	0.65–3.1	4.2–20.0	Stainless	10–60 s	Coppers, brasses
	30–50	2.5–3.1	16.1–51.0	Stainless	5–10 s	Silicon bronze, phosphor bronze
6. 170 mL CrO_3 and 830 mL H_2O	1.5–12	0.95–2.2	F6.1–14.2	Stainless	10–60 s	Brasses
7. 400 mL H_6PO_4 and 600 mL H_2O	1.0–2.0	0.06–0.15	0.39–0.97	Copper or stainless	1–15 min	α, α-β brasses; copper-iron; copper-chromium
8. 30 mL HNO_3, 900 mL methanol, and 300 g $Cu(NO_3)_2$	45–50	1.05–1.25	6.77–8.1	Stainless	15 s	Bronzes (have tendency to etch)
9. 670 mL H_3PO_4, 100 mL H_2SO_4, and 300 mL distilled H_2O	2–3	0.1	0.64	Copper	15 min	Copper; copper-tin up to 6% Sn
10. 470 mL H_3PO_4, 200 mL H_2SO_4, and 400 mL distilled H_2O	2–2.3	0.1	0.64	Copper	10–15 min	Copper-tin up to 9% Sn
11. 350 mL H_3PO_4 and 650 mL ethanol	2–5	0.02–0.07	0.13–0.45	Copper	5–15 min	Copper alloys with high lead (to 30%)
12. 540 mL H_3PO_4 and 460 mL H_2O	2	0.065–0.075	0.4–0.5	Copper	5–15 min	Copper
	2–2.2	0.1–0.15	0.64–0.97			Nickel silver

Source: Ref 87

Table 2.21 Etchants and procedures for microetching coppers and copper alloys

Compositions(a)	Procedure	Copper or copper alloy
1. 20 mL NH_4OH, 0–20 mL H_2O, and 8–20 mL 3% H_2O_2	Immersion or swabbing 1 min; H_2O_2 content varies with copper content of alloy to be etched; use fresh H_2O_2 for best results(b)	Use fresh coppers and copper alloys, particularly Cu-Be; film on etched aluminum bronze can be removed using weak Girard's solution, preferred for brasses
2. 1 g $Fe(NO_3)_3$ and 100 mL H_2O	Immersion	Etching and attack polishing of coppers and alloys
3. 25 mL NH_4OH, 25 mL H_2O, and 50 mL 2.5% $(NH_4)_2S_2O_8$	Immersion	Attack polishing of coppers and some copper alloys
4. 2 g $K_2Cr_2O_7$, 8 mL H_2SO_4, 4 mL NaCl (saturated solution), and 100 mL H_2O	Immersion; NaCl replaceable by 1 drop HCl per 25 mL solution; add just before using; follow with $FeCl_3$ or other contrast etch	Copper; copper alloys of beryllium, manganese, and silicon; nickel silver; bronzes, chromium copper; preferred for all coppers to reveal grain boundaries, grain contrast, and cold deformation
5. CrO_3 (saturated aqueous solution)	Immersion or swabbing	Copper, brasses, bronzes, and nickel silver
6. 50 mL 10–15% CrO_3 and 1–2 drops HCl	Immersion; add HCl at time of use	Same as above; color by electrolytic etching or with $FeCl_3$ etchants
7. 8 g CrO_3, 10 mL HNO_3, 10 mL H_2SO_4, and 200 mL H_2O	Immersion or swabbing	Grain contrast etch for ETP copper, does not dissolve Cu_2O; use after etchant 3 when etching deoxidized microstructure
8. 10 g $(NH_4)_2S_2O_8$ and 90 mL H_2O	Immersion; use cold or boiling	Coppers, brasses, bronzes, nickel silver, and aluminum bronze
9. 10% copper ammonium chloride plus NH_4OH to neutrality or alkalinity	Immersion; wash specimen thoroughly	Copper, brasses, nickel silver; darkens β in α–β brass
10. $FeCl_3$, g; HCl, mL; H_2O, mL 5 50 100 20 5 100(c)(d) 15 25 100 1 20 100 8 25 100 5 10 100(e)(f)	Immersion or swabbing; etch lightly or by successive light etches to required results	Coppers, brasses, bronzes, aluminum bronze; darkens β phase in brass; gives contrast following dichromate and other etches
11. 5 g $FeCl_3$, 100 mL ethanol, and 5–30 mL HCl	Immersion or swabbing for 1 s to several minutes	Copper and copper alloys; darkens β phase in α–β brasses and aluminum brass
12. HNO_3 (various concentrations)	Immersion or swabbing; 0.15–0.3% $AgNO_3$ added to 1:1 solution gives a brilliant, deep etch	Copper and copper alloys
13. NH_4OH (dilute solutions)	Immersion	Attack polishing or brasses and bronzes
14. 50 mL HNO_3, 20 g CrO_3, and 75 mL H_2O	Immersion	Aluminum bronze, free-cutting brass; film from polishing can be removed with 10% HF
15. 5 ml HNO_3, 20 g CrO_3, and 75 mL H_2O	Immersion	Same as above
16. 59 g $FeCl_3$ and 96 mL ethanol	Immersion; heat sample first in hot H_2O	Macro- and microetch for annealed copper-nickel alloys
17. 16 g CrO_3, 1.8 g NH_4Cl, 10 mL HNO_3, and 200 mL H_2O	Immersion	Preferred etch for copper-nickel; preferential attack of copper-rich phase in castings
18. 5 parts HNO_3, 5 parts acetic acid, and 1 part H_3PO_4	Immersion, 3 s	Coppers, brasses
19. Equal parts NH_4Cl and H_2O	Immersion	Coppers and alloys
20. 60 g $FeCl_3$, 20 g $Fe(NO_3)_3$, and 2000 mL H_2O	Immersion	Cu-Ni alloys
21. 1 part acetic acid, 1 part HNO_3, and 2 parts acetone	Immersion	Cu-Ni and Cu-Ni-Al alloys; attacks α-solid solution vigorously
22. 1 g KOH, 20 mL H_2O_2 (3%), 50 mL NH_4OH, and 30 mL H_2O	Dissolve KOH in water, slowly add NH_4OH, then H_2O_2. Use fresh, immerse sample 3–60 s	General-purpose etch
23. 2.0–2.5 N NaOH, 1–2% ethylenediamine and 100g/L $K_3Fe(CN)_6$	Immersion time 1–3 min	Copper and copper alloys
24. Solution A: 3 g CrO_3, 15 mL H_2O, and 1 mL HCl; solution B: 3 mL NH_4OH, and 2 mL H_2O_2 (3%)	Pour solution B into solution A; vigorous reaction, use under hood; swab briskly but lightly 30–40 s; for cold-worked samples, dilute with additional 20% water, etch additional 10 s	Pondo's etch; good for annealed samples; follow procedure for cold-worked material
25. 25 g CrO_3, 2 ml HCl, 2 mL H_2SO_4, and 500 mL H_2O	Immersion or swabbing	Cu-25% Pb-0.75% Ag; for Cu-25% Pb-1% Sn, use 5 mL HCl and 5 mL H_2SO_4

(continued)

(a) The use of concentrated etchants is intended unless otherwise specified. (b) This etchant may be altered with $FeCl_3$. (c) Girard's No. 1 etchant. (d) Plus 1 g CrO_3. (e) Girard's No. 2 etchant. (f) Plus 1 g $CuCl_2$ and 0.05 g $SnCl_2$. Source: Ref 87

Table 2.21 (continued)

Compositions(a)	Procedure	Copper or copper alloy
26. Solution A: 1 g FeCl$_3$, 10 mL HCl, 100 mL H$_2$O; solution B: 5 g FeCl$_3$, 10 or 50 mL HCl, 100 mL H$_2$O; solution C: 8 g FeCl$_3$, 25 ml HCl, 100 mL H$_2$O; solution D: 25 g FeCl$_3$, 25 mL HCl, 10 mL H$_2$O; solution E: 25 g FeCl$_3$, 5 mL HCl, 1 g CrO$_3$, 100 mL H$_2$O; solution F: 5 g FeCl$_3$, 10 mL HCl, 1 g CuCl$_2$, 0.1 g SnCl$_2$, 100 mL H$_2$O; solution G: 5 g FeCl$_3$, 15 mL HCl, 60 mL ethanol; solution H: 2 g FeCl$_3$, 30 mL H$_2$O, 5 mL HCl, 60 mL ethanol; solution I: 2.5 g FeCl$_3$, 1 mL HCl, 100 mL ethanol	Immersion or swabbing; can be used after dichromate etch (No. 4) to develop grain contrast; often used stepwise until proper degree of etch is obtained	A: general purpose; B: use with 50 mL HCl to attack Pb in Cu-Pb alloys, including those with high Sn; G: aluminum bronze; H: α-brass; I: Cu-Sn-As alloys. See also No. 11 for similar uses
27. Solution A: 3% I$_2$ in methanol; solution B: 3 parts methanol, 1 part HNO$_3$	Solution A reveals grain boundaries, solution B develops grain orientation contrast; hold sample 1–2 cm above solution B; vapors of fresh aqua regia produce etching in 10–20 s; do not wash, blow dry	Cu-Au alloys
28. 20 mL H$_2$O$_2$ (30%), 25 mL H$_2$O, 50 mL NH$_4$OH, 5 mL 30% aq. KOH	Add peroxide before use, immerse 2–30 s	Cu-Si alloys
29. 40 mL H$_2$O$_2$ (3%), 58 mL H$_2$O, 2 g KOH, 100 mL NH$_4$OH	Immerse or swab	Cu-Si-Zn alloys
30. 2 g K$_2$Cr$_2$O$_7$, 1 g NaF, 3 mL H$_2$SO$_4$, 100 mL H$_2$O	Immerse or swab	Cromwell's etch for Cu-Sn-Ag dental alloys; attacks β (AgSn) phase
31. 10 g Na$_2$S$_2$O$_3$, 100 mL H$_2$O	Immerse sample 90–150 s with agitation	Aluminum bronze; reveals γ$_2$ but not β
32. 1 part HNO$_3$, 1 part acetic acid, 1 part glycerin	Swab lightly for short time to remove smeared Pb; longer etch reveals structure in Pb	Vilella's etch for Cu-Pb alloys; sharply defines Pb areas
33. 15 mL NH$_4$OH, 15 mL H$_2$O$_2$ (3%), 15 mL H$_2$O, 4 pellets NaOH	Add NaOH last	Cu-Be alloys
34. 5 parts NH$_4$OH, 5 parts H$_2$O, 4 parts H$_2$O$_2$ (3%)	Swab; produces response to polarizad light	High purity Cu-10% Zn
35. 2 g FeCl$_3$, 30 mL H$_2$O, 10 mL HCl, 60 mL ethanol	Immerse 1–2 min; sample responds to polarized light	Cu, Cu-Zn and Cu-Sn alloys
36. 50 mL sat. aq. Na$_2$S$_2$O$_3$, 1 g potassium metabisulfite	Immerse 3 min or more of β brass; up to 60 min for α brass	Tint etch; Klemm's I reagent for brasses
37. 50 mL sat. aq. Na$_2$S$_2$O$_3$	Immerse 6–8 min or more	Tint etch; Klemm's II reagent for α brass
38. 5 mL sat. aq. Na$_2$S$_2$O$_3$, 45 mL H$_2$O, 20 g potassium metabisulfite	Immerse 3–5 minutes	Tint etch; Klemm's III reagent for bronze alloys
39. 240 g Na$_2$S$_2$O$_3$, 30 g citric acid, 24 lead acetate, 1000 mL H$_2$O	Dissolve in given order in water, allowing each to dissolve before adding next. Age in dark bottle for at least 24 h; do not remove precipitate. When stock solution becomes gray or black after prolonged storage, discard. Pre-etch sample with general purpose grain boundary etchant. Immerse sample until surface turns violet to blue. Colors enhanced with polarized light	Beraha's lead sulfide tint etchant for Cu and copper alloys
40. 200 g Cr O$_3$, 20 g anhydrous Na$_2$SO$_4$, 17 mL HCl, 1000 mL H$_2$O	Dip sample 2–20 s, dilute to retard rate of attack	Beraha's tint etch for Cu, brass or bronze alloys
41. 300 mL ethanol, 2 mL HCl, 0.5–1 mL selenic acid	Keep in dark bottle. Pre-etch with grain boundary etchant. Immerse sample until surface is violet to blue. Use plastic tongs; use 0.5 mL selenic acid if color changes too quickly	Beraha's tint etch for brass and Cu-Be alloys
42. 0.1 g AgNO$_2$, 10 mL HNO$_3$, 90 mL H$_2$O	Etch 10 s at 20 °C (68 °F) on electrolytically or chemically polished samples. Use light that is almost cross polarized	de Jong's color etch for OFHC Cu

(a) The use of concentrated etchants is intended unless otherwise specified. (b) This etchant may be altered with FeCl$_3$. (c) Girard's No. 1 etchant. (d) Plus 1 g CrO$_3$. (e) Girard's No. 2 etchant. (f) Plus 1 g CuCl$_2$ and 0.05 g SnCl$_2$. Source: Ref 87

Minimum versus Typical Properties. Because of the imprecision inherent to mechanical properties, producers identify properties either as minimum or typical (also called nominal) values. Minimum values are normally cited in standard product specifications, such as those published by the American Society for Testing and Materials (ASTM).

Typical or nominal properties are most often associated with alloy standards, for example, those that fix recognized composition ranges. North American designations of this type are based on the Unified Numbering System (UNS); among the more widely used compositional standards used in other countries are those of the British Standards Institution (BS) and the Deutsche Institut für Normung (DIN). Efforts to rationalize these standards are being undertaken by such groups as the International Standards Organization (ISO) and the Comité Européen de Normalisation (CEN). Typical mechanical properties can be significantly higher than the minimum values listed in product specifications. While it is possible to use typical properties for many engineering calculations, the properties are given for general information only and cannot be demanded for specific products unless they have been agreed upon by the designer and the alloy producer.

Approximate mechanical properties can also be estimated with reference to standard temper designations, but only for specific alloys. That is, properties associated with temper designations for one alloy do not apply to another. Figure 2.75 illustrates the relationships among hardness, tensile strength, and temper designation for brasses C24000 and C26000. The effect of cold reduction (for example, temper) on tensile strength for several other sheet alloys is shown in Fig. 2.76. Having a similar hardness clearly does not imply similar tensile strengths in different materials. Temper designations are meant to imply a range of mechanical properties that reflect all the variables associated with alloy production and processing. This can be seen in the range of tensile strength data for alloy C26000 shown in Fig. 2.77 (Ref 10).

Yield Strength

Yield strength in copper alloys is defined by a designated point on an engineering stress-strain curve. For some alloys, the linear or Hooke's law region of the curves are quite extensive. In such cases, yield strength is defined as the intersection of the curve with a line having a slope equivalent to that of the elastic modulus but offset by a fixed value of elastic strain, usually 0.2%. Values ob-

Fig. 2.64 C11000 (ETP copper) as cast illustrating the columnar as-cast structure. Source: Ref 91

Table 2.22   Electrolytes and operating conditions for electrolytic etching of copper and copper alloys

Composition	Operating conditions	Copper or copper alloy
1. $5\text{–}14\%\ H_3PO_4$ (8%) and bal H_2O	Voltage range, 1–8 V; etching time 5–10 s	Coppers, cartridge brass, free cutting brass, admiralty, gilding metal
2. 250 mL 8% H_3PO_4, 250 mL 95% ethanol, 500 mL H_2O, 2 mL wetting agent	Voltage range 1–3 V; current density 0.1–0.15 A/cm^2 (0.64–0.97 A/in.^{2}); etching time, 30–60 s	Coppers
3. 30 g $FeSO_2$, 4 g NaOH, 100 mL H_2SO_4, 1900 mL H_2O	0.1 A at 8–10 V for 15 s; do not swab surface after etching	Darkens β phase in brasses and gives contrast after H_2O_2-NH_4OH etch; also for nickel silver and bronzes
4. 1 mL CrO_3 and 99 mL H_2O	6 V; aluminum cathode; etching time, 3–6 s	Beryllium copper and aluminum bronze
5. 5 mL acetic acid (glacial); 10 mL HNO_3, 30 mL H_2O	Voltage range, 0.5–1 V; current density, 0.2–0.5 A/cm^2 (1.3–1.9 A/in.^{2}); etching time, 5–15 s	Copper-nickel alloys; avoiding contrast associated with coring

Source: Ref 87

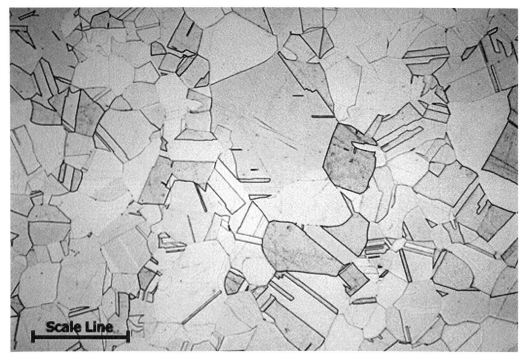

Fig. 2.65 C10100 (OF copper)—cold worked and annealed—illustrating twinned structures. Scale line is approximately 25 μm. Nominal composition is Cu 99.99%. Source: Copper Development Association Inc.

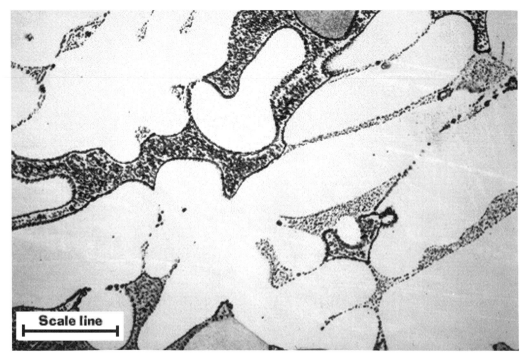

Fig. 2.66 C11000 (ETP copper) showing oxide structure. Scale line is approximately 125 μm. Nominal composition is Cu 99.90%. Source: Copper Development Association Inc.

Fig. 2.67 C26000 (cartridge brass)—cold worked and annealed—illustrating recrystallization. Scale line is approximately 125 μm. Nominal composition is Cu 89.0–90.0, Zn 8.9–11.0, Fe 0.05, and Pb 0.05. Source: Copper Development Association Inc.

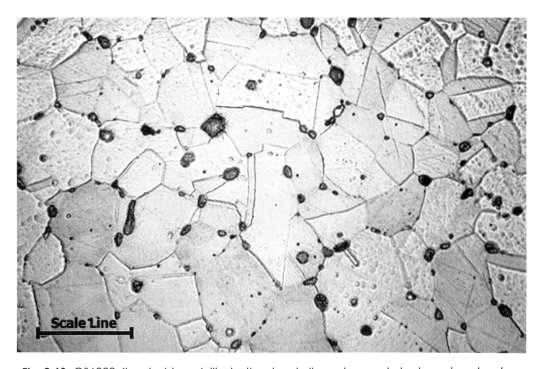

Fig. 2.68 C36000 (leaded brass) illustrating lead dispersion and duplex microstructure. Scale line is approximately 25 μm. Nominal composition is Cu 60.0–63.0, Zn 33.0–37.0, Pb 2.5–3.7, and Fe 0.35. Source: Copper Development Association Inc.

tained in this way are called 0.2% offset yield strengths. For pure copper and many copper al-

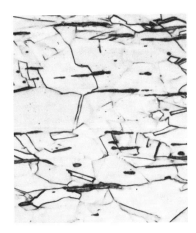

Fig. 2.69 C14500 (DPTE copper)—hot rolled and drawn—illustrating elongated copper tellurides, which improves machinablity.

loys, the linear portion of the stress-strain curve is restricted, making it difficult to construct an offset line with the proper slope. It is, therefore, common practice to define yield strength for such alloys as the ordinate corresponding to a fixed value of total strain, usually 0.5%. Values obtained by this technique are called yield strength at 0.5% extension under load. Because of variations in the shape of stress-strain curves among copper alloys, there is no universal relationship between offset and extension-under-load yield strengths.

Compressive yield strength is also often defined on the basis of a fixed degree of deformation, or "set." Values are typically listed for sets of 0.1%, 1.0%, or 10.0%. Compressive yield strength is higher for higher values of set. It is reasonable to expect that compressive yield strength values are normally different from those measured in a simple tension test. In some cases, they can be significantly lower (Ref 10).

The difference between compressive and tensile yield strength is an important factor in leaf springs, whose opposite surfaces are stressed in

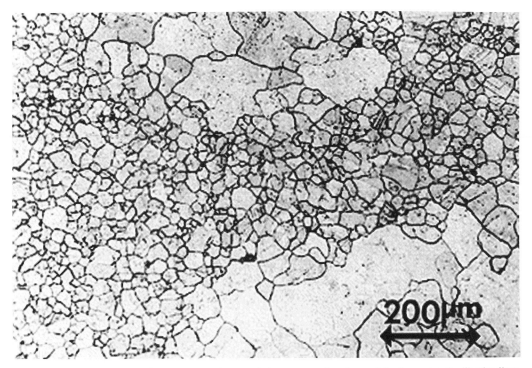

Fig. 2.70 C17200 (beryllium copper)—cast, homogenized, and hot worked—illustrating precipitation hardening. Nominal composition is Be 1.80–2.00, Co + Ni 0.20 min, Co + Ni + Fe 0.6 max, Pb 0.02 max, Cu + sum of named elements 99.5 min. The microstructure shows nonuniform distribution of grain sizes, typical of hot–worked product. Greater uniformity in grain size distribution may be achieved in the finished product by successive cold working and annealing operations. Source: Copper Development Association Inc.

tension and compression during flexure. Such springs are, therefore, designed on the basis of the "flexural yield strength," which is the mathematical average of the tensile and compressive yield strengths. Accurate measurement of flexural yield strength can be difficult in that both tensile and compressive yield strengths are strongly dependent on orientation with respect to the rolling direction. This can be seen in Table 2.23, which illustrates the effect of stress mode and orientation for two common spring materials (Ref 10).

Yield and tensile strengths of wrought copper alloys classified under the UNS are listed in Table 2.24; cast alloys are listed in Table 2.25 (Ref 18, 19).

Hardness

Hardness is a measure of resistance to penetration. It also gives a qualitative indication of the metal's strength and resistance to abrasion. In some cases, hardness can be used to predict certain types of corrosion performance (Ref 10).

For copper alloys, hardness is most often measured by use of the Rockwell hardness test (B and F

scales) and the Brinell hardness test (500 and 1000 kg loads). The Rockwell superficial hardness test (15T, 30T, or 45T scales) should be utilized for thin strip materials, while the Rockwell C scale (diamond penetrator) is necessary for age-

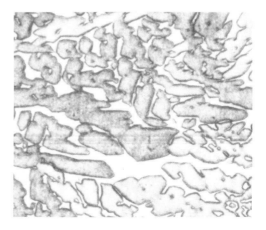

Fig. 2.71 C28000 (Muntz metal)—as cast—illustrating a two-phase structure, α phase in a matrix of β phase.

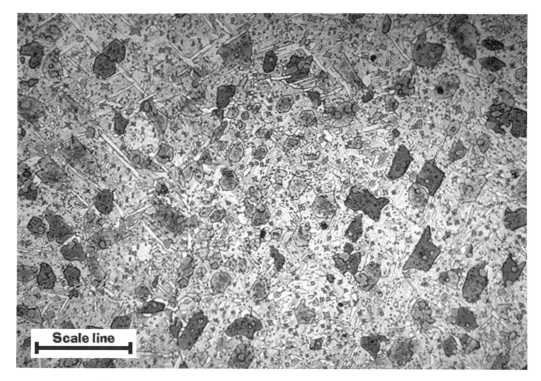

Fig. 2.72 C62500 (aluminum bronze)—a high-aluminum bronze—illustrating complex microstructures, retained beta and eutectoid products. Scale line is approximately 50 μm. Nominal composition is Cu 83.8, Al 12.0, and Fe 4.2. Source: Copper Development Association Inc.

hardened beryllium coppers and some high-strength aluminum bronzes. Hardness values for wrought and cast UNS copper alloys are listed in Tables 2.24 and 2.25.

Fatigue Strength

It has been recognized since the 1870s that stresses much lower than the UTS can lead to fail-

Fig. 2.73 C95400 (aluminum bronze)—solution treated, quenched and tempered—illustrating a martensitic structure.

ure if such stresses are applied repetitively. The phenomenon is known as fatigue, and it commonly constitutes the primary design basis for cyclically loaded products such as springs.

The cyclical stress, S, necessary to cause failure decreases with an increasing number of stress cycles, N. In some metals, a stress level at which failure is independent of N is eventually reached. Below the stress corresponding to this so-called "fatigue life," fatigue failure will not occur. Copper and copper alloys do not necessarily exhibit such a limiting stress, but as a rule of thumb, the endurance limit for copper alloys (and other nonferrous metals) can be taken as 0.35 UTS. A distinction must be made, however, between high-cycle fatigue (high N, lowest stress to failure) and low-cycle fatigue (low N, failure stress approaching the UTS). In both cases, fatigue behavior is strongly dependent on temperature, environment, change in amplitude, prior stress history, shape of the stress-time function, and surface condition (Ref 12).

Representing Fatigue Behavior. The most common way to express fatigue behavior involves measuring the stress required to cause failure at a

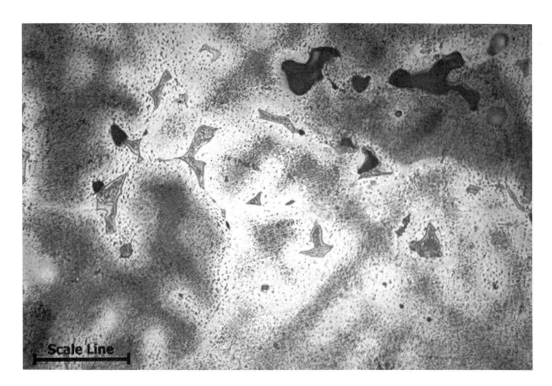

Fig. 2.74 C90700 (tin bronze)—as-cast—illustrating coring. Scale line is approximately 50 μm. Nominal composition is Cu 88–90, Sn 10–12, Pb 0.50, Zn 0.50, Ni 0.50, P 0.30, Sb 0.20, Fe 0.15, S 0.05, and Al 0.005. Source: Copper Development Association Inc.

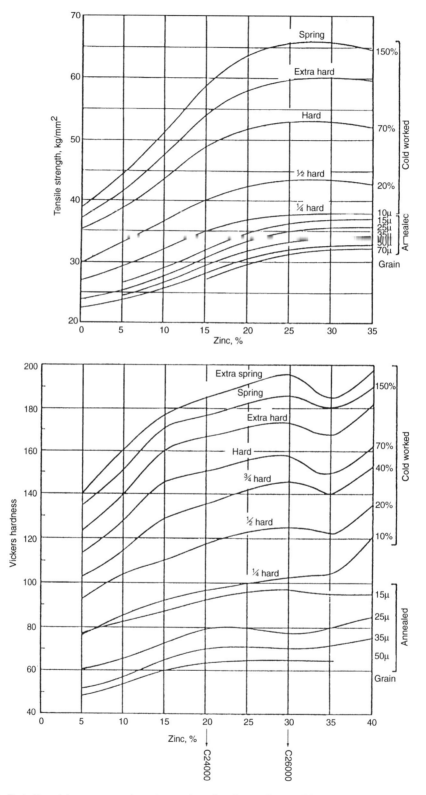

Fig. 2.75 Relationships among hardness, tensile strength, and temper designation for alloys C24000 and C26000.

high number of deflection cycles. Fatigue life can also be expressed in terms of total strain amplitude. Important parameters include the fatigue mode (push-pull, pull-release, etc.); number of cycles, and the stress or strain level corresponding to particular N values. In constant stress amplitude tests, the applied stress at which fracture occurs, N_f, is reported as a function of the number of cycles for which that stress was applied.

Figures 2.78 through 2.81 show S-N curves for several copper-base sheet alloys. Fatigue data for copper alloys used for electrical springs and contacts are compared in Fig. 2.82. Fatigue data for polycrystalline OF copper as a function of strain amplitude are shown in Fig. 2.83. The term, N_i, used in this figure refers to crack initiation life. For unnotched specimens of polycrystalline copper loaded to strain amplitudes between 1×10^{-4} to 8×10^{-4}, failure occurs at a total strain, $\varepsilon_{tot} = 7.2 \times 10^{-4}$ (Fig. 2.84). Fatigue lives of copper single crystals tested in high vacuum are higher by a factor of 15 to 30 than specimens tested in air, suggesting a strong influence of surface condition on fatigue properties (Ref 94–98).

Cyclical stress limits for a given endurance can be plotted against mean applied stress in so-called Goodman diagrams or R-M diagrams. An example of one such diagram for polycrystalline copper is shown in Fig. 2.85. R is the stress range, and M is the mean tensile stress. Fatigue data for several wrought and cast UNS copper alloys are listed in Tables 2.24 and 2.25. When using these data, it is important to make note of the number of cycles for which they apply (Ref 95).

Structure-Property Effects due to Fatigue Loading. Soft recrystallized metals such as copper become increasingly harder during cyclic loading, with hardness eventually reaching a saturation value. Conversely, samples that have been cold worked to relatively high hardnesses soften as N increases. The different behaviors can be explained by the generation of, and/or changes in dislocation structures as fatigue loading progresses. The appearance of crystal (grain) surfaces in copper suggest that cyclical plastic deformation takes place homogeneously only during the initial stages of fatigue life. Thereafter, hardening or softening is apparently linked to specific volume fractions of the material. The affected volumes first take the form of vein-like areas of high dislocation density. The veins are separated by nearly dislocation-free areas called channels. Persistent

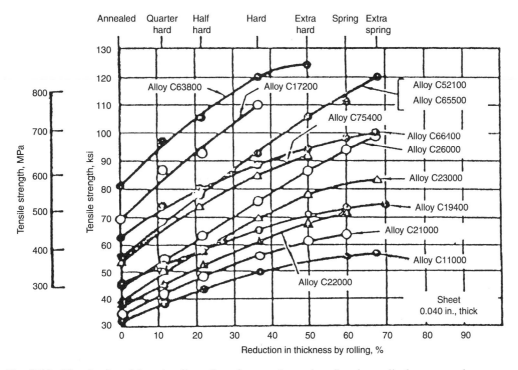

Fig. 2.76 Effect of cold reduction (i.e., temper) on tensile strength for several copper sheet alloys. Source: Ref 10

slip bands, in which strain becomes localized, eventually appear. Fatigue hardening curves for polycrystalline copper in strain-controlled loading with constant applied strain amplitude are shown in Fig. 2.86 (Ref 99, 100).

Fracture

The fracture of a copper alloy will be ductile or brittle depending on the properties of the alloy and the stress state imposed. Ductile fracture is preceded by a high degree of plastic deformation. In extreme cases, specimens stressed in tension may exhibit nearly 100% reduction of area before failure. Brittle fracture is preceded by little if any plastic deformation. Brittle fracture is uncommon in fcc metals such as copper, which have high ductility as a result of their efficient crystallographic slip systems. Amorphous solids, bcc structures, and ionic crystals, in which slip is more difficult and in which cracks can propagate rapidly, are prone to brittle fracture. Pure copper and relatively soft, single-phase copper alloys fail by ductile fracture, whereas hard alloys containing significant amounts of bcc β phase or other hard constituents tend to fail in a brittle or semibrittle

manner. Oxidized copper and harder alloys fail with varying degrees of brittleness (Ref 101).

Brittle Fracture

Brittleness is related to the behavior of dislocations (crystal imperfections) in the region of crack nucleation. In brittle materials, dislocations are practically immobile, and failure proceeds with little or no plastic strain. In semibrittle materials, dislocations are mobile in a limited number of slip planes. In ductile materials such as copper, dislocations can move more freely (Ref 101).

The brittle type of fracture known as cleavage occurs along specific crystallographic planes. Because adjacent grains in polycrystalline materials have different orientations, cleavage cracks change direction when they encounter a grain boundary. Cleavage facets have high reflectivity, which gives the fracture surface a shiny appearance. As a result, brittle fracture is sometimes improperly called crystalline fracture.

Ductile Fracture

Ductile tensile fractures are preceded by local necking. Necking, or plastic instability, occurs when the previously uniform tensile strain becomes localized in one slightly weaker region of the tensile specimen. Within the necked and failed regions, opposing ends of the specimens have cup and cone shapes; these are surrounded by a well-defined conical ring caused by shear failure. The relatively flat central region results from the failure of ligaments between microscopic voids. Such voids are created by dislocation coalescence. In polycrystalline metals, voids can also form around grain boundary triple points. In oxide-bearing copper, they can nucleate when an oxide inclusion or the interface between it and sound metal cracks and separates. The texture of ductile fractures appears dull and is called fibrous. Microscopically, fibrous fracture surfaces appear dimpled, originat-

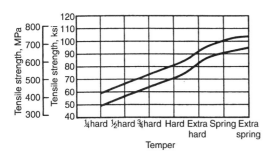

Fig. 2.77 Range of tensile strengths corresponding to rolled tempers in alloy C26000. Source: Ref 10

Table 2.23   Effect of stress mode and orientation on 0.2% offset yield strength of selected copper alloys

Alloy (temper)	Orientation	Tensile mode		Compressive mode	
		MPa	ksi	MPa	ksi
C51000, H04 (hard)	Longitudinal	558	81	427	62
	Transverse	531	77	600	87
C51000, H08 (spring)	Longitudinal	765	111	565	82
	Transverse	738	107	841	122
C68800, H08 (spring)	Longitudinal	772	112	421	61
	Transverse	710	103	827	120

Source: Ref 10

Table 2.24 Mechanical properties of wrought copper tube alloys

Temper	Tensile strength(a), MPa	Yield strength(a), MPa	Elongation(b), %
C10200			
OS050	220	69	45
OS025	235	76	45
H55	275	220	25
H80	380	345	8
C12200			
OS050	220	69	45
OS025	235	76	45
H55	275	220	25
H80	380	345	8
C19200			
H55(c)	290	205(d)	35
C23000			
OS050	275	83	55
OS015	305	125	45
H55	345	275	30
H80	485	400	8
C26000			
OS050	325	105	65
OS025	360	140	55
H80	540	440	8
C33000			
OS050	325	105	60
OS025	360	140	50
H80	515	415	7
C43500			
OS035	315	110	46
H80	515	415	10
C44300, C44400, C44500			
OS025	365	150	65
C46400, C46500, C46600, C46700(e)			
H80	605	455	18
C60800			
OS025	415	185	55
C65100			
OS015	310	140	55
H80	450	275	20
C65500			
OS050	395	...	70
H80	640	...	22
C68700			
OS025	415	185	55
C70600			
OS025	305	110	42
H55	415	395	10
C71500			
OS025	415	170	45

Tube size: 25 mm (1 in.) OD by 1.65 mm (0.065 in.) wall. (a) 0.5% extension under load. (b) In 50 mm (2 in.). (c) Tube size: 4.8 mm (0.1875 in.) OD by 0.76 mm (0.030 in.) wall. (d) 0.2% offset. (e) Tube size: 9.5 mm (0.375 in.) OD by 2.5 mm (0.097 in.) wall. Source: Ref 17

ing around second-phase particles or inclusions (Fig. 2.87). A fractograph of ductile failure in inclusion-free high purity copper is shown in Fig. 2.88 (Ref 101).

Two types of ductile cracking are observed in copper at elevated temperatures. W-type cavitation occurs through stress concentration due to grain boundary sliding. Grain boundary triple points are susceptible to this type of crack nucleation (Fig. 2.89). *R*-type cavitation occurs in copper at grain boundaries oriented normal to the stress axis. It originates at small cavities formed under conditions of low stress and high temperature, as shown in Fig. 2.90 (Ref 39, 101).

Corrosion

Copper is chemically classified as one of the noble metals, and its inherent corrosion resistance is among the metal's best known and commercially most useful attributes. There is ample evidence that its rate of corrosion in natural environments can be extremely low. The fact that metallic, or "native" copper occurs widely in nature is graphic proof of the metal's stability (at least in kinetic, if not thermodynamic terms) in specific geologic environments for periods that, in some cases, exceed 10^9 years. Copper and bronze artifacts have withstood burial in the earth and submersion in the sea for thousands of years, often without serious degradation. Copper roofing in certain rural atmospheres has been found to corrode less than 0.4 mm (0.016 in.) in 200 years. The copper skin of the Statue of Liberty lost just 0.1 mm (0.004 in.) during its first 100 years despite exposure to a marine atmosphere containing a variety of industrial pollutants. In addition to these natural environments, copper and its alloys resist many salts, alkalies, and organic chemicals.

Corrosion resistance is, of course, a relative term, and copper is certainly not chemically inert in all media. Common environments in which copper is susceptible to attack include certain soft, low-pH well waters, oxidizing acids, ammoniacal solutions, amines, cyanides, nitrates and nitrites, oxidizing heavy metal salts, and certain sulfides. The corrosion of copper can also be accelerated by specific microbiologic agents, although copper's natural biostatic nature is of some benefit in this regard (Ref 102, 103).

Given appropriate electrochemical conditions and physical environments, copper metals can be susceptible to most of the familiar forms of corrosion, including general wastage, crevice and pitting attack, dealloying, erosion corrosion, galvanic corrosion, microbially induced corrosion (MIC), and stress corrosion cracking (SCC). However, the fact that copper metals often perform better or more cost-effectively than competing materials is one of the principal reasons for their continued utilization.

Table 2.26 offers a guide to the corrosion resistance of selected wrought copper alloys in various corrosive media. Cast copper alloys are listed in Table 2.27. The tabulated ratings are qualitative because subtle changes in environmental conditions can change the performance of a given alloy significantly. Changes in conditions may also bring about performance with respect to localized corrosion phenomena. In any case, alloy selection is rarely made on the basis of corrosion resistance alone. When choosing between copper alloys and alternative materials, it is necessary to weigh copper's corrosion performance along with its conductivity, strength, and fabricability (Ref 105).

Atmospheric Corrosion

Dry Atmospheres. The most uniform type of general corrosion in copper-base materials occurs in dry air where, in the simplest case, a thin film of cuprous oxide (Cu_2O) forms by direct reaction of copper with oxygen. The oxidation reaction can take place at room temperature, and it is accelerated at elevated temperatures. At moderate temperatures, the oxidation reaction can be written in the form:

$$2Cu + \tfrac{1}{2}O_2 \rightarrow Cu_2O$$

The initial reaction step involves adsorption of a monolayer of oxygen on the clean metallic surface. Electrons then pass from the outermost layer of metal atoms, which become cations to the oxygen atoms, which become anions. If the temperature is sufficiently high to render the ionic species appreciably mobile, a three-dimensional film begins to form. Film growth takes place through the diffusion of metal atoms outward through the film to the surface, where they react with oxygen. Diffusion is aided by the presence of lattice defects in the crystalline oxide film. Cuprous oxide is a p-type semiconductor, meaning that it contains mobile positively charged holes. That is, the Cu_2O lattice contains some vacant Cu^+ sites. Cu^{++} ions in other sites maintain electric neutrality (Ref 106).

If the oxide film is just thick enough to destroy the surface's metallic lustre, the phenomenon is a form of tarnishing. Color is a function of film thickness. Tarnishing is more pronounced in air containing traces of hydrogen sulfide (H_2S), because this leads to the formation of some Cu_2S. Film growth rate also increases when sulfides are present. The Cu_2O film is thermodynamically stable, as shown in the copper-oxygen equilibrium diagram (Fig. 2.9). The thermodynamic driving force is the negative free energy of Cu_2O formation, –35.35 kcal/mol (Ref 32, 107, 108).

From the kinetic standpoint, the oxide film nucleates at active centers such as lattice defects, and after spreading as a result of progressive oxygen adsorption, it thickens very slowly to a low stable value by means of electron transport. At tempera-

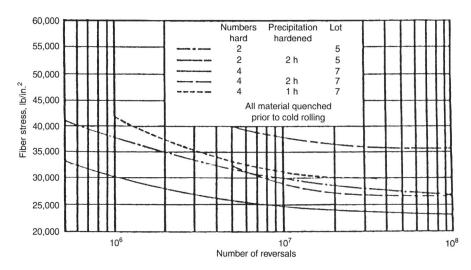

Fig. 2.78 *S-N* fatigue curve for copper alloy C82600. Source: Ref 93

Table 2.25   Mechanical properties of cast copper alloys

UNS number	Tensile strength, MPa	Yield strength(a), MPa	Compressive yield strength(b) , MPa	Elongation, %	Hardness HB(c)	Electrical conductivity, % IACS
ASTM B 22						
C86300	820	468	490	18	225(d)	8.0
C90500	317	152	...	30	75	10.9
C91100	241	172	125 min	2	135(d)	8.5
C91300	241	207	165 min	0.5	170(d)	7.0
ASTM B 61						
C92200	280	110	105	45	64	14.3
ASTM B 62						
C83600	240	105	100	32	62	15.0
ASTM B 66						
C93800	221	110	83	20	58	11.6
C94300	186	90	76	15	48	9.0
C94400	221	110	...	18	55	10.0
C94500	172	83	...	12	50	10.0
ASTM B 67						
C94100	138	97	...	15	44	...
ASTM B 148						
C95200	552	200	207	38	120(d)	12.2
C95300	517	186	138	25	140(d)	15.3
C95400	620	255	...	17	170(d)	13.0
C95400 (HT)(e)	758	317	...	15	195(d)	12.4
C95410	620	255	...	17	170(d)	13.0
C95410 (HT)(e)	793	400	...	12	225(d)	10.2
C95500	703	303	...	12	200(d)	8.8
C95500 (HT)(e)	848	545	...	5	248(d)	8.4
C95600	517	234	...	18	140(d)	8.5
C95700	655	310	...	26	180(d)	3.1
C95800	662	255	241	25	160(d)	7.0
ASTM B 176						
C85700	...	...	...	...	...	...
C85800	380	205(f)	...	15	...	22.0
C86500	...	...	...	...	...	...
C87800	620	205(f)	...	25	...	6.5
C87900	400	205(f)	...	15	...	...
C99700	415	180	...	15	120(d)	3.0
C99750	...	...	...	...	...	...
ASTM B 584						
C83450	255	103	69	34	62	20.0
C83600	241	103	97	32	62	15.1
C83800	241	110	83	28	60	15.3
C84400	234	97	...	28	55	16.8
C84800	262	103	90	37	59	16.4
C85200	262	90	62	40	46	18.6
C85400	234	83	62	37	53	19.6
C85700	352	124	...	43	76	21.8
C86200	662	331	352	20	180(d)	7.4
C86300	820	469	489	18	225(d)	8.0
C86400	448	166	159	20	108(d)	19.3
C86500	489	179	166	40	130(d)	20.5
C86700	586	290	...	20	155(d)	16.7
C87300	400	172	131	35	85	6.1
C87400	379	165	...	30	70	6.7
C87500	469	207	179	17	115	6.1
C87600	456	221	...	20	135(d)	8.0
C87610	400	172	131	35	85	6.1
C90300	310	138	90	30	70	12.4
C90500	317	152	103	30	75	10.9
C92200	283	110	103	45	64	14.3
C92300	290	138	69	32	70	12.3
C92600	303	138	83	30	72	10.0
C93200	262	117	...	30	67	12.4

(continued)

Table 2.25   (continued)

UNS number	Tensile strength, MPa	Yield strength(a), MPa	Compressive yield strength(b), MPa	Elongation, %	Hardness HB(c)	Electrical conductivity, % IACS
C93500	221	110	...	20	60	15.0
C93700	269	124	124	30	67	10.1
C93800	221	110	83	20	58	11.6
C94300	186	90	76	15	48	9.0
C94700	345	159	...	35	85	11.5
C94700 (HT)(g)	620	483	...	10	210(d)	14.8
C94800	310	159	...	35	80	12.0
C94900	262 min	97 min	...	15 min	...	...
C96800	862 min	689 min(f)	...	3 min	...	...
C97300	248	117	...	25	60	5.9
C97600	324	179	159	22	85	4.8
C97800	379	214	...	16	130(d)	4.5

Note: HT indicates alloy in heat-treated condition (a) At 0.5% extension under load. (b) At a permanent set of 0.025 mm (0.001 in.). (c) 500 kgf (110 lbf) load. (d) 3000 kgf (6600 lbf) load. (e) Heat treated at 900 °C (1650 °F), water quenched. (f) At 0.2% offset. (g) Solution anneal of 760 °C (1400 °F) for 4 h, water quenched, and then aged at 315 °C (600 °F) for 5 h and air cooled. Source: Ref 17

tures up to about 100 °C (212 °F), the oxide film grows logarithmically. Oxidation rate is slowed by the presence (doping) of divalent and trivalent alloying elements in the metal and overlying film. Aluminum, beryllium, magnesium, and silicon are seen as effective film stabilizers in that they contribute to the nonconductive nature of the film (Ref 108).

Film thickening rate increases at higher temperatures. For wrought copper alloys, the threshold value for high-temperature behavior is about 150 °C (300 °F) in air and 120 °C (250 °F) in pure oxygen at atmospheric pressure. At higher oxygen partial pressures, for example, up to 1.6 kPa, scaling rate increases rapidly. Above 20 kPa, the rate of increase in thickness is uniform and parabolic, which implies that the oxide film is self-protecting and that the rate of oxidation decreases with time.

Wet Atmospheres. The presence of moisture in the gaseous environment brings about a wholly new corrosion condition provided humidity is greater than the 70 to 80% required for condensation to occur (Ref 105).

In the center of a condensed water droplet, the previously formed Cu_2O oxide film reverts due to oxygen starvation, allowing Cu atoms to be oxidized to the Cu^+ state. The reaction is driven by oxygen reduction at the oxygen-rich periphery of the droplet. The process follows the familiar electrochemical reactions:

Anodic: $Cu^0 = Cu^+ + e$

Cathodic: $\frac{1}{2} O_2 + H_2O + 2e = 2OH^-$

The solubility of Cu^+ ions in the water droplet depends upon the local electrochemical condi-

tions, as illustrated in the Eh-pH equilibrium diagram for the Cu-H_2O system shown in Fig. 2.91. The influence of a species such as chlorine, which complexes with copper and renders in more soluble, is illustrated in Fig. 2.92 (Ref 104, 106).

Electrochemical equilibrium diagrams, also called Eh-pH diagrams or Pourbaix diagrams, are graphical representations of the domain of stability of metal ions, oxides, and other species in solution. The lines that delineate two domains express the value of the equilibrium potential between two species as a function of electrochemical potential, Eh, in mV, and pH.

Surfaces that retain moisture generally corrode more rapidly than surfaces exposed intermittently to rain. Rain (except acidic rain) has a tendency to remove dust particles that can provide sites for crevice corrosion. Exposure of metals in different months of the year can have a pronounced effect on the corrosion rate. Winter exposure is usually the most severe because of an increased concentration of combustion products, notably sulfides, in the air. The presence of SO_2 and other sulfur pollutants, as well as chlorides, creates an especially aggressive environment (Ref 109, 110).

Patination. The initial reaction in a condensed water droplet forms copper hydroxide:

$Cu^+ + OH^- \rightarrow CuOH$, which in neutral solutions precipitates as oxide by hydrolysis:

$$2CuOH + H_2O \rightarrow Cu_2O + 2H_2O$$

This is the physicochemical basis for patination. The familiar range of color seen in patina results from the presence of atmospheric gases, such as carbon dioxide, sulfur dioxide, ammonia, and acetic acid or aerosol suspensions of salts, such as the sodium and/or calcium chloride found in marine atmospheres and road deicing agents. For example, carbon dioxide dissolves in water to form carbonic acid (H_2CO_3), which partially dissociates to carbonate:

$$H_2O + CO_2 \leftrightarrow H_2CO_3$$

$$H^+ + HCO_3^- \leftrightarrow 2H^+ + CO_3^-$$

Sulfur dioxide (SO_2), a component of acidic rain, forms sulfurous acid (H_2SO_3):

$$H_2O + SO_2 \leftrightarrow H_2SO_3$$

$$H_2SO_3 \leftrightarrow H^+ + HSO_3^-$$

and

$$2HSO_3^- + 2H^+ + 2e \leftrightarrow S_2O_4 + 2H_2O$$

The above reactions modify the corrosion product film to produce the familiar colors seen on weathered copper articles. The possible reactions with copper and the corresponding patina colors include:

$$Cu^+ + Cl^- = CuCl$$

$$2CuO + CO_2 + H_2O = CuCO_3 \cdot Cu(OH)_2 \quad \text{green}$$

$$3CuO + CuCl_2 + 3H_2O = CuCl_2 \cdot 3Cu(OH)_2 \quad \text{green}$$

$$CuO + 3H_2SO_3^+ + H_2O = CuS + S_2O_4 + 3H_2O$$
$$\text{black}$$

$$Cu_2O + H^+SO_4 + H_2O = Cu(OH)SO_4 \quad \text{blue}$$

$$\tfrac{1}{2}O_2 + Cu_2O = CuO \quad \text{black}$$

Once formed, the patina film protects the surface from further corrosion while at the same time providing a decorative finish. One of the most common forms of patina has the chemical composition $CuSO_4 \cdot 3Cu(OH)_2$, which is, in fact, the copper mineral brochantite.

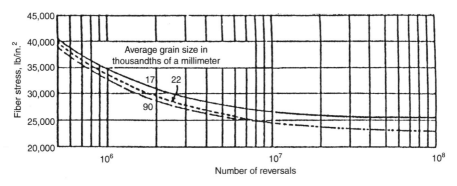

Fig. 2.79 S-N fatigue curve for copper alloy C77000. Source: Ref 93

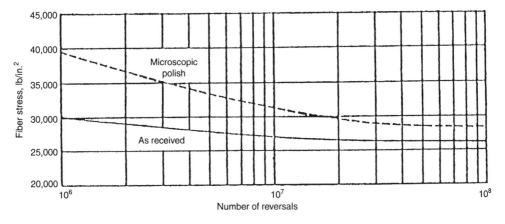

Fig. 2.80 S-N fatigue curve for copper alloy C52100. Source: Ref 93

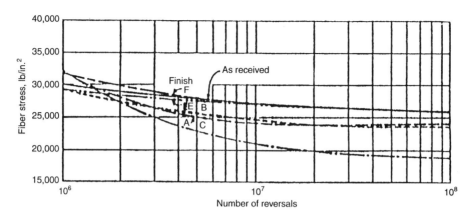

Fig. 2.81 S-N fatigue curve for copper alloy C52100 as function of surface roughness. Source: Ref 93

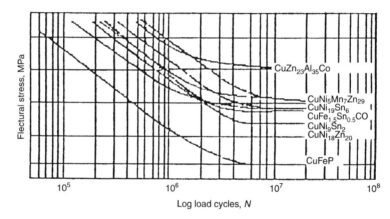

Fig. 2.82 S-N fatigue curves for copper alloys used in low-voltage electrical applications. Source: Ref 96

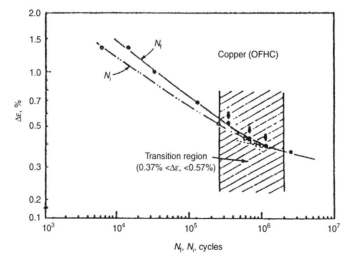

Fig. 2.83 Fatigue curves for polycrystalline OF copper based on strain amplitude. Source: Ref 91

The development of brochantite is a two-step process: a coating of Cu_2O forms first. The cuprous oxide then reacts with oxidized sulfur in the air to form brochantite. Natural patination can require 10 to 20 years of exposure; however, methods have been developed to accelerate the process by preoxidizing the surface (see Chapter 5, "Applications"). Natural patinas vary in color depending on their chemical make-up, and photolysis can affect the stability of colored patina compounds. In the extreme case of high sulfide contamination, the patina will turn black. In regions of high acid rain, the patina slowly dissolves leaving a blue-green coloring over any stone or concrete exposed to runoff water.

Chloride ion, as in marine environments, can also play a role in patina formation through the development of basic copper chloride, $CuCl_2 \cdot 3Cu(OH)_2$. The patina observed on the Statue of Liberty, for example, contains 95% $CuSO_4 \cdot 3Cu(OH)_2$ and 5% $CuCl_2 \cdot 3Cu(OH)_2$. The relative behavior of copper and some copper alloys in three test sites in the United States is shown in Table 2.28 (Ref 110–112).

Galvanic Corrosion

Galvanic corrosion is the accelerated attack that results when one metal makes electrical contact with another in the presence of an electrolyte. Depending on the conditions of exposure, this type of corrosion can take place over wide areas or it can remain localized in the region where the two metals touch.

Chemical elements can be arranged in an electrochemical series according to their electrochemical activity as measured against a standard electrode. Less noble, for example, more active or anodic metals corrode preferentially when placed in contact with more noble or cathodic metals. Thus, iron corrodes preferentially when placed in contact with copper. If the cathodic metal is already in solution in an electrolyte, the anodic metal will displace the cathodic metal in solution, while the cathodic metal simultaneously plates out on the surface of the anodic metal. Metallic copper can, thus, be recovered from solution by passing the solution over iron filings, a commercial process known as "cementation."

When metals or alloys are arranged in order of diminishing negative potential in a practical medium such as seawater, a "galvanic series" is obtained. Galvanic series are specific to the media in which they are determined. Unlike electrochemical series, which are based on thermodynamics, galvanic series often include alloys.

The magnitude of galvanic corrosion damage is a function of current density (amperes/unit area) on the corroding surface. When the surface area of the more noble metal is large in comparison to the more active member, the anodic current density will be unfavorably high. When the area of the more noble metal or alloy is small in comparison to the more active member, only slightly accelerated galvanic corrosion takes place due to the predominant polarization of the more noble material. When copper rivets (small cathodes) were used to

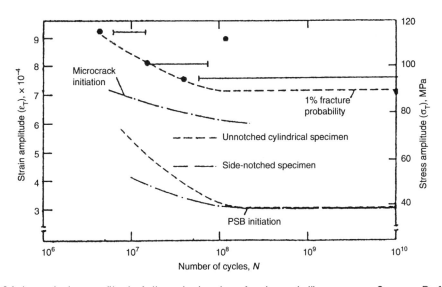

Fig. 2.84 Low-strain amplitude fatigue behavior of polycrystalline copper. Source: Ref 95

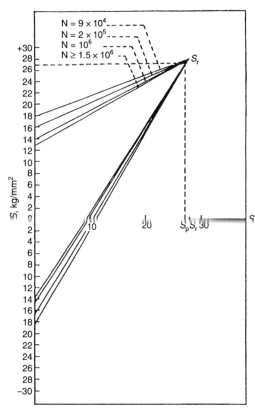

Fig. 2.85 Example of a Goodman diagram for polycrystalline copper. Source: Ref 95

fasten steel plates (large anodes) in seawater, the rivets lasted longer than 1.5 years. Steel rivets (small anodes) used to fasten copper plates (large cathodes) immersed in seawater were completely destroyed during a similar exposure time.

Galvanic corrosion proceeds more rapidly in seawater than in fresh water or the atmosphere because seawater is more electrically conductive. Because copper alloys are widely used in marine applications, particular attention must be paid to possible galvanic couples with less noble metals and alloys including less noble copper alloys. The area relationship is particularly important.

Dealloying

In simplest terms, dealloying is a corrosion process in which the more active metal in a binary alloy is selectively removed from the structure. Dealloying is most widely associated with copper-zinc and copper-aluminum alloys, but it also appears, though less frequently, in copper-nickel and copper-tin alloys. Brasses (copper-zinc) are susceptible to a dealloying process called dezincification, in which zinc is preferentially leached from the alloy, leaving behind a spongy mass of copper and copper oxide. A familiar example is the degradation of the heads of yellow brass screws used to retain faucet washers. All but the so-called inhibited copper-zinc alloys are susceptible to dezincification; however, the process is particularly visible and, therefore, largely identified with brasses containing more than 15% Zn. Dezincification susceptibility increases with zinc content; the beta

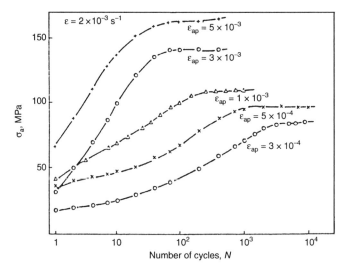

Fig. 2.86 Fatigue-hardening curves for polycrystalline copper under strain-controlled loading. Source: Ref 94

phase found in high-zinc brasses is especially vulnerable to attack.

For many years, there was controversy over whether dezincification took place by direct leaching of the more active component or whether both components dissolved and the more noble component (copper, in the case of brass) redeposited in situ. It is now generally accepted that both mechanisms can operate depending on pH and electrical potential.

Plug-type dealloying refers to attack that is limited to localized areas. Surrounding areas are usually unaffected or only slightly corroded. In layer-type dealloying, the active component of the alloys is removed from a broad area of the surface.

Dezincification of brass is promoted by prolonged contact with waters high in oxygen and carbon dioxide, particularly if the environment is stagnant, for example, under crud deposits. Slightly acidic water that is low in salt content is likely to produce uniform attack at room temperature, whereas neutral or alkaline water, high in salt content and above room temperature, more often produces plug-type attack.

Tin tends to inhibit dezincification, especially in cast alloys. UNS alloys C46400 and C67500, which are duplex ($\alpha + \beta$) brasses containing about 1% tin, are widely used for marine equipment because they exhibit reasonably good resistance to dezincification. Addition of a small amount of phosphorus, arsenic, or antimony to Admiralty Metal (an all-alpha 71Cu-28Zn-1Sn brass) and other copper-zinc alloys inhibit dezincification. Alloys compounded with arsenic, antimony, tin, or phosphorus are commonly known as inhibited brasses. Inhibitors are not effective in α–β brasses because they do not prevent dezincification of the β phase; however, proprietary compositions and heat treatments have been developed to reduce the severity of the problem in such materials by effectively encapsulating the more susceptible beta phase in a matrix of inhibited alpha phase. Where dezincification is a problem, red brass, commercial bronze, inhibited admiralty metal, and inhibited aluminum brass can usually be used safely.

Dealloying occurs in some copper-aluminum alloys, particularly when aluminum content exceeds 8%. It is especially severe in alloys with a continuous γ phase network, and it usually occurs in the plug form. Nickel additions greater than 3.5% or heat treatment to produce an $\alpha + \beta$ microstructure are effectively preventive. Dealloying of nickel in copper-nickels is generally considered negligible; however, in C71500 (30 to 70 copper-nickel), dealloying has been observed at temperatures

above 100 °C (212 °F), low flow conditions, and high local heat flux. Dealloying of tin in cast tin bronzes has been observed, albeit rarely, in hot brine or steam.

Ant-Nest Corrosion

Formates and other organic anions are also involved in a form of corrosion called ant-nest corrosion which is so named because of the resemblance of the corrosion pits to an ant nest. It has been observed in heat exchangers where it has been associated with contamination by breakdown products of trichloroethylene used for degreasing. It has also been observed on copper tubes under urea formaldehyde foam insulation from which formic acid may have been released. It may be related to "bronze disease" described by Evans (Ref 108, 113, 114).

Soil Corrosion

Reports on corrosion resistance of copper conductors and tubes in soils are usually very favorable. Data for exposure times of ten or more years

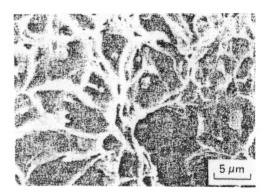

Fig. 2.87 SEM micrograph of ductile-dimple fracture in copper containing oxide inclusions. Source: Ref 98

for buried samples in the United States and in the United Kingdom indicate penetration depths of only a few microns per year. Generally unfavorable results were found for copper imbedded in cinders, acid marsh soil, acid clay loam, and acid humic soils. In such environments, mean depth of penetration rises to 40 µm/year with local penetration (pitting) values as high as 100 to 150 µm/year (0.004 to 0.006 in./year). Corrosion potential values vary between 150 and 300 mV (SHE). Various copper-silicon, copper-zinc, and copper-nickel alloys have also been tested in soil environments. In most soils, maximum penetration depths were not appre-ciable over the exposure period; however, brass samples tended to suffer dezincification (Ref 115, 116).

Corrosion in Potable Water Systems

Corrosion of copper plumbing tube can occur in potable water systems under certain conditions. Basically, there are two types of corrosion that can occur—pitting and generalized—although the details of both types are very complex and both types can occur simultaneously (Ref 115, 116).

Pitting is a form of localized corrosion in which attack is limited to areas ranging from a micro-

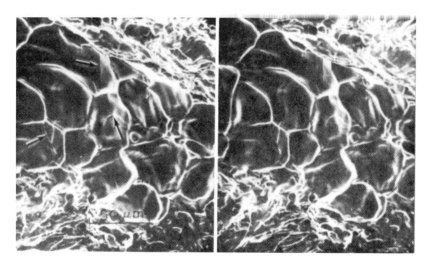

Fig. 2.88 SEM micrograph of semiductile fracture in high-purity copper

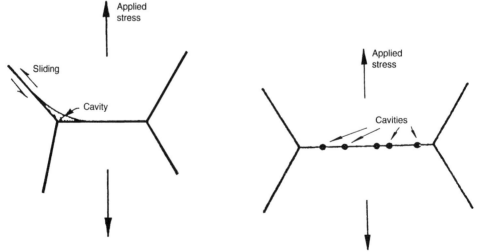

Fig. 2.89 W-type fracture in copper. Fracture initiates at a grain triple point. Source: Ref 98

Fig. 2.90 *R*-type cavitation in copper. Source: Ref 98

Table 2.26   Corrosion resistance of selected wrought copper alloys in various media

	Cu	Low-Zn brass	High-Zn brass	Special brass	Phosphor bronze	Al bronze	Si bronze	Cu-Ni	Ni silver
Acetate solvents	E	E	G	E	E	E	E	E	E
Acetic acid(a)	E	E	P	P	E	E	E	E	G
Acetone	E	E	E	E	E	E	E	E	E
Acetylene(b)	P	P	(b)	P	P	P	P	P	P
Alcohols(a)	E	E	E	E	E	E	E	E	E
Aldehydes	E	E	F	F	E	E	E	E	E
Alkylamines	G	G	G	G	G	G	G	G	G
Alumina	E	E	E	E	E	E	E	E	E
Aluminum chloride	G	G	P	P	G	G	G	G	G
Aluminum hydroxide	E	E	E	E	E	E	E	E	E
Aluminum sulfate and alum	G	G	P	G	G	G	G	E	G
Ammonia, dry	E	E	E	E	E	E	E	E	E
Ammonia, moist(c)	P	P	P	P	P	P	P	F	P
Ammonium chloride(c)	P	P	P	P	P	P	P	F	P
Ammonium hydroxide(c)	P	P	P	P	P	P	P	F	P
Ammonium nitrate(c)	P	P	P	P	P	P	P	F	P
Ammonium sulfate(c)	F	F	P	P	F	F	F	G	F
Aniline and aniline dyes	F	F	F	F	F	F	F	F	F
Asphalt	E	E	E	E	E	E	E	E	E
Atmosphere:									
Industrial(c)	E	E	E	E	E	E	E	E	E
Marine	E	E	E	E	E	E	E	E	E
Rural	E	E	E	E	E	E	E	E	E
Barium carbonate	E	E	E	E	E	E	E	E	E
Barium chloride	G	G	F	F	G	G	G	G	G
Barium hydroxide	E	E	G	E	E	E	E	E	E
Barium sulphate	E	E	E	E	E	E	E	E	E
Beer(a)	E	E	G	E	E	E	E	E	E
Beet-sugar syrup	E	E	G	E	E	E	E	E	E
Benzene, benzine, benzol	E	E	E	E	E	E	E	E	E
Benzoid acid	E	E	E	E	E	E	E	E	E
Black liquor, sulfate process	P	P	P	P	P	P	P	G	P
Bleaching powder (Wet)	G	G	P	G	G	G	G	G	G
Borax	E	E	E	E	E	E	E	E	E
Bordeaux mixture	E	E	G	E	E	E	E	E	E
Boric acid	E	E	G	E	E	E	E	E	E
Brines	G	G	P	G	G	G	G	E	E
Bromine, dry	E	E	E	E	E	E	E	E	E
Bromine, moist	G	G	P	F	G	G	G	G	G
Butane(d)	E	E	E	E	E	E	E	E	E
Calcium bisulfate	G	G	P	G	G	G	G	G	G
Calcium chloride	G	G	F	G	G	G	G	G	G
Calcium hydroxide	E	E	G	E	E	E	E	E	E
Calcium hypochlorite	G	G	P	G	G	G	G	G	G
Cane-sugar syrup(a)	E	E	E	E	E	E	E	E	E
Carbolic acid (phenol)	F	G	P	G	G	G	G	G	G
Carbonated beverages(a)(e)	E	E	E	E	E	E	E	E	E
Carbon dioxide, moist(a)(e)	E	E	E	E	E	E	E	E	E
Carbon tetrachloride, dry	E	E	E	E	E	E	E	E	E
Carbon tetrachloride, moist	G	G	F	G	E	E	E	E	E
Castor oil	E	E	E	E	E	E	E	E	E
Chlorine, dry(f)	E	E	E	E	E	E	E	E	E
Chlorine, moist	F	F	P	F	F	F	F	G	F
Chloroform, dry	E	E	E	E	E	E	E	E	E
Chromic acid	P	P	P	P	P	P	P	P	P
Citric acid(a)	E	E	F	E	E	E	E	E	E
Copper chloride	F	F	P	F	F	F	F	F	F
Copper nitrate	F	F	P	F	F	F	F	F	F
Copper sulfate	G	G	P	G	G	G	G	E	G
Corn oil(a)	E	E	G	E	E	E	E	E	E
Cottonseed oil(a)	E	E	G	E	E	E	E	E	E
Creosote	E	E	G	E	E	E	E	E	E
Dowtherm "A"	E	E	E	E	E	E	E	E	E
Ethanol amine	G	G	G	G	G	G	G	G	G

(continued)

Table 2.26   (continued)

	Cu	Low-Zn brass	High-Zn brass	Special brass	Phosphor bronze	Al bronze	Si bronze	Cu-Ni	Ni silver
Ethers	E	E	E	E	E	E	E	E	E
Ethyls acetate (esters)	E	E	G	E	E	E	E	E	E
Ethylene glycol	E	E	G	E	E	E	E	E	E
Ferric chloride	P	P	P	P	P	P	P	P	P
Ferric sulfate	P	P	P	P	P	P	P	P	P
Ferrous chloride	G	G	G	G	G	G	G	G	G
Ferrous sulfate	G	G	P	G	G	G	G	G	G
Formaldehyde (aldehydes)	E	E	G	E	E	E	E	E	E
Formic acid	G	G	P	F	G	G	G	G	G
Freon, dry	E	E	E	E	E	E	E	E	E
Freon, moist	E	E	E	E	E	E	E	E	E
Fuel or light	E	E	E	E	E	E	E	E	E
Fuel oil, heavy	E	E	G	E	E	E	E	E	E
Furfural	E	E	F	E	E	E	E	E	E
Gasoline	E	E	E	E	E	E	E	E	E
Gelatin(a)	E	E	E	E	E	E	E	E	E
Glucose(a)	E	E	E	E	E	E	E	E	E
Glue	E	E	G	E	E	E	E	E	E
Glycerin	E	E	G	E	E	E	E	E	E
Hydrobromic acid	F	F	P	F	F	F	F	F	F
Hydrocarbons	E	E	E	E	E	E	E	E	E
Hydrochloric acid (muriatic)	F	F	P	F	F	F	F	F	F
Hydrocyanic acid, dry	E	E	E	E	E	E	E	E	E
Hydrocyanic acid, moist	P	P	P	P	P	P	P	P	P
Hydrofluoric acid, anhydrous	G	G	P	G	G	G	G	G	G
Hydrofluoric acid, hydrated	F	F	P	F	F	F	F	F	F
Hydrofluosilicic acid	G	G	P	F	G	G	G	G	G
Hydrogen(d)	E	E	E	E	E	E	E	E	E
Hydrogen peroxide up to 10%	G	G	F	G	G	G	G	G	G
Hydrogen peroxide over 10%	P	P	P	P	P	P	P	P	P
Hydrogen sulfide, dry	E	E	E	E	E	E	E	E	E
Hydrogen sulfide, moist	P	P	F	F	P	P	P	F	F
Kerosine	E	E	E	E	E	E	E	E	E
Ketones	E	E	E	E	E	E	E	E	E
Lacquers	E	E	E	E	E	E	E	E	E
Lacquers thinners (solvents)	E	E	E	E	E	E	E	E	E
Lactic acid(a)	E	E	E	E	E	E	E	E	E
Lime	E	E	E	E	E	E	E	E	E
Lime sulfur	P	P	F	F	P	P	P	F	F
Linseed oil	G	G	G	G	G	G	G	G	G
Lithium compounds	G	G	P	F	G	G	G	E	E
Magnesium chloride	G	G	F	F	G	G	G	G	G
Magnesium hydroxide	E	E	G	E	E	E	E	E	E
Magnesium sulfate	E	E	G	E	E	E	E	E	E
Mercury or mercury salts	P	P	P	P	P	P	P	P	P
Milk(a)	E	E	G	E	E	E	E	E	E
Molasses	E	E	G	E	E	E	E	E	E
Natural gas(d)	E	E	E	E	E	E	E	E	E
Nickel chloride	F	F	P	F	F	F	F	F	F
Nickel sulfate	F	F	P	F	F	F	F	F	F
Nitric acid	P	P	P	P	P	P	P	P	P
Oleic acid	G	G	F	G	G	G	G	G	G
Oxalic acid(g)	E	E	P	P	E	E	E	E	E
Oxygen(h)	E	E	E	E	E	E	E	E	E
Palmitic acid	G	G	F	G	G	G	G	G	G
Paraffin	E	E	E	E	E	E	E	E	E
Phosphoric acid	G	G	P	F	G	G	G	G	G
Picric acid	P	P	P	P	P	P	P	P	P
Potassium carbonate	E	G	E	E	E	E	E	E	E
Potassium chloride	G	G	P	F	G	G	G	E	E
Potassium cyanide	P	P	P	P	P	P	P	P	P

(continued)

Table 2.26 (continued)

	Cu	Low-Zn brass	High-Zn brass	Special brass	Phosphor bronze	Al bronze	Si bronze	Cu-Ni	Ni silver
Potassium dichromate (acid)	P	P	P	P	P	P	P	P	P
Potassium hydroxide	G	G	F	G	G	G	G	E	E
Potassium sulfate	E	E	G	E	E	E	E	E	E
Propane(d)	E	E	E	E	E	E	E	E	E
Rosin	E	E	E	E	E	E	E	E	E
Seawater	G	G	F	E	G	E	G	E	E
Sewage	E	E	F	E	E	E	E	E	E
Silver salts	P	P	P	P	P	P	P	P	P
Soap solution	E	E	E	E	E	E	E	E	E
Sodium bicarbonate	E	E	G	E	E	E	E	E	E
Sodium bisulfate	G	G	F	G	G	G	G	G	G
Sodium carbonate	E	E	G	E	E	E	E	E	E
Sodium chloride	G	G	P	F	G	G	G	E	E
Sodium chromate	E	E	E	E	E	E	E	E	E
Sodium cyanide	P	P	P	P	P	P	P	P	P
Sodium dichromate (acid)	P	P	P	P	P	P	P	P	P
Sodium hydroxide	G	G	F	G	G	G	G	E	E
Sodium hypichlorite	G	G	P	G	G	G	G	G	G
Sodium nitrate	G	G	P	F	G	G	G	E	E
Sodium peroxide	F	F	P	F	F	F	F	G	G
Sodium phosphate	E	E	G	E	E	E	E	E	E
Sodium silicate	E	E	G	E	E	E	E	E	E
Sodium sulfide	P	P	F	F	P	P	P	F	F
Sodium thiosulfate	P	P	F	F	P	P	P	F	F
Steam	E	E	F	E	E	E	F	E	E
Stearic acid	E	E	F	E	E	E	E	E	E
Sugar solutions	E	E	G	E	E	E	E	E	E
Sulfur, solid	G	G	E	G	G	G	G	E	G
Sulfur, molten	P	P	P	P	P	P	P	P	P
Sulfur chloride (dry)	E	E	E	E	E	E	E	E	E
Sulfur dioxide (dry)	E	E	E	E	E	E	E	E	E
Sulfur dioxide (moist)	G	G	P	G	G	G	G	F	F
Sulfur trioxide (dry)	E	E	E	E	E	E	E	E	E
Sulfuric acid 80-95%(j)	G	G	P	F	G	G	G	G	G
Sulfuric acid 40-80%(j)	F	F	F	P	F	F	F	F	F
Sulfuric acid 40%(j)	G	G	P	F	G	G	G	G	G
Sulforous acid	G	G	P	G	G	G	G	F	F
Tannic acid	E	E	E	E	E	E	F	E	E
Tartaric acid(a)	E	E	G	E	E	E	E	E	E
Toluene	E	E	E	E	E	E	E	E	E
Trichloracetic acid	G	G	P	F	G	G	G	G	G
Trichlorethylene (dry)	E	E	E	E	E	E	E	E	E
Trichlorethylene (moist)	G	G	F	G	E	E	E	E	E
Turpentine	E	E	E	E	E	E	E	E	E
Varnish	E	E	E	E	E	E	E	E	E
Vinegar(a)	E	E	P	F	E	E	E	E	G
Water, acidic mine	F	F	P	F	G	F	F	P	F
Water, potable	E	E	G	E	E	E	E	E	E
Water, condensate(c)	E	E	E	E	E	E	E	E	E
Wetting agents(k)	E	E	E	E	E	E	E	E	E
Whiskey(a)	E	E	E	E	E	E	E	E	E
White water	G	G	G	E	E	E	E	E	E
Zinc chloride	G	G	P	G	G	G	G	G	G
Zinc sulfate	E	E	P	E	E	E	E	E	E

Note: The letters E, G, F, and P signify excellent, good, fair, and poor corrosion resistance, respectively. (a) Copper and copper alloys are resistant to corrosion by most food products. Traces of copper may be dissolved and affect taste or color of the products. In such cases, copper alloys are often tin coated. (b) Acetylene forms an explosive compound with copper when moisture or certain impurities are present and the gas is under pressure. Alloys containing less than 65% Cu are satisfactory; when the gas is not under pressure, other copper alloys are satisfactory. (c) Precautions should be taken to avoid SCC. (d) At elevated temperatures, hydrogen will react with tough pitch copper, causing failure by embrittlement. (e) Where air is present, corrosion rate may be increased. (f) Below 150 °C, corrosion rate is very low; above this temperature, corrosion is appreciable and increases rapidly with temperature. (g) Aeration and elevated temperature may increase corrosion rate substantially. (h) Excessive oxidation may begin above 120 °C. If moisture is present, oxidation may begin at lower temperatures. (j) Use of high-zinc brasses, should be avoided in acids because of the likelihood of rapid corrosion by dezincification. Copper, low-zinc brasses, phosphor bronzes, silicon bronzes, aluminum bronzes, and copper nickel offer good resistance to corrosion by hot and cold dilute H_2SO_4 and to corrosion by concentrated H_2SO_4. Intermediate concentrations of H_2SO_4 are sometimes more corrosive to copper alloys than either concentrated or dilute acid. Concentrated H_2SO_4 may be corrosive at elevated temperatures due to breakdown of acid and formation of metallic sulfides and sulfur dioxide, which cause localized pitting. Tests indicate that copper alloys may undergo pitting in 90 to 95% H_2SO_4 at about 50 °C in 80% acid at about 70 °C, and in 60 °C acid at about 100 °C. (k) Wetting agents may increase corrosion rates of copper and copper alloys slightly to substantially when carbon dioxide or oxygen is present by preventing formation of a film on the metal surface and by combining (in some instances) with the dissolved copper to produce a green, insoluble compound. Source: Ref 17

Table 2.27 Corrosion rating of cast copper alloys in various media

	Copper	Sn bronze	Leaded Sn bronze	High-leaded Sn bronze	Leaded red brass	Leaded semi-red brass	Leaded yellow brass	Leaded high-strength brass	High-strength yellow brass	Al bronze	Leaded Ni brass	Leaded Ni bronze	Si bronze	Si brass
Acetate solvents	B	A	A	A	A	A	B	A	A	A	A	A	A	B
Acetic acid														
20%	A	C	B	C	B	C	C	C	C	A	C	A	A	B
50%	A	C	B	C	B	C	C	C	C	A	C	B	A	B
glacial	A	A	A	C	A	C	C	C	C	A	B	B	A	A
Acetone	A	A	A	A	A	A	A	A	A	A	A	A	A	A
Acetylene(a)	C	C	C	C	C	C	C	C	C	C	C	C	C	C
Alcohols(b)	A	A	A	A	A	A	A	A	A	A	A	A	A	A
Aluminum chloride	C	C	C	C	C	C	C	C	C	B	C	C	C	C
Aluminum sulfate	B	B	B	B	B	C	C	C	C	A	C	C	A	A
Ammonia, moist gas	C	C	C	C	C	C	C	C	C	C	C	C	C	C
Ammonia, moisture-free	A	A	A	A	A	A	A	A	A	A	A	A	A	A
Ammonium chloride	C	C	C	C	C	C	C	C	C	C	C	C	C	C
Ammonium hydroxide	C	C	C	C	C	C	C	C	C	C	C	C	C	C
Ammonium nitrate	C	C	C	C	C	C	C	C	C	C	C	C	C	C
Ammonium sulfate	B	B	B	B	B	B	C	C	C	C	C	C	A	A
Aniline and aniline dyes	C	C	C	C	C	C	C	C	C	B	C	C	C	C
Asphalt	A	A	A	A	A	A	A	A	A	A	A	A	A	A
Barium chloride	A	A	A	A	A	C	C	C	C	A	A	A	A	C
Barium sulfide	C	C	C	C	C	C	C	C	B	C	C	C	C	C
Beer(b)	A	A	B	B	B	C	C	C	A	A	C	A	A	B
Beet-sugar syrup	A	A	B	B	B	A	A	A	B	A	A	A	B	B
Benzine	A	A	A	A	A	A	A	A	A	A	A	A	A	A
Benzol	A	A	A	A	A	A	A	A	A	A	A	A	A	A
Boric acid	A	A	A	A	A	A	A	B	A	A	A	A	A	A
Butane	A	A	A	A	A	A	A	A	A	A	A	A	A	A
Calcium bisulfite	A	A	B	B	B	C	C	C	C	A	B	A	A	B
Calcium chloride (acid)	B	B	B	B	B	B	C	C	C	A	C	C	A	C
Calcium chloride (alkaline)	C	C	C	C	C	C	C	C	C	A	C	A	C	B
Calcium hydroxide	C	C	C	C	C	C	C	C	C	B	C	C	C	C
Calcium hypoclorite	C	C	B	B	B	C	C	C	C	B	C	C	C	C
Cane-sugar syrups	A	A	B	A	B	A	A	A	A	A	A	A	A	B
Carbonated beverages(b)	A	C	C	C	C	C	C	C	C	A	C	C	A	C
Carbon dioxide, dry	A	A	A	A	A	A	A	A	A	A	A	A	A	A
Carbon dioxide, moist(b)	B	B	B	C	B	C	C	C	C	A	C	A	A	B
Carbon tetrachloride, dry	A	A	A	A	A	A	A	A	A	A	A	A	A	A
Carbon tetrachloride, moist	B	B	B	B	B	B	B	B	B	B	B	A	A	A
Chlorine, dry	A	A	A	A	A	A	A	A	A	A	A	A	A	A
Chlorine, moist	C	C	B	B	B	C	C	C	C	C	C	C	C	C
Chromic acid	C	C	C	C	C	C	C	C	C	C	C	C	C	C
Citric acid	A	A	A	A	A	A	A	A	A	A	A	A	A	A
Copper sulfate	B	A	A	A	A	C	C	C	C	B	B	B	A	A
Cottonseed oil(b)	A	A	A	A	A	A	A	A	A	A	A	A	A	A
Creosote	B	B	B	B	B	C	C	C	C	A	B	B	B	B
Ethers	A	A	A	A	A	A	A	A	A	A	A	A	A	A
Ethylene glycol	A	A	A	A	A	A	A	A	A	A	A	A	A	A
Ferric, chloride, sulfate	C	C	C	C	C	C	C	C	C	C	C	C	C	C
Ferrous chloride, sulfate	C	C	C	C	C	C	C	C	C	C	C	C	C	C
Formaldehyde	A	A	A	A	A	A	A	A	A	A	A	A	A	A
Formic acid	A	A	A	A	A	B	B	B	B	A	B	B	B	C
Freon	A	A	A	A	A	A	A	A	A	A	A	A	A	B
Fuel oil	A	A	A	A	A	A	A	A	A	A	A	A	A	A
Furfural	A	A	A	A	A	A	A	A	A	A	A	A	A	A
Gasoline	A	A	A	A	A	A	A	A	A	A	A	A	A	A
Gelatin(b)	A	A	A	A	A	A	A	A	A	A	A	A	A	A
Glucose	A	A	A	A	A	A	A	A	A	A	A	A	A	A
Glue	A	A	A	A	A	A	A	A	A	A	A	A	A	A
Glycerin	A	A	A	A	A	A	A	A	A	A	A	A	A	A
Hydrochloric or muriatic acid	C	C	C	C	C	C	C	C	C	B	C	C	C	C

(continued)

Table 2.27 (continued)

	Copper	Sn bronze	Leaded Sn bronze	High-leaded Sn bronze	Leaded red brass	Leaded semi-red brass	Leaded yellow brass	Leaded high-strength brass	High-strength yellow brass	Al bronze	Leaded Ni brass	Leaded Ni bronze	Si bronze	Si brass
Hydrofluoric acid	B	B	B	B	B	B	B	B	B	A	B	B	B	B
Hydrofluosilicic acid	B	B	B	B	B	C	C	C	C	B	C	C	B	C
Hydrogen	A	A	A	A	A	A	A	A	A	A	A	A	A	A
Hydrogen peroxide	C	C	C	C	C	C	C	C	C	C	C	C	C	C
Hydrogen sulfide, dry	C	C	C	C	C	C	C	C	C	C	C	C	B	C
Hydrogen sulfide, moist	C	C	C	C	C	C	C	C	C	B	C	C	C	C
Lacquers	A	A	A	A	A	A	A	A	A	A	A	A	A	A
Lacquers thinners	A	A	A	A	A	A	A	A	A	A	A	A	A	A
Lactic acid	A	A	A	A	A	C	C	C	C	A	C	C	A	C
Linseed oil	A	A	A	A	A	A	A	A	A	A	A	A	A	A
Liquors														
Black liquor	B	B	B	B	B	C	C	C	C	B	C	C	B	B
Green liquor	C	C	C	C	C	C	C	C	C	B	C	C	C	B
White liquor	C	C	C	C	C	C	C	C	C	A	C	C	C	B
Magnesium chloride	A	A	A	A	A	C	C	C	C	A	C	C	A	B
Magnesium hydroxide	B	B	B	B	B	B	B	B	B	A	B	B	B	B
Magnesium sulfate	A	A	A	A	B	C	C	C	C	A	C	B	A	B
Mercury, mercury salts	C	C	C	C	C	C	C	C	C	C	C	C	C	C
Milk(b)	A	A	A	A	A	A	A	A	A	A	A	A	A	A
Molasses(b)	A	A	A	A	A	A	A	A	A	A	A	A	A	A
Natural gas	A	A	A	A	A	A	A	A	A	A	A	A	A	A
Nickel chloride	A	A	A	A	A	C	C	C	C	B	C	C	A	C
Nickel sulfate	A	A	A	A	A	C	C	C	C	B	C	C	A	C
Nitric acid	C	C	C	C	C	C	C	C	C	C	C	C	C	C
Oleic acid	A	A	B	B	B	C	C	C	C	A	C	A	A	B
Oxalic acid	A	A	B	B	B	C	C	C	C	A	C	A	A	B
Phosphoric acid	A	A	A	A	A	C	C	C	C	A	C	A	A	A
Picric acid	C	C	C	C	C	C	C	C	C	C	C	C	C	C
Potassium chloride	A	A	A	A	A	C	C	C	C	A	C	C	A	C
Potassium cyanide	C	C	C	C	C	C	C	C	C	C	C	C	C	C
Potassium hydroxide	C	C	C	C	C	C	C	C	C	A	C	C	C	C
Potassium sulfate	A	A	A	A	A	C	C	C	C	A	C	C	A	C
Propane gas	A	A	A	A	A	A	A	A	A	A	A	A	A	A
Seawater	A	A	A	A	A	C	C	C	C	A	C	C	B	B
Soap solutions	A	A	A	A	B	C	C	C	C	A	C	C	A	C
Sodium bicarbonate	A	A	A	A	A	A	A	A	A	A	A	A	A	B
Sodium bisulfate	C	C	C	C	C	C	C	C	C	A	C	C	C	C
Sodium carbonate	C	A	A	A	A	C	C	C	C	A	C	C	C	A
Sodium chloride	A	A	A	A	A	B	C	C	C	A	C	C	A	C
Sodium cyanide	C	C	C	C	C	C	C	C	C	B	C	C	C	C
Sodium hydroxide	C	C	C	C	C	C	C	C	C	A	C	C	C	C
Sodium hypochlorite	C	C	C	C	C	C	C	C	C	C	C	C	C	C
Sodium nitrate	B	B	B	B	B	B	B	B	B	A	B	B	A	A
Sodium peroxide	B	B	B	B	B	B	B	B	B	B	B	B	B	B
Sodium phosphate	A	A	A	A	A	A	A	A	A	A	A	A	A	A
Sodium sulfate, silicate	A	A	B	B	B	B	C	C	C	A	C	C	A	B
Sodium sulfide, thiosulfate	C	C	C	C	C	C	C	C	C	B	C	C	C	C
Stearic acid	A	A	A	A	A	A	A	A	A	A	A	A	A	A
Sulfur, solid	C	C	C	C	C	C	C	C	C	A	C	C	C	C
Sulfur chloride	C	C	C	C	C	C	C	C	C	C	C	C	C	C
Sulfur dioxide, dry	A	A	A	A	A	A	A	A	A	A	A	A	A	A
Sulfur dioxide, moist	A	A	A	B	B	C	C	C	C	A	C	C	A	B
Sulfur trioxide, dry	A	A	A	A	A	A	A	A	A	A	A	A	A	A
Sulfuric acid														
78% or less	B	B	B	B	B	C	C	C	C	A	C	C	B	B
78% to 90%	C	C	C	C	C	C	C	C	C	B	C	C	C	C
90% to 95%	C	C	C	C	C	C	C	C	C	B	C	C	C	C
Fuming	C	C	C	C	C	C	C	C	C	A	C	C	C	C

(continued)

Table 2.27 (continued)

	Copper	Sn bronze	Leaded Sn bronze	High-leaded Sn bronze	Leaded red brass	Leaded semi-red brass	Leaded yellow brass	Leaded high-strength brass	High-strength yellow brass	Al bronze	Leaded Ni brass	Leaded Ni bronze	Si bronze	Si brass
Tannic acid	A	A	A	A	A	A	A	A	A	A	A	A	A	A
Tartaric acid	B	A	A	A	A	A	A	A	A	A	A	A	A	A
Toluene	B	B	A	A	A	B	B	B	B	B	B	B	B	A
Trichlorethylene, dry	A	A	A	A	A	A	A	A	A	A	A	A	A	A
Trichlorethylene, moist	A	A	A	A	A	A	A	A	A	A	A	A	A	A
Turpentine	A	A	A	A	A	A	A	A	A	A	A	A	A	A
Varnish	A	A	A	A	A	A	A	A	A	A	A	A	A	A
Vinegar	A	A	B	B	B	C	C	C	C	B	C	C	A	B
Water, acid mine	C	C	C	C	C	C	C	C	C	C	C	C	C	C
Water, condensate	A	A	A	A	A	A	A	A	A	A	A	A	A	A
Water, potable	A	A	A	A	A	A	B	B	B	A	A	A	A	A
Whiskey(b)	A	A	C	C	C	C	C	C	C	A	C	C	A	C
Zinc chloride	C	C	C	C	C	C	C	C	C	B	C	C	B	C
Zinc sulfate	A	A	A	A	A	C	C	C	C	B	C	A	A	C

Note: Letters A, B, and C signify recommended, acceptable, and not recommended conditions, respectively. (a) Acetylene forms an explosive compound with copper when moist or when certain impurities are present and the gas is under pressure. Alloys containing less than 65% Cu are satisfactory for this use. When gas is not under pressure, other copper alloys are satisfactory. (b) Copper and copper alloys resist corrosion by most food products. Traces of copper may be dissolved and affect taste or color. In such cases, copper metals are often tin-coated. Source: Ref 106

Table 2.28   Average atmospheric corrosion of copper for 10 and 20 year exposure times at three sites in the United States

New York, NY; urban-industrial		La Jolla, CA; marine		State College, PA; rural	
10 year	20 year	10 year	20 year	10 year	20 year
0.047	0.054	0.052	0.050	0.023	0.017

Note: Corrosion rates are given in mils/year (1 mil/year = 0.025 mm/year). Values cited are for one-half reduction of specimen thickness. Source: Ref 106

scopic point to a spot a few millimeters in diameter. Most conditions of leakage in copper plumbing systems are due to one form or another of pitting corrosion. Pitting attack can occur in tubing carrying cold water with an aggressive chemistry (typically pH 7.0 to 7.7 and dissolved carbon dioxide of at least 25 mg/L). The most cost-effective way of preventing this pitting is by altering the water chemistry by raising the pH and reducing the carbon dioxide content. An excellent summary of pitting corrosion in copper water tube is given by Edwards et al. (Ref 117–122).

Generalized is a form of corrosion in which the attack is more or less uniform. Rather than leakage, generalized corrosion is evidenced by the appearance of "blue water"—a blue-green appearance to the water, some of which settles out on standing as a blue-green precipitate. Another form of generalized corrosion is cuprosolvency, which is a misnomer because metallic copper cannot dissolve per se. The corrosion is due to a chemical attack on the surface that causes the formation of a water-soluble copper species, some of which further reacts in the water to yield a hydrated copper oxide precipitate. Bacteria have been found to be a contributor to cuprosolvency in a form of microbiologically influenced corrosion (MIC) (Ref 105, 107).

In copper plumbing tube, pitting has been classified into three characteristic types. These forms of attack have been found to correlate with surface condition (including the presence of residual soldering flux) and the composition and temperature of the water. A form of MIC pitting has also been observed in some warm water systems (Ref 121).

Type 1 Pitting. Type 1 pits are observed after exposure to moderately hard well water, particularly when the water contains high sulfate and low chloride ion concentrations. It is more likely to affect cold water pipes than hot and usually causes perforation within 1 to 10 years. Aggressive waters typically have a hardness greater than 150 mg/L $CaCO_3$ and a low organic content. Type 1 pits have circular rather than irregular cross sections. They are 2 to 5 mm (0.08 to 0.2 in.) in diameter, although they may be elongated in the direction of water flow. They are found to develop below friable green tubercles or corrosion products, which may consist of basic copper carbonate (malachite) mixed with basic copper sulfate. Green tubercles of basic copper carbonate associated with calcium carbonate have also been observed, and there may also be deposits of basic copper sulfate, cuprous chloride, cuprous oxide, and copper-stained hardness salts (Ref 102, 119, 121).

This type of pitting was once widely associated with galvanic corrosion stemming from residual carbon films arising from drawing lubricants that had been charred during annealing. However, scrupulous cleaning procedures now almost universally used in the plumbing tube industry have eliminated carbon films as a realistic source of attack. Pitting can, however, be caused by the presence of residual, aggressive soldering flux. Nonaggressive soldering fluxes, such as those meeting the requirements of ASTM standard B 813-91, should be chosen to avoid damage through this source (Ref 119–121).

Type 1 pitting can be prevented by raising the pH of the supply water from near neutral to approximately 8.1 to 8.3, for example, with sodium carbonate. This condition should correspond to a free carbon dioxide level approaching zero.

Type 2 Pitting. Type 2 pitting results from soft waters usually having a hardness less than 50 mg/L $CaCaO_3$, a pH below 7.4 and a ratio of bicarbonate to sulfate less than one. It rarely occurs if the water temperature is below 60 °C. Type 2 pits are deep and narrow, usually less than 2mm (0.08 in.) in diameter, filled with hard compact red crystalline cuprous oxide (Cu_2O) covered by a dark green to black layer of a cuprous oxide-basic copper sulfate mixture. The surface between the pits carries a matte black layer of water-formed oxide, often beneath a layer of silt deposited by the water (Ref 119, 121).

Traces of manganese as low as 0.03 mg/L in the water increases the tendency to form Type 2 pitting. A pH between 5 and 7 appears to be a necessary condition, although pitting has also been associated with waters that have been treated to raise the pH above 7.4 (usually to 8.5 to 9). Finally, the HCO_3^-/SO_4^{--} ratio seems to have some influence, because type 2 pitting has not been observed when

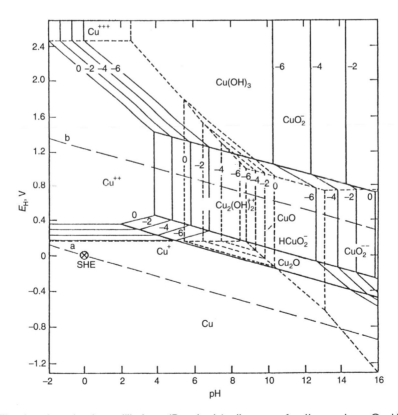

Fig. 2.91 Electrochemical equilibrium (Pourbaix) diagram for the system Cu-H$_2$. Source: Ref 104, 108

this ratio was below 1. Carbon films do not appear to have any bearing on type 2 pitting (Ref 102, 119, 121, 123).

To prevent pitting, it is desirable to have a high bicarbonate concentration. From the corrosion standpoint, water acidification (low pH) is damaging.

Type 3 Pitting. Type 3 pitting was first reported in Germany and other than that instance, it has only been observed in a few sites in Sweden. This type of pitting occurs in pipes carrying cold water with high pH, low hardness, and low mineral and organic contents. Type 3 pits are characterized by the formation of rather broad corroded areas comprising numerous hemispherical pits under a common covering of basic copper sulfate. An oxide membrane extends across all the pits in the group with a perforation above the center of each pit. Pits contain cuprous oxide with up to 1% sulfide. Irregularly shaped pits caused by chloride residues from poorly made solder joints have also been observed. These pits can develop below calcium carbonate-malachite tubercles. It has been sug-gested that carbon films may also play a role in the formation of type 3 pits and it is really a variant of type 1 corrosion in which the nature of the corrosion products reflects the composition of the water (Ref 102, 115, 120, 121, 124, 125).

Microbiologically Influenced Corrosion. In some plumbing systems, pitting has been traced to a microbiological origin. Such corrosion is known as microbiologically influenced corrosion (MIC), and it has also been observed in nickel, stainless steel, and titanium. MIC can be mistaken for any of the three common pitting types. It results from improper water treatment practices and/or building design and operating practices. The phenomenon has been identified in, among other places, Scotland, Saudi Arabia, Germany and the United States. MIC pitting has been reported in cases where the water is in the pH range of 7.4 to 9.3, contains high concentrations of humic substances, and has a low buffering capacity. It also has occurred in instances where the chlorine residual for sanitation purposes has been too low (Ref 102, 103, 105, 126).

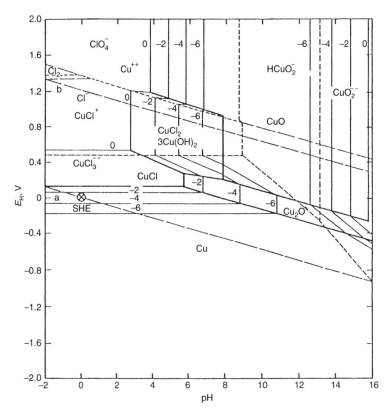

Fig. 2.92 Electrochemical equilibrium (Pourbaix) diagram for the system Cu-Cl-H$_2$O. Source: Ref 105, 106

Under some circumstances, MIC resembles type 3 pitting corrosion in that numerous pits occur beneath a common basic copper sulfate crust with the oxide membrane across the pits being randomly perforated. Under other circumstances, it resembles type 1 pitting in that the pits are hemispherical and contain soft crystalline cuprous oxide with varying amounts of cuprous chloride under a cuprous oxide membrane. It can also resemble type 2 pitting in that the oxide on the surface between the pits is largely cupric oxide. The mounds above the pits are principally basic copper sulfate, often with a deposit of powdery cupric oxide around the periphery and on the deposit itself.

MIC, which also has been referred to as type 1.5 pitting, has been found in hot, cold and warm water systems, but it invariably occurs where ambient temperatures are between 30 and 40 °C (86 and 104 °F) and where water flow rates are low, intermittent, or in some cases, stagnant. Waters typically have a total hardness between 25 and 40 mg/L $CaCO_3$, alkalinity between 10 and 20 mg/L $CaCO_3$, chloride between 15 and 20 mg/L, and sulfate between 10 and 30 mg/L. In most cases, the sulfate content was approximately twice the carbonate content. pH varies widely. Sulfur-reducing bacteria (SRB) thiobacillus thiooxidans and iron sulfate oxidizing bacteria ferrobacillus ferrooxidans are suspected causative agents, although water characteristics and temperature and plumbing system design and installation also seem to play a significant role in system susceptibility to MIC (Ref 105, 126, 127).

Avoidance or correction of MIC conditions must be based on avoidance of the settlement of bacteria and the formation of a biofilm or if encountered, the removal of the biofilm. Fisher et al. discuss the measures taken in large institutional buildings (Ref 128).

Cuprosolvency or "Blue Water" Corrosion. The phenomenon known as "blue water" has been observed in New Zealand, Australia, and elsewhere. Blue water involves the release of solid copper corrosion products into certain soft potable waters. Page (1973) carried out an extensive study of the influence of variables on blue water production finding that copper levels increased at a pH greater than 8.4, and at higher alkalinities and flow rates, but was independent of copper surface condition. The products invariably consist of copper hydroxides and silicates, and there is associated micropitting, indicating a possible continuum between blue water and pitting (Ref 129).

More recently Wells (1994) summarized the earlier work and updated the mechanism from the previous model of a permeable hydrous silica gel layer, which traps copper corrosion products and provides a site for the production of easily scoured flocculent material to a more general one involving the formation of "soft" barriers of either inorganic copper silicate colloidal layers, exo-polysaccharides produced by bacteria, or some combination of the two (Ref 130).

There is anecdotal evidence that bicarbonate dosing of water reduces the incidence of blue water, possibly by providing more effective buffering against the development of high bulk pH from the leaching of lime from mortar-lined mains, and low pH in anodic areas. However, the blue water phenomenon remains unsolved and continues to occur.

Stress-Corrosion Cracking

Stress-corrosion cracking (SCC) is defined as brittle failure under constant stresses in suitably aggressive environments. Properly selected copper alloys possess excellent resistance to SCC in many industrial and chemical environments; however, copper and many copper alloys are susceptible to SCC in a number of media. In some cases, the conditions for cracking are very limited and exist only within a narrow range of pH values and/or a narrow range of potentials. In many cases, the experimental data are limited to a single alloy, and it is not known if the environment is generally deleterious to any copper alloy or to a restricted group of alloys (Ref 108).

Acetate Solutions. Pure copper wire stressed beyond the yield strength was observed to crack in 0.05 N cupric acetate ($Cu(CH_3COO)_2$). Alloy C26000 is susceptible to cracking in the same solution, and the cracking rate under slow strain rate conditions is a function of both pH and applied potential.

Amines. Alloy C26000 is susceptible to cracking in solutions of methyl amine, ethyl amine, and butyl amine when dissolved copper is present in the solution. Susceptibility is a maximum at a potential approximately 50 mV anodic from the rest potential. Tubing fabricated from C68700 exhibited cracks from the steam side of a condenser system after 3048 h of service in a desalination plant.

Ammoniacal Media. All copper-base alloys can be made to crack in NH_3 water vapor, NH_3 solutions, ammonium ion, NH_4^+ solutions and environments in which NH_3 is a reaction product. The rate at which cracks develop is critically dependent on many variables, including stress level, specific alloy, oxygen concentration in the liquid, pH, NH_3 or NH_4^+ concentration, copper ion concentration, and potential.

Early work on the stress corrosion of brass in ammonia provided the following insights:

- SCC occurs in a great variety of brasses that differ widely in compositions, degree of purity, and microstructure.
- Cracking occurs only in objects that are subjected to external or internal tensile stresses.
- Visible corrosion is frequently associated with the effect, but the corrosion may only be superficial.
- Lacquer coatings do not offer complete protection against SCC.
- Sufficient and continuous coatings of a metal, such as nickel, confer complete protection.
- Highly stressed articles may be kept for years in a clean air atmosphere without developing cracks.
- Ammonia and ammonium salts induce cracking.
- Surface defects, which localize stresses, do not appear to contribute to the development of cracks in the absence of an essential corroding agent, such as NH_3.
- Severe corrosion and pitting do not of themselves lead to cracking.
- Cracks often follow an intergranular path, but transgranular cracking is also observed.
- Traces of NH_3 in the environment are an important agent in inducing SCC in atmospheric exposure.
- Ammonia has a specific and selective action on material in the grain boundaries of brass.
- Cracking always begins in surface layers that are under tension, irrespective of whether the stress is applied or residual.
- The behavior of a copper alloy subjected to the combined effect of tensile stress and NH_3 is an index of susceptibility to SCC.
- Susceptibility to SCC in brasses diminishes with increasing copper content.
- Protracted heating of 70Cu-30Zn brass at 100 °C (212 °F) does not develop cracks and does not reduce the internal stress appreciably.

Table 2.29 ranks various copper alloys according to their relative SCC susceptibility in NH_3 environments. Note that cartridge brass, C26000, is assigned the highest susceptibility among common copper alloys.

Chlorate Solutions. Slow strain rate tests, commonly applied to determine susceptibility to SCC, have shown that brass cracks intergranularly and transgranularly when immersed in 0.1 to 5 M sodium chlorate ($NaClO_3$) solutions at pH levels ranging from 3.5 to 9.5.

Chloride Solutions. Copper alloys are not ordinarily susceptible to SCC in chloride solutions; however, the service life of copper alloys under cyclic stress is shorter in chloride solutions than in air. Slow strain rate experiments have also shown that copper and copper alloys have lower fracture stresses in NaCl solutions in the range of anodic potentials that favor SCC (Ref 131).

Citrate Solutions. Alloy C72000 (copper with 15 to 18% nickel) is sensitive to intergranular cracking in citrate solutions containing dissolved copper at pH levels ranging from 7 to 11.

Formate Solutions. Brass is susceptible to SCC in sodium formate ($NaCHO_2$) solutions at pH values exceeding 11 over a considerable range of applied potentials.

Hydroxide Solutions. Brass exhibits increased crack growth rates under slow strain rate conditions when it is exposed to NaOH at a pH between 12 and 13. The rate of crack growth is a function of the applied potential.

Mercury and Mercury Salt Solutions. Brasses crack readily when exposed to metallic mercury or mercury salt solutions that deposit mercury on the surface of the alloy. This high sensitivity to mercury is the basis of an industry test for the detection of residual stresses. The test involves immersing in a solution of mercurous nitrate. This is not to imply that brasses are necessarily sensitive to SCC in such environments because cracking in mercury is the result of liquid metal embrittlement (LME), not SCC.

Nitrate Solutions. Transgranular cracking has been observed in C44300 specimens immersed in naturally aerated 1 N sodium nitrate ($NaNO_3$) at pH 8 and a potential of 0.15 V (SHE). Nickel-brass (Cu-23Zn-12Ni) wires used in telephone equipment were observed to undergo SCC within 2 years. Laboratory tests suggested that nitrate salts were the cause. Cracking did not occur in this alloy in the presence of $(NH_4)_2SO_4$ and ammonium chloride (NH_4Cl) salts.

Nitrite Solutions. Pure copper (99.9 and 99.996% Cu) exhibited transgranular cracking when subjected to a strain rate of 10^{-6}/s while immersed in 1 M sodium nitrite ($NaNO_2$) at a pH of 8.2. Cracking in 1 M $NaNO_2$ was also observed in cartridge brass, C26000, Admiralty brasses, and 90-10 copper-nickel, C70600.

Solders. Some copper alloys are susceptible to cracking in certain solders although the phenomenon would appear to be related to LME rather than SCC. The common lead-free 95Sn-5Sb solder does not produce this form of cracking.

Sulfur Dioxide. Brass is susceptible to SCC in moist air containing 0.05 to 0.5 vol% SO_2. Pre-exposure of the brass to a solution of benzotriazole inhibits the cracking. It is for this reason that brass articles are sometimes packaged in benzotriazole-impregnated paper.

Sulfate Solutions. SCC of C26000 was observed in a solution of 1 M sodium sulfate (Na_2SO_4) and 0.01 M H_2SO_4 when the alloy was polarized at a potential of 0.25 V (SHE) and subjected to a constant strain.

Sulfides. Bronze and other copper-base alloys are generally not acceptable for highly stressed parts in sour gas service because H_2S present in such environments can lead to cracking. Some nickel-copper alloys are considered satisfactory, however (Ref 108).

Tungstate Solutions. Mild transgranular cracking of C44300 was observed in 1 M sodium tungstate (Na_2WO_4) at pH 9.4 and a corrosion potential of 0.080 V (SHE) (Ref 108).

Water. Several cases of SCC in admiralty brass heat-exchanger tubing have been found. The environments in which such SCC was observed included stagnant water, stagnant water contaminated with NH_3 and water accidentally contaminated with nitrate. No cases of SCC in heat-exchanger service were observed in arsenical copper C14200, copper-iron-phosphorus alloy C19400, copper-nickels C70600 and C71500, and aluminum bronze (Ref 108).

Corrosion Fatigue

As its name implies, corrosion fatigue is a progressive mechanical failure mechanism resulting from the combined action of repeated or fluctuating stresses and a corrosive environment. As in non-corrosive environments, corrosion fatigue cracks generally propagate at right angles to the maximum tensile stress in the affected region; however, corrosion accelerates the rate of cracking. Unlike ordinary fatigue failures, which normally exhibit only one failure path, corrosion fatigue failures usually involve several parallel cracks. The appearance of the cracks can be used to differentiate between the two failure modes. Environmental effects can also usually (but not always) be identified by the presence of corrosion

damage or corrosion products on fracture surfaces or within growing cracks.

Copper and copper alloys resist corrosion fatigue in many applications involving repeated stress and environments corrosive to other metals. These applications include such products as springs, switches, diaphragms, bellows, aircraft and automotive gasoline and oil lines, tubes for condensers and heat exchangers, and Fourdrinier wire for the paper industry. Alloys favored for their resistance to corrosion fatigue include beryllium coppers, phosphor bronzes, aluminum bronzes, and copper-nickels (Ref 108).

Erosion-Corrosion and Impingement Attack

Various forms of impingement attack occur where gases, vapors, or liquids impinge on or flow turbulently along metal surfaces at high velocities. The phenomenon is often seen in condensers or heat exchangers and on propellers and impellers. Rapidly moving turbulent water can strip away the protective films from copper alloys, causing the metal to corrode at a more rapid rate. Erosion-corrosion is characterized by undercut grooves, waves, ruts, gullies, and rounded holes. It usually exhibits a directional pattern such that pits are elongated in the direction of flow and are undercut on the downstream side. When this form of corrosion occurs in a condenser tube, it is usually confined to a region near the inlet end of the tube where fluid flow is rapid and turbulent. Partial

Table 2.29 Relative susceptibility to stress-corrosion cracking of some copper alloys in ammonia

Alloy	Susceptibility index(a)
C26000	1000
C35300	1000
C76200	300
C23000	200
C77000	175
C66400	100
C68800	75
C63800	50
C75200	40
C51000	20
C11000	0
C15100	0
C19400	0
C65400	0
C70600	0
C71500	0
C72200	0

(a) 0, essentially immune in SCC under normal service conditions; 1000, highly susceptible to SCC as typified by C26000. Source: Ref 106

tube blockages due to stones or other debris also produce turbulent flow patterns, as do sharp bend radii. Erosion-corrosion is most often associated with waters containing low levels of sulfur compounds and with polluted, contaminated, or silty salt water or brackish water (Ref 108).

Erosion-corrosion can be improved significantly by alloying. Among copper-base materials, the 90-10 copper-nickel alloy, C70600, is perhaps the most widely used material for moderately severe condenser service, particularly in brine-cooled units. In addition to its high erosion-corrosion resistance, the alloy also inhibits biofouling. In small-bore (<25 mm, or 1 in., diameter) tube, the alloy is commonly limited to water velocities below 3.6 m/s (12 ft/s); however, safe limits may be more than twice as high in larger diameter tube. Higher-nickel alloys such as C71500 (30% Ni) are used for more severe service conditions, and an iron-chromium-modified 15%Ni alloy has been shown to be particularly resistive and cost-effective (Ref 132–134).

REFERENCES

1. M.W. Covington, B.R. Cooper, N.L. Church, and W.H. Dresher, Copper, *Encyclopedia of Applied Physics,* Vol 4, VCH Verlagesgesellschaft, GmbH, 1992, p 267–268
2. R.C. Weast, Ed., *Handbook of Chemistry and Physics,* 55th ed., CRC Press, 1974
3. D.W. Wakeman, *The Physical Chemistry of Copper,* Chapter 20, *Copper,* A. Butts, Ed., Reinhold Publishing, 1954
4. J.M. Ziman, *Electrons and Holes: The Theory of Transport,* Oxford University Press, London, England, 1960
5. N.J. Simon, E.S. Drexler, and R.P. Reed, *Properties of Copper and Copper Alloys at Cryogenic Temperatures,* NIST Monograph 177, National Institute of Standards and Technology, U.S. Department of Commerce, Washington, D.C., Feb 1992, p 20.1–20.15
6. Y. Wang, Ed., *Handbook of Radioactive Nuclides,* The Chemical Rubber Co., Cleveland, OH, 1969, p 25
7. R.E. Bolz and G.I. Tuve, Ed., *Handbook of Tables of Applied Engineering Science,* 2nd ed., CRC Press, 1974
8. *Gmelins Handbook of Inorganic Chemistry (Gmelins Handbuch der anorganishen Chemie),* 8th ed., Verlag Chemie, GmbH, 1955
9. *Properties and Selection of Metals,* Vol 1, *Metals Handbook,* 8th ed., American Society for Metals, 1961
10. J.H. Mendenhall, Ed., *Understanding Copper Alloys,* Olin Brass, East Alton, IL, 1977, p 87
11. *Kupfer,* Deutsches Kupfer-Institut E.V., Berlin, Germany, 1982
12. *Properties and Selection: Nonferrous Alloys and Special Purpose Materials,* Vol 2, *ASM Handbook,* ASM International, 1990
13. B. Landolt-Bornstein, III, *Technology, Classification, Value and Behavior of Metallic Materials (Technik, Teil, Stoffwerte, und Verhalten von metallischen Werkstoffen),* Springer-Verlag, 1964
14. Copper and Copper Alloys, *1991 Annual Book of ASTM Standards,* Section 2, Vol 02.01, American Society for Testing and Materials, 1991
15. *Copper Rod Alloys for Machined Products,* Copper Development Association Inc., 1993
16. *Copper Alloy Castings: Alloy Selection and Casting Design,* Copper Development Association Inc., 1994
17. *CopperSelect: Database of Coppers and Copper Alloys,* Copper Development Association Inc., 1994
18. *Alloy Data/2: Wrought Products Standards Handbook,* Copper Development Association Inc., 1994
19. *Alloy Data/7: Cast Products, Standards Handbook,* Copper Development Association Inc., 1994
20. N. Ashcroft and N. Mermin, *Solid State Physics,* Holt Rinehart and Winston, 1976
21. P.B. Coates and J.W. Andrews, A Precise Determination of the Freezing Point of Copper, *J. Phys. F, Met. Phys.,* Vol 8 (No. 2), 1978
22. *American Institute of Physics Handbook,* 3rd ed., McGraw-Hill, 1972
23. B.R. Cooper, H. Ehrenreich, and H.R. Phillipp, *Phys. Rev.,* Vol 138, 1965, p A494–A507
24. H. Ehrenreich and H.R. Phillipp, *Phys. Rev.,* Vol 128, 1962, 1622–1629
25. *Properties and Selection: Nonferrous Alloys and Special-Purpose Materials,* Vol 2, *ASM Handbook,* ASM International, 1990, p 1110–1114
26. A. Butts, Magnetic Properties of Copper and Copper Alloys, *Copper,* Reinhold Publishing Co., 1954
27. C.M. Hurd, A Magnetic Susceptibility Apparatus for Weakly Magnetic Materials and the Susceptibility of Pure Copper in the Range 6–300K, *Cryogenics,* Vol 6 (No. 10), 1966, p 264
28. "OFHC Copper: A Survey of Properties & Applications," brochure, AMAX Copper Inc., New York, NY, 1974, p 88
29. A.R. Kaufmann and C. Starr, Magnetization of Copper-Nickel Alloys, *Phys. Rev.,* Vol 59, 1941, p 690
30. J.W. Fiepke, Permanent Magnet Materials, *Properties and Selection: Nonferrous Alloys and Special-Purpose Materials,* Vol 2, *ASM Handbook,* ASM International, 1990, p 785
31. INCRA, Monograph 13, The Metallurgy of Copper, *The Thermophysical Properties of Liquid Copper and Copper Alloys,* International Copper Research Association Inc., 1989
32. Y.A. Chang and K.-C. Hsieh, *Phase Diagrams of Ternary Copper-Oxygen-Metal Systems,* ASM International, 1989
33. W. Hume-Rothery and K.W. Andrews, Lattice Spacing and Thermal Expansion of Copper, *J. Inst. Met.,* Vol 68, 1942, p 19

34. D. Hull, *Introduction to Dislocations*, 1st ed., Pergamon Press, 1965
35. W. Hume-Rothery, G.W. Mabbot, and K.M. Channel-Evans, *Phil. Trans. Roy. Soc.*, Vol 233, 1934, p 44
36. W. Hume-Rothery, Atomic Theory for Students of Metallurgy, *Inst. of Metals Monograph and Report Series*, No. 3, The Institute of Metals, London, England, 1955
37. W. Hume-Rothery and G. Raynor, *The Structure of Metals and Alloys*, 3rd ed., The Institute of Metals, London, England, 1954
38. B.E. Warren, *X-ray Diffraction*, Addison-Wesley Publishers, 1969
39. B.D. Cullity, *Elements of X-ray Diffraction*, 2nd ed., Addison-Wesley Publishers, 1978
40. W. Hume-Rothery, *The Structure of Metals and Alloys*, 4th ed., The Institute of Metals, London, England, 1962
41. W.L. Johnson, Metallic Glasses, *Properties and Selection: Nonferrous Alloys and Special-Purpose Materials*, Vol 2, *ASM Handbook*, ASM International, 1990, p 804
42. A. Kostam and H. Char, Morphology and Microstructure of Al-Li-Cu Quasicrystals, *J. Mater. Sci.*, Vol 24, 1989, p 1999–2005
43. D. Morris and M. Morris, Microcrystalline or Nanocrystalline Grain Size in Two-Phase Alloys After Mechanical Alloying, *Mater. Sci. Eng.*, Vol A134, 1991, p 1418–1421
44. D. Morris and M. Morris, Microstructure and Strength of Nanocrystalline Copper Alloy Prepared by Mechanical Alloying, *Acta Met.*, Vol 39 (No. 8), 1991, p 1763–1770
45. G.W. Nieman, J.R. Weertman, and R.W. Siegel, Mechanical Behavior of Nanocrystalline Cu, Pd, and Ag Samples, *Microcomposites and Nanophase Materials Symp. Proc.*, The Minerals, Metals and Materials Society, Feb 1991
46. A. Butts, Ed., *Copper*, Reinhold Publishing, 1954
47. R. Hultgren and P. Dasai, Selected Thermodynamic Values and Phase Diagrams for Copper and Some of its Binary Alloys, *INCRA Monograph 1*, International Copper Research Association Inc., 1971
48. Y.A. Chang, J.P. Neumann, A. Mikula, and D. Goldberg, Phase Diagrams and Thermodynamic Properties of Ternary Copper-Metal Systems, *INCRA Monograph 6*, International Copper Research Association Inc., 1979
49. R.M. Brick, Phase Diagrams of Copper Alloy Systems, *Properties and Selection: Nonferrous Alloys and Special-Purpose Materials*, Vol 2, *ASM Handbook*, ASM International, 1990, p 456
50. *Casting Copper-Base Alloys*, American Foundrymen's Society Inc., Des Plaines, IL, 1984
51. J.S. Smart, Jr., The Effect of Impurities in Copper, *Copper*, A Butts, Ed., Reinhold Publishing Co., 1954
52. J.S. Smart, Jr., A.A. Smith, Jr., and A.J. Phillips, *Trans. AIME*, Vol 143, 1941, p 272
53. J.S. Smart, Jr. and A.A. Smith, Jr., *Trans. AIME*, Vol 147, 1942, p 48
54. J.S. Smart, Jr. and A.A. Smith, Jr., *Trans. AIME*, Vol 152, 1943, p 103
55. J.S. Smart, Jr. and A.A. Smith, Jr., *Trans. AIME*, Vol 166, 1946, p 44
56. F. Pawlek and K. Reichel, The Effect of Impurities on the Electrical Conductivity of Copper, *Z. Metallkd.*, Vol 47, 1956, p 347
57. P. Gregory, A.G. Bangay, and T.L. Bird, The Electrical Conductivity of Copper, *Metallurgia*, Vol 71, 1965, p 207
58. L.S. Darken and R.W. Gurry, *Physical Chemistry of Metals*, McGraw-Hill, 1953
59. A.G. Guy, *Essentials of Materials Science*, McGraw-Hill, 1976
60. A. Butts and P.L. Reiber, Jr., Magnetism in Copper Alloys: the Effect of Iron as Impurity, *Proc. ASTM*, Vol 49, American Society for Testing and Materials, 1949, p 857–886
61. F. Bitter and A.R. Kaufmann, Magnetic Studies of Solid Solutions, I: Methods of Observation and Preliminary Results on Precipitation of Iron from Copper, *Phys. Rev.*, Vol 56, 1939, p 1044–1051
62. M. Ohba, H. Abe, and K. Ishida, Contoured Strip Produced by Rolling with a Large Width Increment (Multigage Rolling Process), *J. Jpn. Copper Brass Res. Assn.*, Vol 29, 1990, p 95–100
63. B.F. Decker and D. Harker, Activation Energy for Recrystallization in Rolled Copper, *Trans. AIME*, Vol 188, 1950, p 887
64. D. Bowen, R.R. Eggleston, and R.H. Kropschot, A Study of Annealing Kinetics in Cold Worked Copper, *J. Appl. Phys.*, Vol 23, 1952, p 630
65. F. Haesner and K. Schönborn, *Z. Metallkd.*, Vol 76, 1985, p 3
66. I. Bäckerud, L. Liljenvall, and H. Steen, Solidification Characteristics of Some Copper Alloys, *INCRA Monograph 9*, The Metallurgy of Copper, International Copper Research Association Inc., 1982
67. W. Kurz and D.J. Fisher, *Fundamentals of Solidification*, Trans Tech House, Aedermannsdorf, Switzerland, 1984
68. B. Chalmers, *Principles of Solidification*, John Wiley & Sons, 1967
69. R.W. Cahn and P. Haasen, *Physical Metallurgy*, Elsevier Science, 1983
70. K.A. Warren and R.P. Reed, *Tensile and Impact Properties of Selected Materials from 20 to 300 K*, Monograph 63, National Bureau of Standards, Washington, D.C., 1963
71. D.A. Porter and K.E. Easterling, *Phase Transformations in Metals and Alloys*, Van Nostrand Reinhold Co. Ltd., Berkshire, England, 1981
72. P.G. Shewmon, *Transformations in Metals*, McGraw-Hill, 1969
73. E. Gaffet, C. Louison, M. Harmelin, and F. Faudot, Metastable Phase Transformation Induced by Ball-Milling in the Cu-W System, *Mater. Sci. Eng.*, Vol A134, 1991, p 1380–1384
74. J.W. Cahn, Spinodal Decomposition, *Met. Soc. (Trans. AIME)*, Vol 242, 1968, p 166–182

75. R. Smoluchowski, *Phase Transformations in Solids*, John Wiley & Sons, 1951

76. The Mechanism of Phase Transformations in Metals, *The Institute of Metals Monograph and Report Series, (No. 18) Symp. Proc.*, The Institute of Metals, London, England, 1956, p 346

77. A.J. Ardell, Precipitation Hardening, *Metall. Trans.*, Vol 16A, 1985, p 2131–2165

78. A. Varshavsky, Fatigue-Induced Dissolution of Shearable Particles during Subcritical Crack Growth, *Eng. Fract. Mech.*, Vol 46 (No. 1), 1993, p 151–156

79. F.J.A. den Broeder and S. Nakahara, Diffusion Induced Grain Boundary Migration and Recrystallization in the Cu-Ni System, *Scr. Metall.*, Vol 17, 1983, p 399–404

80. W. Quiming and K. Mokuang, Thermodynamics of the Iron Martensitic Transformation and the M_s Temperature of Iron, *Metall. Trans.*, Vol 22A, 1991, p 1761–1765

81. G. Joseph, J. Anexy, and A. Alarcón, Continuous Cooling Diagram of a Copper-11.3 wt% Aluminum Alloy (Diagrama de Enfriamiento Continuo para una Aleación de Cobre con 11,3% en Peso de Aluminio), *Revista del IDIEM* (Instituto de Investigaciones y Ensayes de Materiales, U. de Chile), Vol 2 (No. 2), Aug 1963

82. P. Tautzenberger, *Shape-Memory-Stellelemente*, Grau GmbH & Co., Ehningen, Germany, 1988

83. L. McDonald Schetky, Shape-Memory Alloys, *Sci. Am.*, Vol 241 (No. 5), 1979, p 74–82

84. J.W. Kim, D.W. Roh, E.S. Lee, and Y.G. Kim, Effects on Microstructure and Tensile Properties of a Zirconium Addition to Cu-Al-Ni Shape Memory, *Metall. Trans.*, Vol 21A, 1990, p 271–274

85. C.M. Wayman, Some Applications of Shape Memory Alloys, *J. Met.*, Vol 32 (No.6), 1980, p 129–137

86. *Materials Selection and Design*, Vol 20, *ASM Handbook*, ASM International, 1987, p 399–414

87. G.L. Kehl, *The Principles of Metallographic Laboratory Practice*, McGraw-Hill, 1949

88. G.F. VanderVoort, *Metallography, Principles and Practice*, McGraw-Hill, 1984

89. J.H. Richardson, *Optical Microscopy for the Materials Sciences*, Marcel Dekker, 1971

90. J.L. McCall and W.M. Mueller, Ed., *Metallographic Specimen Preparation*, Plenum Press, 1974

91. R.E. Rickseeke and T.F. Bower, Copper and Copper Alloys, *Metallography and Microstructures*, Vol 9, *ASM Handbook*, ASM International, 1990, p 399–414

92. *ASM MatDB* (software), ASM International, 1992

93. *CopperSelect: Database*, Copper Development Association Inc., 1991

94. L.C. Lim, Y.K. Tay, and H.S. Fong, Fatigue Damage and Crack Nucleation Mechanisms at Intermediate Strain Amplitudes, *Acta Metall. Mater.*, Vol 38 (No. 4), 1990, p 595–601

95. R. Wang, H. Mughrabi, S. McGovern, and M. Rapp, Fatigue of Copper Single Crystals in Vacuum and in Air, I: Persistent Slip Bands and Dislocation Microstructures, *Mater. Sci. Eng.*, Vol 65, 1984, p 219–233

96. C.H. Greenall and G.R. Gohn, Fatigue Properties of Nonferrous Sheet Metals, *ASTM Proc.*, Vol 37, 1937, p 160–191

97. W. Hessler, H. Müllner, B. Weiss, and R. Stickler, Near Threshold Behavior of Polycrystalline Copper, *Met. Sci.*, Vol 15 (No. 5), 1981, p 225–230

98. J. Rabenschlag and W. Dürrschnabel, Comparison of New Copper Alloys for Low Current Technologies with Respect to Their Behavior during Long-Term Static and Dynamic Loading (Vergleich Neuer Kupferlegierungen für die Schwachstromtechnik Bezüglich Ihres Verhaltens bei Statischer und Dynamischer Dauerbeanspruchung), *Metall., Sonderdruck aus Heft 11*, 1978, (No. 32), p 1123–1125

99. J. Polák, Fatigue Crack Nucleation in Metallic Materials, *Acta Technica*, Vol 35 (No. 2), 1990, p 156–174

100. E.A. Starke, Jr. and G. Lütjering, Cyclic Plastic Deformation and Microstrucure, *Fatigue and Microstructure*, ASM International, 1979

101. A. Meyers and K. Chawla, *Mechanical Metallurgy: Principles and Applications*, Prentice-Hall, 1982

102. P. Angell, H.S. Campbell, and A.H.L. Chamberlain, *Microbial Involvement in Corrosion of Copper in Fresh Water*, ICA Project No. 405, Interim report, International Copper Association Ltd., Aug 1990

103. J.T. Walker, C.W. Keevil, P.J. Dennis, J. McEvoy, and J.S. Colbourne, *The Influence of Water Chemistry and Environmental Conditions on the Microbial Colonization of Copper Tubing Leading to Pitting Corrosion, Especially in Institutional Buildings*, ICA Project No. 407, Final report, International Copper Association Ltd., 1990

104. W. Fischer, H.H. Paradies, D. Wagner, and I. HanBel, Copper Deterioration in a Water Distribution System of a County Hospital in Germany Caused by Microbially Induced Corrosion, Part 1: Description of the Problem, *Werkst. Korros.*, Vol 43, 1982, p 56–62

105. U.R. Evans, *The Corrosion and Oxidation of Metal: Scientific Principles and Practical Application*, Edward Arnold (Publishers) Ltd., London, England, 1960

106. E. King, A. Mah, and L. Pankratz, Thermodynamic Properties of Copper and its Inorganic Compounds, *INCRA Monograph 2, The Metallurgy of Copper*, International Copper Research Association Inc., 1973

107. N.W. Polan, Corrosion of Copper and Copper Alloys, *Corrosion*, Vol 13, *ASM Handbook*, ASM International, 1987, p 610–640

108. P. Duby, The Thermodynamic Properties of Aqueous Inorganic Copper Systems, *INCRA Monograph 4, The Metallurgy of Copper*, International Copper Research Association Inc., 1977

109. M. Pourbaix, *Thermodynamics of Dilute Aqueous Solutions (Thermodynamics des Solutions Aqueses Diluées)*, Centre Belge D'Etude de la Corrosion "CEBELCOR", Brussells, Belgium, 1950

110. W.H.J. Vernon and L. Whitby, The Open Air Corrosion of Copper—A Chemical Study of Surface Patina, *J. Inst. Met.*, Vol 42, 1929, p 389

111. D.H. Osborn, The Truth About Miss Liberty, *Mater. Sci. Eng.*, Vol 57, 1979, p 80–82

112. P.M. Aziz and H.P. Goddard, Mechanism by Which Non-Ferrous Metals Corrode, *Corrosion*, Vol 15 (No. 10), 1959, p 429t–533t

113. D.M.F. Nicholas, "Dezincification of Brass in Potable Water, Urban Water Research of Australia," Report No. 84, Melbourne Water, Aug 1994

114. T. Notoya, Localized "Ant-Nest" Corrosion of Copper Tubing and Preventive Measures, *Mater. Perform.*, Vol 32 (No. 5), 1993, p 53–57

115. O. Von Franqué, *Corrosion Behavior of Copper and Copper Alloys in Soil (Korrosionsverhalten von Kupfer und Kupferlegierungen im Erdboden)*, DKI Special Publication, S. 174, Deutsches Kupfer Institut, Berlin

116. H.H. Paradies, I. HanBel, W. Fischer, and D. Wagner, *Microbiologically Induced Corrosion on Copper Pipes*, INCRA Project No. 404, Final report, International Copper Research Association Inc., 1990

117. M. Romanoff, Underground Corrosion, Circular No. 579, U.S. National Bureau of Standards, Washington, D.C., 1 April 1957

118. V.F. Lucey, Pitting Corrosion of Copper—a Review, *Proc. of the Int. Symp. on Corrosion of Copper and Copper Alloys in Building*, Japan Copper Development Association, Tokyo, 1982, p 1

119. H.S. Campbell, Pitting Corrosion in Copper Water Pipes Caused by Films of Carbonaceous Material, *J. Inst. Met.*, Vol 77, 1950, p 345

120. M. Edwards, J.F. Ferguson, and S.H. Reiber, The Pitting Corrosion of Copper, *J. AWWA*, Vol 86 (No. 7), 1994, p 74–90

121. J.R. Myers and A. Cohen, Pitting Corrosion of Copper in Cold Potable Water Systems, *Mater. Process.*, Vol 34 (No.10), 1995, p 60–62

122. E. Otero, Corrosión Interior de Tuberias en Sistemas Domésticos de Distribución de Agua Potable in J. A. González, *Teoría y Práctica de la Lucha Contra la Corrosión*, Consejo Superior de Investigaciones Científicas, Centro Nacional de Investigaciones Metalúrgicas, Madrid, Espana, 1984, p 129–152

123. V.F. Lucey, Mechanism of Pitting Corrosion of Copper in Supply Waters, *Br. Corros. J.*, Vol 2, 1967, p 175

124. E. Mattsson, Localized Corrosion, *Proc. 6th European Congress on Metallic Corrosion*, London, England, Sept 1977, p 233

125. G. Joseph, R. Perret, M. Arcey, and P. Avila, *Corrosión Localizada en Tubos de Cobre en Villa México*, IDIEM-Universidad de Chile, Technical report, Santiago, Chile, Oct 1973

126. M. Linder, *Influence of Water Composition on Corrosion of Materials in Tap Water*, KI Report, Swedish Corrosion Institute, Stockholm, 1984

127. D.C. Silverman and R.B. Puyear, Effects of Environmental Variables on Aqueous Corrosion, *Corrosion*, Vol 13, *ASM Handbook*, ASM International, 1987, p 37–44

128. W.R. Fisher, D. Wagner, and H. Siedlarek, Microbiologically Influenced Corrosion in Potable Water Installations—An Engineering Approch to Developing Countermeasures, *Mater. Process.*, Vol 34 (No. 10), 1995, p 50–54

129. G.G. Page, Contamination of Drinking Water by Corrosion of Copper Pipes, *N. Z. J. Sci.*, Vol 16, 1974, p 349

130. D.B. Wells, B.J. Webster, P.T. Wilson, and P.J. Bremer, *Proc. Australian Corrosion Association Conf.*, Paper 51, Adelaide, Australia, 1994

131. G. Joseph, *Stress Corrosion Cracking and Corrosion Fatigue Susceptibility if Alpha Aluminum Bronze in Sodium Chloride Solutions*, Universidad de Chile, Santiago, Chile, 1990

132. T.P. May and B.A. Weldon, Copper-Nickel Alloys for Service in Sea Water, *Proc. Int. Congress on Marine Corrosion and Fouling* (Cannes, France), June 1964, p 141

133. K.D. Efird, Effect of Fluid Dynamics on the Corrosion of Copper Base Alloys in Sea Water, *Corrosion*, Vol 33 (No. 1), 1977, p 3

134. W.W. Kirk, "Evaluation of Critical Seawater Hydrodynamic Effects of Erosion-Corrosion of Copper-Nickel Alloys," Final report, International Copper Research Association, New York, NY, June 1987

3

Products

Introduction

Copper and copper-alloy products are commercially classified as refinery shapes, semimanufactured products (commonly called "semis"), and finished products. Refinery shapes are the products of primary copper producers. Along with recycled and re-refined scrap, they are the starting materials for the production of semis. Semis, which are manufactured by wire mills, brass mills, and foundries, are in turn intermediate materials for the manufacture of finished products. This chapter deals only with refinery shapes and semis. Table 3.1 presents an overview of copper products.

Table 3.2 lists the major categories of copper-base materials grouped into compositionally similar categories. The number of copper-alloy families varies among different classification systems, and there are slightly different groupings for wrought and cast products. Wrought alloys classified according to the Unified Numbering System (explained below) include:

- *Coppers:* Unalloyed metals containing at least 99.3% copper
- *High-copper alloys:* Alloys containing between 99.3 and 96% copper
- *Brasses:* Copper alloys containing zinc as the principal addition, including red brass (high in copper), yellow brass (approximately 60 to 75% copper), leaded brass, and tin brass
- *Bronzes:* Copper alloys in which the major alloying element is not zinc or nickel

- *Copper-nickels:* The commercially most important varieties contain 10% and 30% nickel, respectively, along with minor amounts of iron, chromium, and/or other elements
- *Nickel-silvers:* Copper alloys containing zinc and nickel, which together make up about 35 to 40 wt% of the alloy. Leaded (free-cutting) versions of these alloys are also produced
- *Special alloys:* A class of materials with compositions not matching the foregoing categories (Ref 1)

Table 3.1   Classification of copper products

Refinery shapes

    Cathodes
    Rods
    Wirebars
    Ingots

Wire mill products

    Bare wire
    Communication wire and cable
    Building wire
    Magnet wire
    Power cable
    Appliance wire and cordage
    Automotive wire and cable
    Other insulated wire and cable

Brass mill products

    Strip, sheet, and plate
    Mechanical wire (alloy only)
    Rod and bar
    Plumbing tube and pipe
    Commercial tube and pipe

Foundry products

Powder products

Originally, tin was the only principal ingredient when defining a bronze. Today, "bronze" is generally used with a modifying adjective. There are four main families of bronzes: copper-tin-phosphorus alloys (phosphor bronze), copper-tin-phosphorus-lead alloys (leaded phosphor bronze), copper-aluminum alloys (aluminum bronze), and copper-silicon alloys (silicon bronze).

The use of common names for copper alloys is widespread; however, alloy specification in technical and legal documents must follow the accepted designations. For the most part, alloy designations used in this book refer to the Unified Numbering System (UNS), which utilizes five-digit code numbers preceded by the letter "C" to identify copper alloys. The UNS system is used in the United States, Canada, Brazil, and Australia.

Table 3.2   Compositional classification of metallic copper-base materials

Coppers

Wrought coppers	99.3% Cu min.
Cast coppers	99.3% Cu min.

High copper alloys

Wrought	96 to 99.3% Cu
Cast	94% Cu min.

Brasses

Wrought brasses

Copper-zinc alloys (brasses)
Copper-zinc-lead alloys (leaded brasses)
Copper-zinc-tin alloys (tin brasses)

Cast brasses

Copper-zinc-tin alloys (red, semired, and yellow brasses)
Manganese bronze alloys (high-strength yellow brasses)
Leaded manganese bronze alloys (leaded high-strength yellow brasses)
Copper-zinc-silicon alloys (silicon brasses and bronzes)

Bronzes

Wrought bronzes

Copper-tin-phosphorus alloys (phosphorus bronzes)
Copper-tin-lead-phosphorus alloys (leaded phosphorus bronzes)
Copper-aluminum alloys (aluminum bronzes)
Copper-silicon alloys (silicon bronzes)
Copper-zinc-aluminum-iron-manganese alloys (manganese bronzes)

Cast bronzes

Copper-tin alloys (tin bronzes)
Copper-tin-lead alloys (leaded tin bronzes)
Copper-tin-nickel alloys (nickel tin bronzes)
Copper-aluminum alloys (aluminum bronzes)

Copper-nickels

Copper-nickel-zinc-alloys (nickel silvers)

Leaded coppers

Special alloys (wrought and cast)

Copper-iron alloys
Tin brasses

In the United States, the Copper Development Association, Inc. (CDA) administers the classification of copper-bearing materials. New coppers and copper alloys are incorporated into the listing as they come into use, and designations are placed in an inactive status when an alloy ceases to be used commercially. UNS numbers from C10000 through C79999 denote wrought alloys, and cast alloys are numbered from C80000 through C99999.

Technical Requirements

Trade in copper and copper alloys is based on products whose specifications are established jointly by suppliers and consumers through national or international standard-setting organizations. They address qualities such as shape, dimensions, and physical and mechanical properties. Physical and mechanical properties (see Chapter 2) are particularly sensitive to chemical composition, often in the domain of impurity concentrations approaching a few parts per million. While primary copper is a commodity metal, there is product differentiation among the various producers; and refined copper products may contain a trademark to identify origin (Ref 2–5).

National standard-setting organizations for metallic copper and copper-alloy products are listed in Table 3.3. A cross-referenced listing of standards issued by these organizations is given in Table 3.4. Tables 3.5 and 3.6 list compositions, respectively, of wrought and cast coppers and copper alloys available in North America.

Refinery Products

Copper is brought to market in the form of cathodes, ingots, and continuously cast rods. These product forms are collectively known as refinery shapes. Cathodes of pure copper are obtained through electrolytic refining or electrowinning. The cathodes may be traded as-is or remelted and cast into other forms. All copper refinery shapes can be derived from primary copper or re-refined scrap; however, wire rod derived from scrap must be made from Grade 1 scrap if it is to be used for electrical conductors. Chemical compositions of products sold as refinery shapes are included under "Coppers" in Tables 3.5 and 3.6 (Ref 1, 8).

Cathodes

Cathodes (Fig 3.1) are thick sheets of pure copper produced by electrorefining or electrowinning.

Standard cathodes weigh between 90 and 155 kg (200 and 342 lb). Sizes range between 960 and 1240 mm long by 767 to 925 mm wide and 4 to 16 mm thick (roughly 3 ft by 4 ft by $\frac{1}{4}$ in.). Compositional limits for primary cathodes are supervised by the two international copper trading centers, the London Metal Exchange (LME) and the Commodity Exchange (Comex) in New York City. The British Standards Institute (BSI) and the American Society for Testing and Materials (ASTM), among others, have issued corresponding specifications. A ranking of cathode grades has been set up by the LME, and prices are quoted accordingly. Penalties are levied for violations, such as impurities and lack of cleanness (Ref 1, 9).

Compositional limits for refinery shapes under BS DD78 and ASTM B 115, along with two commercial producers' specifications, are shown on Table 3.7. However, refineries tend to offer compositional qualities that are higher than those set forth by market standards. A commercial distinction is made between electrorefined and elec-

trowon cathodes. An extension of the latter case is the designation EW-SX, which denotes use of the solvent extraction (ion exchange) process during refining (Ref 1, 9–11).

Most primary metal cathodes are sold for wire and cable production. They are continuously cast into rod (CCR) as a precursor to wire drawing. Common ingot shapes other than CCR include billets, foundry ingots, and cakes and wire bars. Wire bars have been almost completely replaced by CCR, largely because of the limited unwelded rod length available from wire bars and the generally higher quality of CCR.

Rod

Rods are round, hexagonal, or octagonal sections furnished in coils or straight lengths (Fig. 3.2). Both refineries and wire mills produce CCR. Depending on the process used, as-cast rod may be ready for drawing to wire, or it may require rolling to a suitable diameter. Economics of the wire production process depend on the annealability

Table 3.3   Organizations issuing standards for copper products

Standard	Country	Issuing organization
ABNT	Brazil	Associacao Brasileira de Normas Técnicas, Av. 13 de Maio N°13-28 andar Caixa Postal 1680/CEP, 20.03-Rio de Janeiro, R.J. Brasil
AFNOR	France	Association Française de Normalisation, Tour Europ Cedex 7. 92080 Paris, La Defense, France, Telex: 61 1974 AFNOR F
ARSO/ORAN	Republic of South Africa	African Research Standard Organization, P.O. Box 57363 Nairobi Kenya (Mr. Makane Faye)
ASTM	U.S.A.	American Society for Testing and Materials, 1916 Race Street, Philadelphia, PA 19103, U.S.A.
BSI	United Kingdom	British Standards Institution, 2 Park Street, London W1A 2BS, U.K.
COPANT	Argentina	Pan American Standards Commission, Av. Julio A. Roca 651, Piso 3 Sector 10 (1322), Cable: COPANTE, Buenos Aires, Argentina
COVENIN	Venezuela	Comisión Venezolana de Normas Industriales, Av. Andrés Bello, Edid. Torre Fondo Común, Piso 11, Carcas 1050, Venezuela
CSA	Canada	Canadian Standards Association, 178 Rexdale Blvd., Toronto, Ontario M9W 1R3, Telex: 06-989344 Canada
CSBTS	China	P.O. Box 820, Beijing, China
DGN	México	Dirección General de Normas, Calle Puente de Tecamacalco N°6, Lomas de Talcamachalco Sección Fuentes, Naucapan de Juárez 53950, México
DIN	Germany	Deutsches Institut für Normung e.v. Burggrafenstraße 6, Postfach 1107, D-1000 Berlin 30, Federal Republic of Germany
GOST	CIS	Gosudarstrenny Komitet Standartov, Leninsky Propekt 9, Moskva 17049, URSS, Telex: 411378 gost su
IBN	Belgium	Institut Belge de Normalisation, Av. de la Brabancorne 29, 1040 Bruxelles, Belgium
IRAM	Argentina	Instituto de Racionalización de Materiales, Chile 1192, 1098 Buenos Aires, Argentina
ISI	India	Indian Standards Institution, Manak Bhavan, 6 Banadur Shah Zafar Marg, New Delhi 110002, India Telex: 031-2970
ISO	International	International Organization for Standarization, 1 rue de Verembé (Case Postale 56), CH-1211, Geneve 20, Switzerland
JISC	Japan	Japanese Industrial Standards Committee, Agency of Industrial Ministry of International Trade and Industry, 1-3-1, Kasumigaseki, Chiyoda-Ku, Tokyo 100, Japan
NCh	Chile	Instituto Nacional de Normalización, Matias Causiño 64, Piso 6, Casilla 995, Correo 1, Santiago-Chile
NNI	Norway	Norges Standardeseringsforbund, Kalfjeslaan 2, P.O. Box 5059/2600 GB Delft. Norway, Telex: 19050 nsf n
SABS	South Africa	South Africa Bureau of Standards, Private Bag X191, Pretoria 0001. South Africa, Telex: 3-66 SA
SIS	Sweden	Standardiseringskommissionen Surige Tegnergatan 11, Box 3295 103 66 Stockholm, Telex: 17453 sis S. Sweden
SNV	Switzerland	Schweizerische Normen-Vereinigung, Kirchenweg 4, Postfach, 8032 Zürich, Switzerland
UNE	Spain	Una Norma Española, Calle Fernándes de la Hoz, 52/28010, Madrid, Spain
UNI	Italy	Una Norma Italiana, Ente Nazionale Italiano di Unificazione, Piazza Armando Diaz 2 20123 Milano, Italy, Telex: 312481

Table 3.4(a) Compilation of international standards for refinery products

Product	Standard	Product	Standard
ISO		**BSI**	
Copper refinery shapes	ISO 431-1981	Copper refinery shapes	BS 6017 1981, BS 6019
Tough-pitch refined copper for wrought product and alloy refinery shapes	ISO 1638-1987		
		DIN	
ASTM		Copper cathodes and refinery shapes	DIN 1708 01.73
Coppers	B 224-87	Copper casting alloys; composition of ingot metals	DIN 17656 06.73
Oxygen-free electrolytic copper	B 170-89, F68-82, F96-77	Copper sulfate for electroplating	DIN 50972 12.84
Phosphorized copper	B 379-80(1987)		
Tough-pitch chemically refined copper	B 442-80(1987)	**JISC**	
Tough-pitch refined copper for wrought product and alloy refinery shapes	B 216-89	Tough-pitch copper billets and cakes	H 2123 1971
Tough-pitch refined high-conductivity copper	B 623-80(1987)	Phosphorus deoxidized copper billets and cakes	H 2124 1971
Electrolytic cathode copper	B 115-83a	Oxygen-free electrolytic copper billets and cakes	H 2125 1969
Electrolytic tough-pitch copper	B 5-89	Copper (II) sulfate and pentahydrate	K 8983 1980
Bus bar	B 187-86	Copper (II) sulfate (anhydrous)	K 8984 1980

Table 3.4(b) Compilation of international standards of wrought alloy products

Product	Standard	Product	Standard
ISO		**ISO (continued)**	
Wrought copper-zinc alloys: Chemical composition and forms of wrought products		Drawn square bars: Symmetric plus and minus tolerances on width across flats and form tolerances	ISO 7758 1984
Part 1: Nonleaded and special Cu-Zn alloys	ISO 426-1 1983	Bus bar	ISO 4738 1982
Part 2: Leaded Cu-Zn alloys	ISO 426-2 1983	**ASTM**	
Wrought copper-tin alloys: Composition and forms of wrought products	ISO 427 1983	Requirements for wrought copper and copper-alloy rod, bar, and shapes (metric)	B 249M-86
Wrought copper-aluminum alloys: Composition and forms of wrought products	ISO 428 1983	CuAl	B 150M-86
Wrought copper-nickel alloys: Composition and forms of wrought products	ISO 429 1983	CuBe	B 196M-88
Wrought copper-nickel-zinc alloys: Composition and forms of wrought products	ISO 430 1983	CuCoBe	B 441-85
		CuSi	B 98M-84
Special wrought copper alloys: Composition and forms of wrought products	ISO 1187 1983	Naval brass	B 21M-83b
		Mn bronze	B 138M-84
Copper and copper alloys: Code of designation		CuZnPb	B 140M-85
Part 1: Designation of materials	ISO 1190-1 1982	Bus bars	B 187-86
Part 2: Designation of tempers	ISO 1190-2 1982	CuZn	B 371-84a
Wrought coppers (minimum copper content of 97.5%): Composition and forms of wrought products	ISO 1336 1980	CuNiSi	B 411-85
		CuNiZn	B 151M-89
Wrought coppers (minimum copper content of 99.85%): Composition and forms of wrought products	ISO 1337 1980	Face-centered brass	B 16M-85
		Copper	B 133M-89
		Phosphur bronze	B 139M-83
Wrought copper and copper alloys		Face-centered copper	B 301M-84
Rod and bar: Technical conditions of delivery	ISO 1637 1987	Temper designations for copper and copper alloys, wrought and cast	B 601-88
Extruded round, square, or hexagonal bars: Dimensions and tolerances	ISO 3488 1982	Copper-alloy addition agents	B 644-88
Drawn round bars: All minus tolerances on diameter and form tolerances	ISO 3489 1984	Brasses, special, and bronzes	E 54-80(1984)
		Copper-nickel and copper-nickel-zinc alloys	E 75-76(1984)
Drawn hexagonal bars: All minus tolerances on width across flats and form tolerances	ISO 3490 1984	Copper-beryllium alloys	E 106-83
		Copper-chromium alloys	E 118-83
Drawn square bars: All minus tolerances and form tolerances	ISO 3491 1984	Copper-tellurium alloys	E 121-83
		Copper alloys	E 478-82(1986)
Drawn rectangular bars: Dimensional and form tolerances	ISO 6958 1984	Manganese-copper alloys	E 581-76(1984)
Drawn round bars: Symmetric plus and minus tolerances on width across flats and form tolerances	ISO 7756 1984	**BSI**	
		Copper and copper alloys	BS 2901 1983
		Rods and sections other than forging stock	BS 2874 1969
Drawn hexagonal bars: Symmetric plus and minus tolerances on width across flats and form tolerances	ISO 7757 1984	**DIN**	
		Copper alloys: definitions	DIN 1718 11.59
		Drawn round rod of copper and wrought copper alloys: dimensions	DIN 1756 07.69 (ISO 3489 1984)

(continued)

Table 3.4(b) (continued)

Product	Standard	Product	Standard
DIN (continued)		**DIN (continued)**	
Rectangular bars of copper and wrought copper alloys: dimensions, permissible variations, static values	DIN 1759 06.74 (ISO 6958 1984)	Copper and copper alloys: alternative list	WS 20,000 to 2.1999 03.82
Nonferrous metals: concepts, classification of copper materials	E DIN 17600 (Part 11) 07.87	Nonferrous metals: concepts, materials, explanatory notes	E DIN 17600 (Part 11)
Drawn square rod of copper and wrought copper alloys with sharp edges: dimensions	DIN 1761 07.69 (ISO 3491 1984)	Copper-zinc alloys, brass and high-strength brass	DKI i 5 1985
Drawn hexagonal rod of copper and wrought copper alloys with sharp edges: dimensions	DIN 1763 07.69 (ISO 3490 1984)	Low-alloy copper materials: properties, processes, uses	DKI i 8 1981
Copper master alloys: composition	DIN 17657 03.73	Copper-nickel-zinc alloys, nickel silver: properties, processes, uses	DKI i 13 1980
Wrought copper and copper alloy rod and bar: properties	DIN 17672 (Part 1) 12.83 (ISO 1637 1987)	**JISC**	
Bars of copper and copper alloys: technical conditions of delivery	DIN 17672 (Part 2) 06.74 (ISO 1637 1987)	Phosphor-copper metal	H 2501 1982
		Copper-beryllium master alloys	H 2504 1963
		Copper and copper-alloy covered electrodes	Z 3231 1989
Extruded round rod of copper and wrought copper alloys: dimensions	DIN 1782 07.69	Copper and copper-zinc brazing filler metals	Z 3262 1986
		Aluminum-alloy brazing filler metals and brazing sheets	Z 3263 1980
Forged round bars of copper alloys: dimensions	LN 9468 11.88	Copper-phosphorus brazing filler metal	Z 3264 1985
Drawn round rod of wrought copper alloys: dimensions, weights	LN 1756 06.83	Nickel brazing filler metal	Z 3265 1986
		Gold brazing filler metal	Z 3266 1985
Drawn hexagonal rod of wrought copper alloys with sharp corners: dimensions, weights	LN 1763 06.83	Palladium brazing filler metal	Z 3267 1986
		Vacuum-grade precious brazing filler metals	Z 3268 1988

Table 3.4(c) Compilation of international standards for cast alloy products

Product	Standard	Product	Standard
ASTM		**DIN**	
Bronze castings for bridges and turntables	B 22-89	Copper-tin and copper-tin-zinc casting alloys (cast tin-bronze and gunmetal) castings	DIN 1705 11.81
Bronze castings: valves and fittings	B 61-86, B 62-86	Copper-tin alloy castings, amendment 1	DIN 1705 A 1 06.84
Bronze or ounce metal castings: composition	B 62-86	Copper-zinc alloy castings (brass and special brass castings)	DIN 1709 11.81
Bronze castings in the rough for locomotive wearing parts	B 66-89	Copper-aluminum casting alloys (cast aluminum bronze) castings	DIN 1714 11.81
Bearings: car and journal	B 67-88	Copper-lead-tin casting alloys (cast tin-lead bronze) castings: reference data on mechanical and physical properties	DIN 1716 11.81
Aluminum-bronze sand castings	B 148-88		
Copper-alloy die castings	B 176 88		
Preparing tension test specimens for copper-base alloys for sand castings	B 208-82	Unalloyed and low-alloy copper materials for casting: castings	DIN 17655 11.81
Bronze shapes	B 249-84	Copper casting alloys: ingot materials, composition	DIN 17656 06.73
Copper-base alloy centrifugal castings	B 271-89		
Copper-nickel alloy castings	B 369-87	Copper-nickel casting alloy castings	DIN 17658 06.73
Gear bronze alloy castings	B 427-87	Castings of copper and copper alloys	GDM Copper 02.82
Cast copper-nickel alloys	B 492-82	High-strength copper-base and nickel-copper alloy castings	E 272-75 (1984)
Copper-base alloy continuous castings	B 505-89		
Copper-alloy sand castings for general applications	B 584-89	Tin-bronze castings	E 310-75 (1984)
Copper-alloy sand castings for valve applications	B 763-89	**JISC**	
Copper-beryllium alloy sand castings for general applications	B 770-87	Brass castings	H 5101 1988
		High-strength brass castings	H 5102 1988
BSI		Bronze castings	H 5111 1988
Cast valves and fittings	BS 143, BS 1256, BS 2057	Silicon-bronze castings	H 5112 1988
		Phosphor-bronze castings	H 5113 1988

Table 3.4(d) Compilation of international standards for wire products

Product	Standard	Product	Standard
ISO		**ISO (continued)**	
Wrought copper and copper alloy wire: technical conditions of delivery	ISO 1638 1987	Wrought copper and copper alloy drawn round wire: tolerances on diameter	ISO 3492 1982

(continued)

Table 3.4(d) (continued)

Product	Standard	Product	Standard
ISO (continued)		**BSI (continued)**	
Copper drawing stock (wire rod)	ISO 4738 1982	Hard-drawn copper and copper-cadmium conductors for overhead power transmission purposes	BS 125 1970
ASTM		Copper conductors insulated annealed for electric power and light	BS 152 1922
Hard-drawn copper wire	B 1-95		
Medium-hard-drawn copper wire	B 2-88	Hard-drawn copper and copper-cadmium wire	BS 174 1970, 1986
Bronze trolley wire	B 9-90	Copper binding and jointing wires for telegraph and telephone uses	BS 176 1970
Tinned wire	B 33		
Soft rectangular and square copper wire for electrical conductors	B 48-68(1992)	Cotton-covered copper conductors	BS 1791
Soft or annealed copper wire	B 3-74(1984)	Enameled and cotton-covered copper conductors (oleo-resinous enamel)	BS 1815, BS 3902
Brass wire	B 134-88		
Phosphorus-bronze wire (metric)	B 159M-86A	**DIN**	
Copper-beryllium alloy wire (metric)	B 197M-85	Drawn wire of copper and wrought copper alloys: dimensions	DIN 1757 06.74, 12.83
Copper-nickel-zinc alloy (nickel silver) and copper-nickel alloy wire (metric)	B 206M-87		
General requirements for wrought copper alloy wire (metric)	B 624M-88	Wrought copper and copper alloy wire for general purposes DIN 17677 (Part 1)	
Wire coating continuity	B 290	Drawn spring wire of copper alloys: dimensions	DIN 9421 06.83, LN 9421 06.83
Nickel-coated copper wire	B 355	Copper for electrical purposes, tinned wire: technical delivery conditions	DIN 40500 (Part 5) 06.83
Copper-nickel-silicon alloy wire	B 412-87		
BSI		Hard or soft copper wire: dimensions	DIN 46431
Copper and copper-cadmium trolley and contact wire for electric traction	BS 23 1970	**JISC**	
		Copper and copper alloy wires	H 3260 1986

Table 3.4(e) Compilation of international standards for tubular products

Product	Standard	Product	Standard
ISO		**ASTM (continued)**	
Copper tubes of circular section: dimensions	ISO 274 1975	Seamless and welded copper-nickel tubes for water desalinization	B 552-86
Seamless drawn copper tubes for piping systems	ISO 274 1975		
ASTM		Welded brass tubes	B 587-88
		Welded copper-alloy pipe	B 608-88
Seamless copper pipe, standard sizes	B 42-88	Seamless and welded copper distribution tube (type D)	B 641-88
Seamless red brass pipe, standard sizes	B 43-88		
Seamless copper tube, bright annealed (metric)	B 68M-86	Welded copper-alloy (UNS C21000) water tube	B 642-88
Seamless copper tube (metric)	B 75M-86	Copper-beryllium alloy seamless tube	B 643-83
Seamless copper water tube (metric)	B 88M-88a	Seamless and welded copper and copper-alloy plumbing pipe and tube	B 698-86
Seamless heat exchange tube copper and copper alloy	B 111-88	Seamless copper-alloy (UNS C69100) pipe and tube	B 706-88
Seamless brass tube (metric)	B 133M-86a	Seamless copper water tube, special use (metric)	B 707M-88
Seamless brass tube (metric)	B 135M-86	Welded copper water tube (metric)	B 716M-88
Seamless copper bus pipe and tube	B 188-96	Seamless copper tube in coils	B 743-88
Water-cooled electrical bus bar tube	B 188-96	**BSI**	
Tube and pipe definitions	B 244-87		
General requirements for wrought seamless copper and copper-alloy tube (metric)	B 251M-88	Copper and copper-alloy tube for pressure piping	BS 1306 1975, 1983
Seamless air conditioning and refrigeration tube	B 280-88	High-conductivity copper tubes for electrical purposes	BS 1977 1985
Threadless copper pipe	B 302-88	Copper tubes for water, gas, and sanitation	BS 2871 1971
Copper drainage tube (drainage, waste, and vent systems)	B 306-88	Capillary and compression fittings for copper tubes	BS 4579 (Part 2) 1973, 1985
Surface-enhanced heat exchanger tube	B 359-88	**DIN**	
Hard-drawn copper capillary tube for restrictor applications	B 360-88	Seamless drawn copper tube	
Seamless copper and copper-alloy rectangular waveguide tube	B 372-88	Dimension range and coordination of tolerance	DIN 1754 (Part 1) 08.69
Welded copper tube	B 447-89	Preferred dimensions for general purposes	DIN 1754 (Part 2) 08.69
Seamless copper-nickel pipe and tube (metric)	B 466M-86		
Welded copper-nickel pipe	B 467-88	Preferred dimensions for pipelines	DIN 1754 (Part 3) 04.74
Seamless copper-alloy tubes for pressure applications	B 469-88		
Welded heat exchanger tube	B 543-88	Seamless drawn wrought copper-alloy tubes	

(continued)

Table 3.4(e) (continued)

Product	Standard	Product	Standard
DIN (continued)		**DIN (continued)**	
Dimension range and coordination of tolerance	DIN 1755 (Part 1) 08.69	Seamless drawn wrought copper and copper-alloy tubes for capillary soldering: dimensions	DIN 59753 05.80
Preferred dimensions for general purposes	DIN 1755 (Part 2) 08.69	Welded copper-alloy tubes	DIN 86018 11.72
		Seamless drawn copper-alloy tubes: dimensions	LN 9390 06.83
Preferred dimensions for pipelines	DIN 1755 (Part 3) 08.69	Extruded copper-alloy tubes: dimensions	LN 9391 06.83
Seamless drawn copper tubes for piping systems	DIN 1786 05.80	Seamless drawn copper pipes for gas and water installations: requirements and test specifications, amendments	DVGW GW 392 02.83
Wrought copper and copper-alloy tubes			
Properties	DIN 17671 (Part 1) 12.83	Gilled copper and wrought copper-alloy pipes	VdTUV WB 420/2
Technical conditions of delivery	DIN 17671 (Part 2) 06.69	Gilled pipes with medium rip depth for oxygen-free copper	VdTUV WB 420/2
Extruded dimensions	DIN 59750 06.74	**JISC**	
Seamless drawn tubes and hexagonal hollow sections of wrought copper alloys for freecutting: dimensions	DIN 59752 11.73	Seamless copper and copper-alloy pipes and tubes	H 3300 1986
		Seamless copper-nickel alloy pipes and tubes	H 4551 1977

Table 3.4(f) Compilation of international standards for sheet, strip, and plate products

Product	Standard	Product	Standard
ISO		**ASTM (continued)**	
Plate, sheet, and strip		Copper-aluminum-silicon-cobalt alloy, copper-nickel-aluminum-silicon alloy, and copper-nickel-aluminum-magnesium alloy sheet and strip	B 422-86b
Part 1: Technical conditions of delivery for plate, sheet, and strip for general purposes	ISO 1634-1 1987		
Part 2: Technical conditions of delivery for plate and sheet and for boilers, pressure vessels, and exchangers	ISO 1634-2 1987	Copper and copper-alloy foil	B 451-86
		Copper-zinc-lead alloy (leaded brass) extruded shapes	B 455-89
Wrought copper and copper-alloy cold-rolled flat products delivered in straight lengths (sheet): dimensions and tolerances	ISO 3486 1980	Copper and copper-alloy clad steel plate	B 432-76a
		UNS C26000 brass strip in narrow widths and light gage for heat-exchanger tubing	B 569-86
Wrought copper and copper-alloy cold-rolled flat products in coils or on reels (strip): dimensions and tolerances	ISO 3487 1980	Copper and copper-alloy solar heat absorber panels	B 638-86
		Copper-nickel-tin spinodal alloy strip	B 740-87
		Copper-zirconium alloy sheet and strip	B 747-85
ASTM		Copper-cobalt-beryllium alloy strip and sheet	B 768-88
General requirements for wrought copper and copper-alloy plate, sheet, strip, and rolled bar (metric)	B 248M-86	**DIN**	
		Copper sheet and strip for building purposes: technical delivery conditions	DIN 17650 12.88
Coppers	B 152M-88	Copper and copper-alloy wrought plate more than 10 mm thick: composition, mechanical properties, dimensions and tolerances, comparable ISO designations, requirements for special applications	DIN 17670 (Part 1) A109.88
Cu-Si alloys	B 96M-85		
Brass alloys	B 36-87		
Cu-Zn-Al alloys	B 592-86		
Cu-Al alloys	B 169M-88		
Cartridge brass	B 19-86	Sheet and sheet cut to length	LN 1751 06.83
Cu-Fe alloys	B 465-86	Cold-rolled sheet and cut strip from wrought copper and copper alloys: dimensions	LN 1751 06.83
Cu-Zn-Sn alloys	B 591-86		
Cu-Ni-Sn alloys	B 122-86	Coiled and flat strips from copper alloys for leaf springs	LN 1777 06.83
Leaded brass alloys	B 121-86		
Cu-Co-Be alloys	B 534-88	Cold-rolled coiled and flat strips from copper and copper alloys: dimensions	LN 1791 06.83
Cartridge brass sheet, strip, plate, bar, and disks	B 19-86		
Rolled copper-alloy bearing and expansion plates and sheets for bridge and other structural uses	B 100-86a	**JISC**	
Lead-coated copper sheets	B 101-83	Copper and copper-alloy sheets, plates, and strips	H 3100 1986
Phosphor-bronze plate, sheet, strip, and rolled bar	B 103-88	Phosphor-bronze and nickel-silver sheets, plates, and strips	H 3110 1986
Copper-zinc-manganese alloy (manganese brass) sheet and strip	B 291-86	Copper-beryllium alloy, phosphor-bronze and nickel-silver sheets, plates, and strips for springs	H 3130 1986
Copper sheet and strip for building construction	B 370-88	Copper-nickel alloy strips	H 4555 1978

Table 3.4(g) Compilation of international standards for test methods

Product	Standard	Product	Standard
ISO		**ISO (continued)**	
Unalloyed copper containing not less than 99.90% of copper: determination of copper content, electrolytic method	ISO 1552 1976	Copper alloys: determination of chromium content, titrimetric method	ISO 6437 1984
Wrought and cast copper alloys: determination of copper content, electrolytic method	ISO 1554 1976	Copper alloys: ammonia test for stress-corrosion resistance	ISO 6957 1988
Copper alloys: determination of nickel (low contents) dimethyl-glyoxime spectrometric method	ISO 1810 1976	Copper and copper alloys: determination of sulfur content, combustion titrimetric method	ISO 7266 1984
Chemical analysis of copper and copper alloys: sampling of copper refinery shapes	ISO 1811 1971	**ASTM**	
Copper and copper alloys: selection and preparation of samples for chemical analysis		Expansion (pin test) of copper and copper-alloy pipe and tubing	B 153-85
Part 1: Sampling of cast unwrought products	ISO 1811-1 1988	Mercurous nitrate test for copper and copper alloy	B 154-87
Part 2: Sampling of wrought products and castings	ISO 1811-2 1988	Preparing tension test specimens for copper-base alloys for sand castings	B 208-82
Copper alloys: determination of iron content, 1.10-phenanthroline spectrophotometric method	ISO 1812 1976	Angle of twist in rectangular and square copper and copper-alloy tube	B528-87
Copper and copper alloys: Rockwell superficial hardness test (N and T scales)	ISO 2712 1973	Hydrogen embrittlement of copper	B 577-88
Copper and copper alloys: determination of manganese content, spectrophotometric method	ISO 2543 1973	Conducting bending fatigue tests for copper-alloy spring materials	B 593-85
Copper alloys: determination of aluminum as alloying element, volumetric method	ISO 3110 1975	Determining offset yield strength for copper alloys	B598-86
Copper alloys: determination of tin as alloying element, volumetric method	ISO 3111 1975	Evaluating the corrosivity of solder fluxes for copper tubing systems	B 732-84
Copper and copper alloys: determination of lead-extracting titration method	ISO 3112 1975	Measuring and recording the deviations from flatness in copper and copper-alloy strip	B 754-86
Copper and copper alloys: determination of arsenic-photometric method	ISO 3220 1975	Tension testing of metallic materials (metric)	E 8M-87
Metallic coatings: copper accelerated acetic acid salt spray test (CASS test)	ISO 3770 1976	Sampling wrought nonferrous metals and alloys: determination of chemical composition	E 55-48(1986)
Wrought copper and copper-alloy products: selection and preparation of specimens and test pieces for mechanical testing	ISO 4739 1985	Sampling nonferrous metals and alloys in cast form: determination of chemical composition	E 88-58(1986)
Copper and copper alloys: determination of zinc content, flame atomic absorption spectrometric method	ISO 4740 1985	Determining average grain size	E 112-88
Copper and copper alloys: spectrometric determination of molybdovanadate	ISO 4741 1984	Electromagnetic (eddy-current) testing of seamless copper and copper-alloy tubes	E 243-85
Copper alloys: determination of nickel content, gravimetric method	ISO 4742 1984	**BSI**	
Copper alloys: determination of nickel content, titrimetric method	ISO 4743 1984	Methods for tensile testing of metals	BS 18
Copper and copper alloys: determination of chromium content, flame atomic absorption spectrometric method	ISO 4744 1984	Detection of copper corrosion from petroleum products by the copper strip tarnish test	BS 200 (Part 154) 1982
Copper alloys: determination of iron content, Na2EDTA titrimetric method	ISO 4748 1984	Methods for the analysis of copper alloys	BS 1748
Copper and copper alloys: determination of lead content, flame atomic absorption spectrometric method	ISO 4749 1984	Methods for the determination of copper, lead, iron, aluminum, and nickel in copper alloys	BS 1748 (Parts 1–5) 1961 (1985)
Copper and copper alloys: determination of tin content, spectrometric method	ISO 4751 1984	Method for the determination of phosphorus in copper alloys (photometric method)	BS 1748 (Part 8) 1960 (1985)
Welding and allied processes, assemblies made with soft solders and brazing filler metals: mechanical test methods	ISO 5187 1985	Method for the determination of zinc in copper alloys	BS 1748 (Part 9) 1963 (1985)
Copper and copper alloys: determination of antimony content, rhodamine B spectrometric method	ISO 5956 1984	Methods for the determination of copper and lead in leaded bronze alloys	BS 1748 (Parts 11 and 12) 1964 (1985)
Copper and copper alloys: determination of bismuth content, diethyldithiocarbamate spectrometric method	ISO 5959 1984	Method for the determination of silicon in copper alloys (photometric method)	BS 1748 (Part 78) 1960 (1985)
Copper alloys: determination of cadmium content, flame atomic absorption spectrometric method	ISO 5960 1984	Methods for the analysis of raw copper	BS 1800 1951
		Three methods for copper, two for iron	BS 2690 (Part 1) 1964 (1974)
Liquefied petroleum gases: corrosiveness to copper-copper strip test	ISO 6251 1982	Enameled wire to BS 156: part 1 in the range of bare diameters from 0.0016 to 0.064 in. inclusive with two thicknesses of rayon. Chemical purity of the rayon, increase in diameter due to the rayon, elongation of covered wire, and methods of testing	BS 3902 (Part 1) 1965
		Methods of testing fusion welds in copper and copper alloys	BS 4206 1967
		Determination of acid-insoluble content in iron, copper, tin, and bronze powders	BS 5600 (Part 2, section 2.9) 1980
		Method for hydrogen embrittlement test for copper	BS 5899 (1980)
		Method for determination of resistance to intergranular corrosion of austenitic stainless steels, copper sulfate-sulfuric acid method	BS 5903 1980

(continued)

Table 3.4(g) (continued)

Product	Standard	Product	Standard
BSI (continued)		**JISC (continued)**	
Method for scale adhesion test for oxygen-free copper	BS 5909 1980	Copper-clad laminates for printed wiring boards	C 6481 1986
		Methods for estimating average grain size of wrought copper and copper alloys	H 0501 1986
DIN		Method of eddy-current testing for copper and copper-alloy pipe and tube	H 0502 1986
Testing of solderability for soft soldering, vertical dipping test for specimens of copper alloys: testing, assessment	DIN 32506 (Part 2) 07.86	General rules for chemical analysis of copper products and copper alloys	H 1012 1967
Copper winding wires		Methods for determination of elements in copper and copper alloys	
Enameled round wires	DIN 46453 (Part 1) 04.77	Copper	H 1051 1982
Enameled round wires: determination of limiting temperature	DIN 46453 (Part 2) 04.77	Tin	H 1052 1984
Insulated round wire, bunched enameled braided wires for radio frequency purposes: methods of test	DIN 46453 (Part 4) 10.79	Lead	H 1053 1984
		Iron	H 1054 1984
Enameled round wires: testing of resistance to transformer oil in the presence of water	DIN 46453 (Part 5) 04.77	Manganese	H 1055 1987
		Nickel	H 1056 1987
Rectangular copper or aluminum wires, enameled and/or insulated by covering: methods of test	DIN 46457 07.82	Aluminum	H 1057 1987
		Phosphorus	H 1058 1987
Copper alloys: determination of iron content, Na2 EDTA titrimetric method	DIN ISO 4748 07.86	Arsenic	H 1059 1987
		Cobalt	H 1060 1989
Copper and copper alloys: determination of tin content, spectrometric method	DIN ISO 4751 06.87	Silicon	H 1061 1989
		Methods for emission spectrochemical analysis of electrolytic cathode copper	H 1103 1976
Chemical analysis of copper and copper alloys		Methods of chemical analysis	
Sampling of copper refinery shapes	DIN 50500 11.72 (ISO 1911 1971)	Oxygen-free copper products for electron tubes	H 1202 1982
		Brass	H 1211 1977
Determination of copper in unalloyed copper containing not less than 99.90% of copper	DIN 50502 10.73 (ISO 1553 1976)	Cupro-nickel and nickel silver	H 1231 1977
		Phosphor bronze	H 1241 1977
Determination of copper in wrought and cast copper alloys	DIN 50503 08.80 (ISO 1554 1976)	Bronze	H 1251 1977
Photometric determination of iron in copper and copper alloys containing maximum 0.4% iron	DIN 50504 10.73 (ISO 1812 1976)	Aluminum bronze and special aluminum bronze	H 1252 1977
		Beryllium copper	H 1261 1976
Photometric determination of manganese in copper alloys containing maximum 6% manganese	DIN 50505 10.73	Nickel-copper alloys	H 1271 1962
		Methods for determination of copper in nickel and nickel-alloy castings	H 1272 1988
Photometric determination of nickel in copper alloys containing maximum 2.5% nickel	DIN 50506 10.73 (ISO 1810 1976)	Method for atomic absorption spectrochemical analysis of copper and copper alloys	H 1291 1977
Determination of lead, extracting-titration method	DIN 50507 07.77 (ISO 3112 1975)	Method for fluorescent x-ray analysis of copper alloys	H 1292 1984
Determination of tin as alloying element, volumetric method	DIN 50508 06.77 (ISO 3111 1975)	Methods for determination of copper content	
Determination of aluminum content of copper alloys, volumetric method	DIN 50509 05.77	Magnesium alloys	H 1336 1976
		Aluminum and aluminum alloys	H 1354 1972
Determination of arsenic in copper and copper alloys, photometric method	DIN 50510 04.80 (ISO 3220 1975)	Nickel materials for electron tubes	H 1424 1979
		Methods of chemical analysis of phosphor-copper ingots	H 1552 1976
Determination of phosphorus content, molybdenum blue and molybdovanadate spectrophotometric method	DIN 50511 12.86 (ISO 4741 1984)	Methods of chemical analysis of beryllium-copper master alloys	H 1553 1976
Testing of copper alloys		Methods of chemical analysis of magnesium-copper ingots	H 1554 1976
Mercurous nitrate test	DIN 50911 06.80 (ISO 196 1978)	Method for determination of copper in zirconium and zirconium alloys	H 1657 1985
Stress-corrosion cracking test in ammonia; testing of tubes, rods, and shapes	DIN 50916 (Part 1) 08.76 (ISO 6957 1988)	Methods for determination of copper in tantalum	H 1686 1976
		Method of tension and shear tests for brazed joint	Z 3192 1988
Stress-corrosion cracking test using ammonia, testing of components	DIN 50916 (Part 2) 09.85	Method of wet corrosion test for brazed joint	Z 3195 1971
Corrosion of metals: corrosion behavior of metallic materials against water, scale for evaluation for copper and copper alloys	DIN 50930 (Part 5) 12.80	Method of gaseous corrosion test for brazed joint	Z 3196 1972
		Standard qualification procedure for brazing technique	Z 3891 1977
JISC		Methods for sampling of precious brazing filler metals	Z 3900 1974
Testing methods		Methods for chemical analysis of silver brazing filler metals	Z 3901 1988
Electrical copper and aluminum wires	C 3002 1981	Methods of chemical analysis of copper-phosphorus brazing filler metals	Z 3902 1984
Enameled copper and enameled aluminum wires	C 3003 1984	Methods of chemical analysis of gold brazing filler metals	Z 3904 1979
Rubber or plastic insulated wires and cables	C 3005 1986	Methods of chemical analysis of palladium brazing filler metals	Z 3906 1988
Fiber or paper insulated copper and aluminum winding wires	C 3006 1983		

Table 3.5 Chemical composition of wrought copper and copper alloys

Alloy number (and name)	Nominal composition, %	Commercial forms(a)
C10100 (oxygen-free electronic copper)	99.99 Cu	F,R,W,T,P,S
C10200 (oxygen-free copper)	99.95 Cu	F,R,W,T,P,S
C10300 (oxygen-free extra-low-phosphorus copper)	99.95 Cu, 0.003 P	F,R,T,P,S
C10400	99.95 Cu, 0.025 Ag	F,R,W,S
C10500 (oxygen-free silver-bearing copper)	99.95 Cu, 0.031 Ag	F,R,W,S
C10700 (oxygen-free silver-bearing copper)	99.95 Cu, 0.078 Ag	R,R,W,S
C10800 (oxygen-free low-phosphorus copper)	99.95 Cu, 0.009 P	F,R,T,P
C11000 (electrolytic tough pitch copper)	99.90 Cu, 0.04 O	F,R,W,T,P,S
C11100 (electrolytic tough pitch anneal-resistant copper)	99.90 Cu, 0.04 O, 0.01 Cd	W
C11300 (silver-bearing tough pitch copper)	99.90 Cu, 0.25 Ag	F,R,W,T,S
C11400 (silver-bearing tough pitch copper)	99.90 Cu, 0.31 Ag	F,R,W,T,S
C11500 (silver-bearing tough pitch copper)	99.90 Cu, 0.50 Ag	F,R,W,T,S
C11600 (silver-bearing tough pitch copper)	99.90 Cu, 0.78 Ag	F,R,W,T,S
C12000	99.9 Cu, 0.008 P	F,T,P
C12100	99.9 Cu, 0.008 P, 0.12 Ag	F,T,P
C12200 (high residual phosphorus deoxidized copper)	99.90 Cu, 0.02 P	F,R,T,P
C12500 (fire-refined tough pitch with silver)	99.88 Cu, 0.12 As, 0.03 Sb	
C12700 (fire-refined tough pitch with silver)	99.88 Cu, 0.27 Ag, 0.12 As, 0.03 Sb	
C12800 (fire-refined tough pitch with silver)	99.88 Cu, 0.34 Ag, 0.12 As, 0.03 Sb	
C12900 (fire-refined tough pitch with silver)	99.88 Cu, 0.54 Ag, 0.12 As, 0.03 Sb	
C13000 (fire-refined tough pitch with silver)	99.88 Cu, 0.85 Ag, 0.12 As, 0.03 Sb	
C14200 (phosphorus-deoxidized arsenical copper)	99.68 Cu, 0.3 As, 0.02 P	F,R,W,S
C14300	99.9 Cu, 0.1 Cd	F,R,T
C14310	99.8 Cu, 0.2 Cd	F
C14500 (phosphorus-deoxidized tellurium-bearing copper)	99.5 Cu, 0.50 Te, 0.008 P	F
C14700 (sulfur-bearing copper)	99.6 Cu, 0.40 S	F,R,W,T
C15000 (zirconium copper)	99.8 Cu, 0.15 Zr	R,W
C15100	99.82 Cu, 0.1 Zr	R,W
C15500	99.75 Cu, 0.06 P, 0.11 Mg, 0.26 Ag	F
C15710	99.8 Cu, 0.2 Al_2O_3	F
C15720	99.6 Cu, 0.4 Al_2O_3	R,W
C15735	99.3 Cu, 0.7 Al_2O_3	F,R
C15760	98.9 Cu, 1.1 Al_2O_3	R
C16200 (cadmium-copper)	99.0 Cu, 1.0 Cd	F,R
C16500	98.6 Cu, 0.8 Cd, 0.6 Sn	F,R,W
C17000 (beryllium-copper)	99.5 Cu, 1.7 Be, 0.20 Co	F,R,W
C17200 (beryllium-copper)	99.5 Cu, 1.9 Be, 0.20 Co	F,R,
C17300 (beryllium-copper)	99.5 Cu, 1.9 Be, 0.40 Pb	F,R,W,T,P,S
C17400	99.5 Cu, 0.3 Be, 0.25 Co	R
C17500 (copper-cobalt-beryllium alloy)	99.5 Cu, 2.5 Co, 0.6 Be	F
C18200	99.5 Cu, 0.9 Cr	F,R,
C18400 (chromium-copper)	99.5 Cu, 0.8 Cr	F,W,R,T,S
C18500 (chromium-copper)	99.5 Cu, 0.7 Cr	F,W,R,T,S
C18700 (leaded copper)	99.0 Cu, 1.0 Pb	F,W,R,T,S
C18900	98.75 Cu, 0.75 Sn, 0.3 Si, 0.20 Mn	R
C19000 (copper-nickel-phosphorus alloy)	98.7 Cu, 1.1 Ni, 0.25 P	R,W
C19100 (copper-nickel-phosphorus-tellurium alloy)	98.15 Cu, 1.1 Ni, 0.50 Te, 0.25 P	F,R,W
C19200	98.97 Cu, 1.0 Fe, 0.03 P	R,F
C19400	97.5 Cu, 2.4 Fe, 0.13 Zn, 0.03 P	F,T
C19500	97.0 Cu, 1.5 Fe, 0.6 Sn, 0.10 P, 0.80 Co	F
C19700	99 Cu, 0.6 Fe, 0.2 P, 0.05 Mg	F
C21000 (gilding, 95%)	95.0 Cu, 5.0 Zn	F
C22000 (commercial bronze, 90%)	90.0 Cu, 10.0 Zn	F,W
C22600 (jewelry bronze, 87.5%)	87.5 Cu, 12.5 Zn	F,R,W,T
C23000 (red brass, 85%)	85.0 Cu, 15.0 Zn	F,W
C24000 (low brass, 80%)	80.0 Cu, 20.0 Zn	F,W,T,P
C26000 (cartridge brass, 70%)	70.0 Cu, 30.0 Zn	F,W
C26800, C27000 (yellow brass)	65.0 Cu, 35.0 Zn	F,W,T,P
C28000 (Muntz metal)	60.0 Cu, 40.0 Zn	F,R,W
C31400 (leaded commercial bronze)	89.0 Cu, 1.75 Pb, 9.25 Zn	F,R,T
C31600 (leaded commercial bronze, nickel-bearing)	89.0 Cu, 1.9 Pb, 1.0 Ni, 8.1 Zn	F,R
C33000 (low-leaded brass tube)	66.0 Cu, 0.5 Pb, 33.5 Zn	F,R
C33200 (high-leaded brass tube)	66.0 Cu, 1.6 Pb, 32.4 Zn	T
C33500 (low-leaded brass)	65.0 Cu, 0.5 Pb, 34.5 Zn	T
C34000 (medium-leaded brass)	65.0 Cu, 1.0 Pb, 34.0 Zn	F
C34200 (high-leaded brass)	64.5 Cu, 2.0 Pb, 33.5 Zn	F,R,W,S
C34900	62.2 Cu, 0.35 Pb, 37.45 Zn	F,R
C35000 (medium-leaded brass)	62.5 Cu, 1.1 Pb, 36.4 Zn	R,W
C35300 (high-leaded brass)	62.0 Cu, 1.8 Pb, 36.2 Zn	F,R

(continued)

Table 3.5 (continued)

Alloy number (and name)	Nominal composition, %	Commercial forms(a)
C35600 (extra-high-leaded brass)	63.0 Cu, 2.5 Pb, 34.5 Zn	F,R
C36000 (free-cutting brass)	61.5 Cu, 3.0 Pb, 35.5 Zn	F
C36500 to C36800 (leaded Muntz metal)(b)	60.0 Cu, 0.6 Pb, 39.4 Zn(d)	F,R,S
C37000 (free-cutting Muntz metal)	60.0 Cu, 1.0 Pb, 39.0 Zn	F
C37700 (forging brass)(c)	59.0 Cu, 2.0 Pb, 39.0 Zn	T
C38500 (architectural bronze)(c)	57.0 Cu, 3.0 Pb, 40.0 Zn	R,S
C40500 (high conductivity bronze, penny bronze)	95 Cu, 1 Sn, 4 Zn	R,S
C40800	95 Cu, 2 Sn, 3 Zn	F
C41100 (Lubaloy)	91 Cu, 0.5 Sn, 8.5 Zn	F
C41300	90.0 Cu, 1.0 Sn, 9.0 Zn	F,W
C41500	91 Cu, 1.8 Sn, 7.2 Zn	F,R,W
C42200	87.5 Cu, 1.1 Sn, 11.4 Zn	F
C42500	88.5 Cu, 2.0 Sn, 9.5 Zn	F
C43000	87.0 Cu, 2.2 Sn, 10.8 Zn	F
C43400	85.0 Cu, 0.7 Sn, 14.3 Zn	F
C43500	81.0 Cu, 0.9 Sn, 18.1 Zn	F
C44300, C44400, C44500 (inhibited admiralty)	71.0 Cu, 28.0 Zn, 1.0 Sn	F,T
C46400 to C46700 (naval brass)	60.0 Cu, 39.25 Zn, 0.75 Sn	F,W,T
C48200 (naval brass, medium-leaded)	60.5 Cu, 0.7 Pb, 0.8 Sn, 38.0 Zn	F,R,T,S
C48500 (leaded naval brass)	60.0 Cu, 1.75 Pb, 37.5 Zn, 0.75 Sn	F,R,S
C50500 (phosphor bronze, 1.25% E)	98.75 Cu, 1.25 Sn, trace P	F,R,S
C51000 (phosphor bronze, 5% A)	95.0 Cu, 5.0 Sn, trace P	F,R,W,T
C51100	95.6 Cu, 4.2 Sn, 0.2 P	R,W
C52100 (phosphor bronze, 8% C)	92.0 Cu, 8.0 Sn, trace P	F,R,T,P,S
C52400 (phosphor bronze, 10% D)	90.0 Cu, 10.0 Sn, trace P	F
C54400 (free-cutting phosphor bronze)	88.0 Cu, 4.0 Pb, 4.0 Zn, 4.0 Sn	F,R,W
C60800 (aluminum bronze, 5%)	95.0 Cu, 5.0 Al	F,R,W
C61000	92.0 Cu, 8.0 Al	F,R
C61300	92.65 Cu, 0.35 Sn, 7.0 Al	T
C61400 (aluminum bronze, D)	90.0 Cu, 8.0 Al, 2.0 Ni	F
C61500	89.0 Cu, 1.0 Fe, 10.0 Al	R
C61800	86.5 Cu, 4.0 Fe, 9.5 Al	F
C61900	87.0 Cu, 3.0 Fe, 10.0 Al	F,R
C62300 (aluminum bronze, 9%)	86.0 Cu, 3.0 Fe, 11.0 Al	F,R
C62400 (aluminum bronze, 11%)	82.7 Cu, 4.3 Fe, 13.0 Al	F,R
C62500 (AMPCO 21, Wearite 4-13)(c)	82.0 Cu, 3.0 Fe, 10.0 Al, 5.0 Ni	F,R
C63000 (aluminum bronze, E)	82.0 Cu, 4.0 Fe, 9.0 Al, 5.0 Ni	F,R
C63200	95.5 Cu, 3.5 Al, 1.0 Si	R,W
C63600	95.0 Cu, 2.8 Al, 1.8 Si, 0.40 Co	F
C63800	91.2 Cu, 7.0 Al	F,R
C64200	98.5 Cu, 1.5 Si	R,W,T
C65100 (low-silicon bronze, B)	95.44 Cu, 3 Si, 1.5 Sn, 0.06 Cr	F
C65400	97.0 Cu, 3.0 Si	F,R,W,T
C65500 (high-silicon bronze, A)	70.0 Cu, 28.8 Zn, 1.2 Mn	F,W
C66700 (manganese brass)	58.5 Cu, 36.5 Zn, 1.2 Al, 2.8 Mn, 1.0 Sn	F,R
C67400	58.5 Cu, 1.4 Fe, 39.0 Zn, 1.0 Sn, 0.1 Mn	R,S
C67500 (manganese bronze, A)	77.5 Cu, 20.5 Zn, 2.0 Al, 0.1 As	T
C68700 (aluminum brass, arsenical)	73.5 Cu, 22.7 Zn, 3.4 Al, 0.40 Co	F
C68800 (Alcoloy)	73.3 Cu, 3.4 Al, 0.6 Ni, 22.7 Zn	F
C69000	81.5 Cu, 14.5 Zn, 4.0 Si	R
C69400 (silicon red brass)	96.2 Cu, 3 Ni, 0.65 Si, 0.15 Mg	F
C70250	92.4 Cu, 1.5 Fe, 5.5 Ni, 0.6 Mn	F,T
C70400	88.7 Cu, 1.3 Fe, 10.0 Ni	F,T
C70600 (copper-nickel, 10%)	79.0 Cu, 21.0 Ni	F,W,T
C71000 (copper-nickel, 20%)	75 Cu, 25 Ni	F
C71300	70.0 Cu, 30.0 Ni	F,R,T
C71500 (copper-nickel, 30%)	67.8 Cu, 0.7 Fe, 31.0 Ni, 0.5 Be	F,R,W
C71700	88.2 Cu, 9.5 Ni, 2.3 Sn	F,R,W,T
C72500 (copper-nickel tin bearing)	72.0 Cu, 10.0 Zn, 18.0 Ni	F,R,W,T
C73500	65.0 Cu, 25.0 Zn, 10.0 Ni	F,W
C74500 (nickel silver, 65-10)	65.0 Cu, 17.0 Zn, 18.0 Ni	F,R,W
C75200 (nickel silver, 65-18)	65.0 Cu, 20.0 Zn, 15.0 Ni	F
C75400 (nickel silver, 65-15)	65.0 Cu, 23.0 Zn, 12.0 Ni	R,W
C75700 (nickel silver, 65-12)	59.0 Cu, 29.0 Zn, 12.0 Ni	F,T
C76200	55.0 Cu, 27.0 Zn, 18.0 Ni	F,R,W
C77000 (nickel silver, 55-18)	82.0 Cu, 16.0 Ni, 0.5 Cr, 0.8 Fe, 0.5 Mn	F,T
C77200	65.0 Cu, 2.0 Pb, 25.0 Zn, 8.0 Ni	F
C78200 (leaded nickel silver, 65-8-2)		

(a) F, flat products; R, rod; W, wire; T, tube; P, pipe; S, shapes. Ranges are from softest to hardest commercial forms. The strength of the standard copper alloy depends on the temper (annealed grain size or degree of cold work) and the section thickness of the mill product. Ranges cover standard tempers for each alloy. (b) Values are for as-hot-rolled material. (c) Values are for as-extruded material. (d) Rod 61.0 Cu min. Source: Ref 6, 7

Table 3.6 Chemical composition of cast copper and copper alloys

UNS designation	Nominal composition(a), %
Coppers	
C80100	99.95 Cu + Ag min, 0.05 other max
C80300	99.95 Cu + Ag min, 0.034 Ag min, 0.05 other max
C80500	99.75 Cu + Ag min, 0.034 Ag min, 0.02 B max, 0.23 other max
C80700	99.75 Cu + Ag min, 0.02 B max, 0.23 other max
C80900	99.70 Cu + Ag min, 0.034 Ag min, 0.30 other max
C81100	99.70 Cu + Ag min, 0.30 other max
High-copper alloys	
C81300	98.5 Cu min, 0.06 Be, 0.80 Co, 0.49 other max
C81400	98.5 Cu min, 0.06 Be, 0.80 Cr, 0.40 other max
C81500	98.0 Cu min, 1.0 Cr, 0.50 other max
C81700	94.25 Cu min, 1.0 Ag, 0.4 Be, 0.9 Co, 0.9 Ni
C81800	95.6 Cu min, 1.0 Ag, 0.4 Be, 1.6 Co
C82000	96.8 Cu, 0.6 Be, 2.6 Co
C82100	97.7 Cu, 0.5 Be, 0.9 Co, 0.9 Ni
C82200	96.5 Cu min, 0.6 Be, 1.5 Ni
C82400	96.4 Cu min, 1.70 Be, 0.25 Co
C82500	97.2 Cu, 2.0 Be, 0.5 Co, 0.25 Si
C82600	95.2 Cu min, 2.3 Be, 0.5 Co, 0.25 Si
C82700	96.3 Cu, 2.45 Be, 1.25 Ni
C82800	96.6 Cu, 2.6 Be, 0.5 Co, 0.25 Si
Red brasses and leaded red brasses	
C83300	93 Cu, 1.5 Sn, 1.5 Pb, 4 Zn
C83400	90 Cu, 10 Zn
C83600	85 Cu, 5 Sn, 5 Pb, 5 Zn
C83800	83 Cu, 4 Sn, 6 Pb, 7 Zn
Semired brasses and leaded semired brasses	
C84200	80 Cu, 5 Sn, 2.5 Pb, 12.5 Zn
C84400	81 Cu, 3 Sn, 7 Pb, 9 Zn
C84500	78 Cu, 3 Sn, 7 Pb, 12 Zn
C84800	76 Cu, 3 Sn, 6 Pb, 15 Zn
Yellow brasses and leaded yellow brasses	
C85200	72 Cu, 1 Sn, 3 Pb, 24 Zn
C85400	67 Cu, 1 Sn, 3 Pb, 29 Zn
C85500	61 Cu, 0.8 Al, bal Zn
C85700	63 Cu, 1 Sn, 1 Pb, 34.7 Zn, 0.3 Al
C85800	58 Cu, 1 Sn, 1 Pb, 40 Zn
Manganese and leaded manganese-bronze alloys	
C86100	67 Cu, 21 Zn, 3 Fe, 5 Al, 4 Mn
C86200	64 Cu, 26 Zn, 3 Fe, 4 Al, 3 Mn
C86300	63 Cu, 25 Zn, 3 Fe, 6 Al, 3 Mn
C86400	59 Cu, 1 Pb, 40 Zn
C86500	58 Cu, 0.5 Sn, 39.5 Zn, 1 Fe, 1 Al
C86700	58 Cu, 1 Pb, 41 Zn
C86800	55 Cu, 37 Zn, 3 Ni, 2 Fe, 3 Mn
Silicon bronzes and silicon brasses	
C87200	89 Cu min, 4 Si
C87400	83 Cu, 14 Zn, 3 Si
C87500	82 Cu, 14 Zn, 4 Si
C87600	90 Cu, 5.5 Zn, 4.5 Si
C87800	82 Cu, 14 Zn, 4 Si
C87900	65 Cu, 34 Zn, 1 Si
Tin bronzes	
C90200	93 Cu, 7 Sn
C90300	88 Cu, 8 Sn, 4 Zn
C90500	88 Cu, 10 Sn, 2 Zn
C90700	89 Cu, 11 Sn
C90800	87 Cu, 12 Sn
C90900	87 Cu, 13 Sn
C91000	85 Cu, 14 Sn, 1 Zn
C91100	84 Cu, 16 Sn
C91300	81 Cu, 19 Sn

UNS designation	Nominal composition(a), %
Tin bronzes (continued)	
C91600	88 Cu, 10.5 Sn, 1.5 Ni
C91700	86.5 Cu, 12 Sn, 1.5 Ni
Leaded tin bronzes	
C92200	88 Cu, 6 Sn, 1.5 Pb, 4.5 Zn
C92300	87 Cu, 8 Sn, 4 Zn
C92400	88 Cu, 10 Sn, 2 Pb, 2 Zn
C92500	87 Cu, 11 Sn, 1 Pb, 1 Ni
C92600	87 Cu, 10 Sn, 1 Pb, 2 Zn
C92700	88 Cu, 10 Sn, 2 Pb
C92800	79 Cu, 16 Sn, 5 Pb
C92900	84 Cu, 10 Sn, 2.5 Pb, 3.5 Ni
High-leaded tin bronzes	
C93200	83 Cu, 7 Sn, 7 Pb, 3 Zn
C93400	84 Cu, 8 Sn, 8 Pb
C93500	85 Cu, 5 Sn, 9 Pb
C93700	80 Cu, 10 Sn, 10 Pb
C93800	78 Cu, 7 Sn, 15 Pb
C93900	79 Cu, 6 Sn, 15 Pb
C94000	70.5 Cu, 13 Sn, 15 Pb, 0.50 Zn, 0.75 Ni, 0.25 Fe, 0.05 P, 0.35 Sb
C94100	70.0 Cu, 5.5 Sn, 18.5 Pb, 3.0 Zn, 1.0 other max
C94300	70 Cu, 5 Sn, 25 Pb
C94400	81 Cu, 8 Sn, 11 Pb
C94500	73 Cu, 7 Sn, 20 Pb
Nickel-tin bronzes	
C94700	88 Cu, 5 Sn, 2 Zn, 5 Ni
C94800	87 Cu, 5 Sn, 5 Ni
C94900	80 Cu, 5 Sn, 5 Pb, 5 Zn, 5 Ni
Aluminum bronzes	
C95200	88 Cu, 3 Fe, 9 Al
C95300	89 Cu, 1 Fe, 10 Al
C95400	85 Cu, 4 Fe, 11 Al
C95410	85 Cu, 4 Fe, 11 Al, 2 Ni
C95500	81 Cu, 4 Ni, 4 Fe, 11 Al
C95600	91 Cu, 7 Al, 2 Si
C95700	75 Cu, 2 Ni, 3 Fe, 8 Al, 12 Mn
C95800	81 Cu, 5 Ni, 4 Fe, 9 Al, 1 Mn
Copper-nickels	
C96200	88.6 Cu, 10 Ni, 1.4 Fe
C96300	79.3 Cu, 20 Ni, 0.7 Fe
C96400	69.1 Cu, 30 Ni, 0.9 Fe
C96600	68.5 Cu, 30 Ni, 1 Fe, 0.5 Be
C96700	67.6 Cu, 30 Ni, 0.9 Fe, 1.15 Be, 0.15 Zr, 0.15 Ti
Nickel silvers	
C97300	56 Cu, 2 Sn, 10 Pb, 12 Ni, 20 Zn
C97400	59 Cu, 3 Sn, 5 Pb, 17 Ni, 16 Zn
C97600	64 Cu, 4 Sn, 4 Pb, 20 Ni, 8 Zn
C97800	66 Cu, 5 Sn, 2 Pb, 25 Ni, 2 Zn
Leaded coppers	
C98200	76.0 Cu, 24.0 Pb
C98400	70.5 Cu, 28.5 Pb, 1.5 Ag
C98600	65.0 Cu, 35.0 Pb, 1.5 Ag
C98800	59.5 Cu, 40.0 Pb, 5.5 Ag
Special alloys	
C99300	71.8 Cu, 15 Ni, 0.7 Fe, 11 Al, 1.5 Co
C99400	90.4 Cu, 2.2 Ni, 2.0 Fe, 1.2 Al, 1.2 Si, 3.0 Zn
C99500	87.9 Cu, 4.5 Ni, 4.0 Fe, 1.2 Al, 1.2 Si, 1.2 Zn
C99600	58 Cu, 2 Al, 40 Mn
C99700	56.5 Cu, 1 Al, 1.5 Pb, 12 Mn, 5 Ni, 24 Zn
C99750	58 Cu, 1 Al, 1 Pb, 20 Mn, 20 Zn

(a) Nominal composition unless otherwise noted. For seldom used alloy, only compositions are available. Source: Ref 6, 7

and drawability of the rod. These properties are extremely sensitive to factors, such as dissolved impurities; refractory, oxide, or slag inclusions; and cavities caused by porosity, seams, surface defects, or piping. The quality of CCR exceeds that of wire bars with respect to all of these factors. Rod must display a drawability index between 130 and 270×10^3 at 5,000 to 10,000 kg (11,000 to 22,000 lb) (Ref 12–14).

Wire Bars

Wire bars deserve mention because they constituted the most important standard unit of trade for raw copper for many years. Wire bars are cigar-shaped ingots with roughly square cross sections

Fig. 3.1 Copper cathodes, as produced in an electrolytic refinery. This refinery shape remains a standard of copper trade contracts. Source: ASARCO, Inc.

and tapered ends. Quality requirements are spelled out in ASTM B 5. Along with chemical composition, physical properties, weights, workmanship, finish, and appearance, the standard specifies two standard size ranges: 91 to 104 kg (200 to 230 lb) and 109 to 136 kg (240 to 300 lb). Both types are 137.2 cm (54 in.) long, and the heavier version are slightly wider. Other sizes are traded by agreement between manufacturers and purchasers (Ref 1).

Billets

Billets are made by continuous casting. Billets for extrusion are cast as solid cylinders up to 300 mm (1 ft) in diameter and weigh between 45 and 180 kg (100 and 400 lb). For some applications, billets are cast as rectangular shapes up to 7.3 m (20 ft) long and weigh up to 11 Mt (12.5 ton).

Ingots

Ingots are chill cast in metallic molds. Shapes vary from simple oblongs (for remelting by foundries) to cakes (large slabs suitable for rolling to plate or sheet). Ingots are also continuously and semicontinuously cast. Dimensions are not standardized but fixed by refiners or requested by customers. Cakes can be up to 300 mm (1 ft) thick and weigh as much as 27 Mt (30 ton) (Ref 1).

Miscellaneous Products

In addition to these standard refinery shapes, some refineries also supply copper powder, copper sulfate, and copper anodes for electroplating.

Quality Requirements

Where appropriate, both composition and electric conductivity can be specified. The maximum mass resistivity for wire bars, cakes, slabs, and billets for electrical use is specified by ASTM B 5 at 0.15328 $\Omega g/m^2$, which corresponds to a volume resistivity of 0.017241 $\Omega mm^2/m$. These values are equivalent to a conductivity of 100% of the International Annealed Copper Standard (IACS) at 20 °C (68 °F). The maximum mass resistivity for other than electrical uses is established by ASTM B 5 as 0.15694 $\Omega g/m^2$ (conductivity, 97.66% IACS) at 20 °C (Ref 1, 15).

Under ASTM B 5, variations of ±5% in weight or 6.4 mm (1/4 in.) in any dimension from the manufacturer's published list or the purchaser's specified size are permitted. Wire bars may vary in length ±1% from the listed or specified value. Cakes are allowed to vary ±3% from the listed or specified size in any dimension greater than 203 mm (8 in.). The weight of copper ingots and ingot bars should not exceed those specified by more than 10%, but weight variations are otherwise not considered important (Ref 1).

Wire bars, cakes, slabs, and billets should be free from molding imperfections, such as shrinkage holes, cold sets, pits, sloppy edges, concave tops, and similar defects. This requirement does not apply to ingots or ingot bars in which physical defects are of no consequence.

When mechanical properties are specified, North American practice generally further identifies alloys by the temper, or mechanical/metallurgical state. Temper designations codified under ASTM B 601 are listed in Table 3.8.

Table 3.7   Composition limits for refinery shapes

Element group	Element	BS DD 78 Max. conc., ppm	BS DD 78 Group total, ppm	ASTM B 115 Max. conc., ppm	ASTM B 115 Group total, ppm	LME Cathodes HML	LME Cathodes HMG-s	LME Wire-bar, HMG	ENAMI Cathodes, ER	SMP Cathodes, EW
...	Cu	...	...	...	...	99.95	99.99	99.90	...	...
1	Se	2.0	3.0	4.0	3.0	<1.0	<1.0	<1.0	0.5	...
	Te	2.0	3.0	2.0	3.0	<0.5	<0.5	<0.5	0.5	1.0
	Bi	2.0	3.0	2.0	3.0	<1.0	<1.0	<1.0	0.1	1.0
	Cr	...	15.0	...	...	...	...	...	...	...
	Mn	...	15.0	...	...	...	...	...	...	...
	Sb	4.0	15.0	5.0	...	<2.0	<1.0	<1.0	1.0	1.0
	Cd	...	15.0	...	...	...	...	...	...	...
	As	5.0	15.0	5.0	...	<2.0	<1.0	<1.0	0.8	1.0
	P	...	15.0	...	...	...	...	...	...	...
3	Pb	5.0	5.0	8.0	8.0	2–4	2–4	3/5	0.5	1.0
4	S	15.0	15.0	25.0	25.0	8–15	8–12	5–10	5.9	5.0
5	Sn	...	20.0	10.0	...	<1.0	<1.0	<1.0	0.2	1.0
	Ni	...	20.0	10.0	...	2–5	1–4	3–5	0.8	1.0
	Fe	10.0	20.0	15.0	...	4–8	2–4	10–15	0.9	1.0
	Si	...	20.0	...	...	...	...	...	...	...
	Zn	...	20.0	...	...	1–3	1–3	2–5	...	...
	Co	...	20.0	...	...	...	...	...	...	...
6	Ag	25.0	25.0	25.0	25.0	12–20	10–15	10–20	8.0	1.0
7	O	...	...	200.0	200	...	...	200–300	45	70
Total		65.0		90.0						

Fig. 3.2   Copper rod—the principal intermediate product for wire manufacture. Source: ASARCO, Inc.

Table 3.8 Temper designations for copper and copper alloys as defined in ASTM B 601

Temper designation	Temper name or material condition	Temper designation	Temper name or material condition
Cold-worked tempers		**Annealed tempers(a) (continued)**	
H00	$\frac{1}{8}$ hard	O82	Annealed to temper: $\frac{1}{2}$ hard
H01	$\frac{1}{4}$ hard	**Annealed tempers(c)**	
H02	$\frac{1}{2}$ hard	OS050	Average grain size 0.005 mm
H03	$\frac{3}{4}$ hard	OS010	Average grain size 0.010 mm
H04	Hard	OSO15	Average grain size 0.015 mm
H06	Extra hard	OS025	Average grain size 0.025 mm
H08	Spring	OS035	Average grain size 0.035 mm
H10	Extra spring	OS050	Average grain size 0.050 mm
H12	Special spring	OS070	Average grain size 0.070 mm
H13	Ultra spring	OS100	Average grain size 0.100 mm
H14	Super spring	OS120	Average grain size 0.120 mm
H50	Extruded and drawn	OS150	Average grain size 0.150 mm
H52	Pierced and drawn	OS200	Average grain size 0.200 mm
H55	Light drawn: light cold rolled	**Solution-treated temper**	
H58	Drawn general purpose	TB00	Solution heat treated
H60	Cold heading; forming	**Solution-treated and cold-worked tempers**	
H63	Rivet	TD00	TB00 cold worked to $\frac{1}{8}$ hard
H64	Screw	TD01	TB00 cold worked to $\frac{1}{4}$ hard
H66	Bolt	TD02	TB00 cold worked to $\frac{1}{2}$ hard
H70	Bending	TD03	TB00 cold worked to $\frac{3}{4}$ hard
H80	Hard drawn	TD04	TB00 cold worked to full hard
H85	Medium-hard-drawn electric wire	**Precipitation-hardened temper**	
H86	Hard-drawn electrical wire	TF00	TB00 and precipitation hardened
Cold-worked and stress-relieved tempers		**Cold-worked and precipitation-hardened tempers**	
HR01	H01 and stress relieved	TH01	TD01 and precipitation hardened
HR02	H02 and stress relieved	TH02	TD02 and precipitation hardened
HR04	H04 and stress relieved	TH03	TD03 and precipitation hardened
HR06	H06 and stress relieved	TH04	TD04 and precipitation hardened
HR08	H08 and stress relieved	**Precipitation-hardened and cold-worked tempers**	
HR10	H10 and stress relieved	TL00	TF00 cold worked to $\frac{1}{8}$ hard
HR50	Drawn and stress relieved	TL01	TF00 cold worked to $\frac{1}{4}$ hard
Cold-worked and order strengthened tempers		TL02	TF00 cold worked to $\frac{1}{2}$ hard
HT04	H04 and order heat treated	TL04	TF00 cold worked to full hard
HT06	H06 and order heat treated	TL08	TF00 cold worked to spring
HT08	H08 and order heat treated	TL10	TF00 cold worked to extra spring
As manufactured tempers		TR01	TL01 and stress relieved
M01	As sand cast	TR02	TL02 and stress relieved
M02	As centrifugal cast	TR04	TL04 and stress relieved
M03	As plaster cast	**Mill-hardened tempers**	
M04	As pressure die cast	TM00	AM
M05	As permanent mold cast	TM01	$\frac{1}{4}$ HM
M06	As investment cast	TM02	$\frac{1}{2}$ HM
M07	As continuous cast	TM04	HM
M10	As hot forged and air cooled	TM06	XHM
M11	As hot forged and quenched	TM08	XHMS
M20	As hot rolled	**Quench-hardened tempers**	
M30	As hot extruded	TQ00	Quench hardened
M40	As hot pierced	TQ50	Quench hardened and temper annealed
M45	As hot pierced and re-rolled	TQ75	Interrupted quench hardened
Annealed tempers(a)		**Tempers of welded tubing(d)**	
O10	Cast and annealed(b)	WH00	Welded and drawn to $\frac{1}{8}$ hard
O20	Hot forged and annealed	WH01	Welded and drawn to $\frac{1}{4}$ hard
O25	Hot rolled and annealed	WM00	As welded from H00 strip
O30	Hot extruded and annealed	WM01	As welded from H01 strip
O40	Hot pierced and annealed	WM02	As welded from H02 strip
O50	Light annealed	WM03	As welded from H03 strip
O60	Soft annealed	WM04	As welded from H04 strip
O61	Annealed	WM06	As welded from H06 strip
O65	Drawing annealed	WM08	As welded from H08 strip
O68	Deep drawing annealed	WM15	WM50 and stress relieved
O70	Deep soft annealed		
O80	Annealed to temper: $\frac{1}{8}$ hard		
O81	Annealed to temper: $\frac{1}{4}$ hard		

(continued)

Table 3.8 (continued)

Temper designation	Temper name or material condition	Temper designation	Temper name or material condition
Temper of welded tubing(d)		**Temper of welded tubing(d)**	
WM20	WM00 and stress relieved	WO50	Welded and light annealed
WM21	WM01 and stress relieved	WR00	WM00; drawn and stress relieved
WM22	WM02 and stress relieved	WR01	WM01; drawn and stress relieved
WM50	As welded from O60 strip		

(a) To produce specified mechanical properties. (b) Homogenization anneal. (c) To produce prescribed average grain size. (d) Tempers of fully finished tubing that has been drawn or annealed to produce specified mechanical properties or that has been annealed to produce a prescribed average grain size are commonly identified by the appropriate H, O, or OS temper designation. Source: Ref 7, 16

Wire Mill Products

Wire mills produce electrical conductors. Products include round and flat wires, cables, straps, square and rectangular bars, angle shapes, channels, and special profiles. Wires can be in the form of single filaments, multiple filaments, or cable. Insulated and uninsulated wire can be fabricated into cables for various uses in a wide variety of forms and shapes (Ref 17, 18).

Copper-Base Conducting Materials

For power transmission, power distribution, and telecommunications, the most important conducting material is high-purity copper normally corresponding to electrolytic tough-pitch copper, C11000, or, for critical applications, one of the oxygen-free coppers, C10100 through C10800.

Copper Conductors. Copper wire and stranded conductors are normally supplied in hard-drawn or medium-hard-drawn tempers in sizes 4 AWG and larger. Twelve-wire strands are used occasionally in some sizes to obtain flexibility in larger diameters and to minimize corona effects. Three-wire strands are used extensively to obtain both practical wire sizes and create the triangular configuration that increases resistance to vibration-induced fatigue. Overhead conductors, which require higher strength than pure copper provides, can be made from copper-clad steel.

Copper-alloy conductors provide higher strengths, as well as improved abrasion and corrosion resistance, than are attainable in pure copper. The tradeoff is conductivity, which is lower than that of unalloyed copper. Trolley wire, used for railroad electrification in 2/0 to 4/0 AWG sizes, is the most familiar example of a copper-alloy conductor. Cadmium coppers, such as C16200 and C16500, are usually chosen for this application. The electrical resistivity of trolley wire ranges from 0.166 Ω/km and 0.267 Ω/km depending on diameter. Grooved forms of trolley wire are covered under ASTM B 9 (Ref 1, 15).

Bus bar, rod, and shape specifications are given in ASTM B 187; those for bronze shapes are given in ASTM B 249M. In the United States, copper-base bus bars are made from coppers C10100 through C12000 (Ref 1, 19).

Wire is a single-strand conductor. It may be bare or insulated, as shown in Fig. 3.3. The cross section of electrical wires is nearly always circular, but wires can also be flattened. Special cross sections are used for trolley wire (Ref 15, 17).

Wire Gages. Circular wire diameters are designated by gage numbers in commercial practice. The most common system used in the United States is the American Wire Gage (AWG) system, formerly known as the Brown and Sharpe Gage (B & S). The AWG sizes are defined by a geometric progression based on the ratio of the diameters of any two consecutive sizes in imperial (fps) units. The diameters follow the sequence of drawing dies used in wire manufacture. Thirty-nine sizes are defined (Ref 20), ranging between No. 36 (0.0050 in.) and No. 0000 or 4/0 (0.46 in.). Therefore, the ratio between consecutive sizes is:

$$\sqrt[39]{(0.460/0.0050)} = 1.12293$$

A few approximate but useful rules pertaining to the AWG include:

- An increase of three gage numbers (e.g., from No. 10 to 7) doubles the area and weight and consequently halves the dc resistance.
- An increase of six gage numbers (e.g., from No. 10 to 4) doubles the diameter.
- An increase of 10 gage numbers (e.g., from No. 10 to 1/0) multiplies the area and weight by 10 and divides the resistance by 10.
- A No. 10 AWG wire has a diameter of about 0.10 in. (2.54 mm), an area of about 10,000 circular mils, and a resistance of approximately 1.0 Ω per 1000 ft (302 m) for standard annealed copper at 20 °C.
- The weight of No. 2 AWG copper wire is very close to 200 lb per 1000 ft (268 kg per km).

The metric or millimeter wire gage assigns progressive numbers to increasing sizes. Thus, 0.1 mm (0.004 in.) diameter is No. 1, 0.2 mm is No. 2, etc. The German wire gage contains 25 sizes of diameters and thicknesses expressed in millimeters (Table 3.9).

Stranded Conductors

Stranded conductors are composed of a group of wires usually braided or twisted together and not insulated from one another.

A cable is composed of one or more stranded conductors. Cables can be bare, sheathed, or covered with a polymeric material, with or without a semiconducting or metallic shielding. They may also be armored with braided steel for mechanical protection.

Commercial cable designations have been issued under the following standards-based designations: ASTM, BSI (U.K.), Insulated Cable Engineers' Association (ICEA), International Electrotechnical Commission (IEC), Military (MIL), National Electric Code (NEC) (National Fire Protection Agency, Quincy, MA), National Electrical Manufacturers' Association (NEMA), Rural Electrification Administration (REA), Underwriters Laboratories (UL), and Verband Deutscher Electrotechniker (VDE) in Germany.

Stranded Concentric or Random Conductors. Concentric strands are composed of a central core surrounded by one or more layers of helically laid wires. If the wires are laid randomly, the cable is called "bunch cable" or toron. A toron whose strands are very fine and flexible is called a cord. When the outer layers have been smoothed, as by drawing, the cable is said to be compressed.

Mixed Conductors. Stranded conductors can be made from two different types of wire. Such cables are generally designed for physical and electrical characteristics different from those found in homogeneous materials. Cables combining copper and steel strands, for example, can be stretched over long spans, where greater than average strength must be combined with high conductivity. Such so-called copper cable steel reinforced (CCSR) products are made in a large variety of steel/copper combinations.

A similar product is made from copper-clad steel, marketed as Copperweld (Copperweld Fayetteville Division, Fayetteville, TN). In this instance, a copper sheath is metallurgically bonded to a steel core, forming a composite conductor. Such cables are available as individual filaments 0.2 mm (0.008 in.) and larger, containing a single heat-treated steel core, or as stranded conductors made up of these wires. Such products are also used for overhead lines, where strength, corrosion resistance, and conductance are important requirements. They can be used in combination with ordinary copper wires in a wide range of strengths and conductances. Typical uses include conducting guy wires; overhead conductors for telephone, telegraph, and signal lines; messenger wires for aerial cable; catenary cables for railroad electrification; and radio antennas.

Bare Wire and Cable

Bare wire and cables are used for overhead lines and grounding connections to keep cost and operating efficiency at optimal levels. Table 3.10 lists frequently used bare copper wire sizes. Among the

Cross sections of copper conductors

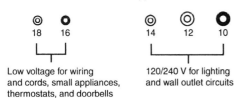

Low voltage for wiring and cords, small appliances, thermostats, and doorbells — 120/240 V for lighting and wall outlet circuits

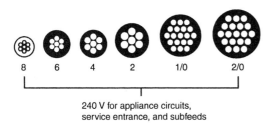

240 V for appliance circuits, service entrance, and subfeeds

Fig. 3.3 Sizes of copper conductors

Table 3.9 German wire gages

No.	Diameter, mm	No.	Diameter, mm
1	5.50	14	1.75
2	5.00	15	1.50
3	4.50	16	1.375
4	4.25	17	1.250
5	4.00	18	1.125
6	3.75	19	1.000
7	3.50	20	0.875
8	3.25	21	0.750
9	3.00	22	0.650
10	2.75	23	0.562
11	2.50	24	0.500
12	2.25	25	0.438
13	2.00		

Source: Ref 17

technical standards applicable to these products are ASTM B 1, B 2, B 3, and IEC 228 (Ref 21).

Magnet (Winding) Wire

Magnet wire is fine copper wire that has been coated with a polymeric insulation, which is usually called "enamel," although the polymer may not actually be an enamel in the chemical sense of the word. Maximum service temperature depends on the type of insulating polymer used (Table 3.11). Specifications for magnet wires are given in standards such as JIS C3103; NEMA MW-1000, MW-35C, and MW-15. The most common sizes range between 23 and 44 AWG. High-purity oxygen-free (OF) copper grades are used for high-temperature applications in reducing atmospheres or in motors/generators cooled by hydrogen gas. The OF coppers are immune to hydrogen embrittlement, which can arise when gaseous hydrogen reacts with the copper oxides found in tough-pitch coppers to form water vapor filled voids (Ref 21–23).

Telecommunication Cable

Telephone connections require wire pairs. Single pair, three wire, and multipair cables are manufactured in a variety of designs. All telephone cables use colored polyvinyl chloride or polypropylene insulation to identify the individual wires in the cable.

Multipair cables use a polyethylene sheath that can be shielded with copolymer-coated aluminum foil. The sheath or foil is sometimes placed over a nonhygroscopic tape wrapping. Some multipair cables are bound to a galvanized steel supporting cable; others incorporate a galvanized steel messenger wire. For waterproofing, petroleum jelly may be injected to fill interstices.

Building Wire and Cable

Building wire is used to distribute electric power inside residential, commercial, or industrial buildings at potentials of 600 V or less. Copper building wire sizes generally range from No. 14 AWG through 750 kcmil. Solid wire is used from No. 14 through 10 AWG. Beginning with No. 8 AWG, the wire is stranded to provide flexibility. Selection of building wires is based on safe operating temperatures. Maximum currents (ampacities) for insulated wires are listed in Table 3.12.

For distribution mains inside buildings, the NEC requires conductors containing insulated wires for neutral and hot phases plus a bare wire for grounding (Fig. 3.1). Wires frequently incorporate color coding systems for ready identification: grounding conductors are normally green or natural gray (or bare), "hot" phases are red and black, and neutral conductors may be white or gray. The following is a general description of conductor types as published by the Copper Development Association (Ref 20).

Nonmetallic (NM, NM-B) and NM Corrosion Resistant (NMC, NMC-B) Cable. Nonmetallic sheathed cable is a factory assembly of two or

Table 3.10 Common bare copper wire sizes

Gage, AWG	Section, mm²	Diameter, mm	Weight, kg/km	Ampacities(a)
12	3.31	2.05	29.4	45
...	4.00	2.25	35.3	50
10	5.26	2.59	46.8	61
...	6.00	2.76	53.2	64
9	6.63	2.91	59.0	69
8	8.37	3.26	74.4	81
...	10.00	3.57	89.0	89
7	10.55	3.66	93.8	93
6	13.30	4.11	118.2	108
...	16.00	4.50	141.4	118
5	16.77	4.62	149.1	125
4	21.15	5.19	188.0	145

(a) Ampacities of bare or covered conductors based on 40 °C (104 °F) ambient temperature, 75 °C (167 °F) total conductor temperature, 61 cm/s (24 in./s) wind speed. Source: Ref 21

Table 3.11 Characteristics of common magnet wires

Designation	Description	Operating temperature, max	Applications
COSOL	Copper, polymethane	105 °C	Recommended use electromagnetic coils, regulator coils, and when many welds are required as there is no need for insulation removal for welding
BONDEZEM	Polyester, copper, cementable layer	150 °C	Electric motors, ballast coils, and applications where ribbons or impregnating varnishes for coil support can be eliminated
HTH	Polyester, copper, amideimide	200 °C	Motors, dry transformers, and electrical equipment operating up to 200 °C
HHTH	Polyester, copper, amideimide	200 °C	Air tight motors for refrigeration, using Freon 12 or 22, and applications similar to HTH

Source: Ref 22

more insulated conductors with a nonmetallic outer sheath that is both moisture resistant and flame retardant. The cable may have an insulated or bare conductor for equipment grounding purposes. To conform to NEC requirements, the cable must be approved type NM or NMC; conductor sizes range from No. 14 through 2 AWG. Type NM cable is used in normally dry locations. It can be placed in air voids of masonry block or tile walls where the walls are not exposed to excessive moisture or dampness. However, it should not be embedded in masonry, concrete, fill, or plaster. Type NMC is selected for applications that may be moist, damp, or corrosive (Ref 20).

Armored cable (AC) is a fabricated assembly of insulated conductors in a flexible metallic en-

Table 3.12   Ampacities of insulated conductors rated 0 to 2000 V

Wire size, AWG or kcmil	Ampacity, A, of copper insulated conductors				Ampacity, A, of aluminum or copper-clad aluminum insulated conductors			
	At 60 °C, types TW(a), UF(a)	At 75 °C, types FEPW(a), RH(a), RHW(a), THHW(a), THWN(a) XHHW(a), USE, ZW(a)	At 85 °C, type V	At 90 °C, types TA, TBS, SA, SIS, FEP(a) FEPB(a), RHH(a), THHN(a), THHW(a)	At 60 °C, types TW(a), UF(a)	At 75 °C, types RH(a), RHW(a), THHW(a), THW(a), THWN(a), XHHW(a), USE(a)	At 85 °C, type V	At 90 °C, types TA, TBS, SA, SIS, RHH(a), THHW(a), THHN(a)
18	...	...	...	14	...	...	...	...
16	...	...	18	18	...	...	...	...
14	20(a)	20(a)	25	25(a)	...	...	...	...
12	25(a)	25(a)	30	30(a)	20(a)	20(a)	25	25(a)
10	30	35(a)	40	40(a)	25	30(a)	30	35(a)
8	40	50(a)	55	55	30	40	40	45
6	55	65	70	75	40	50	55	60
4	70	85	95	95	55	65	75	75
3	85	100	110	110	65	75	85	85
2	95	115	125	130	75	90	100	100
1	110	130	145	150	85	100	110	115
1/0	125	150	165	170	100	120	130	135
2/0	145	175	190	195	115	130	145	150
3/0	165	200	215	225	130	155	170	175
4/0	195	230	250	260	150	180	195	205
250	215	255	275	290	170	205	220	230
300	240	285	310	320	190	230	250	255
350	260	310	340	350	210	250	270	280
400	280	335	365	380	225	270	295	305
500	320	380	415	430	260	310	335	350
600	355	420	460	475	285	340	370	385
700	385	460	500	520	310	375	405	420
750	400	475	515	535	320	385	420	435
800	410	490	535	555	330	395	430	450
900	435	520	565	585	355	425	465	480
1000	455	545	590	615	375	445	485	500
1250	495	590	640	665	405	485	525	545
1500	520	625	680	705	435	520	565	585
1750	545	650	705	735	455	545	595	615
2000	560	665	725	750	470	560	610	630
Ambient temperature(b), °C				**Ampacity correction factors**				
21–25	1.08	1.05	1.04	1.04	1.08	1.05	1.04	1.04
26–30	1.00	1.00	1.00	1.00	1.00	1.00	1.00	1.00
31–35	0.91	0.94	0.95	0.96	0.91	0.94	0.95	0.96
36–40	0.82	0.88	0.90	0.91	0.82	0.88	0.90	0.91
41–45	0.71	0.82	0.85	0.87	0.71	0.82	0.85	0.87
46–50	0.58	0.80	0.80	0.82	0.58	0.75	0.80	0.82
51–55	0.41	0.74	0.74	0.76	0.41	0.67	0.74	0.76
56–60	...	0.67	0.67	0.71	...	0.58	0.67	0.71
61–70	...	0.52	0.52	0.58	...	0.33	0.52	0.58
71–80	...	0.30	0.30	0.41	...	...	0.30	0.41

Not more than three conductors in raceway or cable or earth (directly buried) (a) Unless otherwise specifically permitted elsewhere in this code, the overcurrent protection shall not exceed 15 A for 14 AWG, 20 A for 12 AWG, and 30 A for 10 AWG copper; or 15 A for 12 AWG and 25 A for 10 AWG aluminum and copper-clad aluminum after any correction factors for ambient temperature and number of conductors have been applied. (b) For ambient temperatures other than 30 °C, multiply the ampacities by the appropriate factor.

closure. The armored casing provides mechanical protection and can also serve as equipment-grounding conductor (although separate copper grounding conductors are now preferred). Type AC conductors have a moisture-resistant and fire-retardant fibrous covering. Type AC cables, except ACL (lead), have an integral bonding strip of copper or aluminum that is in intimate contact with the armor for its entire length. It is listed for 600 V applications, primarily branch circuits and feeders; copper conductor sizes are No. 14 through 1 AWG (Ref 20).

Metal-clad (MC) cable is a cable with individually insulated conductors enclosed in metallic sheath, which may be in the form of an interlocking tape or simply a tube. Its sturdy construction allows it to be used in a wide variety of building wire applications (Ref 20).

Service entrance (SE) or underground SE (USE) cable is a moisture resistant but not flame retarding jacketed product. An SE cable can be used for interior wiring as if it were NM if all the conductors have either rubber or thermoplastic insulation. Type USE is a cable with moisture resistant but not flame retarding insulation. It is primarily used underground (Ref 20).

Underground feeder (UF) cable is made with a flame-retardant covering, but it is also moisture-,

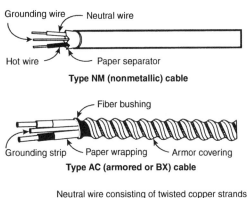

Type NM (nonmetallic) cable

Type AC (armored or BX) cable

Type SE (service entrance) cable

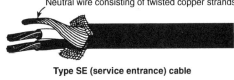

Type UF (underground feeder) cable

Fig. 3.4 Types of cables

fungus- and corrosion-resistant, making it suitable for direct burial. It may contain an insulated or bare conductor for equipment grounding. Type UF is not normally used for applications inside buildings (Ref 20).

TC cable is a factory assembly of two or more insulated conductors in a nonmetallic sheath. It may be supplied with or without grounding conductors. The grounding conductors may be bare or insulated. Type TC ranges from No. 18 AWG through 1000 kcmil copper. It is used for power, lighting, control and signal circuits and is installed in cable trays or raceways. It may also be supported by messenger wire.

Fire-Resistant Cables. Polyvinyl chloride (PVC), neoprene, and other polymers produce halogen gases (usually containing chlorine) when burned or heated above dissociation temperatures. Smoke containing these gases impairs human breathing and visibility. It can also have a corrosive effect on buildings and equipment. Fire-safe insulation materials that do not produce hazardous gases include ethylvinylacetate (EVA) and carbopolyamine. Features of fire-safe sheathed cables are specified in the following standards: ICEA S-61-402; IEC 227, 228, 502, and 754-1; VDE 0207-23,24; BS 6899-9 and 2782-1; IEC 332-3-82; and ASTM E 62-83 and D 2863-77 (Ref 20).

Mineral insulated (MI) cables consist of high-conductivity copper conductors embedded in a densely compacted mineral insulator of magnesium oxide. They are enclosed by a robust yet ductile solid drawn copper, copper alloy, or stainless steel sheath, which is both liquid and gas tight. Mineral insulated cable is labeled by Underwriters Laboratories for sizes No. 16 AWG to 250 kcmil for one conductor, No. 16 to 4 AWG for two and three conductors, No. 16 to 6 AWG for four conductors, and No. 16 to 10 AWG for seven conductors. The cable is rated for 600 V (Ref 24, 25).

Power Cable

The general term "power cable" applies to products used primarily for the transmission and distribution of power. Power cables can be used inside buildings for local distribution, but they are more commonly used outdoors, under or above ground.

Insulated power distribution cables are insulated to avoid accidental contact, to prevent grounding, and to reduce or eliminate corona losses, inductive influence, and mechanical damage. The several types of cable sheathing are classified as follows:

- *Integral insulation:* The conductor has only one covering (Fig. 3.5) that acts both as electrical and mechanical insulation (Ref 23).
- *Coating:* The conductor (Fig. 3.6) has a polymeric protecting film over the first layer of insulation (Ref 21).
- *Armor:* The conductor is protected from mechanical damage (Fig. 3.7) by a metallic cover in addition to the polymeric insulation (Ref 21).
- *Screen or shield:* The conductor receives additional electrical protection (Fig. 3.8) from a layer of copper or aluminum foil or copper wire mesh or strip (Ref 21).
- *Semiconducting sheathing:* The conductor is covered with semiconducting wrappings that alternate with ordinary polymeric sheathing (Ref 21).

Cables for low voltage (up to 1 kV) distribution lines are frequently insulated with cross-linked polyethylene (XLPE), with or without a PVC cover. An EVA insulation is used for applications requiring low smoke and nontoxic gas emission in the event of fire. Such fire-safe cables are available in singlestrand and multistrand configurations. Examples are listed in Table 3.13 (Ref 22).

Insulated High-Voltage Distribution Cable. Armored cables with XLPE insulation and rubber, PVC, or polyurethane sheathing, with or without polypropylene fill, are used for underground power distribution and mining applications. Such cables operate at up to 15 kV. Cables have laminated (paper wrapped) insulation. They may contain several layers of conductors and insulation and are often fitted with a copper wire mesh or foil shielding (known as a screen) for grounding, protection against voltage transients, and eddy current transport. Sometimes they also contain a steel wire mesh for mechanical protection. Examples are listed in Table 3.13 (Ref 21–23).

Appliance Wire and Cordage

Cordage for domestic appliances and instruments is manufactured from No. 34 AWG bunch-stranded soft wire, which, in some cases, is tinned for ease of soldering or phase identification. Insulation is most commonly PVC, but neoprene, ethylene propylene, and proprietary compositions are also used. For higher electrical loads, polypropylene filler may be used. Cords for electric irons or stoves may have cotton filling. Lamp cord utilizes soft copper No. 34 AWG conductors insulated with PVC (Ref 21, 23).

Cords for heavy-duty equipment, such as welding machines, must be very flexible and able to

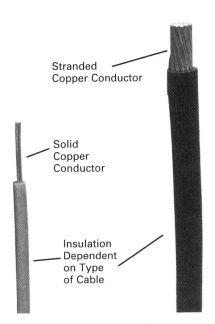

Fig. 3.5 Typical conductors with integral insulation. Source: Copper and Brass Development Association, Canada

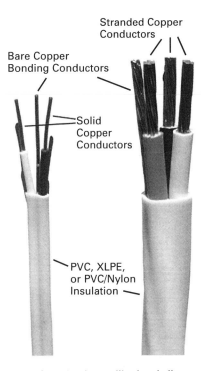

Fig. 3.6 Conductor with insulation and exterior coating. Source: Copper and Brass Development Association, Canada

transmit high current and withstand mechanical stress. Some types must also resist flames, outdoor exposure, acids, oil, grease, abrasion, or welding sparks. Conductors consist of stranded No. 34 AWG soft copper. Strands may be covered with polyester film and protected with heavy neoprene insulation, with or without cotton filling.

Automotive Wire and Cable

Automotive wire and cable requires insulation that is resistant to elevated temperatures, petroleum products, humidity, fire, and chemicals. PVC, neoprene, and polyethylene are most com-

Copper Conductor

Conductor Shield

XLPE or EPR Insulation

Insulation Shield

Copper Bonding Conductor

Copper Tape Shield

Fillers

Binder Tape

PVC Inner Jacket

Interlocking Aluminum or Galvanized Steel Armour

PVC Outer Jacket

Fig. 3.7 Armored cable as used in mines and industrial applications. Source: Copper and Brass Development Association, Canada

mon. Potentials range from 12 V for electrical systems to between 300 and 15,000 V for instrument, lighting, and ignition systems. In the United States, specifications for automotive cables are standardized by the Society of Automotive Engineers (SAE) and by individual automobile manufacturers. Automotive wire types and sizes corresponding to SAE J558 and SAE J557 are listed in Table 3.14 (Ref 21).

Brass Mill Products

Brass mills are the principal producers of wrought copper-alloy products. The semis they produce are used by other industries to make finished goods. Semis are classified according to physical shape.

Sheet, Strip, and Plate

Sheets are flat-rolled products up to and including approximately 4.8 mm (0.188 in.) thick and over 500 mm (20 in.) wide. Strip is defined as any flat product other than flat wire. Thickness follows the range for sheet, although strip is furnished in widths between 32 and 305 mm (1.25 and 12 in.) with drawn or rolled edges. Plate is defined as flat product that is more than 4.8 mm (0.188 in.) thick and over 300 mm (12 in.) wide. Dimensional ranges for flat-rolled products are given in Table 3.15 (Ref 19, 26).

Edge condition is an important consideration in flat-rolled brass mill products. Sawed or slit edges, for example, may not be suitable for flexural applications, because they can initiate fatigue failures. Product with drawn, machined, or rolled edges can be used as supplied. Edge conditions available in flat-rolled products are given in Table 3.16 (Ref 26, 27).

Flat-rolled coppers are used in the manufacture of products ranging from roofing sheet to coinage and electrical components. Product is supplied in the annealed condition and in a range of as-rolled tempers. Tempers based on grain size are also available. The choice of grade and temper depends on the intended application. For example, OF coppers, C10100 to C10800, can be welded and brazed without risk of hydrogen embrittlement, unlike tough-pitch coppers, such as C11000. On the other hand, pure coppers anneal readily during brazing or soldering, while deoxidized coppers, grades containing small amounts of silver, and high-copper alloys retain as-rolled tempers (Fig. 3.9) to elevated temperatures (Ref 26, 27).

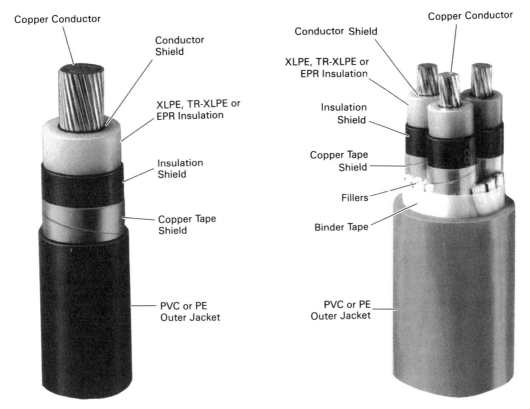

Fig. 3.8 Typical shielded conductor and shielded power cable. Source: Copper and Brass Development Association, Canada

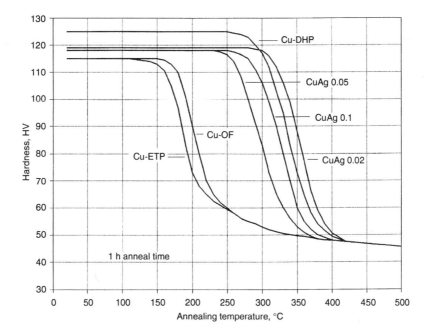

Fig. 3.9 Softening curves for various coppers

Table 3.13   Specifications of power distribution cables

Type	Sheathing	Installation	Standards	Standing design	Voltage	Service temperature	Gage	Nominal section, mm²	Conductor	Emergency temperature
XTU	XLPE + PVC	O, U, W	ICEA S-66-524	Single	600 V	90 °C	14–4/0 AWG 250–1000 MCM	2.08–506.7	Soft copper cable compacted over 6 AWG	130 °C
XTMU	XLPE with PVC filling and cover	O, U, W	ICEA S-66-524	Multi	600 V	90 °C	14–4/0 AWG 250–500 MCM	…	Soft copper wire or cable compacted over 6 AWG	130 °C
XU	XLPE	O, U	ICEA S-66-524	Single	600 V	90 °C	8–4/0 AWG 250e1000 MCM	8.37–506.7	Soft copper cable compacted over 6 AWG	130 °C
EVALEX(a), -S, -M	EVA insulation and cover	Public buildings, hospitals, mines	IEC 502, IEC 228, IEC 332-3, IEC 754-1 ASTM Fireproof, E662-79	Single, multi	0.6 A, 1.0 kV	90 °C		1.5–630	Soft copper wire and cable. Only cable if total section larger than 6 mm²	130 °C
PT	PE + PVC cover	O, U	ICEA S-61-402	Single	600 V	75 °C	8–4/0 AWG 250–1000 MCM	8.30–506.7	Soft copper wire or cable with wires concentrically distributed	…
NSYA	PVC	O	VDE 0250, IEC 228	Single	1000 V	70 °C	…	1.5–2.5 4.0–9.5	Soft copper wire or cable; individually insulated cables	…
PI	PE	…	ANSI-C8-35	…	1000 V	75 °C	12–4/0 AWC 250–500 MCM	…	Hard copper wire or cable	…
NYA	PVC	…	VDE 0250	…	1000 V	70 °C			Soft copper wire	…
NYAF	PE	…	VDE 0250	…	1000 V	70 °C			Soft copper flexible cable	…
NSYAF	PE	…	VDE 0250	…	1000 V	70 °C			Soft copper flexible wire or cable	…
THW	PE	…	UL-83	…	600 V	75 °C			Soft copper cable	…
THHN and THWN	XPE	…	UL-83, ANS-C83-80	…	600 V	90 °C			Electrolytic soft copper wire or cable	…
TPS	3 wires with PVC, shielded with PVC	…	VDE 0250	…	600 V	75 °C			Soft copper wire	…
USE-RHH	Ethylene-propylene, neoprene cover	…	UL-854 ICEA S-66-524	…	600 V	90 °C	12–3/0 AWC 4/0–1000 MCM		Soft copper cable	…
USE-RHHM	Ethylene-propylene, neoprene cover	…	UL-854 ICEA S-66-524	…	600 V	90 °C	…	3.301–107	Soft copper cable	…
E	EPDM	…	ICEA S-68-516	…	600 V	90 °C		6–240	Copper cable	…
X	XLPE	…	ICEA S-68-524	…	600 V	90 °C	…	6–240	Copper cable	…
ET	EPDM	…	ICEA S-68-515	…	0.6–5.8–15 kV 25–35 kV	90 °C	12–4/0 AWG 250–1000 MCM		Copper cable	…
XT	XLPE	…	ICEA S-66-524	…	0.6–5.8–15 kV 25–35 kV	90 °C	…	…	…	…
NYY-O	PVC thermoplastic insulation plus filling and cover	U, W	VDE 0271	Multi	1000 V	70 °C	…	…	…	…

O is overhead; U is underground; W is underwater. (a) Trademark of Cocesa Chile, low smoke and polluting gases emission in fire emergencies. Source: Ref 21, 23

Copper-alloy strip is also manufactured in a range of tempers from soft to extra spring. Again, the choice of temper depends on the application and the degree of deformation required to manufacture the finished product. Simple electrical springs, for example, may be stamped from hard temper strip, while complex-shaped connectors and leadframes may require softer starting material. Table 3.17 lists common electrical spring alloys according to both UNS and DIN designations. Other products of this type are listed in Table 3.15. It should be noted that compositional tolerances called out by the two standards systems do not always match exactly (Ref 28, 29).

Copper foil is a special form of sheet and strip (Table 3.18). It is produced both by rolling and by electrodeposition. Both forms are used in the manufacture of electronic circuit boards. In general, the rolled product is used in applications where flexibility is required, while the electrodeposited product is used in rigid products. Copper foil that is nominally less than 0.51 mm (0.02 in.) thick is covered under ASTM B 451 (Ref 16).

Electrodeposited copper foil is formed on rotating stainless steel drums from which it is continuously stripped. Foil can also be deposited directly on printed circuit boards (PCBs) using patented electrolytic or electroless plating processes. Foil thicknesses for PCBs have fallen from 70 to 35 μm and as little as 12 μm ($\frac{1}{3}$ ounce per ft^2) as a result of component miniaturization. Copper laminations as thin as 5 to 10 μm are formed on aluminum, which is removed by etching once the copper has bonded to the substrate (Ref 30).

Electronic Strip. Copper and copper-alloy electronic strip, which is normally produced by rolling but can also be electrodeposited, is typically 0.5 mm (0.02 in.) or more in thickness and up to few millimeters wide. The strip is used in electronic connectors. Popular connector alloys are listed in Table 3.19 (Ref 28, 31).

Table 3.14   Specifications of automotive wires

Type	Max. service temperature, °C	Max. voltage ampacity, V/A	AWG	Standard
TAC	105	300 V	10–20	SAE
PVIC-7	75	15,000 V	16	SAE
VBC	105	300 V	2–6	SAE
PRT	30	30–37 A	10–20	SAE J558
IT-7	…	…	16	SAE J557
SGT	40	189–192 A	1–6	SAE J558
SGR	40	223–109 A	1–6	SAE J558

Source: Ref 21

Table 3.15   Flat products (including rectangles and squares) furnished in rolls or in straight lengths (in British units)

| Thickness, in. | Products available at a given width | | | |
	≤1.25 in.	>1.25 to 12 in.	>12 to 24 in.	>24 in.
≤188	Strip(a)	…	…	…
	Flat wire(b) (including square wire)	Strip	…	Sheet
>188	Bar(c)	…	Plate	…

(a) Product originally produced with slit, sheared, or sawed edges, whether or not such edges are subsequently rolled or drawn. (b) Product with all surfaces rolled or drawn, without previously having been slit, sheared, or sawed. (c) When bar is ordered, it is particularly desirable that the type of edge be specified. Source: Ref 26

Table 3.16   Typical edges available with flat products

| Edge available | Product | | | | |
	Flat and square wire	Strip	Sheet	Plate	Bar
Drawn flat products					
Drawn	•	•		•	
Slit		•	•	•	•
Sheared, sawed		•	•	•	•
Rolled flat products					
Machined		•	•	•	•
Rolled	•	•		•	

Note: The listing of a product in the above diagram does not necessarily signify that the edge indicated is commercially available in the complete size range for that product. Source: Ref 26

Table 3.17   Copper alloy spring materials

UNS(a)	DIN
C50500	CuSn2
C51100	CuSn4
C51000	CuSn5
C51900	CuSn6
C52100	CuSn8
C42500	CuSn3Zn9
C23000	CuZn15
C26000	CuZn30
C27000	CuZn36
C27200	CuZn36
C27200	CuZn37
C27400	CuZn37
C68800	CuZn23Al3Co
C72500	CuNi9Sn2
C75700	CuNi12Zn24
C76400	CuNi18Zn20
C77000	CuNi18Zn27
C19400	CuFe2P
C17000	CuBe1.7
C17200	CuBe2
C17500	CuCo2Be
C17510	CuNi2Be

(a) Tolerance domains for UNS classified alloys do not always overlap with DIN standards.

Mechanical Wire

Copper and copper-alloy wire used for other than purely electrical applications (but may include electrical or electronic uses) is known as mechanical wire. Such wire has a variety of uses in fasteners and small machined or cold-headed components.

Plain drawn wire is supplied in hard and soft tempers in sizes from 0.03 to 3.0 mm (0.001 to 0.12 in.). It is produced in C10200 and C11000 coppers and in alloy C16200 (cadmium copper). Standards for the product are addressed by DIN 46 431, DIN 40 500 Part 4, and ASTM B 1 and B 3.

Tinned copper wire is available with either electrolytic or hot-dipped tin coatings. Properties are covered by DIN 40 500 Part 5 (Class 1–5) and ASTM B 33. Hot-dip tinned wires are drawn to size before tinning and then passed through a bath of molten tin. Electrolytically tinned wires are produced by plating tin onto a relatively large diameter wire, which is then drawn to the desired size (Ref 28).

Solder-Coated Wire. The common commercial types of solder- and tin-coated wire are described in Table 3.20. Coating thicknesses are listed in Table 3.21.

Silver-plated copper wires are available in a range of sizes and qualities. Coating thickness, 1.02 μm (40 μin.) minimum, and continuity are

Table 3.18 Typical copper foil properties

Temper designation	Material and condition	Nominal weight, g/m²	Tensile strength, min, MPa	Elongation in 50 mm, min, %
...	Electrodeposited	153	105	3
		305 and over	205	3
O61	Rolled and	153	105	5
	annealed	305	140	10
		610 and over	170	20
H00	Light cold rolled	610 and over	220	5
H08	As rolled	All weights	345	...

Source: Ref 16

Table 3.19 Preferred copper alloys for electrical and electronic connectors

Alloy	Applications
Brass	
C230	Automotive connectors,
C260	household recepticals,
	appliance terminals
Phosphor bronze	
C510	Telecommunications, computers
C521	
Beryllium copper	
C170	Military, aerospace,
C172	computer, telecommunications
C173	
C174	

Fig. 3.10 Bar and rod products

Table 3.20 Solder- and tin-plated copper wires

Type of plating	Standard composition, %	Features
Standard, solder-plated wire	65Pb, 35Sn	Solder-plated wire for general uses has adequate heat resistance. Melting point is about 230 °C (446 °F). Has excellent soldering properties.
Heat-resistant type, solder-plated wire	75Pb, 25Sn	Excellent soldering qualities despite its comparatively high melting temperature (melting point about 260 to 270 °C, or 500 to 518 °F). Suitable for parts molded at a high temperature.
Eutectic type, solder-plated wire	40Pb, 60Sn	Has the lowest solder melting point, about 185 °C (365 °F). Excellent solderability and soldering properties. Suitable for parts in which heat-resistance is not essential.
Tin-plated wire	100Sn	Like solder-plated wires, this type has excellent soldering properties. Highly resistant to corrosion, retains its luster and good soldering qualities.

measured in accordance with ASTM B 298. The wire is normally insulated with Teflon (E.I. Du Pont de Nemours & Co., Inc., Wilmington, DE).

Nickel-Plated Wire. Nickel is plated on copper wire to a minimum thickness of 1.27 µm (54 µin.). Coating continuity is specified by ASTM B 355.

Rod, Bar, and Shapes

For the copper metals, *rod* is defined as a round, hexagonal, or octagonal product. *Bar* refers to square or rectangular cross sections, while *shapes* can have oval, half-round, geometric, or custom-ordered profiles. Examples of rod and bar products are shown in Fig. 3.10, and a variety of specialty shapes are shown in Fig. 3.11. The three basic product forms are differentiated from wire in that they are sold in straight lengths, whereas wire is sold in coils (Ref 1, 26).

Specifications for copper shapes are given in ASTM B 187; those for bronze shapes are given in ASTM B 249M. Compositions conform to grades listed in Table 3.5; grades and densities (used by machinists to calculate part weights) are listed in Table 3.22. Shapes are normally supplied in either

the O60 (soft annealed) or H04 (hard) tempers. Dimensional tolerances, as specified by ASTM B 249 and 249M, are very specific. Manufacturers can make distinctions between shape (referring to the inside cross sectional pattern) and profile (by the outside pattern). Products of this type are intended for use in architecture as window mullions, balustrades, doors, etc. (Ref 1, 19, 32).

Tube and Pipe

Copper fabricators make a distinction between tube and pipe as these terms are defined under ASTM B 224. In a generic sense, hollow copper products are always referred to in the singular. *Tube* is a unidirectional elongated hollow product of uniform round or other cross section having a

Table 3.21 Standard thicknesses of electroplate on solder-plated and tin-plated wires

Classification of coating	Thickness of coating, µm
Extra heavy coating	14 ± 2
Heavy coating	10 ± 2
Medium coating	6 ± 1
Thin coating	2 ± 1

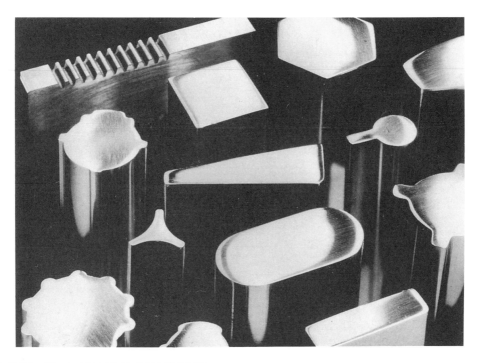

Fig. 3.11 Examples of custom shapes

Table 3.22 Designations and densities of wrought copper and copper alloys

ASTM designation	Material	Copper or copper alloy, UNS No.	Density, g/cm^3
B 16	Free-cutting brass	C36000	8.50
B 21	Naval brass	C46200	8.44
		C46400	8.41
		C48200	8.44
		C48500	8.44
B 98	Copper-silicon alloy	C65100	8.75
		C65500	8.53
		C65800	8.53
		C66100	8.53
B 124	Copper	C11000	8.94
	Copper-tellurium	C14500	8.94
	Copper-sulfur	C14700	8.94
	Forging brass	C37700	8.44
	Naval brass	C46400	8.41
	Medium leaded naval brass	C48200	8.44
	Leaded naval brass	C48500	8.44
	Aluminum-bronze	C61900	7.5
	Aluminum-bronze, 9%	C62300	7.65
	Aluminum-nickel-bronze	C63000	7.58
	Aluminum-silicon-bronze	C64200	7.69
	Aluminum-silicon-bronze, 6.7%	C64210	7.69
	High-silicon-bronze (A)	C65500	8.53
	Manganese-bronze (A)	C67500	8.36
	Nickel-silver, 45–10	C77400	8.47
B 133	Copper:		
	deoxidized and oxygen-free	…	8.94
	other classifications	…	8.89
B 138	Manganese-bronze	C67000	7.92
B 139	Phosphor-bronze	C51000	8.86
		C52100	8.80
		C52400	8.77
		C53400	8.91
		C54400	8.86
B 140	Leaded red brass	C31400	8.83
		C31600	8.86
		C32000	8.77
B 150	Aluminum-bronze	C61300	7.89
	Aluminum-bronze	C61400	7.89
	Aluminum-bronze	C61900	7.5
	Aluminum-bronze, 9%	C62300	7.65
	Aluminum-bronze	C62400	7.45
	Aluminum-nickel-bronze	C63000	7.58
	Aluminum-nickel-bronze	C63200	7.64
	Aluminum-silicon-bronze	C64200	7.69
	Aluminum-silicon-bronze, 6.7%	C64210	7.69
B 151	Copper-nickel-zinc alloy (nickel-silver) and copper-nickel alloy	C70600	8.94
		C71500	8.94
		C72000	8.94
		C74500	8.66
		C75200	8.77
		C75700	8.69
		C76400	8.72
		C77000	8.69
		C79200	8.69
		C79400	8.77
B 196	Copper-beryllium alloy	C17000	8.22
		C17200	8.22
		C17300	8.22
B 301	Free-cutting copper	C14500	8.94
		C14700	8.94
		C14710	8.94
		C14720	8.94
		C18700	9.94
B 371	Copper-zinc-silicon-alloy	C69400	8.19
		C69700	8.30
B 411	Copper-nickel-silicon alloy	C64700	8.91
B 441	Copper-cobalt-beryllium	C17500	8.75

(continued)

Table 3.22 (continued)

ASTM designation	Material	Copper or copper alloy, UNS No.	Density, g/cm^3
	Copper-nickel-beryllium	C17510	8.75
B 453	Copper-zinc-lead (leaded brass)	C33500	8.47
		C34000	8.47
		C34500	8.47
		C35000	8.44
		C35300	8.47
		C35600	8.50
B 455	Copper-zinc-lead (leaded brass)	C38000	8.44
		C38500	8.47

Source: Ref 16

continuous periphery. *Pipe* is a tube conforming to the particular dimensions commercially known as "standard pipe sizes" (Ref 33–35).

Copper and copper-alloy tube is available in seamless and welded forms. Plumbing tube and pipe is used for potable water and drain, waste, and vent service. It may be seamless or welded, and it may be made from either copper or copper alloys. Commercial tube and pipe is used for air conditioning, refrigeration, and other applications. Common couplings are fittings manufactured from tube, some having wrought or cast components (Fig. 3.12).

Plumbing Tube. Table 3.23 presents an overview of copper tube and pipe types, specifications, uses, and applicable UNS grades. Letter designations for copper tube refer to wall thicknesses. Thus, type K is heaviest; type L is standard; and type M is lightest (Table 3.24). Type DWV refers to drainage, waste, and ventilation, for which this grade is used. Types K, L, and M are called A, B, and C in ASTM B 88M, the metric specification for seamless copper water tube. For the four types of tube, the outside diameter is always 3.2 mm (0.125 in.) larger than the nominal or standard size. The inside diameters, therefore, depend on tube size and wall thickness (Ref 1, 35).

Types K, L, M, and DWV are supplied in hard (cold drawn) temper in straight lengths ordinarily 6.1 m (20 ft) long. Products are also available in the soft (annealed) temper as straight lengths or coils. In some countries, drawn plumbing tube is also furnished in a bending temper. As its name implies, this light-drawn temper of intermediate strength and hardness is intended for applications requiring bending.

Hard-temper tube can be joined by soldering, brazing, or adhesion using capillary fittings, or by welding. Tube in the bending and soft tempers can be joined by these means and by compression- or flare-type fittings. It is also possible to expand the end of one tube so that it can be joined to another by soldering or brazing without a capillary fitting, a procedure that can be efficient and economical in many installations.

All tube applied to ASTM standards is manufactured with a minimum 99.9% copper content (including silver). Analyses correspond generally to CDA/ASTM alloy UNS C12200, DHP (deoxidized high-phosphorus copper) and UNS C12000, DLP (deoxidized low phosphorus copper). Deoxidation is required because the tube is commonly joined by soldering. Table 3.25 lists specifications, grades, and uses for other copper plumbing tube products (Ref 1, 16).

Distribution tube (type D) is a seamless or welded tube suitable for the conveyance of fluids in hot and cold water distribution systems, wet heating and cooling systems, and condensate drain lines installed above ground. It is joined with capillary type solder fittings. As specified under ASTM B 641, distribution tube is available in grades C10100, C10200, C10300, C10800, C12000, and C12200.

Threadless pipe is a seamless tube conforming to particular dimensions. It is supplied in straight lengths for piping systems that are assembled with brazed joint pipe fittings. Threadless pipe is covered by ASTM B 302, which lists grades C10300, C10800, C12000, and C12200.

Seamless copper pipe is suitable for use in plumbing, boiler feed lines, and similar applications. Covered under ASTM B 42, the product is available in grades C10200, C10300, C10800, C12000, and C12200.

Seamless Red Brass Pipe. In the annealed temper, seamless red brass pipe is suitable for use in plumbing, boiler feed lines, and similar applications. Seamless red brass pipe is supplied in alloy C23000 and is covered under ASTM B 43.

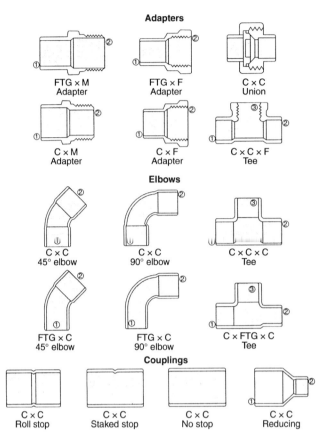

Fig. 3.12 Selected pressure fittings. Source: Ref 35

Table 3.23   Copper tube alloys and their typical applications

UNS No.	Alloy type	ASTM specifications	Typical uses
C10200	Oxygen-free copper	B 68, B 75, B 88, B 111, B 188, B2 80, B 359, B 372, B 395 , B447	Bus tube, conductors, wave guides
C12200	Phosphorus-deoxidized copper	B 68, B 75, B 88, B 111, B 280, B 306, B 359, B 360, B 395, B 447, FB 543	Water tubes; condenser, evaporator, and heat exchanger tubes; air conditioning and refrigeration, gas, heater and oil burner lines; plumbing pipe and steam tubes; brewery and distillery tubes; gasoline, hydraulic and oil lines; rotating bands
C19200	Copper-iron	B 111, B 359, B 395, B 469	Automotive hydraulic brake lines; flexible hose
C23000	Red brass, 85%	B 111, B 135, B 359, B 395, B 543	Condenser and heat exchanger tubes; flexible hose; plumbing pipe; pump lines; plumbing brass goods
C26000	Cartridge brass, 70%	B 135	Pump and power cylinders and liners; plumbing brass goods
C33000	Low-leaded brass (tube)	B 135	Screw machine parts; plumbing goods
C36000	Free-cutting brass	...	Bourdon tubes; musical instruments
C43500	Tin brass	...	Condenser, evaporator and heat exchanger tubes; distiller tubes
C44300	Inhibited admiralty metal	B 111, B 359, B 395	Heat exchanger tubes; electrical conduits; condenser, evaporator and heat exchanger tubes; distiller tubes
C60800	Aluminum bronze, 5%	B 111, B 359, B 395	Heater exchanger tubes; electrical conduits
C65100	Silicon-bronze B	B 315	Chemical equipment, heat exchanger tubes; piston rings
C65500	Silicon-bronze A	B 315	Condenser, evaporator and heat exchanger tubes; distiller tubes
C68700	Arsenical aluminum brass	B 111, B 359, B 395	Condenser, evaporator, and heat exchanger tubes; salt water piping; distiller tubes
C70600	Copper-nickel, 10%	B 111, B 359, B 395, B 466, B 467, FB 543, B 552	Condenser, evaporator and heat exchanger tubes; distiller tubes; salt water piping
C71500	Copper-nickel, 30%	B 111, B 359, B 395, B 466, B 467, B 543, B 552	...

Source: Ref 7

Table 3.24 Physical characteristics of copper tube types K, L, M, and DWV

Size	Outside diameter	Inside diameter	Wall thickness	Cross sectional area of bore, in.^{2}	External surface, ft^2/linear ft	Internal surface, ft^2/linear ft	Weight, lb/linear ft
Type K							
¼	0.375	0.305	0.035	0.073	0.098	0.080	0.145
⅜	0.500	0.402	0.049	0.127	0.131	0.105	0.269
½	0.625	0.527	0.049	0.218	0.164	0.138	0.344
⅝	0.750	0.652	0.049	0.334	0.196	0.171	0.418
¾	0.875	0.745	0.065	0.436	0.229	0.195	0.641
1	1.125	0.995	0.065	0.778	0.294	0.261	0.839
1¼	1.375	1.245	0.065	1.22	0.360	0.326	1.04
1½	1.625	1.481	0.072	1.72	0.425	0.388	1.36
2	2.125	1.959	0.083	3.01	0.556	0.513	2.06
2½	2.625	2.435	0.095	4.66	0.687	0.638	2.93
3	3.125	2.907	0.109	6.64	0.818	0.761	4.00
3½	3.625	3.385	0.120	9.00	0.949	0.886	5.12
4	4.125	3.857	0.134	11.7	1.08	1.01	6.51
5	5.125	4.805	0.160	18.1	1.34	1.26	9.67
6	6.125	5.741	0.192	25.9	1.60	1.50	13.9
8	8.125	7.583	0.271	45.2	2.13	1.98	25.9
10	10.125	9.449	0.338	70.2	2.65	2.47	40.3
12	12.125	11.315	0.405	101.0	3.17	2.96	57.8
Type L							
¼	0.375	0.315	0.030	0.078	0.098	0.082	0.126
⅜	0.500	0.430	0.035	0.145	0.131	0.113	0.198
½	0.625	0.545	0.040	0.233	0.164	0.143	0.285
⅝	0.750	0.666	0.042	0.348	0.196	0.174	0.362
¾	0.875	0.785	0.045	0.484	0.229	0.206	0.455
1	1.125	1.025	0.050	0.825	0.294	0.268	0.655
1¼	1.375	1.265	0.055	1.26	0.360	0.331	0.884
1½	1.625	1.505	0.060	1.78	0.425	0.394	1.14
2	2.125	1.985	0.070	3.09	0.556	0.520	1.75
2½	2.625	2.465	0.080	4.77	0.687	0.645	2.48
3	3.125	2.945	0.090	6.81	0.818	0.771	3.33
3½	3.625	3.425	0.100	9.21	0.949	0.897	4.29
4	4.125	3.905	0.110	12.0	1.08	1.02	5.38
5	5.125	4.875	0.125	18.7	1.34	1.28	7.61
6	6.125	5.845	0.140	26.8	1.60	1.53	10.2
8	8.125	7.725	0.200	46.9	2.13	2.02	19.3
10	10.125	9.625	0.250	72.8	2.65	2.53	30.1
12	12.125	11.565	0.280	105.0	3.17	3.03	40.4
Type M							
⅜	0.500	0.450	0.025	0.159	0.131	0.118	0.145
½	0.625	0.569	0.028	0.254	0.164	0.149	0.204
¾	0.875	0.811	0.032	0.517	0.229	0.212	0.328
1	1.125	1.055	0.035	0.874	0.294	0.276	0.465
1¼	1.375	1.291	0.042	1.31	0.360	0.338	0.682
1½	1.625	1.527	0.049	1.83	0.425	0.400	0.940
2	2.125	2.009	0.058	3.17	0.556	0.526	1.46
2½	2.625	2.495	0.065	4.89	0.687	0.653	2.03
3	3.125	2.981	0.072	6.98	0.818	0.780	2.68
3½	3.625	3.459	0.083	9.40	0.949	0.906	3.58
4	4.125	3.935	0.095	12.2	1.08	1.03	4.66
5	5.125	4.907	0.109	18.9	1.34	1.28	6.66
6	6.125	5.881	0.122	27.2	1.60	1.54	8.92
8	8.125	7.785	0.170	47.6	2.13	2.04	16.5
10	10.125	9.701	0.212	73.9	2.65	2.54	25.6
12	12.125	11.617	0.254	106.0	3.17	3.04	36.7
Type DWV							
1¼	1.375	1.295	0.040	1.32	0.360	0.339	0.65
1½	1.625	1.541	0.042	1.87	0.425	0.403	0.81
2	2.125	2.041	0.042	3.27	0.556	0.534	1.07
3	3.125	3.030	0.045	7.21	0.818	0.793	1.69
4	4.125	4.009	0.058	12.6	1.08	1.05	2.87
5	5.125	4.981	0.072	19.5	1.34	1.30	4.43
6	6.125	5.959	0.083	27.9	1.60	1.56	6.10
8	8.125	7.907	0.109	49.1	2.13	2.07	10.6

According to standard (in B.S. units). Source: Ref 35

Welded tube is available in alloy C21000 and is covered under ASTM B 642. It is supplied in straight lengths. The tube is most commonly used with solder-type fittings and is not intended for bending or flaring. Welded tube is suitable for general applications; however, in some countries, building codes do not permit its use in plumbing systems.

Commercial tube and pipe is differentiated from common plumbing tube by dimension (it is generally more robust than common tube) and special features.

Air Conditioning and Refrigeration Tube. Copper tube for air conditioning and refrigeration field service is designated by its actual outside diameter. Specifications for this product are given in Table 3.26. It is available in soft annealed coils or hard-drawn lengths (Ref 35).

Condenser and Heat Exchanger Tube. Copper materials for condenser and heat exchanger use range from OF copper qualities through an extensive range of alloys. Some typical applications, along with corresponding ASTM standards, are listed in Table 3.25. Corrosion and particularly erosion-corrosion resistance are the most important factors to be considered when selecting materials for heat exchanger service. Corrosion response varies considerably among the copper alloys; relevant data can be found in Chapter 2 (Ref 36).

Seamless Condenser Tube and Ferrule Stock. Seamless tube is used in surface condensers,

Table 3.25 Overview of standard plumbing tube per ASTM B 698

ASTM Specification	Type	Designation	Application	Standard mill sizes, in.	UNS alloy availability C10100	C10200	C10300	C10800	C12000	C12200	C21000	C23000
B 42	Seamless	Copper pipe	Plumbing and feed lines	$\frac{1}{8}$–12		•	•	•	•	•		
B 43	Seamless	Red brass pipe	Plumbing and feed lines	$\frac{1}{8}$–12								•
B 88	Seamless	K, L, M	General plumbing water tube	$\frac{1}{4}$–12		•	•	•	•	•		
B 302	Seamless	TP	Plumbing and feed lines	$\frac{1}{4}$–12			•	•	•	•		
B 306	Seamless	DWV	Drainage, waste, vent	$\frac{1}{4}$–8			•	•	•	•		
B 641	Seamless	D	Distribution	$\frac{1}{4}$–3	•	•	•	•	•	•		
B 641	Welded	D	Distribution	$\frac{1}{4}$–3	•	•	•	•	•	•		
B 642	Welded	Welded alloy water tube	General plumbing	$\frac{1}{4}$–3							•	

Source: Ref 16

Table 3.26 Physical characteristics of copper tube air conditioning and refrigeration tube for field service

Size	Nominal dimensions, in. Outside diameter	Inside diameter	Wall thickness	Calculated values based on nominal dimensions Cross-sectional area of bore, in.^{2}	External surface, ft^2/linear ft	Internal surface, ft^2/linear ft	Weight, lb/linear ft
$\frac{2}{3}$	0.125	0.065	0.030	0.00332	0.0327	0.0170	0.0347
$\frac{3}{16}$	0.188	0.128	0.030	0.0129	0.0492	0.0335	0.0577
$\frac{1}{4}$	0.250	0.190	0.030	0.0284	0.0655	0.0497	0.0804
$\frac{5}{16}$	0.312	0.248	0.032	0.0483	0.0817	0.0649	0.109
$\frac{3}{8}$	0.375	0.315	0.030	0.0780	0.0982	0.0821	0.126
$\frac{3}{8}$	0.375	0.311	0.032	0.0760	0.0982	0.0814	0.134
$\frac{1}{2}$	0.500	0.436	0.032	0.149	0.131	0.114	0.182
$\frac{1}{2}$	0.500	0.130	0.430	0.145	0.131	0.113	0.198
$\frac{5}{8}$	0.625	0.555	0.035	0.242	0.164	0.145	0.251
$\frac{5}{8}$	0.625	0.545	0.040	0.233	0.164	0.143	0.285
$\frac{3}{4}$	0.750	0.666	0.042	0.348	0.196	0.174	0.362
$\frac{7}{8}$	0.875	0.785	0.045	0.484	0.229	0.206	0.455
$1\frac{1}{8}$	1.125	1.025	0.050	0.825	0.294	0.268	0.655
$1\frac{3}{8}$	1.375	1.265	0.055	1.26	0.360	0.331	0.884
$1\frac{5}{8}$	1.625	1.505	0.060	1.78	0.425	0.394	1.14
$2\frac{1}{8}$	2.125	1.985	0.070	3.09	0.556	0.520	1.75
$2\frac{5}{8}$	2.625	2.465	0.080	4.77	0.687	0.645	2.48
$3\frac{1}{8}$	3.125	2.945	0.090	6.81	0.818	0.771	3.33
$3\frac{5}{8}$	3.625	3.425	0.100	9.21	0.949	0.897	4.29
$4\frac{1}{8}$	4.125	3.905	0.110	12.0	1.08	1.02	5.38

Source: Ref 35

evaporators, and heat exchangers. A ferrule tube is a tube from which metal rings or collars (ferrules) are made for use when installing condenser tubes. Seamless copper and copper-alloy tube is covered under ASTM B 111.

Welded heat exchanger tube is also used in surface condensers, evaporators, and heat exchangers, as well as general engineering applications. Welded tube is made from clean strip that is formed into a tube and seam welded longitudinally using forge welding, fusion, or induction methods. Such tube is covered under ASTM B 543.

Water desalination tube can be either seamless or welded. Preferred materials are copper-nickels, such as C70600, C71500, C71640, and C72200. The relevant standard is ASTM B 552.

Enhanced surface heat exchanger tubes can be made with internal and/or external profiles, whose patterns are designed to maximize the type of heat exchange for which the tubes are intended. Enhanced surfaces are created by cold forming. A large variety of tubes are now manufactured, and several examples of tubes with internal enhanced surfaces are shown in Fig. 3.13. These patterns can be either parallel to the longitudinal axis or circumferentially extend from the tube. They serve to increase the effective surface area for heat transfer and improve the film coefficient at the tube interior (Fig. 3.14). Tubes with external enhanced surfaces are shown in Fig. 3.15 (Ref 37).

Composite tubes with mechanically attached aluminum fins, used for residential and commercial heating, are also available.

Special-Purpose Tube. In addition to the above categories, a number of uncommon, proprietary, and special-purpose tubular products should be mentioned.

Leak Detecting Tubes. Special tubes have been designed for nuclear power plant and other safety-critical installations. These are compound structures incorporating two axially symmetric tubes, where the outer is at least internally finned.

Bourdon tube with an oval cross section, which causes the tube to flex under changing pressure conditions, is useful in instruments, such as pressure gages.

Pitting Corrosion Resistant Tube. Intended for central heating and sanitary applications, these tubes are claimed to exhibit improved aqueous corrosion resistance. Such tubes, manufactured in Europe and elsewhere to meet DIN 1786, are furnished as coils and straight lengths measuring 6 m by 1 cm (20 ft by 0.4 in.) up to 267 m by 3 cm (876 ft by 1.2 in.) (Ref 38, 39).

Insulated Tube. Tubes with coextruded external PVC insulation are made for hot and cold water service. Such tubes are also used in under floor heating systems where direct contact of the copper with concrete may be undesirable.

Waveguide Tube. Seamless rectangular tube is commonly used for electronic waveguides. Such tube is ordinarily made from OF coppers (C10200 and C10300), but deoxidized copper (C12000) and commercial bronze (C22000) can also be used. Waveguide tube is made in accordance with ASTM B 372-93 (Ref 1).

Bus Pipe or Tube. A high-conductivity copper tubular product is manufactured under standard specification ASTM B 188-88 for use as water-cooled electrical bus bars.

Foundry Products

One of the most important characteristics of copper is the ease with which it can be cast. Indeed, cast copper and bronze artifacts are among the oldest metallic items manufactured by man. The fact that thousands of such products have survived since the Bronze Age also attests to the corrosion resistance of copper. Early bronze and brass was relatively simple in composition, although quite similar alloys are still cast today. On the other hand, modern metallurgy and casting technology have brought the number of commercial copper casting alloys into the hundreds.

Table 3.6 lists copper-base casting alloys recognized by CDA and classified under the Unified Numbering System. The list applies mainly to North America, but similar alloys can be found in other national and international classification systems.

Cast copper alloys are used in virtually every industrial market category from ordinary plumbing goods to precision electronic components and state-of-the-art marine and nuclear equipment. Their favorable properties are often available in useful combinations, which is an important feature for products that must satisfy several requirements simultaneously.

The ability to withstand corrosive environments is the most important and best known characteristic of cast copper alloys. Not surprisingly, water handling equipment of one form or another constitutes the largest single market. Copper-alloy castings are also widely used to handle corrosive industrial and process chemicals, and they are well known in the food, beverage, and dairy industries.

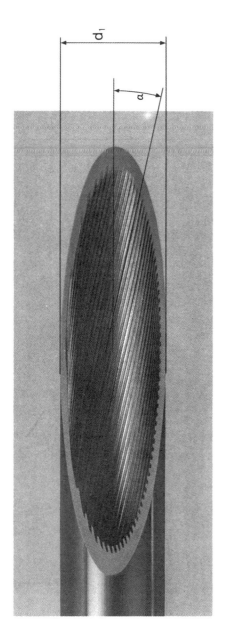

d₁ OD
α twist of fins
H fin height
wᵣ root wall

Fig. 3.13   High-performance, internally grooved, enhanced heat-transfer tube. Source: Ref 37

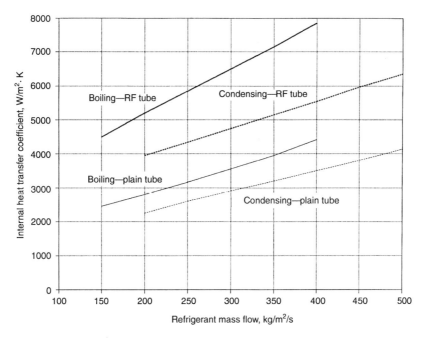

Fig. 3.14 Heat-transfer performance for enhanced surface tube in Fig. 3.13

Fig. 3.15 High-performance, externally finned, enhanced heat transfer tubes. Source: Ref 37

Some cast copper alloys exhibit strengths that rival quenched and tempered steels. Unlike many steels, almost all copper alloys retain their mechanical properties, including impact toughness, at very low temperatures. Copper alloys tend to lose strength at elevated temperatures, but some alloys are used routinely at temperatures as high as 425 °C (800 °F). No class of engineering materials can match the combination of strength, corrosion resistance, and conductivity of copper alloys over such a broad temperature range.

Cast sleeve bearings are an important application for copper alloys largely because of excellent tribological properties. For sleeve bearings, no material of comparable strength can match high-leaded bronze in terms of low wear rates against steel. For worm gears, nickel bronzes and tin bronzes are unchallenged industry standards.

Copper and its alloys resist seawater corrosion. Equally important, they inhibit biofouling, the growth of algae, and other marine organisms. As a result, copper-alloy castings have long been used for products such as seawater piping, pump and valve fittings, and marine hardware.

The thermal and electrical conductivities of copper are higher than any other metal except silver. Even copper alloys with conductivities as low as 20% that of pure copper conduct heat and electricity better than stainless steel and titanium. Unlike most other metals, the thermal conductivity of many cast copper alloys increases with rising temperature. As a result of favorable conductivities, copper-alloy castings find many uses in electrical and heat transfer equipment.

The most common cast copper alloys are compositions based on copper-zinc, copper-tin, copper-zinc-tin, copper-zinc-tin-lead, copper-aluminum, copper-aluminum-nickel, copper-aluminum-nickel-iron, and copper-nickel. Many of these basic alloys also contain small amounts of iron, silicon, manganese, and other elements to improve properties or to make them easier to cast or fabricate (Ref 40, 41–43).

Standards and Specifications

Standards and specifications address the quality and properties of the many classes of cast products. Examples of the more important cast copper-base products, along with corresponding ASTM standards, include (Ref 1, 44, 45):

- *Components of valves, flanges, and fittings:* ASTM B 61-86, B 62-86; DIN V86086; and BS 143, 1256, 2051

- *Unfinished bronze castings:* Furnished in the rough for locomotive components: ASTM B 66
- *Railroad journal bearings:* ASTM B 67
- *Aluminum bronze sand castings:* ASTM B 148
- *Copper-alloy die castings:* ASTM B 176
- *Copper-base centrifugal castings:* ASTM B 271
- *Gear bronze sand castings and static and centrifugal chill castings:* ASTM B 427
- *Centrifugally cast alloy C96300 (80-20 copper-nickel) tail shaft sleeves:* ASTM B 492
- *Continuously cast rod, bar, tube, and shapes:* ASTM B 505
- *Sand castings for general applications:* ASTM B 584
- *Sand castings for valve applications:* ASTM B 763
- *Copper-beryllium sand castings for general applications (as-cast or solution annealed and aged):* ASTM B 770

Powder Products

Powder metallurgy (P/M) offers the ability to produce products having nearly net shapes, thereby reducing or eliminating the need for expensive machining. Powder metallurgy also offers the possibility of producing composite materials that cannot be made by conventional cast or wrought processes. Copper-base P/M products comprise a small fraction of total copper usage; however, within the P/M market, copper and copper alloys rank second only to iron-base materials in terms of tonnage consumed.

Powders are produced mainly by air or water atomization (pouring a stream of molten metal through a blast of air or water) and by oxide reduction. These processes yield powders whose shapes favor good compressibility and high green (unsintered) strength. Copper powder can also be made by chemical and electrolytic processes, but the resulting powder has inferior properties. A selection of particle shapes produced by various powder manufacturing methods is shown in Fig. 3.16, and powder properties are listed in Table 3.27.

Powder metallurgy products are made by pressing powders in a die cavity whose shape approximates the contours of the finished part. Powder may be prealloyed and homogeneous, or it may be in the form of a suitably proportioned mixture of the constituent elements of the desired alloy. Oxide dispersion strengthened (ODS) coppers, another class of P/M materials, are metal-matrix composites in which improved room- and elevated-temperature mechanical properties derive from fine particles of alumina dispersed in essentially pure copper.

Compaction pressures needed to form P/M products are typically in the several hundred MPa range. After pressing, the so-called green compacts are sintered from 815 to 925 °C (1500 to 1700 °F) for copper alloys and at temperatures approaching the melting point (1083 °C, or 1981 °F) for pure copper. Sintering bonds the compressed powder particles metallurgically and, in the case of mixed powders, drives diffusion to form the alloy. Parts may be sized to final dimensions in a second die after sintering. Sizing raises density to near theoretical values. It also imparts cold work, which strengthens the parts.

Bearings

Self-lubricated sintered bronze bearings constitute the largest fraction of the copper-base P/M market. Sintered bearings are very widely used in light and medium duty applications, such as the fractional horsepower electric motors found in automotive vehicles, agricultural equipment, business machines, machine tools, appliances, and consumer products. The bearings are intentionally pressed to leave interconnecting channels between the bronze particles. The channels are then vacuum-impregnated with oil. In use, the oil seeps out of the channels to lubricate the bearing surface. As the bearing operates, the oil circulates continuously between the bearing surface and the interparticle reservoir.

The most widely used sintered bronze bearing alloy is 90%Cu-10%Sn, although brasses and nickel-silvers are also known. Bronzes may contain up to a few percent graphite or molybdenum disulfide for improved lubricity. A selection of sintered bronze bearings is shown in Fig. 3.17.

Structural Products

Copper-base P/M structural parts are made from a large variety of alloys and composites. Copper, high-copper alloys, brasses, bronzes, and nickel-silvers are most popular. Powder metallurgy parts are generally relatively small, although products up to several inches in cross section are produced. Because the major rationale for using P/M processing is to avoid the cost of machining, many P/M parts resemble machined items: gears and gear segments, actuators, slides, etc. Such products are normally made from copper alloys (Fig. 3.18).

Parts that require improved strength at elevated temperatures may be made from ODS copper. Electrical and electronic P/M products can also be specified in ODS copper, although pure copper is

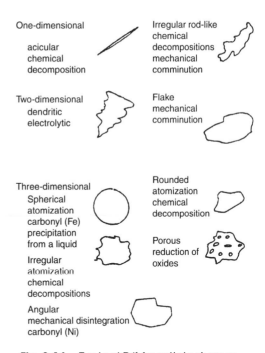

Fig. 3.16 Typical P/M particle shapes

Table 3.27 Typical properties of copper powder produced by various methods

Property	Atomized	Electrolytic	Hydrometallurgy	Solid state reduction
Copper, %	99–99.5	99–99.5	99–99.5	98–99
Weight loss in H_2, %	0.1–0.75	0.1–0.75	0.1–0.75	0.1–0.75
Acid insoluble, %	0.5–0.1 max	0.03 max	0.03 max	0.03 max
Apparent density, g/cm^3	2–4	1.5–4	1.5–2.5	2–4
Flow, s/50 g	20–35	30–40	none	2–35
Green strength, MPa (psi)	0–17.2 (0–2500)	2.8–41.3 (400–6000)	0–68.9 (0–10,000)	0–17.2 (0–2500)
–325 mesh, %	25–80	5–90	60–95	25–50

Source: Ref 46

more common. Purity is important when the highest electrical conductivity must be preserved.

Frictional Products

Products, such as clutch faces and brake linings, have been made from copper-base materials for many years (Fig. 3.19). Although copper-base frictional materials have been replaced to some extent by organic materials, copper-base products continue to be used in moderate to heavy duty applications, such as trucks, machinery, and aircraft. Copper-base friction elements are best used wet (submerged in oil) or under semifluid conditions.

Friction materials are composites, mixtures of several materials, each of which contributes to the function of the whole. Sintered copper or copper-alloy powder is typically the matrix within which the frictional materials themselves are embedded. Copper is used because it transmits heat well, allowing the device to operate cooler. Admixtures of silica, mullite, alumina, and silicon nitride provide friction. Other hard materials are added to impart wear resistance, which is also aided by the addition of lubricants, such as graphite or molybdenum disulfide.

Filters

Filters constitute an important portion of the copper-base P/M market. Powder metallurgy filters are generally made from atomized prealloyed powder, because the particles are spherical and have readily controllable sizes and size ranges. This in turn helps control the pore size of the filter. Compacts are gravity sintered (not pressed) to give a range of pore sizes between 5 to 125 μm (20 to 500 μin.).

Powder metallurgy filters are made from a variety of ferrous and nonferrous metals. Tin bronzes dominate among the copper-base metals, although nickel-silvers and copper-tin-nickel alloys can also be found. The major advantage of copper-base filter materials over other materials, such as stainless steel, is the combination of relatively low cost and good corrosion resistance.

Fig. 3.17   Sintered bronze bearings

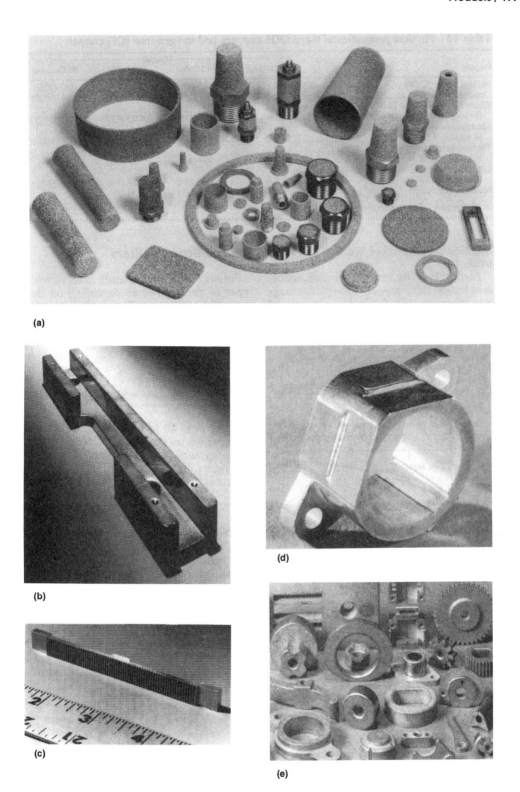

Fig. 3.18 Copper-base P/M products

Electrical, Electronic, and Other P/M Products

Copper-graphite composites are the familiar brushes found in electric motors. Sintered copper-tungsten, copper-tungsten carbide and copper-molybdenum composites are used for high-current,

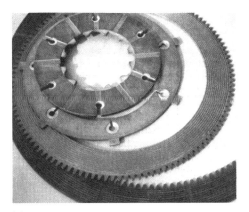

(a)

(b)

(c)

Fig. 3.19 Copper-base P/M friction elements. (a) Grooved P/M friction elements for wet applications. (b) Copper-base P/M clutch plates (280 to 500 mm OD) used in power-shift transmissions for tractors. (c) Copper-base P/M friction pad

high-voltage electrical contacts and in spark gaps. Copper-tungsten composites have demonstrated low wear rates when used as gas metal arc welding (metal inert gas) and submerged-arc welding contact tips. Similar materials may hold promise as high-performance resistance welding electrodes. Other important examples of copper-base P/M products include copper-nickel coinage, tokens, and medallions. Producing these from sintered powder compacts eliminates the need and cost of producing strip as an intermediate product.

Forgings

Copper-base forgings represent a relatively small but important class of products (Fig. 3.20). Forgings are typically moderate-size products, such as valves, fittings, mechanical devices, and architectural hardware, rarely exceeding 90 kg (200 lb). Because forgings tend to be somewhat more costly than comparably sized castings, forged products are usually reserved for applications in which special qualities are needed.

Strong, Tough Structures. Forging of copper metals is performed hot, and the severe deformation involved produces a dense, fibrous grain structure that gives the products excellent mechanical properties. Forgings are therefore preferred for thin-walled pressure-retaining devices, such as valves and fittings.

Fine Surface Finishes. Copper-base forgings can be expected to have surface finishes as fine as at least 32 μm (125 μin.). Finer finishes are possible in many cases, but this depends very much on the size and shape of the product. In general, better surface finishes are easier to obtain on smaller products.

Consistent Dimensions, Close Tolerances. Copper-base forgings can be made to precise dimensions and to sections thinner than 3.2 mm (0.125 in.); however, section sizes are normally limited by the features of the part in question. Typical commercial tolerances fall between ±0.2 and ±0.4 mm (±0.008 and ±0.015 in.), depending on configuration, in forgings weighing less than about 0.9 kg (2 lb). Tolerances are slightly wider in heavier forgings, but dimensions can be held as tight as ±0.025 mm (±0.001 in.) in special cases. Flatness tolerances are typically on the order of 0.12 mm/25 mm (0.005 in./in.) for the first 25 mm and 0.075 mm/25 mm (0.003 in./in.) thereafter.

Intricate detail and sharp lettering makes forging the preferred method of manufacture for decorative and architectural products, such as doorplates.

Lower Environmental Risk. Unlike sand casting, forging produces neither hazardous fumes nor residues that require expensive clean up or special disposal. There is no waste, and all unused metal is recycled to make new alloy.

Cost Considerations. Forgings are usually more costly than castings, but there are exceptions. Forging dies cost about one-half as much as dies for pressure die casting (a competing process). Also, forging dies are usually a one-time expense to the customer, whereas the maintenance, repair, and replacement of casting dies are usually the customer's responsibility. Finally, forgings use significantly less metal per part than castings or screw machine products because forged products can be made with thinner walls and lighter sections. Forging also generates less runaround scrap, thereby reducing energy consumption.

Materials. Ideal forging characteristics include low force requirements, little tendency to crack, and good surface finishes. The UNS copper alloys with acceptable forgeability are listed in Table 3.28. Among these alloys, forging brass, C37700, is by far the most commonly used. It is a leaded yellow brass containing sufficient beta phase to provide good high-temperature ductility. The alloy contains approximately 2% lead to make it free cutting. Other commonly forged copper alloys include naval brass, C46400, and an inhibited aluminum bronze, C64200. These alloys are used in place of C37700 when higher strength and/or corrosion resistance is required.

Chemical Products

Copper chemicals have found a wide variety of applications, many of which are related to the benefits of copper on health and nutrition. The Western pharmaceutical industry uses copper chemicals as a dietary supplement in baby food and in various vitamin and mineral tonics. For example, copper sulfate is added to the feed of pigs as a growth stimulant. Inhabitants of Asia and Africa use copper sulfate for curing sores and skin diseases, a practice that dates to earliest recorded history. Copper salts prevent algae growth in ponds, potable water tanks, and swimming pools. In some countries, copper sulfate is used to control such water-borne diseases as schistosomiasis and bilharzia in humans and liver fluke in animals. Less than 1 ppm (part per million) $CuSO_4$ in water is capable of controlling the snails that transmit such diseases. Copper salts also have a wide spectrum of applications in the prevention of biofouling in seawater, and they are commonly used as preservatives for wood and fabrics. A description of the biological effects of copper on plants, animals, and humans is given in Chapter 6.

A new field was opened for copper compounds in the second half of the 20th century with the discovery of the advantages of copper-base organometallic compounds. These compounds are replacing Grignard reagents (magnesium organometallic compounds) in organic synthesis for polymer production. Copper compounds are also used as catalysts. An overview of applications of copper compounds is given in Chapter 5.

Table 3.28   Common copper-base forging alloys

UNS No.	Alloy name, nominal composition	UNS No.	Alloy name, nominal composition
C10200	Oxygen-free Cu; 99.95 Cu	C48500(a)	High leaded naval brass; 60 Cu, 1.8 Pb, 0.7 Sn, 37.5 Zn
C10400	Oxygen-free Cu with Ag; 99.95 Cu	C62300	Al-bronze, 9%; 9 Al, 3.0 Fe, Bal Cu
C11000	Electrolytic tough pitch Cu; 99.90 Cu	C62400	Al-Bronze 10 $\frac{1}{2}$%; 10.5 Al, 3.0 Fe, Bal Cu
C11300	Tough-pitch Cu with Ag; 99.90 Cu	C63000	Al-bronze, 10% Al, 3 Fe, 5 Ni
C12200	Phos.-deoxidized, low residual phosphorus Cu, 99.90 Cu, 0.02 P	C63200	Ni-Al bronze, 9%; 82 Cu, 9 Al, 4 Fe, 5 Ni
C14500	Te-bearing Cu; 99.5 Cu, 0.008 P, 0.55 Te	C64200	Arsenical Si-Al bronze; 91.2 Cu, 7.0 Al, 1.8 Si, 0.15 As
C14700	S-bearing Cu; 99.90 Cu, 0.35 S	C67000	Mn brass; 70 Cu, 28.8 Zn, 1.2 Mn
C15000	Zr-copper; 99.80 Cu, 0.15 Zr	C67300	Leaded Si-Mn brass; 60 Cu, 2.5 Pb, 3 Mn, 1 Si
C17000	Be-Copper; 98.3 Be, 1.7 Be	C67400(a)	High Mn-bronze; 58.5 Cu, 1.20 Al, 2.8 Mn, 0.50 Si
C18200	Cr-copper; 99.1 Cu, 0.9 Cr	C67500(a)	Mn-bronze A; 58.5 Cu, 1,4 Fe; 1.0 Sn, 0.1 Mn; 39.0 Zn
C36500(a)	Leaded muntz metal; 59.5 Cu, 0.5 Pb, 40 Zn	C67600(a)	Leaded Mn-bronze; 59 Cu, 0.15 Mn, 1.0 Si, 1.0 Fe, 0.75 Pb, bal Zn
C37700(a)	Forging brass, 59.5 Cu, 2 Pb, 38 Zn	C70600	Copper-nickel, 10%; 88.8 Cu, 10 Ni, 3 Fe
C38500(a)	Architectural bronze; 57 Cu, 3 Pb, 40 Zn	C71500	Copper-nickel, 30%; 69.5 Cu, 30 Ni, 0.5 Fe
C46400(a)	Naval brass; 60 Cu, 0.8 Sn, 39.2 Zn		

(a) Most widely used alloys

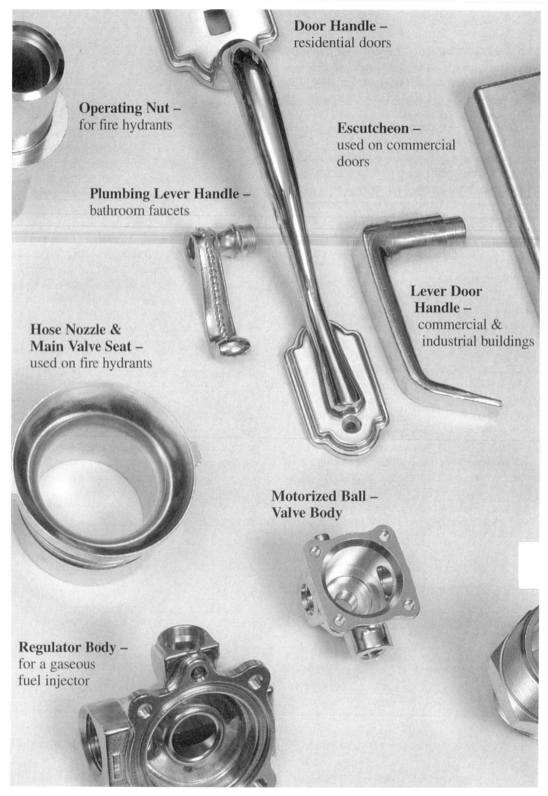

Fig. 3.20 Some typical bronze forgings. Source: Copper and Brass Development Association, Canada

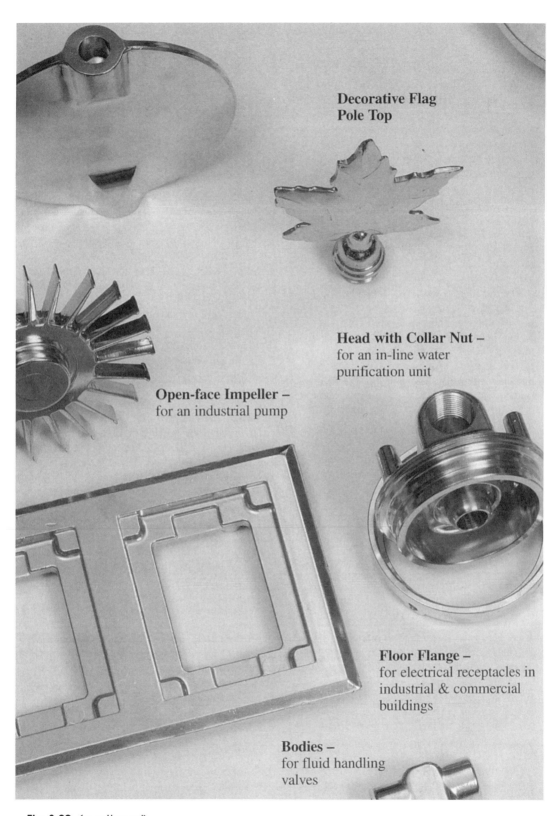

Decorative Flag Pole Top

Head with Collar Nut – for an in-line water purification unit

Open-face Impeller – for an industrial pump

Floor Flange – for electrical receptacles in industrial & commercial buildings

Bodies – for fluid handling valves

Fig. 3.20 (continued)

Common Copper Chemicals

The number of copper chemicals in industrial use is too large to address completely in this text. The following descriptions contain the copper chemicals that are in relatively common use or whose use has particular commercial importance.

Copper Acetates. Cupric acetate (normal cupric acetate, neutral verdigris), $Cu(C_2H_3O_2)_2H_2O$, formula weight 199.67, forms a green powder or dark green efflorescent crystals of the monoclinic system. It has a melting point of 115 °C (239 °F), a boiling point of 240 °C (464 °F), and a specific gravity of 1.882 g/cm^3. It is soluble in water and alcohol and slightly soluble in ether. Cupric acetate is prepared by dissolving cupric oxide, verdigris, or a copper carbonate in acetic acid. It may also be prepared by treating a copper sulfate solution with lead acetate or by the action of acetic acid on copper with subsequent crystallization. Cupric acetate is used as a basic raw material for the manufacture of cupric acetoarsenide (Paris green).

Basic cupric acetate (blue verdigris, French verdigris), $Cu(C_2H_3O_2)_2 \cdot CuO \cdot 6H_2O$, formula weight 369.33, when mixed with green verdigris, $2Cu(C_2H_3O_2)_2 \cdot CuO \cdot 6H_2O$, formula weight 549.00, is known as common verdigris. Basic cupric acetate is manufactured by the simultaneous action of acetic acid and air on copper turnings and scrap. The copper is generally packed into a tower, acetic acid is sprayed over the top, and a countercurrent of warm air is passed through the mass. Blue verdigris forms a light blue powder or silky blue crystalline needles or scales, which decompose to a green mass through the loss of water at 60 °C (140 °F). Basic copper acetate is slightly soluble in cold water and ammonium hydroxide and soluble in dilute acid. The green coating on uncleaned copper articles is sometimes incorrectly called verdigris; it consists of copper carbonates, not acetates. Basic copper acetate is an important ingredient of catalytic preparations for making a large number of organic compounds, such as acetaldehyde and acetic acid from ethyl alcohol; propionic acid, butyric acid, higher alcohols, ketones, and acids by the reduction of carbon dioxide or carbon monoxide; and for numerous other reactions.

Other important applications for copper acetates include use as fungicides, catalysts for rubber vulcanization, textile dyes, ceramic pigments, and impregnants in antitarnish wrappings for silverware.

Copper cyanide, CuCN, forms colorless crystals. It exists as an independent compound, but it is also found in combined form as cuprocyanides (e.g., $M[Cu(CN)_x]$). Among its many uses, cuprous cyanide is a common electroplating electrolyte. It has also been used or investigated as a catalyst, insecticide, antifouling agent, and as an ingredient of ointments for the treatment of trachoma and conjunctivitis.

Copper Acetylides. Either one or both of the hydrogen atoms in the acetylene molecule can be replaced by metals to form acetylides. Copper, silver, and mercury form acetylides that, when dry, are highly explosive. Copper acetylide is used in very small quantities as an ingredient of explosives.

Copper Arsenate. This chemical is light blue, blue, or bluish-green in color. It is a powder of variable composition containing approximately 33% Cu and 29% As. It is soluble in dilute acids and ammonium hydroxide but insoluble in alcohol and water. Copper arsenate is used in agriculture as an insecticide and fungicide.

Copper bromide, CuBr or Cu_2Br_2, formula weight 143.49 or 286.97, is a white crystalline (tetrahedral) powder. It forms a grayish-brown or greenish-brown mass when molten. It has a melting point of 504 °C (1004 °F), a boiling point of 1345 °C (2453 °F), and a specific gravity of 4.72 g/cm^3. It is slightly soluble in cold water and decomposes in hot water. It is soluble in hydrogen bromide, hydrochloric acid, and ammonium hydroxide and is insoluble in acetic acid and boiling concentrated sulfuric acid. The white powder becomes dark blue when exposed to sunlight. Cuprous bromide is prepared by heating cupric bromide. A mixture of cuprous and cupric bromides is formed when copper reacts with bromine vapor at dark red heat. If an excess of bromine is used, the cupric form is obtained. This reaction is very vigorous.

Cupric bromide, $CuBr_2$, formula weight 223.40, forms black prismatic crystals of the monoclinic system. It has a melting point of 498 °C (928 °F). It is very soluble in water; soluble in alcohol, acetone, ammonia, and pyridine; and insoluble in benzene. Rectangular crystals or greenish-yellow needles of the dihydrate are formed by evaporation of a solution of cupric bromide. The tetrahydrate results when the anhydrous form is evaporated below 29 to 30.5 °C. This form has long green needlelike monoclinic crystals. Cupric bromide is prepared as stated earlier or by the action of bromine water on copper or of hydrobromic acid on cupric oxide. Cupric bromide is used in photography as an intensifier and as a brominating agent in organic syntheses.

Copper Carbonates. The normal cupric or cuprous carbonates have not been isolated, although there are some questions concerning the existence of the cuprous form, Cu_2CO_3. Numerous basic cupric carbonates have been reported having ratios of $CuO:CO_2:H_2O$ of 10:1:6, 8:1:5, 6:1:0, 3:1:2, 8:3:6, 5:2:6, 2:1:0, 2:1:(1 or 2), 5:3:0, 8:5:7, and 3:2:1. Several of these occur as minerals; the most important copper carbonate minerals are malachite and azurite.

The commercial grade of basic cupric carbonate, $CuCO_3Cu(OH)_2$, formula weight 221.17, specific gravity 4.0 g/cm^3, is in the form of dark green monoclinic crystals or powder. The American Institute of Homeopathy and the Association of Official Agricultural Chemists have developed specifications, standards, and methods of analysis. Copper carbonate is used as a pigment in the ceramics, paint, and varnish industries and as a fungicide for treating wheat and sorghum seed. It is also used as an intermediate material in the production of other copper compounds, as an ingredient in pyrotechnic compositions, and as a pickling agent for imparting a black color to brass.

Copper Chlorides. Cupric chloride, $CuCl_2$, formula weight 134.48, appears in the form of a yellowish brown, deliquescent powder or brown sublimate. It has a melting point of 498 °C (928 °F), a boiling point of 993 °C (1819 °F), and a specific gravity of 3.05 g/cm^3. The dihydrate occurs in long green deliquescent needles of the rhombic system; it has a melting point of 110 °C (with loss of water) and a specific gravity of 2.38 g/cm^3 (Ref 49).

Cupric chloride is used as a mordant in the dyeing and printing of textile fabrics in the metallurgical industry for refining copper, gold, and silver and in the wet process for recovering mercury from its ores. It is also used in pyrotechnics, photography, and the petroleum industry for sweetening (deodorizing) or catalytic oxidation of mercaptan.

Cuprous chloride, $CuCl$ or Cu_2Cl_2, formula weight 99.03 or 198.05, forms a white powder composed of small hexakistetrahedral crystals belonging to the cubic system. It has a melting point of 422 °C (792 °F), a boiling point of 1366 °C (2491 °F), and a specific gravity of 3.53 g/cm^3. Exposure to air and sunlight of the moist powder results in color changes to yellow, dirty violet, and finally blue-black. Cuprous chloride is used to oxidize and sweeten mercaptan. Ammoniacal solutions of cuprous chloride are employed for the absorption of any carbon monoxide that may be present in a gas as an impurity. It is also used as oxygen absorbent; a catalyst in organic and inorganic chemical manufacture; a condensing agent for soaps, fats, and oils; and for denitration of nitrocellulose (Ref 49).

Copper fluoborate is used in copper plating baths. It permits use of higher current densities when other conditions of deposition are similar to those maintained for a copper sulfate bath.

Copper gluconate, $Cu(C_6H_{11}O_7)_2H_2O$, is an odorless, light blue, fine crystalline powder. It is soluble in water but insoluble in acetone, alcohol, and ether. It is a pharmaceutical used in medicine and as a food additive.

Cupric hydroxide, $Cu(OH)_2$, formula weight 97.59, is formed as a blue hydrogel when an alkali is added to a cupric solution. The hydroxide is used as a mordant and pigment, in the manufacture of many copper salts, and for staining paper.

Copper Iodides. Cuprous iodide, CuI or Cu_2I_2, formula weight 190.49 or 380.98, occurs as a white or brownish-white crystalline powder with cubic or hemihedral crystals. It has a melting point of 605 °C (1121 °F), a boiling point of 759 to 772 °C (1398 to 1422 °F), and a specific gravity of 5.62 g/cm^3. It is insoluble in water and soluble in sodium hydroxide, dilute potassium iodide, potassium cyanide, and hot concentrated hydrochloric acid. Concentrated nitric and sulfuric acids, alkalis, and alkali carbonates decompose cuprous iodide. It is less sensitive to light than cuprous bromide and cuprous chloride. The iodide is prepared by the reaction of iodine with hot, finely divided copper or by treating a solution of cupric sulfate with potassium iodide. Cupric iodide, CuI_2, formula weight 317.41, has not been isolated in the solid form. Copper tetraiodide (CuI_4), copper hexaiodide (CuI_6), and copper nonaiodide (CuI_9) have been reported to be formed in attempts to prepare cupric iodide.

Cupric nitrates, $Cu(NO_3)_2$, formula weight 187.59, is known in trihydrated and hexahydrated forms. The trihydrate occurs as blue deliquescent prismatic crystals. It has a melting point of 114.5 °C (238 °F), a boiling point of 170 °C (338 °F), and a specific gravity of 2.047 g/cm^3. Basic cupric nitrate, $Cu(NO_3)_23CuO_3H_2O$, formula weight 480.34, forms a green powder or bluish crystals and decomposes at high temperatures. Cupric nitrate is used in electroplating solutions for obtaining a burnishing effect on iron and in the preparation of oxide or metal catalysts. It is also used in ceramics, in dyeing as a mordant, in fireworks, and in photography (Ref 49).

Copper Oxides. Cuprous oxide, Cu_2O, occurs as the natural mineral cuprite, but the industrial chemical is normally synthesized. It has a formula weight of 143.14 and exists as dark red crystals or granular powder. It has a melting point of 1235 °C (2255 °F), a boiling point of 1800 °C (3272 °F), and a specific gravity of 6.0 g/cm^3. U.S. Navy specification 52C4c addresses two grades of the compound in terms of its reducing power and total copper content. Grade I must contain a minimum of 97% Cu_2O, 86% total copper, and 97% reducing power as Cu_2O. For Grade II, the corresponding figures are 70%, 80%, and 90%. Cuprous oxide is used as a fungicide and seed dressing (together with $CuSO_4$), as an ingredient of antifouling marine paints, as a stain to produce copper ruby glass, as a component of Cu_2O/Cu rectifiers, and in photoelectric cells. The oxide is also utilized as a catalyst (Ref 49, 50).

Cupric oxide (black copper oxide), CuO, formula weight 79.57, is found as a brownish black amorphous powder. It has a melting point of 1026 °C (1879 °F) and a specific gravity of 6.45 g/cm^3. It occurs naturally as paramelaconite and tenorite. It is synthetically produced by any of several methods. Industrial applications include ceramic glazes (blue-green and red tints), chemical intermediates (e.g., the production of synthetic fibers), animal dietary supplements, reagent for organic and gas analysis, and as a catalyst in the reduction of organic compounds (Ref 49, 51).

Copper oxychlorides encompass a range of chemicals best defined by the molecular ratios $CuO:CuCl_2:H_2O$ (i.e., various oxychlorides in anhydrous and hydrated forms). Copper oxychlorides are used as agricultural fungicides. Examples include Cupro-K and copper compound A (tetracalcium copper oxychloride). A hydrated oxychloride (Brunswick green) is used as a pigment and as a component of artificial patina (Ref 49–51).

Copper soaps or fatty acid salts are very important fungicides. During World War II, oxidation-resistant copper naphthenate was used in huge quantities to treat sandbags, ropes, tarpaulins, and similar materials. Copper oleate, stearate, and tallate were used when naphthenic acids became scarce. Copper naphthenate and resinate have also been used to pressure-treat wood. The choice of vehicle in which the fungicide is dissolved depends on the application. For textiles, copper soaps can be dissolved in volatile solvents, such as mineral spirits; for wooden railroad ties or telephone poles, cheap nonvolatile oils, such as gas oils, can be used; while fungus-resistant paints utilize linseed oil/solvent mixtures.

Copper Sulfate. Penthydrated copper sulfate, $CuSO_4 \cdot 5H_2O$, also known as blue vitriol and blue stone, is the best known and most widely used copper compound. Uses range from agricultural biocides to electroplating solutions. Commercial copper sulfate contains 25% copper and is sold with a guaranteed minimum purity of 98% copper sulfate. It is sold as large, small, and granulated crystals and as a powder.

Specialty Coppers and Copper Alloys

Copper-base metals can be considered in terms of alloys that have a wide range of uses and whose applications demand a unique combination of properties. The former includes such generic materials as the tough pitch and deoxidized coppers as well as the common brasses, bronzes, and copper-nickels. However, while it would be difficult to find a copper metal with only one use, several coppers and copper alloys offer such a unique combination of properties that they are considered specialty materials.

Oxygen-Free (OF) Coppers

The designation "oxygen-free copper" refers to a group of highly refined coppers that are processed under protective atmospheres and slags such that they contain very little dissolved oxygen. Oxygen exists in copper as copper oxide and is visible in the microstructure if enough is present. The oxide has only a small effect on conductivity and mechanical properties, but it markedly degrades weldability. Copper oxide is reduced to copper and oxygen under the heat of welding. Hydrogen present in atmospheric moisture or acetylene combustion products combines with the liberated oxygen to form water, which appears as vapor-filled voids. The voids embrittle the weld metal; hence, they cause poor weldability. Products to be welded must therefore be made from either OF copper or copper in which the oxygen is combined in a stable combined form, such as phosphorus deoxidized copper (Ref 52, 53).

Oxygen-free coppers have slightly higher electrical and thermal conductivities than deoxidized and tough-pitch coppers. The International Annealed Copper Standard (IACS) for conductivity, established in 1913, is based on a copper, which under standard conditions exhibits a volume resistivity of 0.017241 $\Omega mm^2/m$, or a conductivity of

100% IACS. While it is not unusual for modern deoxidized and tough-pitch coppers to exhibit conductivities greater than 100% IACS (on account of high purity), OF coppers do so consistently. Products, such as electronic components, bus bars, and waveguides, that require the absolutely highest conductivity are made, therefore, from one of the OF coppers.

Oxygen-free coppers also form strong glass-to-metal seals (provided the glass is of the proper type). The seal actually forms against the copper oxide on the surface of the metal; and in the case of high-purity OF copper, the oxide is particularly adherent. Therefore, OF coppers are used in electron tubes and similar devices.

The most common OF copper (under the UNS numbering system) is the wrought grade, UNS C10200, which contains a maximum of 10 ppm oxygen (0.0010%). This copper is referred to as OF copper, high-conductivity OF copper, and OFHC copper, which is a registered trade name that has fallen into generic usage. Copper C10200 is immune to hydrogen embrittlement, it is readily weldable, and is suitable for all electrical and most electronic uses (Ref 54).

UNS C10100 is known as oxygen-free electronic (OFE) copper. It contains only 5 ppm oxygen and less than 100 ppm (0.010%) total impurities. In other words, it is certified to be at least 99.99% pure copper. As such, it is the purest commercial metal and one of the purest materials of any type available in the world. In addition to the properties it shares with C10200 and other OF coppers, C10100 exudes no volatiles in high vacua and elevated temperatures. It forms a strongly adherent oxide film. It is used in electronics, glass-to-metal seals, electron tubes, and rectifiers, and as a matrix in superconducting cables (see Chapter 4).

Silver-Bearing Copper

Silver-bearing coppers are often (but not always) coppers in which silver, an "impurity" in the ore, is intentionally not removed from the metal during refining. Silver has very little deleterious effect on conductivity when present in the small concentrations associated with the silver-bearing grades of copper. Silver also imparts a modest degree of annealing resistance. This feature permits cold-rolled or cold-drawn (and therefore hard) copper to be soldered or brazed and still retain mechanical properties. Silver-bearing copper was at one time promoted for use in the manufacture of automotive radiators, which are assembled by soldering and/or brazing.

Because silver is relatively benign toward electrical conductivity, standard compositions (as in the UNS system) list copper and silver together, such as Cu (including Ag, % min). This format is followed for all coppers, although standards do list silver content limits separately for the designated silver-bearing grades. Silver-bearing coppers are available in oxygen-free, tough-pitch (both electrolytic and fire-refined), and deoxidized qualities.

There are twelve UNS designated silver-bearing coppers; numerous others, some equivalent to UNS grades, are recognized in international classification systems. Among the more common grades are C10400, which is equivalent to CuAg 0.03(OF); C10500, equivalent to CuAg 0.05; C10700, equivalent to CuAg 0.1(OF), and CuAg 0.2(OF), an OF grade for which there is no corresponding UNS quality.

Because of its rather high silver content, CuAg 0.2(OF) is more creep- and anneal-resistant than the leaner grades, and it can sustain stress at temperatures up to 400 °C (752 °F). Along with other silver-bearing coppers, it is used for hollow conductors (for gas-cooled generator windings, induction furnace coils, and similar applications), commutator switches, and heavy duty electrical equipment (Ref 54).

Other High-Strength, High-Conductivity Alloys

The addition of alloying elements is a common means to achieve higher strength in copper-base metals, but improvements come at the cost of decreased thermal and electrical conductivity. Creating both high strength and acceptably high conductivity often involves a compromise. Beryllium coppers are perhaps the best known examples of high-strength, high-conductivity copper alloys, yet even these important materials are subclassified by whichever of the two properties has been optimized.

The zirconium coppers comprise another interesting group of alloys. A North American version bears the UNS designation C51000 (99.8% Cu + Ag, 0.15 to 0.20% Zr), but other grades are common in Europe and elsewhere. The alloy can be brazed and electron-beam welded. It is used for power semiconductor bases, leadframes, and a variety of welding products, as well as commutators, automotive radiator strip, and continuous casting molds (Ref 53, 55–64).

Specialty Product Forms

In addition to specialty grades, a number of new product forms have been developed in response to customer demand. Such demands have originated mainly from the electronic and automotive markets. Examples of the latter include extra wide brass foil, 38 μm (0.0015 in.) thick copper radiator strip (which is also available in copper-tin alloys), and extra-thin brass strip for the manufacture of radiator tube.

Advanced Technology Products

A number of copper-base materials are finding uses in what can generally be described as advanced technologies. Better-known examples include superconducting cable, superconducting ceramics, metallic glasses, and shape memory effect alloys. All of these materials are accepted commercial products; but while they show great promise for expanded copper usage in the future, at present they do not represent a significant fraction of the copper market.

Superconductors

Low-temperature (i.e., operating in liquid helium at 4.2 K) superconducting (SC) wire and tape have been available for many years. Nevertheless, they can be considered advanced-technology materials because of the considerable amount of alloy and process development that continues to be devoted to them.

Copper is not a superconductor, although its conductivity at cryogenic temperatures exceeds 1000% IACS. Copper and tin bronzes, however, are important components of many types of commercial SC wire and cable. The most important and best known applications for such conductors are windings for the high-field electromagnets found in magnetic resonance imaging (MRI) equipment and particle accelerators (see Chapter 5).

SC products are composite structures in which filaments of a type II superconductor, such as

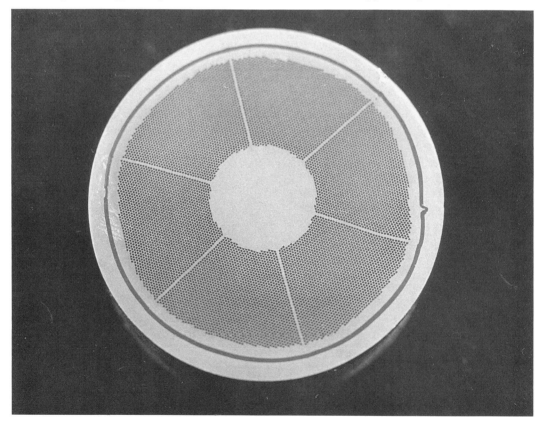

Fig. 3.21 A typical hot-pressed compact of superconducting filaments in a copper matrix. Source: Outokumpu Copper Partner AB

Nb$_3$Sn (among scores of others), are embedded in a copper-base matrix/stabilizer. Copper acts as a thermal sink and an electrical shunt in the event of momentary and local loss of superconductivity. Bronze acts as a source of tin for the in situ formation of the niobium-tin intermetallic. Diffusion barriers are normally placed or formed between the SC and matrix/stabilizer materials to prevent the superconductor from changing composition over time.

Individual superconducting filaments range from about 6 to 8.3 μm (236 to 326 μin.) in diameter to much larger sizes, depending on the intended application. Superconducting wire is offered in a variety of sizes, configurations, and copper to superconductor (Cu:Sc) ratios. The product is formed as a compact, which is extruded and drawn to its final dimensions. Figure 3.21 shows a compact before extrusion, and Fig. 3.22 shows a variety of hot-pressed compacts and superconducting cables obtained from them (Ref 65).

A series of high-temperature (high T_c) superconducting materials has been under development since their discovery in the early 1980s. The superconductors are ceramics; the best known of which is a mixed copper-rich oxide with the composition YBa$_2$Cu$_3$O$_7$. The potential value of high T_c superconductors is that they can operate at liquid nitrogen temperatures, a much less costly regime than the liquid helium required for conventional superconductors. Large-scale commercial applications for the materials had not been commercially developed by the mid-1990s, but initial uses are expected in computers and electronic devices rather than in applications requiring a large current flow. As with conventional superconductors, copper may find use as a stabilizer or matrix material, because the conductors themselves are quite brittle. Understandably, a large number of patents have been registered for these ceramic materials. The patents relate to the products themselves and to their use in bulk, thin layers, films, and transmission lines (Ref 67–70).

Metallic Glasses

As the name implies, metallic glasses are metallic materials that have no fixed crystalline struc-

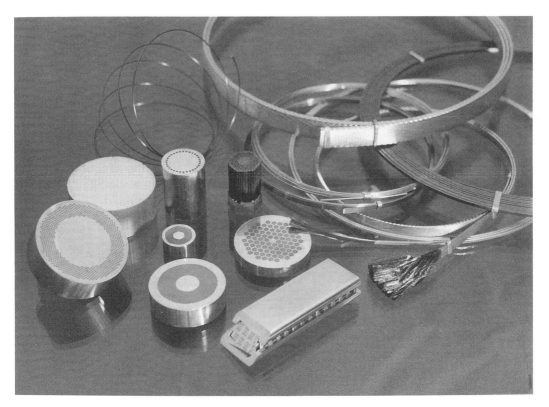

Fig. 3.22 A variety of hot-pressed compacts and the superconducting cables made from them by extrusion. Source: Outokumpu Copper Partner AB

ture at low temperatures (i.e., in the "solid" state). They have a number of very useful properties, including high corrosion resistance, low melting points (glass transition temperatures), and low coercive forces (see Chapter 2).

Metallic glasses are obtained by the extremely rapid solidification of a molten alloy of suitable composition. The glassy state can also be obtained by vapor deposition, sputter deposition, electroplating, electroless plating, and chemical vapor deposition. There are two types of metallic glasses that refer to the atomic nature of the constituent elements: metal-metalloid and metal-metal. Many varieties of both types have been developed since the metallic glass phenomenon was first studied in the 1950s. One reported copper-containing metal-metalloid glass-forming composition is designated $Pd_{77.3}Cu_{6.0}Si_{16.5}$. In the glassy state, the material has a room temperature hardness of 500 HV, which corresponds to a yield strength of 1570 MPa (227 ksi) and an elastic modulus of 88.2 Gpa (12.8 MPa). Examples of copper-containing metal-metal metallic glasses include $Zr_{50}Cu_{50}$, $Zr_{35}Cu_{65}$, and $Cu_{60}Zr_{40}$. Copper-tin, copper-gold, and copper-lead alloys, as well as copper itself, can be splat-cooled to a glassy state at cooling rates on the order of 10^6 K/s (Ref 71, 72).

The only commercial application for copper-base metallic glasses to date has been as low-melting-point brazing alloys. In this instance, metallic glass is supplied in the form of thin ribbons, which are fitted between the surfaces to be brazed. The alloys become fluid upon heating to the glass transition temperature (which can be considerably below the liquidus temperature of the corresponding alloy), after which they freeze as ordinary crystalline solids. It has been suggested that metallic glasses can also find use as protective coatings (possibly by way of plasma or thermal spray deposition) and as nonmagnetic electrical devices (Ref 71, 72).

Shape Memory Effect Alloys

Shape memory effect (SME) alloys are capable of returning to a preset shape or configuration upon heating to a particular temperature. That is, they "remember" the shape they held before deformation. The driving force is a reversible martensitic reaction that gives rise to stresses that are high enough to cause elastic strains of considerable magnitude. The materials have also been called "marmem" alloys to highlight the microstructural basis for the memory effect.

Several alloy systems exhibit SME. The effect has been observed in a number of copper-base alloy systems, including Cu-Zn, Cu-Sn, Cu-Al-Ni, and Cu-Zn-X (where X = Ga, Si, Sn, Au). To date, only Cu-14/14.5%Al-3/4.5%Ni and Cu-10/30%Zn-4/10%Al alloys have been found to exhibit SME-based stresses or strains large enough for commercial exploitation (Ref 73, 74).

Examples of devices in which SME alloys are used include heat shrinkable pipe and flange couplings, clamps, orthodontal devices, greenhouse window controls, automotive clutch fans, variable-orifice carburetors, thermostatic radiator valves, and temperature controllers and activators. Perhaps the most spectacular use of a SME to date is as the driver for a spontaneously unfolding space craft antenna (Ref 75, 76).

REFERENCES

1. *1989 Annual Book of ASTM Standards, Nonferrous Metal Products: Copper and Copper Alloys,* Vol 02.01, American Society for Testing of Materials, Philadelphia, PA, 1989
2. *Wrought Mill Products,* Copper/Brass/Bronze, Standards Handbook, CDA Pub. No. 104/8, Copper Development Association Inc., New York, NY, 1968
3. "Production of Refined Metal, Production, and Consumption—Forming, Copper and Alloys" ("Fabrication de Métal Raffiné Productions et Consommations—Mises À Forme. Cuivre et Alliages: Principles Qualités Rencontrées Principales Propriétés et Utilisations") brochure, CIMNF Cu 031, Centre D'Information des Metaux non Ferreux, Le Cuivre, Février, Belgium, 1986
4. "Specifications for Copper Refinery Shapes," British Standard Institution, BS 6017, 1981
5. W.M. Tuddenham and R.J. Hibbeln, Ed., Sampling and Analysis of Copper Cathodes, STP 831, ASTM, Baltimore, MD, March 1984
6. *Wrought and Cast Products,* Copper/Brass/Bronze, Standards Handbook, CDA Pub. No. 101/8, Copper Development Association Inc., Greenwich, CT, 1988
7. *Properties and Selection: Nonferrous Alloys and Pure Metals,* Vol 2, *Metals Handbook,* 9th ed., American Society for Metals, 1979, p 239–249
8. "Outokumpu Superconductors," brochure, Outokumpu Oy, Copper Division, Pori, Finland
9. "Specification for Higher Purity Copper Cathode," British Standards Institution, DD 78, 1981
10. "OD Electrorefined Copper Cathode Typical Analysis," brochure, Olympic Dam Marketing Pty. Ltd., Parkside, South Australia, July 1990
11. W.C. Cooper, Impurity Behavior and Control in Copper Electrorefining, *Proc. of Copper '87,* Vol 3, Chilean Institute of Mining Engineers, 1987, p 451–465
12. R.E. Lee and S.J. Harris, Reinforcement of Composites with Continuous Fibers, *J. Mater. Sci.,* Vol 9, 1974, p 359–368

13. R.G. Austey, Drawability Index of Copper Wire, *Wire Ind.*, Sept 1989, p 580–582

14. R.D. Adams, A. Cofer, and U. Sinha, "Southwire Continuous Rod," Paper 15, presented at Symp. Cu Forever (Bombay, India), Indian Copper Development Centre, 1987

15. *Stranded Conductors for Electrical Purposes*, Drahtwerk Waidhaus, Germany

16. *1991 Annual Book of ASTM Standards: Copper and Copper Alloys*, Vol 02.01, ASTM, Philadelphia, PA, 1991

17. A. Knowlton, Ed., *Standard Handbook for Electrical Engineers*, McGraw-Hill Book Co., New York, NY, 1949

18. D.E. Tyler and J. Crane, High Performance Copper Alloys for Electrical and Electronic Applications—an Integrated Approach to Meet Customer Needs, *Proc. Copper '90* (Ottawa, Canada), Vol 1, Pergamon Press, 1990, p 367–377

19. "Catalog: Tube and Plate Plant" ("Catálogo Planta de Tubos y Planchas"), MADECO, Santiago, Chile, March 1991

20. *Copper Building Wire*, Copper/Brass/Bronze Product Handbook, CDA Pub. 601/0, Copper Development Association Inc., Greenwich, CT

21. "Catalog: Electrical Wire and Cable" ("Catálogo Alambres y Cables Eléctricos"), brochure 0001240, Cobre Cerrillos S.A., Santiago, Chile

22. "Cocessa Technical Bulletin" ("Boletines Técnicos Cocesa"), brochure, Cobre Cerrillos S.A., Santiago, Chile

23. "Electrical Conductor Catalog" ("Catálogo Conductores Eléctricos"), Catálog 751, MADECO, Santiago, Chile, 1990

24. M. Earley, R. Murray, and J. Caloggero, *The National Electrical Code 1990 Handbook*, National Fire Protection Association, Quincy, MA, 1990

25. B.K. Hay, Mineral Insulated Copper Sheathed Wiring Cables, *ATB Metal.*, Vol 16 (No. 2), 1976, p 130–132

26. Wrought Mill Products, Copper/Brass/Bronze, *Standards Handbook*, CDA Pub. 102/3, Copper Development Association Inc., New York, NY, 1983

27. "Wire and Extruded Products, Alloyed Strip," brochure, Outokumpu OY, Copper Products Division, Porie, Finland, 1988

28. "Copper Strip and Wire for Structural Elements in Electronics" ("Bänder und Drähte aus Kupferwerkstoffen für Bauelemente der Elektrotechnik und der Elektronik"), DKI Information Pub. 1020, Deutshes Kupfer Institut, Berlin

29. "Schwarzwald Metal Works" ("Metallwerke Schwarzwald"), brochure, Metallwerke Schwarzwald, GmbH, Schwenningen, Alemania

30. L. Betty, "Process and Apparatus for Electroplating Copper Foil," U.S. Patent 4,898,647, 2 June 1990

31. "Copper, Food, Health" ("Kupfer Lebensmittel Gesundheit"), brochure, DKI Information Publication, Bestell-Nr 1,019, Deutsches Kupfer Institut, E.V., Berlin

32. "Stainless Steel & Bronze Sections," brochure, Drawn Metal Ltd., Swinnow Lane, Bramley

33. A. Cohen, Copper for Hot and Cold Potable Water Systems, *Heat/Piping/Air Cond.*, Vol 50 (No. 5), May 1978, p 81–87

34. A. Cohen, Copper and Copper Alloys in the Great Pipe Debate, *ASTM Stand. News*, Vol 15 (No. 9), 1987, p 54–59

35. *The Copper Tube Handbook*, Copper Development Association Inc., New York, NY, 1995

36. A.H. Tuthill, Practical Guide for Selecting Metals for Heat Exchanger Tubes, *Mat. Perf.*, Vol 29 (No. 11), 1990, p 56–59

37. "Ripple-Fin Tubes, GEWA Finned Tubes, Type D," brochure, Wieland-Werke AG, Ulm, Germany

38. "Copper Tube Systems for Sanitation and Heating" ("Kupferrohrsysteme für Sanitär und Heizung"), Wieland-Werke AG, Ulm, Germany

39. "La Vostra Fiducia Nella Qualità", Centro Informazioni SANCO, Milan, Italy

40. Cast Copper and Copper Alloy Products, Part 7, Alloy Data, *Standards Handbook*, Copper Development Association Inc., New York, NY, 1978

41. "The Aluminum Bronzes" ("Die Aluminium Bronzen"), brochure, Deutshes Kupfer Institut, Berlin, 1958

42. H.J. Bargel and G. Schulze, Ed., *Materials News (Werkstoffkunde)*, VDI-Verlag GmbH, 1983

43. B. Landolt-Bornstein III, *Technology, Classification, Value and Behavior of Metallic Materials (Band Technik Teil Stoffwerte und Verhalten von Metallischen Werkstoffen)*, Springer-Verlag, 1964

44. "Copper-Aluminum Alloys: Properties, Production, Fabrication, Use" ("Kupfer-Aluminium Legierungen Eigenschaften Herstellung Verarbeitung Verwendung"), brochure, Deutsches Kupfer Institut, Berlin

45. "Copper-Nickel Alloys," brochure, Bulletin TN30, Copper Development Association, Potters Bar, Herts, U.K., Sept 1982

46. J.L. Everhart, Copper and Copper Alloy Powder Metallurgy Properties and Applications, Technical Report, Copper/Brass/Bronze, *Powder Metals Properties and Applications*, CDA Pub. 129/6, Copper Development Association Inc., New York, NY

47. "Forgings," Copper/Brass/Bronze, Design Guide, CDA Pub. 705/5, Copper Development Association Inc., Greenwich, CT

48. *Properties and Selection of Metals*, Vol 1, *Metals Handbook*, 8th ed., American Society for Metals, 1961

49. R.E. Kirk and D.F. Othmer, *Encyclopedia of Chemical Technology*, Vol 15, 4th ed., John Wiley & Sons, New York, NY, 1991

50. *Gmelins Handbook of Inorganic Chemistry, Copper: Part A, Section 1 (Gmelins Handbuch der Anorganischen Chemie, Kupfer: Teila - Lieferung 1)*, Verlag Chemie, GmbH, Weinheim/Bergstrasse, 1955

51. "HCOKOF Copper, Regular Grade," ASTM 10200 OF, brochure, Outokumpu OY, Copper Products Division, Porie, Finland, 1988
52. "Uses of Copper Compounds," brochure, Bulletin TN11, Copper Development Association, Potters Bar, Herts, U.K., Feb 1974
53. "High Conductivity Coppers," brochure, Properties and Applications Bulletin T29, Copper Development Association, Potters Bar, Herts, U.K., Nov 1981
54. "Wire and Extruded Products, Oxygen-Free Copper and Silver-Bearing Copper," brochure, Outokumpu OY, Copper Products Division, Porie, Finland, 1988
55. "Zirconium Copper," brochure, Outokumpu OY, Copper Products Division, Porie, Finland, 1987
56. "Chromium, Zirconium Copper Welding Tips," brochure, Outokumpu OY, Copper Products Division, Porie, Finland, 1988
57. "Wire and Extruded Products, Copper Hollow Conductors for Magnet Windings," brochure, Outokumpu OY, Copper Products Division, Porie, Finland, 1988
58. "Drawn Products, Hollow Conductors, Magnet Tubes, Rotor Tubes, Stator Tubes, Sections for Special Applications, List of Sizes for Which Tools are Available," brochure, Outokumpu OY, Copper Products Division, Porie, Finland, Feb 1989
59. "Wire and Extruded Products, Copper for Electric Generators," brochure, Outokumpu OY, Copper Products Division, Porie, Finland, 1988
60. "Tubes for MIG/MAG Welding," brochure, Outokumpu OY, Copper Products Division, Porie, Finland, 1988
61. "Wire and Extruded Products, Hot Tinned Strip," brochure, Outokumpu OY, Copper Products Division, Porie, Finland, 1988
62. "Hitachi Oxygen-Free Copper (Hitachi OFC)," brochure, Catalogue EA-101E, Hitachi Cable Co. Ltd., Tokyo, 1985
63. "Hitachi Oxygen-Free Copper Strip and Tape," brochure, Catalog EA-121D, Hitachi Cable Co. Ltd., Tokyo, 1989
64. "Hitachi Oxygen-Free Copper for Automotive Radiators," brochure, Catalog EA-131, Hitachi Cable Co. Ltd., Tokyo, 1985
65. "Superconductors," brochure, Catalog EA-600C, Hitachi Cable Co. Ltd., Tokyo, 1989
66. Superconducting Wire; Cable Direct Lines to the Future; and Oxygen-Free Copper—The Ultimate Foundry Product, *Outokumpu News,* Vol 27 (No. 2), 1990
67. M.W. Dew, L. Jolla, and R.I. Creedon, "Superconducting Transmission Line System," U.S. Patent 4,947,007, 7 Aug 1990
68. W.C. Keur, C.A.H.A. Mutsaers, and H.A.M. Van Hall, "Superconductive Thin Layer," U.S. Patent 4,948,779, 14 Aug 1990
69. R.E. Soltis, E.M. Logothetis, and M. Aslam, "Preparation of Superconducting Oxide Films Using a Pre-Oxygen Nitrogen Anneal," U.S. Patent 4,943,558, 24 July 1990
70. H.C. Ling, "Method of Making High Density YBaCu$_3$O$_x$ Superconducting Material," U.S. Patent 4,943,557, 24 July 1990
71. P. Chaudhari, B.C. Giessen, and D. Turnbull, Metallic Glasses (Vidrios Metélicos), *Investigations and Science (Investigacíon y Ciencia),* Vol 45 (No. 6), 1980, p 58–72
72. J.D. Streeter, "Introduction to Thermal Spraying" ("Introduccion al Rociado Termico"), brochure, Eutectic + Castolin, Santiago, Chile, 1988
73. J. Perkins and R.O. Sponholz, Stress-Induced Martensite Transformation Cycling and Two-Way Shape Memory Training in Cu-Zn-Al Alloys, *Metall. Trans. A.,* Vol 15 (No. 2), 1984, p 313–321
74. R.L. Mannheim, Basic Facts in the Technology of Beta-Brass Production for Shape Memory Effect, *Proc. Copper '87* (Santiago, Chile), Vol 1, Chilean Institute of Mining Enginners, Nov 1987, p 339–346
75. B.J. Spalding, New Uses for Materials with Memories, *Chemical Week,* 12 Feb 1988
76. C.M. Wayman, Some Applications of Shape-Memory Alloys, *JOM,* Vol 32 (No. 6), p 129–137

4

Fabrication

Melting Technology

All copper products, whether cast or wrought, pure or alloyed, begin as molten metal. The starting material may be primary copper, for example, that which has been refined from ore, or remelted and/or re-refined scrap. See Chapter 1 for a discussion of these terms. The metal is processed in an appropriately sized furnace capable of furnishing liquid metal of sufficient quality for the type of alloy to be produced and the nature of the casting operation. This chapter deals with primary copper and pure alloying elements and does not include the rather complex copper scrap cycle (Ref 1).

The choice of melting furnace strongly influences the cost of the melting operation and the quality of the metal produced. The relevant parameters include:

- The melting point of the alloy
- The danger of contamination of the molten metal by the furnace atmosphere and/or refractories
- Furnace capacity, melting and throughput rates
- Power source (electricity, oil, gas), fuel cost and availability
- Heat recovery, if any

Most modern copper-melting furnaces incorporate heat recovery systems to reduce energy consumption (Ref 2).

Fuel-Fired Furnaces

ASARCO furnaces, named for the American Smelting and Refining Company, are commonly used for melting pure copper cathodes and clean scrap. The product is tough pitch copper, which is normally fed to wire rod casting machines. The first ASARCO furnaces were put in operation in the late 1950s and have since been built in a range of sizes (Ref 3).

An ASARCO furnace is a shaft furnace (Fig. 4.1), internally shaped like an inverted cone, about one-half as wide at the bottom as at the top. At the charge level, the diameter ranges from 1.5 to 1.8 m (4.9 to 5.9 ft). By adjusting the fuel/air mixture, the furnace atmosphere is kept slightly reducing so as not to oxidize the molten metal. Fuels include desulfurized natural gas, propane, butane, and naphtha. The charge is melted by radiation from the side walls and by conduction and convection from the molten metal in the bottom of the furnace. Energy consumption is 10^5 kcal per ton of cathode (expressed as 100% Cu) (Ref 3).

Reverberatory furnaces, also called open hearth furnaces, are rectangular in plan, with burners situated in the two shorter furnace walls. The shallow molten bath is heated by flame and hot gases passing overhead and by radiation from the arched refractory roof. Hot (1100 °C, or 2012 °F) exiting gases are directed through a heat recovery boiler before being released to the stack (Ref 4).

Reverberatory furnaces are slower and thermally less efficient than ASARCO furnaces, requiring up to 24 h to melt, hold, and cast the charge, which may be as large as 400 tons of copper. Despite protective slags, the charge can pick up sulfur, and the metal is normally further refined by air oxidation and poling (Fig. 4.2). Before the advent of the ASARCO furnace, fuel-fired reverberatory furnaces were used almost exclusively to melt cathode for the production of tough pitch copper. The furnaces are still used in those few cases where wire bars are cast from cathode copper in copper refineries. Reverberatory furnaces are sometimes used for brass melting but almost never for melting other alloys (Ref 1, 3).

Cupola smelting furnaces resemble blast furnaces, in that the charge is in direct contact with the solid fuel, normally coke. High-pressure air is injected through tuyères located above the crucible at the bottom of the furnace. They are among the cheapest and simplest of all metallurgical furnaces, but they tend to produce significant levels of air pollution (Ref 5).

Cupola furnaces were originally used to smelt copper, a practice since discontinued. By using special low-sulfur and low-ash coke, they can be used to melt brass, but the process is difficult because the furnaces tend to oxidize and volatilize zinc, as shown in Fig. 4.3 (Ref 1).

Crucible furnaces are the simplest melting devices for small and intermittent production (Fig. 4.4, 4.5, and 4.6). They consist of a small vertical cylindrical chamber inside which heat is transmitted to a crucible by direct flame, resistance elements, or

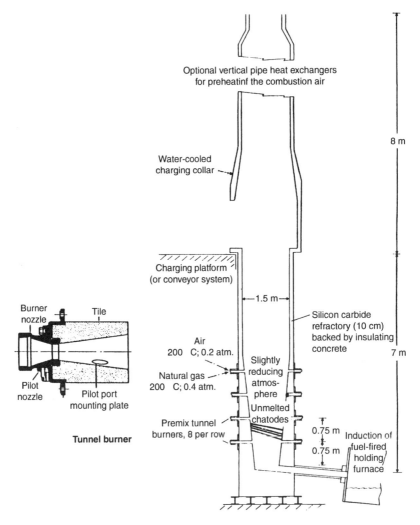

Fig. 4.1 ASARCO shaft furnace for melting cathodes. The insert shows a premix tunnel burner in which combustion is complete before the gases enter the shaft (Ref 3).

induction coils. Crucibles are made from clay-agglomerated graphite, carbon-agglomerated silicon carbide, or other refractory compositions. The furnaces may be covered or open, fixed or tilting, and may even be portable. Charges of 100 to 500 kg (220 to 1100 lb) of brass can be melted in 30 to 120 min. Crucible furnaces are very well suited for small alloy batches and hence are favored for artistic work (Ref 6).

Copper converters are refractory-lined cylindrical vessels, oriented horizontally (Fig. 4.7). The charge is fed through, and melt and slag tapped from, an opening in the furnace wall at mid-length. Charging and tapping are accomplished by rotating the vessel about its longitudinal axis. Converters are autogenously fired; that is, air, injected through tuyères, burns sulfur and iron in the charge, known as "matte," producing heat. Molten combustion products are slagged off periodically (Ref 7).

Rotary furnaces, also known as drum or trommel furnaces, consist of horizontal cylinders terminating in cones at both ends (Fig. 4.8). Gas, oil, or pulverized coal burners are mounted at one end, and the opposite end receives the charge. The tap hole is located at the center of the side wall, as in a converter, and like converters, the furnaces rotate about their horizontal axis. The furnace is continuously rotated during its operation, causing the entire heated lining to come into contact with the charge. For better thermal efficiency, combustion air is heated by exchange with the exhaust gases (Ref 2, 8).

Rotary furnaces range between 50 and 6000 kg (110 and 13,200 lb) in capacity. They have a fast melting rate, low melting losses, and produce metal with low gas content and uniform composition and temperature. The furnaces have low fuel consumption and allow complete control of temperature and atmosphere. On the other hand, rotary furnaces do not allow easy purging with inert gases because of the shallow bath. They also do not permit fast charging of new materials, and they carry a severe air pollution penalty. They are primarily used to pretreat scrap to be charged into a melting furnace (Ref 2).

Electric Furnaces

The three types of electric furnaces used in copper industry include direct and indirect arc furnaces and several styles of induction furnaces (Fig. 4.9).

Direct or submerged arc furnaces (Fig. 4.10) have been used to smelt copper matte in the United States, Japan and Scandinavia. They serve well in this role, but they cannot be used as simple melting furnaces, especially for alloys, because the extremely high temperature at the arc makes composition control difficult.

Indirect Arc Furnaces. Indirect arc, or "Detroit," furnaces are usually constructed in the form of refractory-lined horizontal steel cylinders in which two carbon electrodes are fitted axially (Fig. 4.11). Heating is by radiative heat transfer from an arc struck between the electrodes, for example, the arc does not directly impinge on the charge. Detroit furnaces have proven useful in small foundries, but they are rarely used in large operations. They can be built in a wide range of capacities.

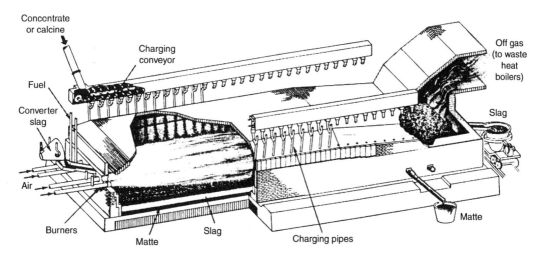

Fig. 4.2 Reverberatory copper smelting furnace (Ref 4)

Core-Type, Low-Frequency Induction Furnaces. Low-frequency (50 to 60 Hz), core-type induction furnaces operate as short-circuited transformers. The primary coil and an iron core are embedded in the refractory of the furnace's inductor channels. Heat is generated in the molten metal in the channels, which act as the "transformer's" secondary circuit. Electromagnetic forces circulate the molten metal and provide a uniform temperature and composition throughout the melt. Despite the chance for gas pickup in the constantly circulating metal, induction furnace metals are generally cleaner than those from either reverberatory or fuel-fired crucible-type furnaces. Core-type, low-frequency induction furnaces are available in sizes ranging from 20 to 200 kW (Fig. 4.12). They are used both as melting devices and as hot metal reservoirs for continuous casting machines, in which case they are used in series with separate melting furnaces (Ref 9).

Coreless low-frequency induction furnaces, "channel" or "crucible furnaces," consist of a cylindrical, refractory-lined melting chamber surrounded by a water-cooled primary coil. In this case, the induced current melts the charge directly. Unlike core-type furnaces, coreless units can be charged with cold metal, and they are, in fact, mainly used for scrap melting. Sizes range from

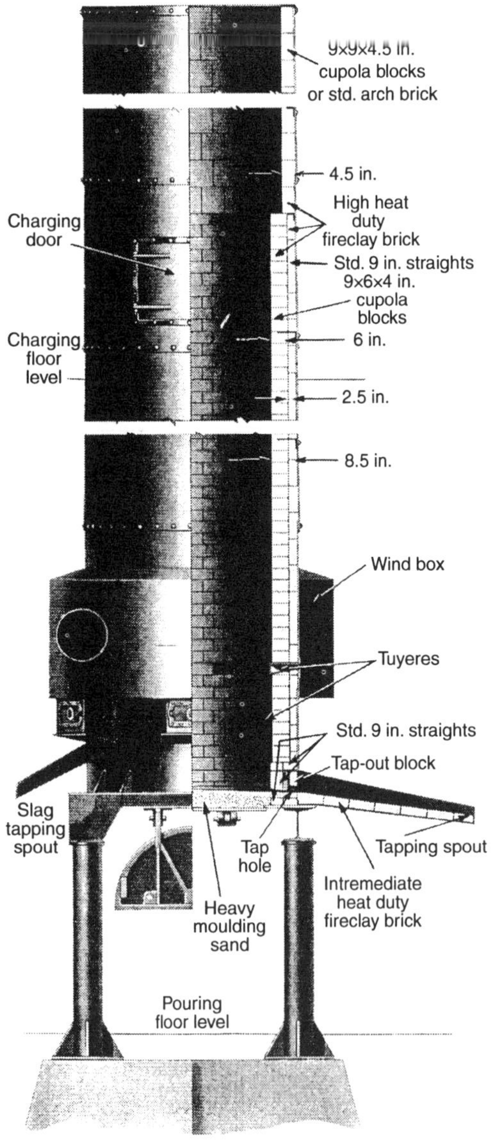

Fig. 4.3 Section of typical cupola furnace

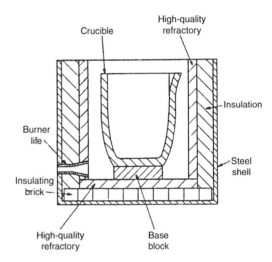

Fig. 4.4 Crucible furnace

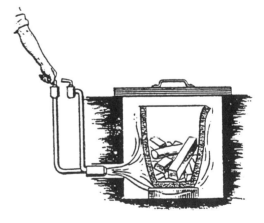

Fig. 4.5 Heating a crucible furnace with a gas heater

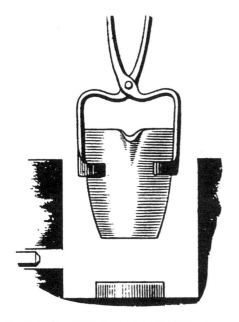

Fig. 4.6 Crucible furnace discharge

75 to 700 kW, with capacities from kilos to many tonnes.

High-frequency induction furnaces have limited commercial application because of their relatively small capacity and their inability to melt fines. They are otherwise similar to coreless low-frequency units. Capacities range from grams to several hundred kilos (Ref 10).

Refractories

Copper processing furnaces are constructed from a variety of refractory matcrials, whose selection depends on the metallurgical process involved and on the technical requirements of the equipment in question.

Shaft Furnaces. For example, gas-fired shaft furnaces are lined with nitride-bonded silicon carbide refractories to resist abrasion erosion and impact as well as thermal shock and alkali attack. Such refractories have rather high thermal conductivity, and furnaces are insulated with firebrick or

Fig. 4.7 Group of Peirce-Smith copper converters in operation, each with one cylindrical holding furnace

insulating castables. Properties of one type of agglomerated silicon carbide shaft furnace refractory are given in Table 4.1; those for insulating brick and firebrick are listed in Tables 4.2 and 4.3, respectively (Ref 10–13).

Channel-type induction furnaces utilize silicon carbide bricks below the liquid level. Above that, as well as for backup, 45% silica-alumina brickwork is used. Properties are given in Table 4.4. Furnaces may also be entirely lined with high alumina-phosphate agglomerated brick and castable or rammed refractories. Similar combinations are also used in Detroit-type furnaces and in reverberatory and rotary drum furnaces (Ref 13).

Lining Life and Contamination. The life of refractory linings depends on furnace practice (continuous or intermittent operation) and on the type of refractory. Silicon carbide linings can never be allowed to cool down, whereas silica-alumina bricks allow some flexibility in operation. Wear is always more pronounced in contact with solid charges or at the metal slag interface. In rotary furnaces, this problem is much less important. As a general rule, all brickwork damage should be repaired after each pour in tilting and shaft furnaces.

To minimize contamination, all refractories that contact molten metal should contain less than 0.5% iron. Otherwise, contamination of the melt by refractories is not a serious problem. Reactions such as silica reduction in aluminum melts do not occur in copper; however, slag erosion has been observed in chrome magnesite bricks.

Refractory coatings and gunning mixes for brickwork repair are available in silico-aluminum fire clay from 40% through 60% Al_2O_3, and as silicon-carbide alumina gunning mixes. Compositions are listed in Table 4.5.

Raw Materials

Charge Preparation. Copper products are generally considered high-value goods, and appropriate care is taken in their production, including melting. The degree of care can vary slightly, however, depending on the nature of the product in question. For example, because of the detrimental effect of impurities on electrical conductivity, charge materials for cast electrical products must be as pure as possible. For casting general-purpose alloys, it is important that the charge should be as homogeneous as possible and that its composition be within specified limits. During charging, it is helpful that the melting temperatures of the charge components be as close as possible. This avoids volatilization and oxidation of low melting point elements, and it also helps

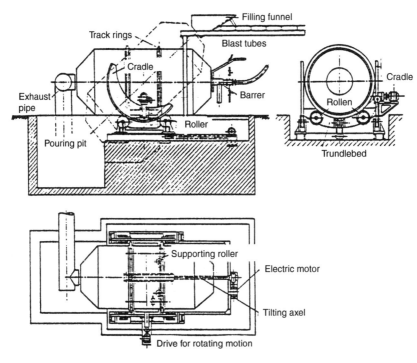

Fig. 4.8 Rotatory furnace for copper refining

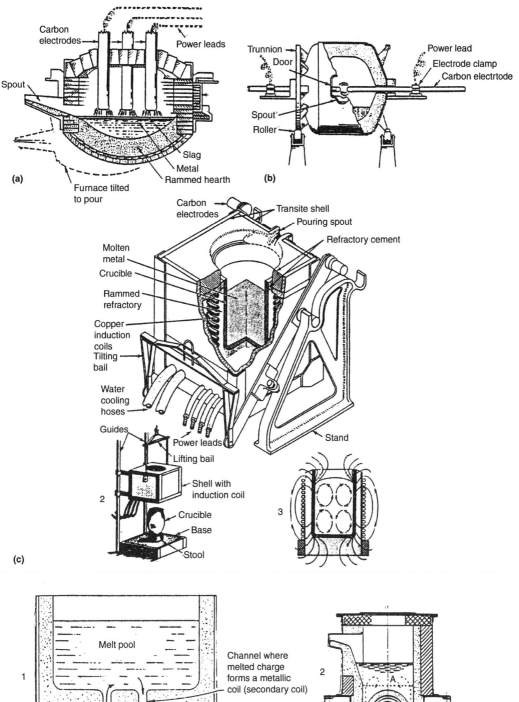

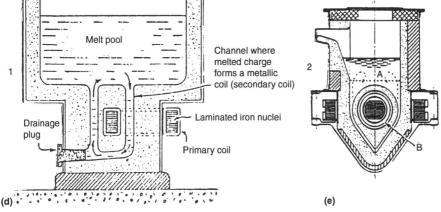

Fig. 4.9 Electric-arc furnaces. (a) Direct arc-furnace (three phase). (b) Indirect arc-furnace—"rocking arc." (c) High frequency electric induction furnaces. (1) Tilting furnace. (2) Lift coil furnace. (3) Lines of magnetic force and stirring action on the molten-metal bath. (d) Low-frequency electric induction furnace. (1) Static furnace. (2) Tilting furnace

maintain homogeneity in the melt and in the finished casting. Common practice, therefore, is to utilize so-called master alloys rather than individual elements when making up a charge. Such alloys, which are usually supplied to foundries in ingot form, are uniform in composition and melting point.

Scrap Recycling. Scrap constitutes nearly one-half of the world's copper supply, making recycling an extremely important element in the economics of copper utilization (See Chapter 1). The majority of the world's copper-base scrap is consumed by brass mills, where it typically consti-

tutes 50 to 75% of the raw materials charge (Ref 14).

Copper scrap can be classified in four categories, and the recycling method and degree of refining varies accordingly. Cathode-quality scrap from manufacturing processes (rejected rod, bare wire, molds) is recycled in-house in ASARCO shaft furnaces along with the normal charge of cathode copper. Scrap contaminated by other miscellaneous metals is melted in rotary or reverberatory furnaces and cast into anodes for electrolytic refining. Most impurities (Al, Fe, Zn, Si, and Sn)

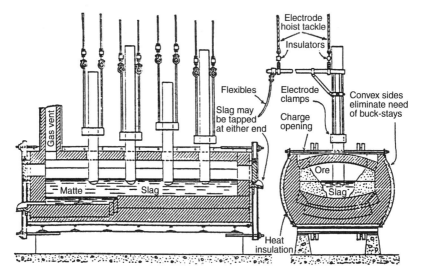

Fig. 4.10 Diagrammatic sketch of the early Westly electric smelting furnace

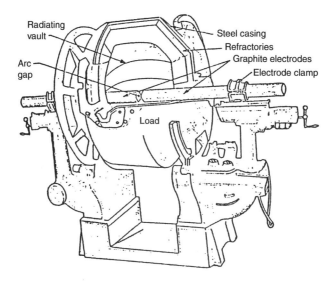

Fig. 4.11 Half-section view of an indirect arc-furnace

can be removed by slagging or by air oxidation and fire refining, although pollution concerns now limit this practice. Fire-refined scrap is suitable for alloying purposes. Alloy scrap consists mainly of brass, plus some bronzes and copper-nickels. Such scrap may be classified as "new" (manufacturing wastes) or "old" (obsolete, discarded products). It is remelted in rotary, hearth, or induction furnaces, with some refining and composition adjustment and cast as remelt ingots or new semifinished products. Low-grade scrap of variable composition (12 to 95% Cu) is treated as a copper-rich ore or concentrate and is processed in a smelter (Ref 3).

Recycled scrap is heated in two stages. The first stage dries the scrap and burns away organic matter, such as grease and oil; the second stage melts the metal and provides opportunity to remove metallic impurities in a cleansing slag.

Melting Practice

Gases constitute the major impurities to which cast copper metals are susceptible, and of the gases, hydrogen is the most detrimental. Hydrogen solubility rises with temperature in both solid and liquid metal, although there is a large decrease in solubility upon solidification (Fig. 4.13). Gas liberated during solidification forms bubbles which, if they remain entrapped, appear as porosity. Hydrogen and oxygen present together combine to form water vapor porosity—a phenomena known as steam porosity, the "hydrogen disease"

Table 4.1 Properties of agglomerated silicon carbide brick for copper shaft furnaces

Property	Value
Bulk density, g/mL	2.56–2.66
Apparent porosity, %	11–15
Modulus of rupture, kg/cm^2	176–246
Crushing strength, kg/cm^2	634–987
Permanent lineal change after preheating to 1400 °C (2550 °F)	–0.1 to +0.1
Approximate chemical analysis	
Silica (SiO$_2$)	7.3
Alumina oxide (Al$_2$O$_3$)	2.1
Ferric oxide (Fe$_2$O$_3$)	1.1
Silicon carbide (SiC)	89.2

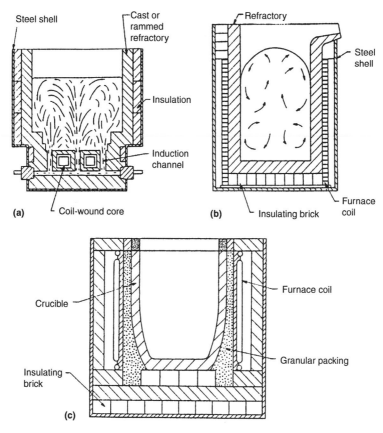

Fig. 4.12 Induction furnaces. (a) 60-cycle core type. (b) 60-cycle coreless. (c) High-frequency

or hydrogen embrittlement. Sulfur and oxygen can cause similar porosity through the evolution of sulfur dioxide. Hydrogen-oxygen equilibria are shown in Fig. 4.14 and 4.15 (Ref 15, 16).

The main source of hydrogen in molten metal is the dissociation of water vapor at the bath surface. Reducing gas or oil flames can also contribute, as can moisture in ladle linings and hydrated corrosion products. Sulfur and oxygen are generally introduced in charge components. Good melting practice, therefore, includes maintaining an oxidizing furnace atmosphere, rapid melting to minimize contact with the atmosphere, and taking care to charge only dry materials (Ref 17).

One time-tested method to produce sound copper castings is known as oxidation-reduction. The melt is first oxidized by the furnace atmosphere and oxidizing flames. The high-oxygen content drives off the hydrogen, after which excess oxygen is removed by sparging the melt with a hydro-

carbon such as natural gas or by inserting green wood poles (hence, poling) into the melt. The use of green wood poles has long been known as a simple and safe, although inefficient, method to deoxidize copper, both in the smelter and in the brass mill (Ref 15).

Copper castings for other than electrical uses are deoxidized with phosphorus, added as a master alloy containing 10 to 15% phosphorus. Any excess phosphorus improves the metal's castability and is not considered harmful. Although seldom used, lithium, magnesium, and calcium (as calcium boride) are also very effective deoxidizers. Lithium offers the additional benefit that it can combine directly with hydrogen.

Castings for electrical applications are more difficult to produce in phosphorus-deoxidized (PDO) copper because even as little as 0.05% phosphorus lowers electrical conductivity by 72%. One alternative is to use copper-phosphorus as the primary

Table 4.2 Insulating fireclay brick for copper shaft furnaces

Type	K^R-20	K-23	K-25	K-26 LI	K-28	K-30	K-30000	Insalcor
Color code	Green	Red	Blue	Brown	Orange	Black	Purple	...
Service temperature, °C	1100	1260	1371	1430	1540	1600	1650	1790
Bulk densities (per ASTM C 134-70)								
g/22.86 cm^3	771	816	998	1270	1360	1360	1542	2180
kg/m^3	465	496	609	770	817	817	929	1314
Melting point, °C (°F)	1510 (2750)	1510 (2750)	1538 (2786)	1755 (3191)	1755 (3191)	1755 (3191)	1843 (3350)	1843 (3350)
Rupture modulus (per ASTM C 93-83), kg/cm^2	7.7	8.4	9.8	11.2	14.7	16.8	18.2	35.0
Cold crushing strength (per ASTM C 93-83), kg/cm^2	7.7	10.2	15.4	11.9	15.4	20.7	19.3	70.0
Permanent lineal change (ASTM C 210-68), %								
1065 °C (1950 °F)	0	...	...	...	...	...	...	...
1230 °C (2245 °F)	...	0	...	...	...	...	...	...
1343 °C (2450 °F)	...	...	−0.1	...	...	...	...	...
1400 °C (2550 °F)	...	...	...	−0.1	...	...	...	...
1510 °C (2750 °F)	...	...	...	...	−0.6	...	...	...
1540 °C (2805 °F)	...	...	...	...	...	−0.5	...	...
1620 °C (2950 °F)	...	...	...	...	...	...	−0.6	...
1790 °C (3255 °F)	...	...	...	...	...	...	...	+0.4
Thermal conductivity (ASTM C 182-88), W/m · K								
260 °C (500 °F)	0.12	0.13	0.14	0.25	0.26	0.29	0.30	0.87
540 °C (1005 °F)	0.14	0.16	0.17	0.27	0.30	0.35	0.35	0.85
815 °C (1500 °F)	0.17	0.19	0.20	0.35	0.36	0.43	0.40	0.91
1095 °C (2000 °F)	...	0.22	0.23	0.43	0.46	0.55	0.49	1.05
1315 °C (2400 °F)	...	...	...	...	...	...	0.58	1.21
Deformation under 69 kPa hot load (per ASTM C 16-81), %								
1095 °C (1000 °F)	0	...	...	...	...	...	...	...
1205 °C (2200 °F)	...	0.1	0.1	0.2	0.2	0.2	...	...
1450 °C (2640 °F)	...	...	...	...	...	...	0.5	...
12.5 psi, 1½ h, 1500 °C (2730 °F)	...	...	...	...	...	...	...	0.1
Chemical analysis ASTM (C 573-70), %								
Alumina (Al$_2$O$_3$)	39	39	45	45	45	46	61	77
Silica (SiO$_2$)	44	44	38	52	52	52	36.5	21
Ferric oxide (Fe$_2$O$_3$)	0.4	0.4	0.2	1.0	1.0	0.9	0.4	0.4
Titanium oxide (TiO$_2$)	1.5	1.5	1.4	2.0	2.0	2.0	1.4	0.6
Calcium oxide (CaO)	15.4	16.0	14.5	0.3	0.3	0.5	0.4	0.1
Magnesium oxide (MgO)	0.1	0.1	0.1	0.1	0.1	0.1	Trace	0.1
Alkalis as (Na$_2$O)	0.3	0.4	0.5	0.3	0.3	0.4	0.1	0.3
Reversible coefficient of thermal expansion, °C^{-1} (°F^{-1})	5.4×10^{-6} (3×10^{-6})	5.4×10^{-6} (3×10^{-6})	5.8×10^{-6} (3.2×10^{-6})	5.2×10^{-6} (2.9×10^{-6})	5.2×10^{-6} (2.9×10^{-6})	5.2×10^{-6} (2.9×10^{-6})	5.2×10^{-6} (2.9×10^{-6})	5.2×10^{-6} (2.9×10^{-6})

deoxidizer, followed by an addition of calcium to combine excess phosphorus as harmless calcium phosphide, which slags off. Hydrogen, along with small amounts of oxides (slag inclusions, for example) can be removed by bubbling a dry inert gas, such as nitrogen, carbon dioxide, or argon, through the molten metal (Ref 17).

Melt Shop Quality Control. Among the quality control instrumentation now in use in the copper industry are electronic load cell weighing systems, which permit automated furnace charging, digitized power and flow meters and high-speed emission spectrography and x-ray fluorescence for chemical analysis (see Chapter 3). Solid-state oxygen sensors (oxygen probes) are replacing traditional sample extraction techniques in, among other places, ASARCO furnaces. Furnace and waste gases are now controlled using such techniques as paramagnetic oxygen analysis, infrared analysis, and solid-state electrochemical anlysis (Ref 18).

Casting

Copper alloys are known for their good castability. As with other metals, however, the production of sound, defect-free copper castings depends primarily on careful control of alloy composition, melt quality (for example, clean and gas-free metal), temperature, pouring and casting techniques, mold design, and mold materials.

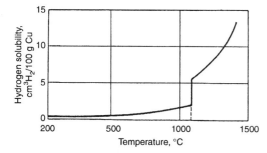

Fig. 4.13 Solubility of hydrogen in pure copper as a function of temperature

Table 4.3 Insulating castables for copper shaft furnace linings

Brand	RESCA(a) HW 40–66
Max. service temperature, °C	1400
Water required for 100 kg of dry refractory	25–30
Dry material required, kg/m^3	1350
Approx. bulk density after drying at 110 °C, kg/cm^3	1440
Rupture modules after drying at 110 °C, kg/cm^3	12–19
Crushing strength after drying at 110 °C, kg/cm^2	38–59
Permanent linear change at 1500 °C	
after drying	Negotiable
after heating	0.2 to –0.6

(a) RESCA, Refractorios Chilenos S.A.

Table 4.4 Compositions and properties of silica-alumina bricks

Property	Product		
	KUME super-conductivity fire bricks	Coral-P-600 high alumina phosphate agglom.	Korundal-XD high alumina brick
SiO$_3$, %	52.5	8.8	9.2
Al$_2$O$_3$, %	41.4	87.1	90.3
Fe$_2$O$_3$, %	2.4	1.4	0.2
CaO, %	0.3	...	0.1
Norton cone equivalent	33	40	41
Temperature, °C	1740	1885	1970
Bulk density, kg/m^3	2160–2245	2835–2963	2900–2963
Apparent porosity, %	19–22	14–18	14–18
Modulus of rupture, kg/cm^2	56–72	135–253	154–210
Crushing strength, kg/cm^3	220–290	540–1000	528–840
Permanent linear change after heating to 1725 °C, %	–1.0 to 1.0	–0.5 to –1.0	0.5 to 1.5

Table 4.5 Composition of gunning mixes for copper melting furnaces

Name of refractory	Composition, %							
	SiC	Al$_2$O$_3$	SiO$_2$	Fe$_2$O$_3$	CaO	MgO	TiO$_2$	Na$_2$O + K$_2$O
Supercote 40	...	36–38	57–59	1.0–2.0	1.0–2.0	0.1–0.6	1.0–2.0	0.5–2.0
Supercote 60	...	60–62	60–62	1.5–2.5	0.5–1.5	0.1–0.4	1.0–2.0	1.0–2.0
Alkagun	...	35.7	45.9	1.6	13.7	0.7	...	...
Hot Gun 28	...	43.8	48.5	1.9	1.9	...	...	...
Hargun C	...	47.2	44.1	1.0	4.7	...	...	...
Alusa Gun Mix	...	59.9	33.1	1.1	3.3	...	...	...
Gunning Mix HW 20	...	36.0	34.2	2.5	20.2	...	...	...
One Shot Mix	...	44.2	43.0	1.1	7.7	...	...	...
Harbide Gun Mix 80	...	16.6	1.7	1.0	3.7	...	...	...
Gunning Mix 23/67	70	17.0	7.8	22.3	0.2	11.3	...	Cr$_2$O$_3$ 38.2

With regard to cleanliness, copper casting alloys can be categorized as those which exhibit a tendency to form continuous oxide films on the surface of the melt (type I) and those that do not (type II). Copper alloys belonging to the oxide film-forming group include aluminum bronzes, silicon bronzes, chromium copper, beryllium coppers, and high zinc brasses. Type II alloys include copper itself, copper-nickel and many copper-tin, copper-zinc, and copper-zinc-lead alloys (Ref 19).

When casting type I alloys, it is important to avoid strongly agitating the bath, as during induction melting, while skimming or when making solid ladle additions. During pouring, the distance from furnace to ladle or ladle to mold should be kept short. Sprues, gates, and tundishes should be close to one side of the mold to maintain steady metal flow through the oxide film during pouring.

Oxidation of melts of type II alloys can cause two effects. A considerable amount of copper oxide may form on the surface and immediately dissolve in the bath. The oxides of the alloying elements zinc, tin and lead, which are insoluble in the melt, tend to form dry flakes or dross. Dross can be dissolved in protective fluxes and removed as a slag.

Dross-forming alloys are easier to cast than oxide film-forming alloys. It is important, however, to avoid entrapping dross or slag particles in the mold. When pouring, it is beneficial to fill the pouring sprue quickly. The use of tundishes, conical sprues, and wedge-shaped runners and gates is recommended.

Foundry Practice

Foundry practice involves all process steps and associated equipment necessary to transport molten metal from the furnace to the mold cavity, where it solidifies into the rough casting. The aim throughout is to maintain the metal at the proper temperature, free from contamination by gases, tramp metal, and refractories. The following is a brief description of some of the equipment, processes, and technical considerations that go into the manufacture of quality copper-base castings.

Casting ladles are refractory-lined steel vessels typically capable of holding up to about 200 kg (400 to 450 lb) of metal (Fig. 4.16), although they can be much larger. Ladles may pour through an ordinary spout or through a syphon, better to retain dross and slag (Fig. 4.17). When pouring, the spout should not be higher than 150 mm (6 in.) above the sprue.

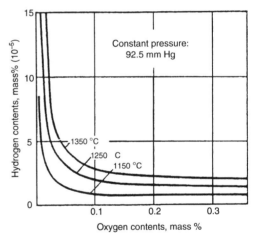

Fig. 4.15 Equilibrium between oxygen and hydrogen in the molten copper as a function of temperature

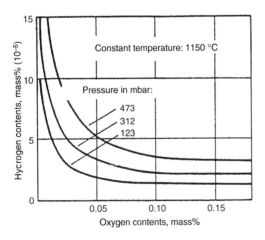

Fig. 4.14 Equilibrium between oxygen and hydrogen contents in the molten copper as a function of pressure

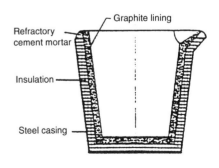

Fig. 4.16 Pouring ladle with insulation and graphite lining

Pouring basins, or tundishes, are recommended for castings requiring more than 15 kg (33 lb) of metal. Tundishes facilitate smooth feeding of melt and avoid interruptions in pouring. They also enable risers to be placed more expeditiously. A typical section through a sand mold with a pouring basin is shown in Fig. 4.18.

Sprues and Risers. Metal enters the mold cavity through a funnel-shaped vertical channel called a sprue. Risers are disposable extensions of castings, which serve to contain sufficient molten metal to fill difficult portions of the mold. Risers also help direct heat flow and the progress of solidification. Sprue and riser dimensions are important

elements of mold design. Most molds for copper alloy castings, requiring less than 15 kg (33 lb) of metal, use copes (the upper half of the mold) between about 100 and 150 mm (4 and 6 in.) high. The sprue, in the upper third of the cope, should be conical, 50 mm (2 in.) in diameter at the top and 19 mm (0.75 in.) at the base. Risers should also be conical, as shown in Fig. 4.19. The lower diameter of the sprue actually depends on the casting rate, which in turn is fixed by the metallostatic height of the riser plus the pouring basin. Figures 4.20 and 4.21 give the relevant relationships.

Mass flow rate, expressed in kilograms per second, depends on the size of the casting size and

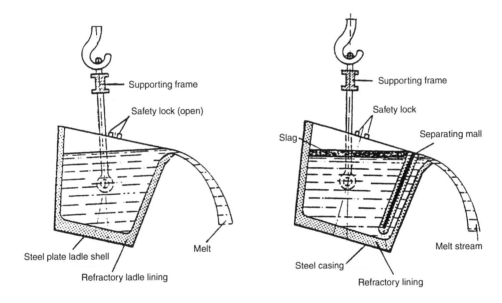

Fig. 4.17 Section through tilting pouring ladles. (a) Ordinary construction. (b) Syphon type ladle

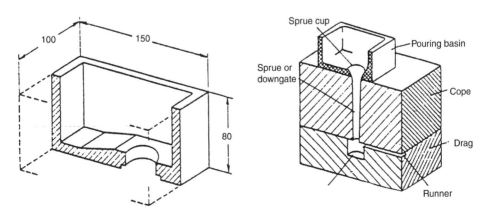

Fig. 4.18 Section through sand mold with pouring basin

the breadth of the alloy's solidification interval. Alloys with solidification intervals narrower than 150 °C (270 °F), such as copper-tin and copper-tin-zinc alloys (UNS C90200 to C91700, copper-tin-lead and copper-lead-tin alloys (C92800 to 94500), and leaded red brasses (C83600 to C84800), when hand poured as castings smaller than 15 kg (33 lb), can satisfactorily accommodate casting rates between 1.6 to 2.0 kg/s (3.5 to 4.4 lb/s). Small simple parts such as medals or plaques can be cast at 4.0 to 4.5 kg/s (8.8 to 10 lb/s). Many automatic casting machines operate at casting rates between 3.5 and 4.5 kg/s.

Alloys, such as the aluminum bronzes (C95200 to C95800), brasses (C85200 to C85800), aluminum brasses (C86100 to C86800), pure copper (e.g., C80200), copper-chromium alloys (C81100 to C81500), and manganese-bearing nickel-silver (C99700), exhibit solidification intervals narrower than 50 °C (90 °F). For these Type I alloys, castings smaller than 15 kg (33 lb) should be poured at rates between 1 and 2 kg/s (2.2 and 4.4 lb/s) to avoid vortex formation, which might drag oxides into the mold cavity.

Gating. For Type I copper alloys with solidification intervals up to 110 °C (200 °F) and Type II alloys with intervals between 50 and 110 °C (90 and 200 °F), it is good practice to place the gates in the drag and main volume of the casting in the cope. Gates should be rectangular and their cross sections should be between two and four times that of the lower riser section. When fed horizontally, the cross section of gates should be six times that of the smallest cross section of the risers and sprues. For vertical feeding, which is preferable for these alloys, the cross-section relationship for riser (or sprue) to runner to gate ranges between 1 to 4 to 4 and 1 to 2 to 2 (Ref 19, 20).

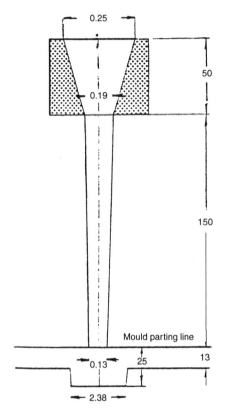

Fig. 4.19 Downgate with superposed funnel to lessen flow turbulence

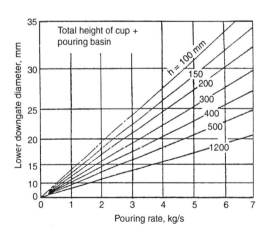

Fig. 4.20 Pouring rate of copper alloys with conical gates

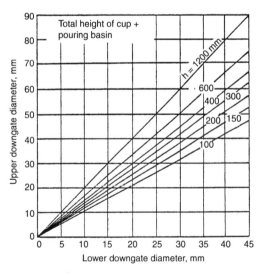

Fig. 4.21 Determination of upper gate diameter

Solidification, Shrinkage and Porosity. The solidification of copper alloys is discussed in detail in the Chapter 2. Solidification modes, which define the so-called Group I and Group II alloys, are summarized in Table 4.6. Short-freezing Group I alloys solidify by a process of skin formation, with freezing proceeding monotonically inward from the casting surface. Group II alloys solidify over wide freezing ranges, passing through an extended mushy stage before freezing completely (Fig. 4.22a–c). Figure 4.23 shows an intermediate structure, and Fig. 4.24 shows an equiaxed structure typical of Group II solidification. Group I alloys tend to form open funnel-shaped shrinkage pipes and centerline cavities at hot spots, as shown in Fig. 4.25. Group II alloys tend to form microporosity in slowly cooled sections, as shown in Fig. 4.26 (Ref 19).

Solidification Rates. In Group I alloys, the skin formation rate can be expressed as:

$$s = qt^{1/2}$$

where s is shell thickness after pouring time, t and q is a solidification constant unique to each alloy. For sand-cast coppers with a narrow solidification interval, q is 6.86 mm/min$^{1/2}$.

For Group II alloys, the total solidification time, t, is:

$$t = \frac{1(V)^2}{q^2 (A)^2}$$

where V is casting volume, A is surface area, and $1/q^2$ is approximately 5.39 s/cm. V/A is M and is also called the solidification modulus.

Riser size can be calculated as follows:

$$Relative\,volume = \frac{Volume\,of\,casting}{Volume\,of\,risers}$$

$$Shape\,factor = \frac{Length\,of\,casting + width\,of\,casting}{Thickness\,of\,casting}$$

Table 4.7 gives minimum riser volumes for copper castings. For practical purposes, the solidification time for the riser, t_s, should be longer than that of the casting, t_G; therefore,

$$\frac{t_s}{t_G} = F^2 > 1$$

Table 4.6 Solidification behavior of copper casting alloys

Group I, shell-type solidification, pure copper or alloys with small solidification interval

Deoxidized copper
Oxygen-free copper
Copper-cadmium alloys
Copper-chromium alloys
Copper-beryllium alloys(a)
Copper-zinc alloys(a)
Aluminum and manganese containing copper-zinc alloys
Copper-aluminum alloys
Copper-silicon alloys
Copper-nickel alloys

Group II, mushy solidification, alloys with wide freezing range

Copper-nickel-zinc alloys
Copper-tin alloys
Copper-lead alloys
Copper-tin-zinc (lead) alloys

(a) Classification depends on alloy compositions and freezing conditions. Alloys with longer freezing times show tendency to mushy-type solidification.

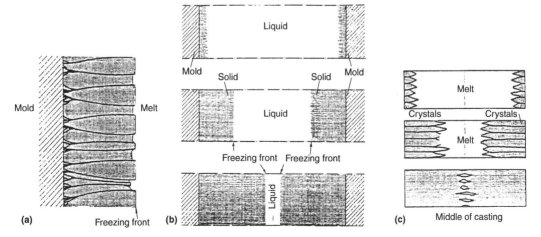

Fig. 4.22 (a) Freezing front in high purity copper. (b) Smoothwall solidification process in a section of a plate shaped casting of high-purity copper. (c) Roughwall solidification process for low alloy coppers

where F is the constant which for optimum solidification in copper alloys is 1.3.

Batch Casting

Ingot/Anode Casting. Machines known as Walker wheels are used to cast anodes of refined blister copper. Steel anode molds are placed horizontally on a rotating table about 7 m (23 ft) in diameter and fed from an electrically operated (tilting) tundish or ladle. When the newly cast anode is three index positions away from the tundish, cold water is sprayed downward onto the copper and upward from underneath the mold for cooling and to form steam, which offsets shrink-

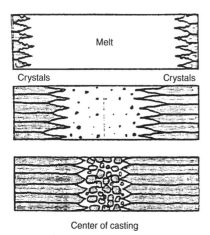

Fig. 4.23 Mixed equiaxed-columnar solidification of a copper alloy

age (Fig. 4.27). PDO and OF coppers do not evolve steam and cannot be cast in horizontal molds because they do not produce a flat surface when they solidify. These coppers are usually cast in vertical water-cooled molds mounted on a rotating wheel. Molten copper is fed into the shrinkage pipe as it forms. To avoid oxygen pickup, coke is placed in the tundish and runners and carbonaceous dressings are used in the molds. Slabs and billets of tough pitch copper for rolling and extrusion are also cast in vertical molds. Vertical casting on rotating wheels can be carried out at production rates of up to 15 tons/h (Ref 3).

Sand casting represents about 85% of copper and copper-alloy castings produced in the United States. Sand molds can be classified according to the binder used, as described below.

Green sand molding, still one of the most widely used processes, utilizes clay and water to form a semi-rigid sand structure. Silica (quartz) sand is most common, but zircon, olivine, a magnesium-iron-orthosilicate $(Mg, Fe)_2SiO_4$, a mixture of fosterite (Mg_2SiO_4), and fayalite (Fe_2SiO_4) are also used. Binding clays adsorb moisture, which acts to hold sand grains together. Kaolinite (fireclay) and montmorillonite (bentonite) are the most important foundry clays (Ref 21).

Surface- or skin-dried sand molds are sometimes preferable to green sand molds. Their dry surfaces, free from gas-forming humidity, have improved erosion and thermal shock resistance compared with green sand molds (Ref 19).

Dry sand molds use oil as a binder and plasticizer. So-called waterfree molding sand for copper-

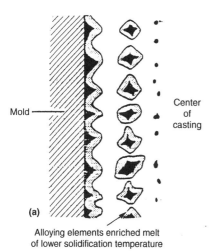

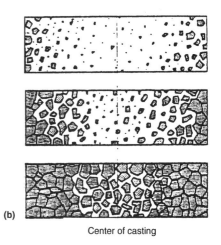

Fig. 4.24 (a) Schematic illustration of short-time blocking of crystal growth. (b) Equiaxed solidification process of copper alloys with high alloy contents

alloy castings is composed of dry quartz sand, about 5% bentone (clay treated to swell in oil), and 2% oil. Dry sand molds are noted for their strength (Ref 19).

Portland Cement-Bonded Sand Molds. Sands with a Portland cement binder are used only for casting very large products such as ship propellers. Molds are air-dried but may require a higher drying temperature than clay-bonded molds. Such molds attain a dry strength high enough to allow transport by cranes. Cement sand molds tend not to form gases during casting. They also chill the solidifying metal; this helps produce sound structures. However, such molds are inflexible, and they cannot be used where restrain may cause hot tearing.

Chemically Bonded Sand Molds. Chemical, or polymeric, bonding produces higher strength molds than are possible with sand casting methods. Both organic and inorganic binders are used. Hardening processes typically involve chemical reactions; a process based on the reaction between sodium silicate (water glass) and CO_2 was one of

the first successful waterless bonding systems. Many other waterless "no-bake" processes are now available (Ref 19).

Shell molding, Croning process (Fig. 4.28), involves the preparation of thin, bonded shell-like molds, formed by pressing resin-impregnated sands against heated patterns. After the bonding resins have set, mold halves are clamped together to form a cavity. The shell molding process is capable of producing castings with fine surface detail and good dimensional fidelity; however, shell molding is somewhat labor-intensive and its cost must be justified by the resulting product quality. Most copper alloys can be cast by the shell molding process (Ref 22).

Permanent mold casting (gravity die casting) is conducted in split molds made from steel, cast iron, or other suitable materials. Its main distinction from pressure die casting is that it relies on gravity alone to distribute metal within the mold. Between 5000 and 20,000 castings, sometimes more, can be produced by one mold (Ref 19, 22).

The most common copper permanent mold and die casting alloys are yellow brasses, silicon

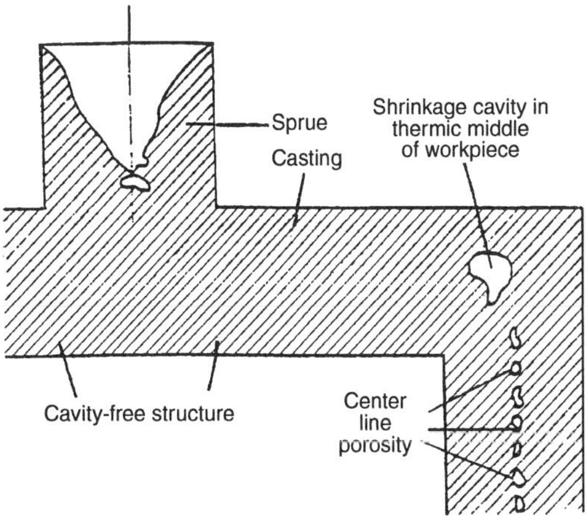

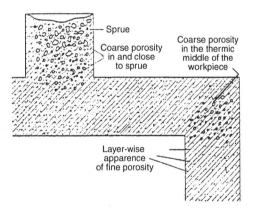

Fig. 4.25 Schematic representation of shrinkage cavities in castings of shell-type solidifying copper-base materials

Fig. 4.26 Schematic representation of porosity formation in castings of copper-base alloys with broad solidification interval

Table 4.7 Required minimum riser volumes

	Minimum values V_S/V_G, %			
	Insulated riser		Sand riser	
Type of casting(a)	H/D = 1:1	H/D = 2:1	H/D = 1:1	H/D = 2:1
Bulky castings (cube, etc): dimensional ratios 1:1:1 to 1:1. 33:1	32	40	140	198
Bulky castings: dimensional ratios up to 1:2:4	26	32	106	140
Average-size castings: dimensional ratios up to 1:3:9	19	22	58	75
Thin-walled castings: dimensional ratios up to 1:10:10	13	15	30	38
Castings with very thin walls: dimensional ratios up to 1:15:30 or larger	8	9	12	14

(a) Dimensional ratio = thickness:width:length

brasses, and the 10 to 11% aluminum bronzes C95200 and C95300 (GK CuAl10Fe, DIN 1714). These alloys solidify in such a fashion that the castings can withstand shrinkage stresses that arise in the unyielding metal molds (Ref 19, 22).

Dimensional tolerances for permanent-mold and other casting processes are listed in Table 4.8. It can be seen that tolerances for the permanent mold process are closer than those for sand casting. Surface qualities are excellent, especially when graphite mold coatings are used. Castings have a fine grain size because of the chill effect of the metal mold, and they are free from porosity owing to the metallostatic head and the close control over solidification that the process offers. Permanent-mold castings are normally considered to be superior

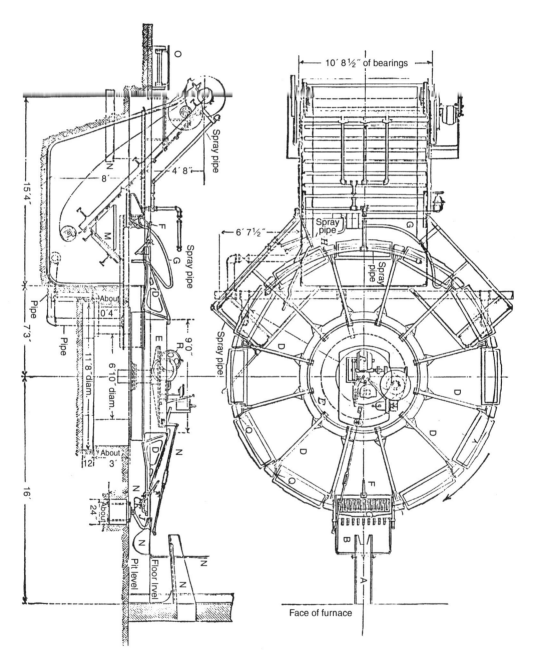

Fig. 4.27 Walker casting machine for copper anodes or other shapes and ingots

to sand castings and pressure die castings, although the other casting processes have their own advantages.

Plaster Casting. Copper alloys are occasionally cast in plaster molds. Such castings have smooth surfaces, good dimensional accuracy, and, if properly made, few if any defects. Molds are made from plaster of Paris, often containing agents that promote porosity to vent gases. The so-called Antioch process is used primarily for plaques and statuary art works. This process utilizes plaster molds composed of 50% sand, 40% plaster, 8% talc and 2% other substances. Plaster molding materials are also applicable to investment casting (Ref 19).

Plaster molds can be used to produce electrical and electronic products and components, which must be defect-free. They can also be used for large turbine or pump runners requiring very smooth internal and external surfaces. Other products that are sometimes cast in plaster molds include gear blanks, splined rod, non-sparking tools, and decorative hardware.

Die casting (pressure die casting) involves forcing molten metal into a split steel mold (Fig. 4.29). Copper alloys are cast in so-called cold

chamber machines, in which mold temperatures are in the vicinity of 500 °C (900 °F). Dies must be made from steels that resist thermal shock well to maintain surface finishes that are free from heat checks and cracks.

In Europe and the United Kingdom, pressure die casting of copper alloys is commonly used for the manufacture of plumbing fittings and other hardware. With its ability to offer high production rates, good surface finish, and freedom from the environmental problems associated with disposing of spent foundry sand, die casting is gaining popularity in North America as well. The most significant limitation of the process is the size of the parts it can produce. Tool wear, brought about by the relatively high operating temperatures, is another major concern (Ref 19, 22).

Table 4.8 Dimensional tolerances for castings produced by various methods

Permanent mold casting, mm (in.)	0.25–1.3 (0.010–0.050)
Die casting, mm/mm (in./in.)	0.03–0.3 (0.001–0.003)
Investment casting, mm/mm (in./in.)	
For extreme precision	±0.006 (±0.00025)
For average work	±0.1 (±0.004)
Plaster-mold casting	Good dimensional accuracy
Shell-mold casting, mm/mm (in./in.)	0.05–0.4 (0.002–0.015)

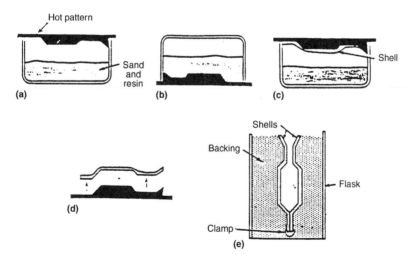

Fig. 4.28 Shell molding. (a) A heated metal pattern is clamped over a box containing sand mixed with thermosetting resin. (b) Box and pattern are inverted for a short time. Heat in the pattern gives an initial set to resin next to pattern. (c) When box and pattern are returned to upward position, shell of resin-bonded sand is retained on pattern surface, while unaffected sand falls into box. The shell on the pattern is placed in an oven and baked to thoroughly set the resin. The temperature of the oven and the time required to give the shell its final set varies with resin and other factors. (d) The shell is ejected from the pattern usually by ejector pins built into the pattern. (e) Two shells are assembled with clamps and supported in a flask with metal shot or other backing material. The shell mold is now ready to receive molten metal

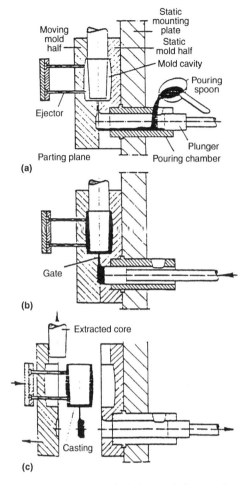

Fig. 4.29 Simplified picture of die casting in cold chamber machines. (a) Filling of casting chamber. (b) Pressure feeding of mold. (c) Ejection of finished casting after mold opening

Copper alloys suitable for die casting are listed in Table 4.9; their properties are given in Chapter 2. The most important physical property required for a die casting alloy is a relatively low melting temperature. Copper-zinc alloys (yellow brasses) are the most common materials. Fittings and other plumbing hardware should be made from dezincification-inhibited brasses, which differ from ordinary brasses only by the addition of small amounts of antimony, arsenic, phosphorus, or tin. Some proprietary die casting brasses employ heat treatment to reduce sensitivity to dezincification.

Investment casting, also known as precision casting, lost-wax, or (in recent years) lost-foam casting, has been used by artisans and sculptors for centuries. The process is still used to cast bronze statuary, plaques, and artwork, but it is also extensively employed in the manufacture of precision parts that require little or no machining. A wax or foam polymer pattern is immersed in a plaster or refractory slurry, which is allowed to harden. The pattern is then melted or burned out, leaving a precisely formed mold cavity (Fig. 4.30). The process is expensive and time-consuming, but dimensional accuracy is better than that of all other casting methods. In general, copper alloys applicable to die and permanent-mold processes can also be investment cast successfully.

A proprietary modification of the investment method known as the "Shaw Process" is occasionally utilized for copper. This process makes use of refractory molds. The molds can be maintained at high temperatures, making possible the casting of thin, complex sections (Ref 19, 22).

Centrifugal casting consists of pouring molten metal into a rapidly spinning mold and allowing it to solidify. The three types of centrifugal-casting are illustrated in Fig. 4.31.

Centrifugal castings are extremely clean because nonmetallic impurities, which are lighter than metal, concentrate on the inner surface,

Table 4.9 Nominal compositions of copper die-casting alloys

Alloy number or name	Composition, %										
	Cu	Sn	Pb	Zn	Fe	Al	Si	Mn	Ni	Sb	Other
C85800	58.0	1.00	1.00	bal	< 0.50	< 0.25	< 0.25	...	...	...	< 0.50
C86200	64.0	< 0.20	< 0.20	bal	3.00	4.00	...	3.0	...	...	...
C86500	58.0	0.50	< 0.40	bal	1.00	1.00	...	1.00	< 0.50	...	...
C87800	82.0	< 0.25	0.15	bal	< 0.15	< 0.15	4.00	< 0.15	...	...	< 0.25
C87900	65.0	< 0.25	0.25	bal	< 0.15	< 0.15	1.00	< 0.15	...	...	< 0.50
OM-Metal(a)	63.5	0.50	2.00	bal	< 0.50	0.40	0.70	0.50	...	0.50	< 0.50
A-Metal(b)	64.5	1.0	2.50	bal	< 0.15	0.30	0.50	0.30	...	...	...
C99700	58.0	...	< 1.00	bal	...	< 1.00	...	13.00	5.00	...	...
C99750	59.5	< 0.50	1.50	bal	< 0.25	1.25	< 0.50	19.50	...	...	< 0.50

(a) Commercial name of Olaf Manner AG, Sweden. (b) Commercial name of Anderson Co. AB, Sweden

where they can be machined away. The castings are also very sound because centrifugal force exerts high pressure on the liquid metal as it is fed to the solidification interface. Gas bubbles, like impurities, are removed by flotation. Dimensional tolerances and surface quality are quite good.

Several alloys can be cast sequentially to provide layered structures. For example, the prototype-spent nuclear fuel container section shown in Fig. 4.32 incorporates an inner shell of aluminum bronze (C95800) for strength and an outer layer of pure copper for predictable long-term corrosion performance.

Copper alloys most frequently cast by the centrifugal process are red, semi-red, and yellow brasses; tin bronzes; aluminum bronzes; and nickel-aluminum bronzes (Ref 22, 23).

Cleaning and Inspection. Castings must be cleaned and trimmed to remove sand, gates, risers, and other protuberances not part of the final product. Copper alloys present no extraordinary prob-

lems during cleaning, although some have a tendency to display hot shortness (high-temperature brittleness) and these can only be shaken out of the mold after they have thoroughly cooled (Ref 22).

Like other metals, copper-alloy castings are tested both destructively (proof tests of simultaneously cast test bars and hydrotests for pressure vessels) and nondestructively. Nondestructive tests include visual inspection and examinations using dye penetrants, x-rays, gamma rays, ultrasonics, and eddy currents.

Alloys containing lead or tin can, under certain conditions, exude these metals during solidification, producing defects known as lead or tin "sweat." With these exceptions, however, defects found in copper castings differ little from those found in other metals.

Continuous Casting

The simplest form of continuous casting is that in which metal is poured into and through an

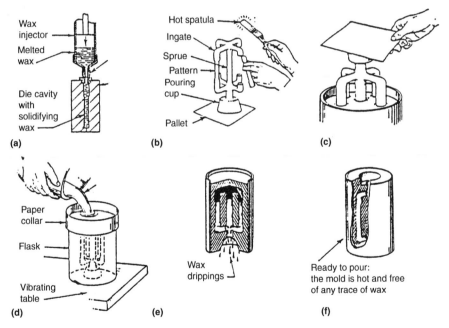

Fig. 4.30 Preparing a mold for investment casting (the lost wax or precision-casting process). (a) Wax is melted and injected into a metal die to form the disposable pattern. (b) Patterns are welded to wax gates and runners to form a "tree." (c) The tree is precoated by dipping in a refractory slurry and then dusted with refractory sand. (d) A metal flask is next placed around the tree and sealed to the pallet. Then the investment, a coarser refractory in a more viscous slurry, is poured around the precoated tree. (e) When the investment has set, the mold is placed in an oven at 93 °C to dry the investment and melt out the wax pattern. (f) Finally, before casting, the mold is placed in a furnace and carefully fired to 700 to1400 °C to remove all wax residue and reach the temperature at which it will receive the molten metal.

open-ended mold cavity having water-cooled sides and a movable bottom. As hot metal is fed to the top of the cavity, it freezes from the walls inward. When a sufficient shell has solidified, the bottom is withdrawn at a rate equal to the molten metal feed rate. The result is a continuous strand having the external shape of the mold cavity. The shape may be round, as in the casting of brass extrusion billets or copper wire rod; hollow, as with bronze bearing and blanks, or profiled, as for gear blanks. Two types of wire rod casting machines, horizontal and vertical, are illustrated in Fig. 4.33. Figure 4.34 shows an outline of a continuous tube shell casting machine.

ASARCO shaft furnaces are frequently used as sources of hot metal for continuous cast wire rod, while hearth-type or induction furnaces are more commonly used for copper alloys.

Semicontinuous direct chill casting is used to produce rectangular slabs for subsequent rolling to sheet and cylindrical billets for forging and extrusion. This method is applicable to both copper and copper alloys. The melt is poured through a graphite-lined, water-cooled mold, which is in the form of a collar, approximately 30 cm (1 ft) long. The bottom of the mold cavity is initially capped by a chilled steel block. Rapid chilling action causes the sides and bottom of the casting to solid-

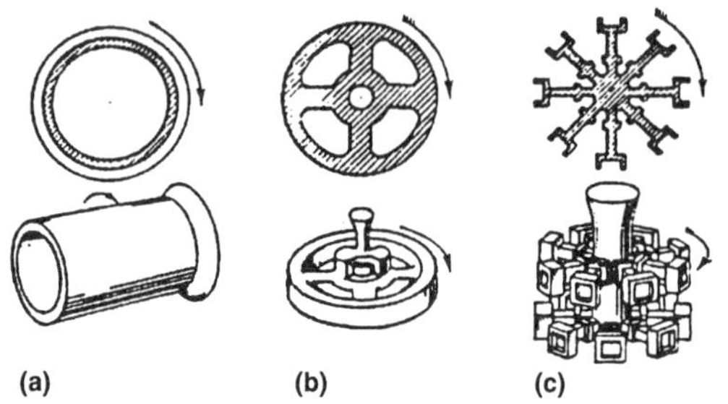

Fig. 4.32 Centrifugally cast test ring for a nuclear fuel element cannister, comprised of multiple layers of dissimilar metals. The inner shell is Ni-Al bronze and the outer shell is pure copper. Source: Copper Development Association Inc.

(a)   (b)   (c)

Fig. 4.31 Centrifugal casting, the three principal categories. (a) True centrifugal. (b) Semicentrifugal. (c) Centrifuged (for example, cast in a mold cavity spun around the axis of rotation)

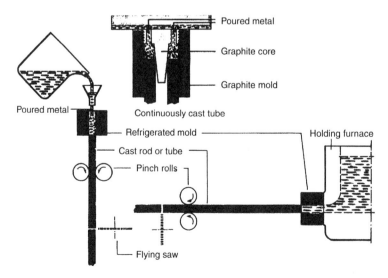

Fig. 4.33 Horizontal and vertical continuous casting schematic. Source: DKI (Germany)

ify. Once a solid plug has formed, the bottom of the mold is slowly withdrawn. The water-cooled molds vibrate with a vertical deflection of 0.5 to 5.0 cm (0.2 to 2 in) at 20 to 30 strokes/min. Casting rates in commercial practice are approximately 4 to 5 mT/h.

Continuous Casting Systems. Several integrated systems are now used to produce copper wire rod—the refinery shape from which copper conductors are drawn.

The Southwire and Properzi Wheel and Band Processes. A diagram of a typical Southwire rod casting process is shown in Fig. 4.35 and 4.36(a) and that of a Properzi process in Fig. 4.36(b). Copper is melted in an ASARCO shaft furnace, cast on a moving wheel, shaped on a series of rolling mills, and coiled for shipment. In both the Southwire and Properzi processes, copper is cast into the gap between a copper-covered continuously turning wheel and a steel band moving together at the same speed. Metal solidifies through the cooling action of water sprays directed at the rear surface of the band. After the wheel and band move together for some distance, the band is pulled off onto an idler wheel. The cast bar is then pulled under slight tension through a continuous rolling mill delivering rod 0.6 to 1.5 cm (0.24 to 0.6 in.) in diameter suitable for drawing. Casting rates are on the order of 20 to 50 mT/h (Ref 3, 24–27.

Hazelett Twin Band and Contirod-Hazelett Processes. In these methods, copper is cast be-

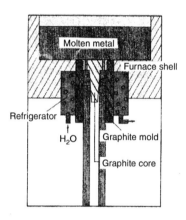

Fig. 4.34 Outline of a continuous tube casting installation

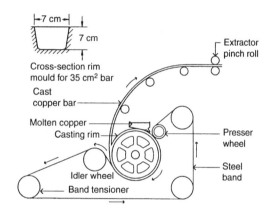

Fig. 4.35 Layout of the Southwire continuous rod casting process. Inset shows cross-section of casting (Ref 26).

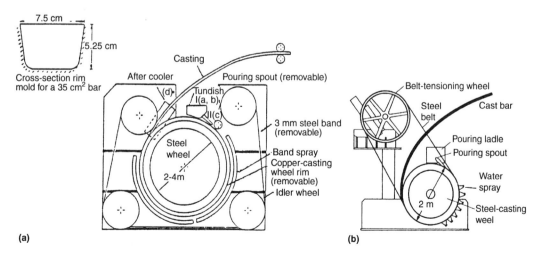

(a) (b)

Fig. 4.36 (a) Cross-section of the Southwire wheel and band continuous casting system. (b) Schematic diagram of the casting details in a Properzi casting machine (Ref 28).

tween two moving sloped steel bands, each 1.1 mm (0.04 in.) thick. The bands are held 5 cm (2 in.) apart by flexible dams at the side of the casting cavity. The horizontal position of the dams is adjustable to obtain any desired bar width. Hazelett plants can operate efficiently at 7 mT/h. The Contirod-Hazelett system incorporates an ASARCO furnace, the Hazelett twin band machine and a continuous, automatically controlled rod-rolling mill. Figure 4.37 shows a schematic picture of the casting part of this system. Here the rod is cast between two moving belts. The process produces copper with a fine-grained and laminar structure. Typical cross sections range from 120 mm by 50 mm (4.7 by 2 in.) to 140 mm by 70 mm (5.5 by 2.75 in.). When operating at the smallest section size, the machine produces up to 40 tons/h (Ref 3, 28).

The General Electric dip-forming process produces rod by pulling a freshly shaved, 1 cm (0.4 in.) copper rod rapidly upward through a bath of molten OF copper held at 1150 °C. The enlarged rod passes into a rolling mill in an inert gas-filled chamber, where it is rolled to its original size. A portion (about one-third) of the rolled rod is recycled back to become the core for the next batch and the remainder is shipped as predraw rod. The production rate of the dip-forming process is 5 to 10 tons/h (Ref 27).

Outokumpu Upcast. In the Outokumpu upcast process, metal is pulled upward directly from a molten bath through a vertical water-cooled, copper-jacketed graphite die as shown in Fig. 4.38. To assure good surface quality, the graphite die is pulsed longitudinally at 100 to 200 strokes/min. The average casting speed is 1 m/min, (3.3 ft/min). For extended mold life, the copper to be cast must be oxygen-free; however, upcasters are also used to produce copper alloy rod (Ref 29–31).

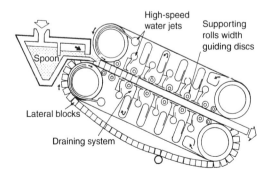

High-speed water jets

Supporting rolls width guiding discs

Spoon

Lateral blocks

Draining system

Fig. 4.37 Hazelett machine schematic

Horizontal Casters. Copper alloy rod is produced in horizontal continuous casting machines. Vertical machines, which produce rod of limited lengths and are also used with copper alloys, are semi-continuous. Horizontal machines utilize continuous melting furnaces, usually of the channel induction type. They feed holding furnaces, the function of which is to maintain a nearly constant metallostatic head to the water-cooled, graphite-lined casting dies. Large conventional casters of this type produce billets or "logs" typically around 230 mm (9 in.) in diameter. Pinch rolls withdraw the solidified billets with a slow pulsating motion; travelling saws cut them to approximately 750 mm (30 in.) lengths suitable for extrusion presses.

Small horizontal casters operate much the same way as larger systems. In this case, the holding furnace and casting crucible form one unit consisting of a solid, self-draining graphite crucible, a submerged casting die, and a low-intensity electric heater. Metal in the crucible is protected by an inert gas atmosphere. Such casters can be fed with molten metal, scrap, swarf, ingots, or virgin metal. They generally have multistrand casting dies. Figure 4.39 illustrates a typical multistrand unit, and Fig. 4.40 shows the schematic layout of a casting machine and its associated coiling unit. Plant capacities claimed by one manufacturer are listed in Table 4.10. Casting rates range from 20 to 600 kg/h (44 to 1320 lb/h) (Ref 32).

Rolling

Continuous cast wire rod is ordinarily rolled to intermediate diameters before it is drawn to wire. Rolling is also used to produce copper alloy rod, plate, sheet, foil, and strip.

Cold rolling is one of the principal means by which copper alloys in sheet, strip, and plate forms are hardened to useful tempers. Temper describes an alloy's physical state in any of several ways, one of which being the strength derived from work hardening. A century-old but still commonly used scheme for relating temper and reduction is that upon which the so-called Brown & Sharpe gage numbers are based. The gage numbers define the progression of deformation between successive cold reduction steps.

Rolling Equipment

Figure 4.41 shows typical roll arrangements in different rolling mills. Very sturdy two-high roll stands are used to hot roll strip. The rolls are reversible to permit rolling in forward and reverse directions. Large horizontal rolls reduce the strip's thickness, and a pair of vertical edging rolls maintain the proper width. Rolls are water cooled to avoid overheating, and a polishing stone continuously dresses the rolls as they operate. Figure 4.42 is an illustration of a rolling mill stand showing the principal items of equipment (Ref 33, 34).

Depending on the type of mill, roll setting gear may take one of two forms. In single-stand, two-high reversing mills, the top roll is adjusted by electrically driven screws to give the required setting for each pass. Upward thrust is applied to the lower chock of the top roll by hydraulic rams (Fig. 4.43). Continuous or reversing mills require no adjustment from pass to pass because section thickness is governed by appropriate pass selection for a given rolling campaign. In two high stands, either one or both rolls may be adjustable. In three high stands, the middle roll is usually fixed and the top and bottom rolls are adjustable (Fig. 4.44 and 4.45).

Rolls used with copper and copper alloys are made from gray iron, chilled cast iron, alloy cast iron, alloy gray iron, forged steel, cast steel, alloy steel, and very high carbon steel.

Ancillary Equipment. The payout side of a (reversing) hot rolling stand typically feeds to a 110 m (350 ft) runout table. High-pressure water sprays provide temperature control and scale removal. Alligator shears, a coiler, and a transfer buggy complete the mill system.

Scale and Oxide Film Removal. Most applications require a rolled surface with a bright, oxide film-free surface. Oxide films form through atmospheric oxidation during hot rolling. One recently developed method to prevent surface oxidation is to perform rolling in a totally enclosed mill, and to provide cooling with a chemically reducing medium, then immediately pickling and waxing the product after the last rolling stage (Ref 35).

Alternatively, sheet and strip surfaces can be mechanically cleaned or scalped by a machining operation after the oxide film has been broken by rolling. A specially designed in-line milling machine performs this operation. After inspection, scalped bars are ready for rolling or drawing to final gauge, temper, and width.

Horizontal continuously cast bars are annealed prior to rolling to maximize their ductility. Rods are then cold rolled to an extent sufficient to produce the desired grain size and structure after a second annealing operation, after which the bars are scalped.

Hot Rolling

Hot rolling, the quickest and most economical size reduction method, is defined as deformation above an alloy's recrystallization temperature.

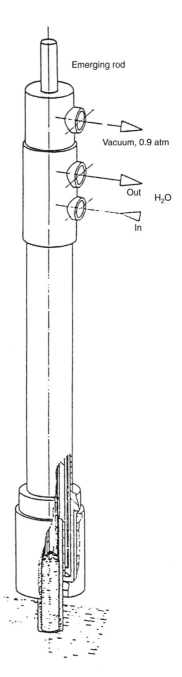

Fig. 4.38 Upcast continuous casting machine and detail of water-cooled graphite die through which metal rod is pulled

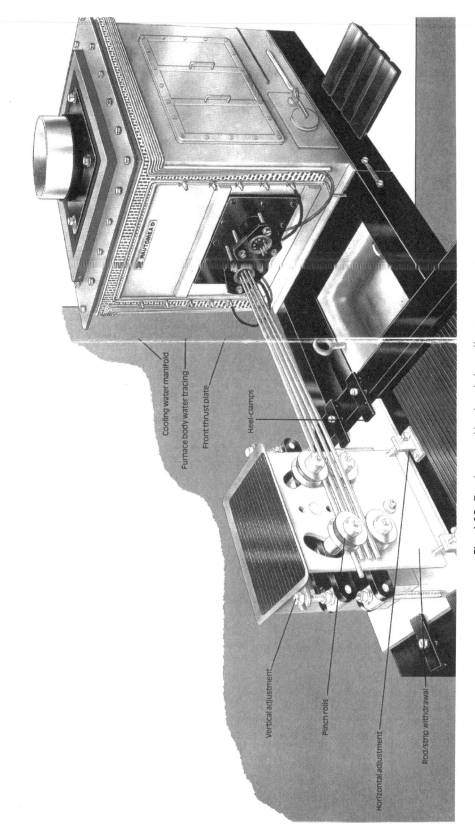

cooling water manifold

Furnace body water tracing

Front thrust plate

Heel clamps

vertical adjustment

Pinch rolls

Horizontal adjustment

Rod/strip withdrawal

Fig. 4.39 Rautomead horizontal continuous proce :s

Recrystallization temperature is the point at which deformed, for example, rolled metal grains containing high levels of residual stress spontaneously recrystallize into new, stress-free structures. The temperature is a function of the alloy's composition. The rolling schedule must be such that the desired thickness is reached before the metal cools below this temperature. Hot working characteristics of copper and copper alloys are listed in Chapter 3. Modern roll stands are of the horizontal or vertical cantilevered type, equipped with fast-response electric sensors to monitor and control mill operations (Ref 24).

Rolling mills work in line with pickling plants. Rolled product is first cooled with a isopropyl alcohol in water, then, in the case of wire products, delivered as ready-to-draw, bright pickled wire in coils as heavy as 1451.5 kg (3190 lb), without welds.

Rolling Schedules. The sequence of deformations during hot rolling is known as the rolling schedule. The heaviest deformation usually occurs halfway through the schedule, and the finishing pass serves to true up the gage in the case of flat products or shape for rod.

Different alloys behave in distinct ways during rolling, and it is necessary to plan rolling schedules accordingly. The thermal profile of the operation must first be calculated based on the alloy's allowable hot rolling range and the product's finished dimensions. Allowance must be made for the mill's stiffness, or elastic compliance, and the rolling characteristics of the alloy. Consideration must be given to the product's size and shape, so that reductions can be allocated properly among the rolling passes (Ref 33).

Rod Rolling. In the past, copper rod was commonly produced by rolling wirebars in a mill arrangement known as the Belgian layout. As wire rod replaced wirebars, continuous mill layouts incorporating several strands became the industry norm. Five to ten stands of two-high rolls are common. Twist in the rod is avoided by mounting the rolls alternately vertical and horizontal. The grooved surfaces of the rolls allow a pair of rolls to process a progression of sections without changes in roll setting. Continuous mills have stands placed one after another, and the speed of successive stands increases progressively to compensate for the decreasing rod diameter. In mills fed by a continuous casting machine, input section sizes range from 17.42 to 51.6 cm^2 (2.7 to 8 in.^{2}) fed at speeds between 12 and 18 m/min (39 and 59 ft/min). It is important that the correct relationship be maintained between amount of reduction and stock thickness. Stock that is far too large will simply not enter the rolls, and stock that overfills the roll gap results in the formation of ribs and fins.

Flat Products. Starting materials for sheet and strip rolling are supplied in the form of slabs. So-called book molds, used for many years to cast individual slabs, have now been replaced by semi-continuous or continuous cast product (Fig. 4.46). Such slabs are 50 to 100 mm (2 to 4 in.) thick, 610 to 915 mm (2 to 2.5 ft) wide and about 2400 mm (8 ft) long. Continuous cast strip is typically 250 to 610 mm (10 to 24 in.) wide and from about 9 to 16 mm (0.625 to 0.875 in.) thick (Ref 35).

Preparation for hot rolling strip begins by trimming the as-cast slabs and heating them to a uniform temperature. Temperature control is very important. If hot rolling is attempted at too high or too low temperatures, the stock may be damaged and may have to be scrapped, although slabs must remain sufficiently hot throughout the rolling process. The slab temperature and the degree of deformation in the last rolling pass determines the

Table 4.10 Continuous caster performance data

Machine model	Output kg/h	Crucible capacity, kg	Power rating(a), KVA	Typical strand geometry, dimensions, m			
				Flats		Wire rod	Hollow section, mm
				Single	Twin		
RX 2200	600	2200	360	1000	450	32	6 @ 80
RX 1700	500	1700	300	950	400	30	6 @ 80
RX 1100	400	1100	240	900	360	24	5 @ 80
RT 850	250	850	170	400	150	12	2 @ 80
RT 650	175	650	105	400	150	8	2 @ 80
RT 400	135	400	80·	400	150	8	2 @ 80
RC 650	175	650	105	150	50	4	1 @ 80
RC 500	125	500	75	150	50	4	1 @ 80
RMT 150	90	150	100	200	75	8	1 @ 80
RM 050	60	50	75	150	50	4	1 @ 80
RMJ 005	20	10	30	75	25	2	1 @ 25

(a) Assumes integrated melting and casting. Output will be greater and power required will be less, if melting is separate

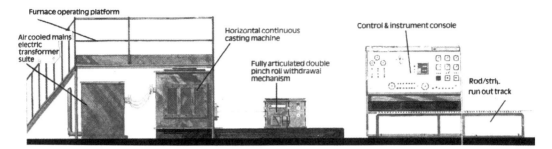

Fig. 4.40 Schematic layout of a multistrand horizontal casting machine and two associated coiling units

grain size of the metal. This is an important parameter, because later processing to meet specified properties depends strongly upon it (Ref 35).

For precipitation hardening alloys (chromium coppers, beryllium coppers), hot rolling must be carried out at a temperature above that at which precipitation occurs. Quenching with high-pressure water sprays immediately after the last hot rolling pass leaves the product in the solution heat-treated condition, thus avoiding the need for re-annealing prior to heat treatment.

Cold Rolling to Final Gage

Precise control of final product thickness is accomplished by cold rolling. The goal is to achieve final gage and flatness in the smallest number of steps so as to minimize costs. Uniformity throughout the length of the coil is essential. So-called "good shape" material is that which is flat without humps, waves, or buckles.

Continuous casting of thin slabs has obviated the need for hot rolling at some mills, allowing substantial cost reductions. The direct cold rolling of cast strip is particularly advantageous for alloys that are difficult to hot roll.

Because the degree of reduction per pass depends on the unit forces transmitted by the rolls, small diameter rolls are needed to effect reasonable reductions in thickness on thin strip. However, because small diameter rolls lack the stiffness necessary to assure uniform thickness across the width of the strip, brass mills generally use four-high and cluster mills. The small diameter working rolls in such mills are supported by a series of backup rolls. Sendzimir or "Z" mills (Fig. 4.46), are used for very thin strip.

Surface quality requirements necessitate the use of hardened steel rolls for cold finishing. Lubricants reduce friction between the rolls and the strip and protect against scratches. Lubricants must be sulfur-free to avoid staining the metal.

Texture and Grain Size. Cold-finishing imparts texture and work hardens the strip. The degree of work hardening, along with annealing, if used, gives the strip its final temper. Large amounts of cold reduction prior to annealing may develop texture, defined as a variation in properties with respect to the rolling direction. Such directionality adversely affects formability and appearance. The degree of reduction during cold rolling strongly influences grain size in strip that is subsequently annealed.

Forging

Forging, also called hot stamping, produces products of near-net shape by deforming metal under impact or pressure in a suitably shaped die. Starting material is usually in the form of round billets, although ingots, slabs, and extruded shapes are also used. Forging is performed on hammers and presses. Forging hammers, which typically produce high-impact velocities, are effective on thin stock. Presses are more effective than hammers for deforming relatively thick sections (Ref 34).

Forging temperatures, which vary with alloy composition, range between the recrystallization temperature of the alloy and the temperature at which incipient fusion or hot shortness occurs. Maximum forging temperatures are also limited by oxidation, loss of alloy components through sublimation and excessive grain growth. For brasses and bronzes, forging temperatures range from 643 to about 927 °C (1100 to about 1700 °F) (Ref 36).

The copper-zinc equilibrium shown in Fig. 4.47 delineates forging temperature ranges for several brasses. The beta phase enhances hot forgeability,

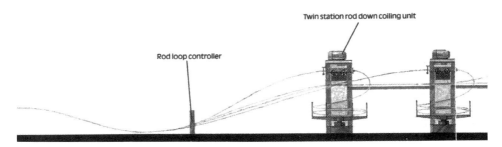

Rod loop controller

Twin station rod down coiling unit

Fig. 4.40 (continued)

therefore, high-zinc, for example, all-beta alloys can be forged more readily than all alpha and duplex alloys. On the other hand, duplex and beta brasses can suffer embrittling grain growth during excessively long heating in the beta region. Alloying elements other than zinc can be considered in terms of their "zinc equivalency," for example, how much they (like zinc) promote the formation of beta. Low zinc (alpha) brasses can only be forged if they are free from impurities. In addition, they require higher stresses than duplex or beta brasses for equivalent deformations (Ref 37, 38).

Forging Alloys

Common copper-base forging alloys are listed in Table 4.11. Those designated by a superscript "b" account for 90% of all U.S. commercial copper-base forgings. Forging brass, C37700, the most common forging alloy, is regarded as the standard and is given an arbitrary forgeability rating of 100; however, all alloys listed in the table support appreciable hot deformation without cracking (Ref 37).

Copper-nickels, silicon bronzes, and other copper alloys are also forged; however, these alloys

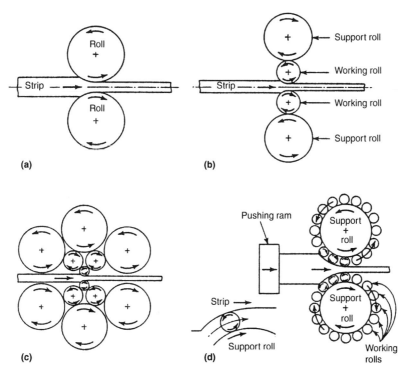

Fig. 4.41 Typical roll arrangement in different rolling mills. (a) Two-high rolling mill. (b) Four-high rolling mill. (c) One-, two-, three-cluster mill. (d) Planetary rolling mill

are more difficult, and more expensive, to forge than the brasses. Copper-nickels, which have high forging temperatures, should be heated in a controlled atmosphere, which complicates the process. Silicon bronzes require both high forging temperatures and high forging stresses. They tend to cause more rapid die deterioration than the common forging alloys.

Minor amounts of addition elements can influence forging behavior. If they are insoluble at forging temperatures, they can cause hot shortness. Lead, for example, is soluble up to 2.0% in beta brass at all temperatures, and lead contents as high as 2.5% are permissible in Cu40% Zn duplex brasses. On the other hand, more than 0.10% Pb in a Cu-30% Zn alpha brass can lead to catastrophic high-temperature cracking. As a rule of thumb, leaded brasses show better forgeability if their beta contents are greater than 50%.

Forging Equipment

Most copper-base, closed-die forgings are produced in crank-type mechanical presses. Unlike the blow of a forging hammer, a press blow is more a squeeze than an impact, and it is delivered in a stroke of uniform length. The principal components of a mechanical forging press are described in Fig. 4.48.

Forging Dies. Die designs for copper and copper alloys are different from those used to forge steel. Draft angles for brass can be decreased to between 0 and 5 degrees, depending on configuration. Because of differences in thermal expansion of steel and copper between room temperature and the forging range, die cavities used with copper alloys are usually machined 0.005 in. smaller than those for forging steels. Finally, die cavities in brass-forging dies are usually polished to finer

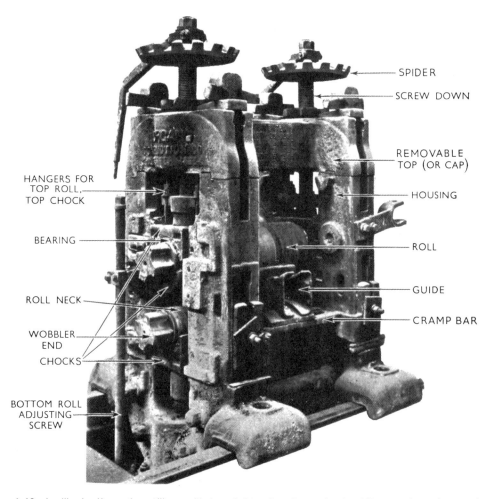

Fig. 4.42 An illustration of a rolling mill stand showing the principal items of equipment

surface finish for forging copper and copper alloys.

Die materials for the hot forging of copper alloys depend on the configuration of the part and on the number of parts to be produced. Die steels commonly used for copper metals include H11 or H12 hot work steels, or L6 die steel with hot work steel inserts, depending on the size of the die. Die steels recommended for forging copper metals include AISI grades H10–H14, H19, H21–26 and ASM 6G, 6F2, and 6F3.

Forging Process Steps

An estimated 90% of brass forgings are produced hot, in closed dies, in a single stroke. Complex or large shapes may require multi-step operations in which the work piece is progressively deformed to its final configuration. In such cases, a "fuller" is an impression used to reduce the cross section and to lengthen the forging stock. The fuller is usually elliptical in the longitudinal section. A "blocker" serves to prepare the shape of the stock before it is forged to final shape in the finisher. The "finisher impression" gives the final overall shape of the workpiece. In this impression, excess metal is forced out into the flash. Flash is trimmed and recycled.

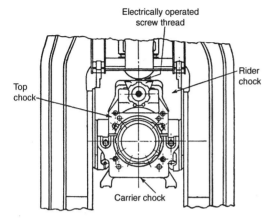

Fig. 4.43 Arrangement of chocks for the top roll of a cogging mill. The rider chock is under a constant hydraulic pressure through an attachment, which is not shown

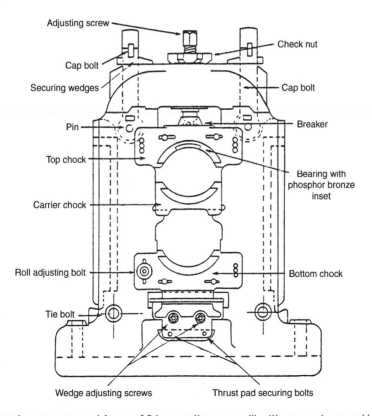

Fig. 4.44 Stand arrangement for an 18 in. continuous mill with open-topped housing

Open-die forging is also known as hand, smith, hammer, or flat-die forging. Swaging is a form of open-die forging. Open-die forging is used when the product is too large to be produced in closed dies; when required mechanical properties cannot be developed by other deformation processes; when the cost of closed dies cannot be justified; or when closed dies cannot be manufactured in time.

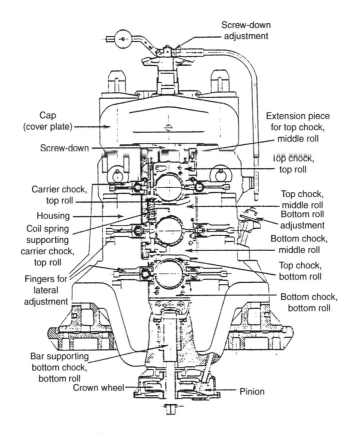

Fig. 4.45 Three-high rolling mill

Table 4.11   Copper-base forging alloys: forgeability ratings and forging temperatures

Alloy	Nominal composition	Relative forgeabiliy(a), %	Forging temperature, °C (°F)
C10200	99.95 Cu min	65	730–845 (1350–1550)
C10400	Cu-0.027 Ag	65	730–845 (1350–1550)
C11000	99.99 Cu min	65	730–845 (1350–1550)
C11300	Cu-0.027 Ag + O	65	730–845 (1350–1550)
C14500	Cu-0.65Te-0.90Cr-0.10	65	730–845 (1350–1550)
C18200	Cu-0.10Fe-0.90Cr-0.10	80	730–845 (1350–1550)
C35300(b)	Cu-36Zn (Sb)	~50	750–800 (1380–1450)
C37700(b)	Cu-38Zn-2Pb	100	650–760 (1200–1380)
C46400(b)	Cu-39.2Zn-0.85Sn	90	600–700 (1100–1300)
C48200	Cu-38Zn-0.7Pb	90	650–760 (1200–1400)
C48500	Cu-37.5Zn-1.8Pb-0.7Sn	90	650–760 (1200–1480)
C62300	Cu-10Al-3Fe	75	700–875 (1300–1600)
C63000	Cu-10Al-5Ni-3Fe	75	800–925 (1450–1700)
C63200	Cu-9Al-5Ni-4Fe	70	825–900 (1500–1650)
C64200	Cu-7Al-1.8Si	80	700–870 (1300–1600)
C65500	Cu-3Si	40	700–875 (1300–1600)
C67500(b)	Cu-39Zn-1.4Fe-1Si-0.1Mn	80	625–750 (1150–1450)
C71500(b)	Cu-30Ni-0.5Fe	60	675–800 (1250–1450)

(a) Takes into consideration such factors as pressure, die wear, and hot plasticity. (b) Commercially important alloy

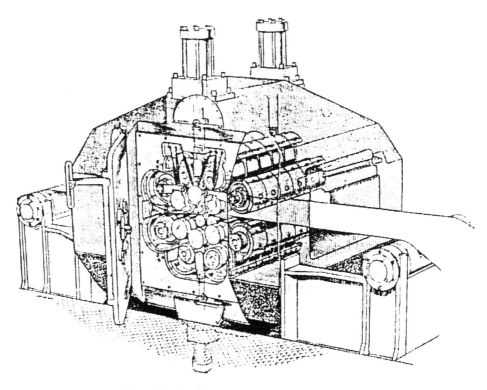

Fig. 4.46 Phantom view of Sendzimir mill

Cleaning. Forging dies are normally lubricated with colloidal graphite suspended in water. The graphite must be completely removed before placing the forging in service because it strongly promotes galvanic corrosion. Brasses and most other copper alloy forgings are commonly pickled in hot dilute sulfuric acid, although hydrochloric acid can also be used. The tough, adherent, alumina-rich oxide film that forms on aluminum bronzes can be removed by a 2 to 6 min immersion in 10 wt% solution of sodium hydroxide in water at 75 °C (167 °F). A thorough water rinse is then applied, after which the forgings are pickled in acid solution in the same way as brasses. The silicate-rich scales on silicon-containing alloys are removed in hydrofluoric acid or proprietary fluorine-containing compounds. Alloys containing nickel should be heated in a controlled atmosphere to minimize the oxide film prior to conventional brass-type pickling. The surface finish of cleaned forgings should be 5 μm (200 μin.) or better. By more precise control, a finish of 2.5 μm (100 μin.) or better is attainable. Neither the type of alloy nor the amount of draft have much influence on surface finish (Ref 38).

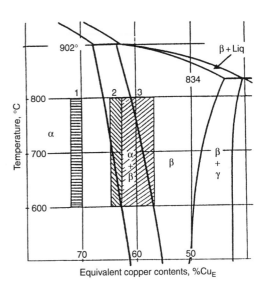

Fig. 4.47 High-temperature structure at forging temperatures. 1. Military brass. 2. Not dezincifying brass (C99400-C99500). 3. Forging brass (C37700)

Cold Forging

With the exception of the heat treatable (age-hardening) grades, copper alloys can only be strengthened appreciably by cold deformation, as in rolling, drawing, or, in this case, cold forging. Copper-base materials are, in general, rather ductile at room temperature; therefore, it is not surprising that among common engineering metals, the cold forgeability of copper alloys is rated higher than all but aluminum and magnesium alloys (Ref 38).

Upset forging is seldom applied to copper alloys because the materials are so easily extruded. A copper nail can be made by extruding the shaft from the head, whereas steel requires upsetting the head from the shaft. Some products do benefit from upsetting operations, and such operations can be applied to copper alloys. Forging brass can be upset as severely as three times the starting

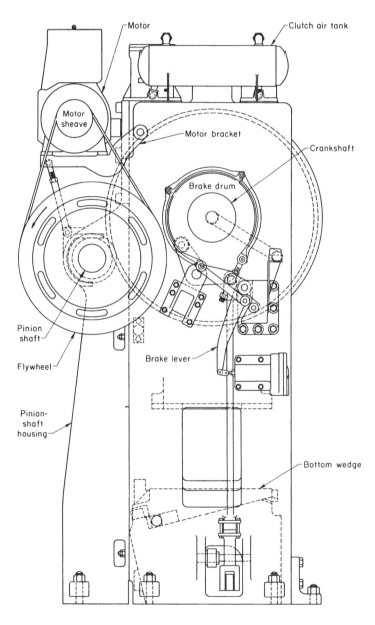

Fig. 4.48 Principal components of a mechanical forging press

diameter, although the allowable upset for other copper alloys is somewhat less.

Forging Products

Copper-base forgings exhibit high strength as a result of their fibrous texture, fine grain size, and structure (Fig. 4.49). They can be made to closer tolerances and with finer surface finishes than sand castings, and, while forgings are somewhat more expensive than sand castings, their cost can be justified in light of their soundness and generally better properties. Forged and sand-cast microstructures are compared in Fig. 4.50. An overview of forging tolerances is given in Table 4.12. Tolerances in closed die forgings can be as close as ±0.25 mm (0.01 in.), sometimes better, for small-to-medium sized forgings. Small draft angles can easily be accommodated within these tolerance limits (Ref 39).

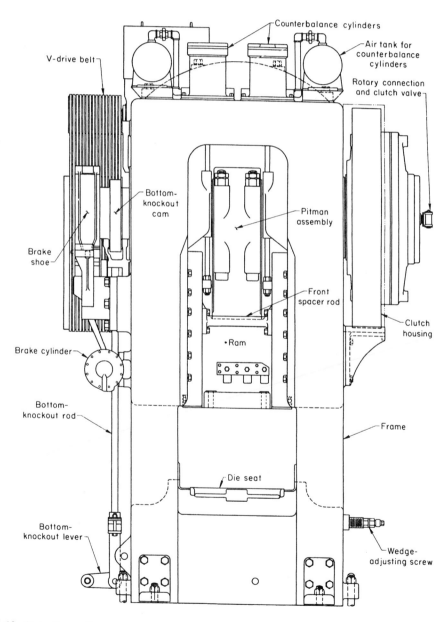

Fig. 4.48 (continued)

Brass forgings are commonly used in valves, fittings, refrigeration components, and other low- and high-pressure gas and liquid handling products. Industrial and decorative hardware products are also frequently forged. High-strength bronze forgings are used for mechanical products, such as gears, bearings and hydraulic pumps.

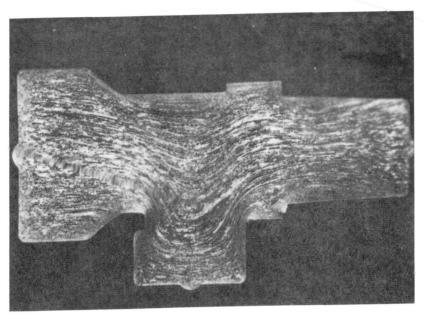

Fig. 4.49 Macrophotographic print of a longitudinal section of a work piece of CuZn39Pb2, C37700 alloy. Picture shows fibers orientation

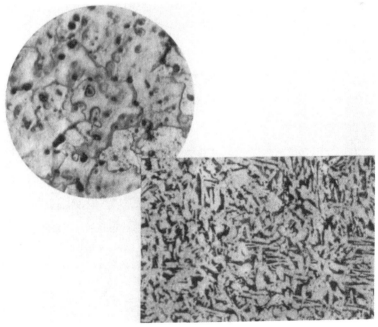

Fig. 4.50 Forged (left) and sand cast (right) microstructures. 250×

Extrusion

During extrusion, a billet of appropriate size is squeezed under high pressure from a closed container through a die to yield a product with a smaller diameter and/or desired shape. Copper alloys are extruded by the three methods described in Fig. 4.51.

Direct, or forward, extrusion is characterized by the continuous motion of the billet within the extrusion chamber. In the case of indirect, or backward, extrusion, the billet remains motionless as the die penetrates into the extrusion chamber. The extruded product exits through the hollow ram, which also holds the die. Hydrostatic extrusion, in which pressurized fluid takes the place of the ram, is a particular case of direct extrusion. Both direct and indirect extrusion can be performed hot or cold, with or without lubrication. Indirect extrusion is conceded to produce product of higher quality than direct extrusion, which can generate centerline defects, among other problems (Ref 40).

Figure 4.52 illustrates a typical extrusion press. The complete extrusion plant also includes a billet heater and a billet feed and transport system, a discard separator, a die changing mechanism, a runout table with quenching apparatus, a cross-transfer conveyor, roller conveyors, and a stretcher. Copper and copper alloys are extruded to rod, section, strip, and wire in horizontal presses. Tubes and hollow products can also be produced in such equipment by extruding over a mandrel or, in certain cases, a spider die (Ref 40).

Extrudability of Copper Alloys

Extruded materials can be classified on the basis of the temperature range over which they can be extruded. The minimum forging temperature is that at which the flow stress of the material is low enough to ensure easy deformation, and the maximum of the range is the temperature above which hot shortness occurs. Copper alloys can be extruded at room temperature, but most extrusion is performed between 600 and 1000 °C (1100 and 1830 °F). On this basis, therefore, copper alloys would be considered to have moderately good extrudability. Commercial extrusion processes transform round billets into various shapes. A variety of hollow sections are pictured in Fig. 4.53 (Ref 41–44).

Among other factors, extrudability varies with alloy composition. Pure copper and beta brasses are the most easily extruded. Other copper alloys of this group comprise cadmium copper, silver-bearing copper, chromium copper, zirconium copper, aluminum bronzes up to 5%Al, tin bronzes with 2% Sn and duplex $(\alpha + \beta)$ brasses. High-alloy aluminum bronzes (over 8% Al or complex bronzes) are considered very difficult to extrude because they require relatively high specific pressures. High tin bronzes such as DIN CuSn8 and CuNiFe alloys belong in this group, as well. Alloys considered moderately difficult to extrude

Table 4.12 Tolerances for small copper-base forgings with weights up to 1 kg (2 lb)

	Tolerances, mm (in.) or degrees for alloy(a):			
Forging types	C36500, C37700, C38500, C46400, C48200, C48500, C67500	C10200, C10400, C11000, C11300, C14500, C14700, C15000, C16200, C17000, C18200	C62300, C64200	C63000, C63200, C65500, C67500
Solid	0.2 (0.008)	0.25 (0.010)	0.25 (0.010)	0.3 (0.012)
Solid, with symmetrical cavity	0.2 (0.008)	0.25 (0.10)	0.25 (0.10)	0.3 (0.012)
Solid, with eccentric cavity	0.2 (0.008)	0.3 (0.012)	0.3 (0.012)	0.3 (0.012)
Solid, deep extrusion	0.25 (0.010)	0.3 (0.012)	0.3 (0.012)	0.36 (0.014)
Hollow, deep extrusion	0.25 (0.010)	0.3 (0.012)	0.3 (0.012)	0.36 (0.014)
Thin section, short (up to 150 mm incl.)	0.25 (0.010)	0.3 (0.012)	0.3 (0.012)	0.36 (0.014)
Thin section, long (over 150 mm incl.)	0.4 (0.015)	0.4 (0.015)	0.4 (0.015)	0.5 (0.020)
Thin section, round	0.25 (0.010)	0.3 (0.012)	0.3 (0.012)	0.36 (0.014)
Draft angles (outside and inside 1° to 5°)	$\frac{1}{2}$°	$\frac{1}{2}$°	$\frac{1}{2}$°	$\frac{1}{2}$°
Matching allowance (min. on one surface)	0.8 ($\frac{1}{32}$)	0.8 ($\frac{1}{32}$)	0.8 ($\frac{1}{32}$)	0.8 ($\frac{1}{32}$)
Flatness (max deviation per 25 mm)	0.13 (0.005)	0.13 (0.005)	0.13 (0.005)	0.13 (0.005)
Concentricity (total indicator reading)	0.5 (0.020)	0.76 (0.030)	0.76 (0.030)	0.76 (0.030)
Nominal web thickness and tolerance, mm (in.)	3.2 ($\frac{1}{8}$)	4.0 ($\frac{5}{32}$)	4.0 ($\frac{5}{32}$)	4.8 ($\frac{3}{16}$)
	0.4 ($\frac{1}{64}$)	0.4 ($\frac{1}{64}$)	0.4 ($\frac{1}{64}$)	0.4 ($\frac{1}{64}$)
Nominal fillet and radius	1.6 ($\frac{1}{16}$)	2.4 ($\frac{3}{32}$)	2.4 ($\frac{3}{32}$)	3.2 ($\frac{1}{8}$)
	0.4 ($\frac{1}{64}$)	0.4 ($\frac{1}{64}$)	0.4 ($\frac{1}{64}$)	0.4 ($\frac{1}{64}$)
Approximate flash thickness	1.2 ($\frac{3}{64}$)	1.6 ($\frac{1}{16}$)	1.6 ($\frac{1}{16}$)	2.0 ($\frac{5}{64}$)

(a) Tolerances should be understood as plus and minus; if tolerances all plus or all minus are desired, value is double of that given

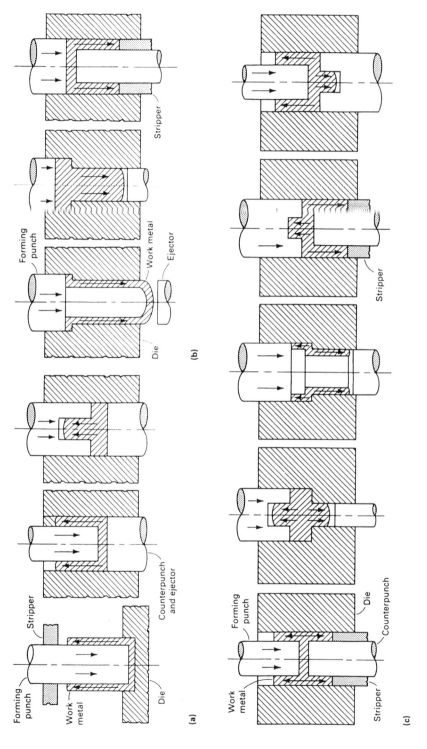

Fig. 4.51 Displacement of metal in cold extrusion. (a) Backward extrusion. (b) Forward extrusion. (c) Combined backward and forward extrusion

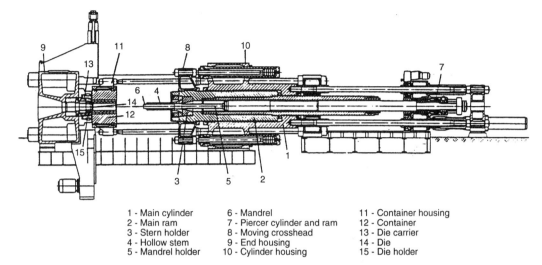

1 - Main cylinder		6 - Mandrel		11 - Container housing	
2 - Main ram		7 - Piercer cylinder and ram		12 - Container	
3 - Stern holder		8 - Moving crosshead		13 - Die carrier	
4 - Hollow stem		9 - End housing		14 - Die	
5 - Mandrel holder		10 - Cylinder housing		15 - Die holder	

Fig. 4.52 Cross-section through a horizontal rod and tube press with an external piercer

Fig. 4.53 Examples of extruded hollow copper sections

include leaded nickel silvers, which tend to show hot shortness; alpha brasses; $\alpha + \beta$ special brasses; aluminum bronzes containing from 5 to 8% Al; tin bronzes ranging from 4 to 8% Sn, and copper-nickels with 1 to 2% Si. Other difficult-to-extrude alloys are those that display hot shortness, such as silicon bronzes and leaded nickel silver. Unlike its action in forging, lead does not limit extrudability appreciably. Highly leaded alloys such as free-cutting brass are routinely extruded in large quantities.

The reduction and shaping parameters to consider are:

- The extrusion temperature and its relationship with the flow stress of the corresponding alloy. Table 4.13 lists approximate values of the flow stress, k_f, for copper and a number of copper alloys at the stated extrusion temperatures. Table 4.14 lists values for deformation resistance, k_w, and deformation efficiency,

Table 4.13 Approximate values of the flow stress k_f for the most common working temperatures at extrusion ratio $\varphi = \ln (A_0/A_1) = 2.5$

UNS number	DIN designation	Temperature, °C (°F)	Strain rate φ', s^{-1}	k_f, N/mm^2
C11000	E-Cu	690 (1275)	0.0006	85
C11100	SE-Cu	850 (1560)	3.3	56
		950 (1740)	3.8	37
C11100	SF-Cu	500 (930)	1.23	148
		700 (1290)	1.23	78
		900 (1650)	1.23	42
C22000	CuZn10	600 (1110)	2.5	186
		800 (1470)	2.5	88
C24000	CuZn20	600 (1110)	2.5	197
		800 (1470)	2.5	83
C26000	CuZn30	600 (1110)	2.5	183
		800 (1470)	2.5	65
C26800, C27000	CuZn35	600 (1110)	2.5	155
		800 (1470)	2.5	40
C28000	CuZn37	600 (1110)	2.5	62
		800 (1470)	2.5	12
C44300, C44000, C44500	CuZn35Ni	700 (1290)	3	140
		800 (1470)	3	85
C37700	CuZn38Pb1	500 (930)	0.1	86
		600 (1110)	0.1	50
C38500	CuZn39Pb3	600 (1110)	3	70
		700 (1290)	3	30
C50500	CuSn2	600 (1110)	2.5	260
		800 (1470)	2.5	140
C51000	CuSn5	600 (1110)	2.5	270
		800 (1470)	2.5	145
C51100	CuSn6	600 (1110)	2.5	310
		800 (1470)	2.5	120
C52100	CuSn8	800 (1470)	2.9	125
C52400	CuSn10	600 (1110)	2.5	275
		700 (1290)	2.5	190
C70250	CuNi5Fe	800 (1470)	3	230
		900 (1950)	3	145
C96200	CuNi10Fe	800 (1470)	3	125
		1000 (1830)	3	67
C75200	CuNi17Zn20	800 (1470)	3	153
		1000 (1830)	3	68
C71000	CuNi20	920 (1690)	1.9	103
C71500	CuNi30	1000 (830)	2	142
C71700	CuNi30Fe	1000 (1830)	2.8	120
		1100 (2010)	3.1	97
C71900	CuNi30Cr3	1000 (1830)	0.9	113
C60800	CuAl5	700 (12900)	3	190
		800 (1470)	3	130
C61400	CuAl8Fe	900 (1950)	3	75
C95300	CuAl10Fe	650 (1200)	3	100
		750 (1380)	3	45
C17300	CuBe2	700 (1290)	10	215
		800 (1470)	10	150
C80100	CuAg	950 (1740)	3.3	48

D_F in extrusion with and without lubricants for the same materials.

- The mechanical workability of the material at the corresponding temperatures. Workability is defined by the mathematical relationship: ratio of shear deformation at fracture-through-torsion to flow stress is φ_{Fr}/k_f. A correlation between workability and extrudability values is shown in Table 4.15.
- The maximum extrusion ratio
- The permitted temperature range
- The load or specific pressure required within that range
- The exit speed at a constant extrusion load
- The maximum extrusion speed to the onset of hot shortness

Extrusion speeds for copper alloys are higher than those encountered in aluminum extrusion. This may introduce the requirement to use hydraulic accumulator drive. High extrusion speeds minimize the fall in temperature of the billet during the "push" cycle specially when temperatures required for extrusion are more elevated. Billet temperatures for copper and copper alloy extrusion are usually between 600 and 1000 °C (1112 and 1832 °F). The container temperature normally cannot exceed 500 °C (932 °F). The large heat loss this limitation causes results in undesirable cooling toward the end of the push. This heat loss is compensated in part by the heat of deformation, which causes a temperature increase toward the end of the cycle. An upper limit on operating speed is usually set by the capacity of the runout equipment. Exit speeds in practice are more than 300 m/min (18 km/h) or (984 ft/s) (Ref 40).

Extrusion Methods

Cold Extrusion. During cold extrusion, the slug or preform enters the extrusion die at room temperature. The workpiece temperature may rise by several hundred degrees as work of deformation is converted to heat. Like hot extrusion, cold extrusion can be performed backward (indirect), forward (direct), or a combination of the two. The advantages that cold extrusion offers over hot extrusion include a lack of oxidation, higher mechanical properties in the extruded product, closer tolerances, better surface finishes, and higher production rates (Ref 40).

Copper and its alloys rank second among common metals in cold extrudability; however, on a tonnage basis, cold extrusion is far less important than hot extrusion. Among the copper metals, OF copper (C10200) is the most readily extruded. Harder copper alloys such as aluminum-silicon bronze and nickel silver are more difficult to extrude cold than softer, more ductile alloys such as cartridge brass (C26000), which can satisfactorily withstand cold reduction of up to 90% between anneals. Alloys containing as much as 1.25%Pb can be successfully cold-extruded if the amount of upset is small and the workpiece is in compression at all times during metal flow. Copper alloys containing more than 1.25%Pb are likely to fracture (Ref 38, 40).

Hot extrusion is widely used in the production of wire, pipe, and tube, in which it is an intermediate process, and rod, bar, and shapes, for which it can be used to produce finished products. An example of the latter are architectural shapes such as banister rails, which are extruded from brass. Billet temperatures vary from about 595 to 995 °C (1100 to 1825 °F). Copper and brasses extrude readily, with ram speeds ranging from 50 to 400 mm/s (2 to 16 ft/s). Stiffer alloys require high pressures greater than 690 MPa (100 ksi) (Ref 38).

Table 4.14   Deformation resistance k_w and deformation efficiency factor, D_F, in extrusion with and without lubricant

Material	Temperature, °C (°F)	Billet lubricant	k_w, N/mm^2	D_F
C37700	650 (1200)	Oil/graphite	102–166	0.44–0.27
	700 (1290)	Oil/graphite	79–98	0.32–0.26
	750 (1300)	Oil/graphite	79–98	0.23–0.21
	800 (1470)	Oil/graphite	46–79	0.33–0.19
	800 (1470)	None	39–48	0.39–0.31
C28000	800 (1470)	None	35–46	0.51–0.39
	800 (1470)	Rough surface	29–34	0.69–0.59
C27000	750 (1300)	Oil/graphite	180–186	0.28–0.27
	800 (1470)	Oil/graphite	104–110	0.34–0.32
	850 (1560)	Oil/graphite	72–79	0.35–0.32
C11000	800 (1470)	Oil/graphite	159–166	0.37–0.35
	900 (1650)	Oil/graphite	106–109	0.41–0.39
	900 (1650)	Rough surface	106–148	0.41–0.29
	1000 (1830)	Oil/graphite	93–94	0.32

The sequence of operations for unlubricated forward extrusion of a copper or brass billet is shown in Fig. 4.54.

1. The heated billet and the dummy block are loaded into the container.

Table 4.15   Workability and extrudability of copper alloys

$W = \varphi_{F} / k_{f}$ (N/mm^2)

Workability (W)	Extrudability
<2	Poor
2–4	Average
4–15	Good
>15	Very good

2. The billet is extruded by the force of the ram.
3. The container is separated from the die, the extruded section with the butt, and the dummy block.
4. The butt is sheared off.
5. The shear die, the container, and the ram are returned to their initial (loading) positions (Ref 40).

The sequence of operations for backward extrusion is essentially the same as that for forward extrusion. Copper alloy rod, bar, and shapes are commercially extruded by both the forward and back extrusion processes.

Hydrostatic Extrusion. In hydrostatic extrusion, the billet is forced through the die by fluid

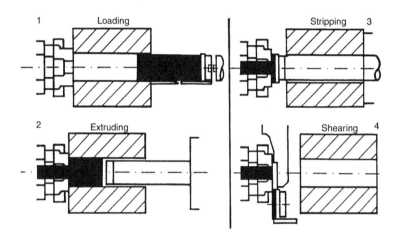

Fig. 4.54 Sequence of operations in direct extrusion (schematic)

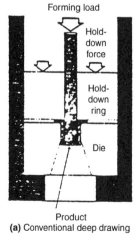

(a) Conventional deep drawing

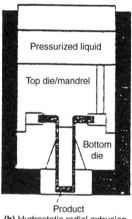

(b) Hydrostatic radial extrusion

Fig. 4.55 Comparison of tooling for conventional deep drawing and hydrostatic radial extrusion

pressure in the extrusion chamber. There is no ram as such. The process is seldom used for the commercial extrusion of copper alloys.

Combined Extrusion/Drawing

In fine wire extrusion, the extruded product is drawn through the die and is concurrently forced through by hydrostatic pressure in the extrusion chamber. It can be used to produce extremely fine wires from a coiled "billet." Pressure is maintained until the billet material is exhausted. Extrusion is stopped by a knot at the end of the billet wire; this blocks the die and prevents escape of the high pressure fluid. Using this process, copper wires have been produced in sizes down to 0.0152 mm (0.0006 in.) (Ref 45).

Hydrostatic radial or lateral extrusion is illustrated in Fig. 4.55. This offers the ability to produce very deep cups or tubes with complex cross-sectional shapes in one operation rather than the multiple stages needed in conventional deep drawing.

Helical extrusion combines hydrostatic extrusion, conventional extrusion, and an intermediate stage that is similar to the turning process performed on a lathe. Metal is first extruded through a conventional die, proceeding immediately over the conical tip of a piercing mandrel to produce a tubular form. As the annulus forms, material is circumferentially sheared from it by a rotating cutting tool. The sheared material is extruded through a die hole situated immediately in front of the collecting tool and rotating with it (Fig. 4.56). The process has only been applied to copper. It promises the ability to produce wire directly from billet, thereby eliminating several costly process steps.

The Conform process is a continuous extrusion process in which the extrusion container is a groove in a steel wheel. The groove is closed by a shoe covering a quadrant of the wheel and is blocked at one end by an abutment. The extrusion wheel rotates continuously. Feedstock in the form of continuously cast rod or clean particulate material is fed into the groove, where frictional forces push it against the abutment and radially out through a die. Figure 4.57 schematically illustrates one- and two-wheel systems (Ref 46, 47).

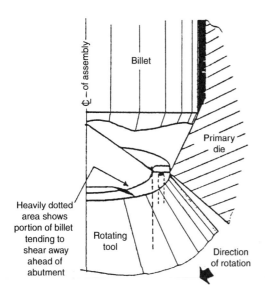

Fig. 4.56 Helical extrusion tool action

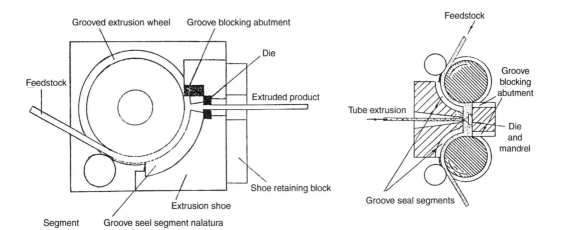

Fig. 4.57 One- and two-wheel conform processes

Cross-Sectional Shape and Size

If extrusion is followed by cold drawing, the dimensions of the extrusion must be coordinated with those of the final product to control the degree of cold work and the mechanical properties of the final product. Process economics also play a role: it is sometimes possible to obtain the desired mechanical properties in a single draw. This is not difficult with large sections but is limited with bar dimensions less than 10 mm (0.4 in.) or flat bar and sections less than 5 mm (0.2 in.) thick because the extrusion load needed for such small sections exceeds the capacity of most presses. Thus, very small final dimensions are produced by several cold working operations with intermediate annealing, which raises production considerably (Ref 40, 48).

Small-diameter bar and sections are cooled and coiled immediately after extrusion. The coils of round or flat product, and occasionally profiled sections, are next cold drawn on drum-type drawing machines known as bull blocks. Intermediate annealing between drawing stages may be required, and the degree of drawing thereafter plus the final anneal, if applied, sets the temper of the product. Finished product is sold in coils or is straightened and cut to lengths. Hollow copper sections, such as water-cooled bus bars, are extruded from hollow billets over a stationary mandrel.

Impact Extrusion Processes

Impact extrusion, which is normally performed cold, is similar to indirect extrusion, and can be used to provide a variety of internal and external surfaces as shown in Fig. 4.58. The punch descends at a high speed, striking the blank or slug, which is then extruded upward, forming a tube or shaft, depending on whether the extruded metal travels outside or within the punch. Copper and alpha brasses, among other metals, are routinely extruded at impact speeds up to 11 m/s (36.5 ft/s) (Ref 48–50).

Mechanical Properties. Copper recrystallizes when impact extruded at ratios greater than 8:1, and rod formed in this manner exhibits the same mechanical properties as annealed material. Alpha brass recrystallizes at extrusion ratios greater than 23 to 1 at an impact velocity of 11 m/s (36.5 ft/s). Rod extruded at a ratio of 8 to 1 shows some evidence of recrystallization; however, structures retain the directionality of extruded material (Ref 50).

Forming of Plate, Sheet, and Foil

Blanking, Cutting, Piercing, and Related Operations

Material Characteristics. Copper and copper alloys are readily blanked and pierced; however, it is important that strip be flat and of the correct dimensions and temper so as to ensure the proper shape and properties in the finished product. Flatness and dimensions are direct consequences of the care taken to produce the strip. Mechanical properties and, for example, shear-to-break characteristics, are determined by the composition and temper of the strip (Ref 38).

The quality of blanked edges is determined jointly by die clearance and material properties. Burr-free and distortion-free parts can be cut from annealed copper alloy strip at die clearances up to about 5% of the strip thickness. Unalloyed coppers such as C10100 and C10200 require smaller clearances to produce burr-free edges, even in rolled tempers. Copper alloys containing second-phase particles, for example, C19400, ones that have relatively high solute contents, C26000 and

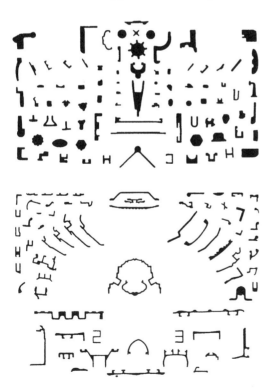

Fig. 4.58 Examples of impact extruded shapes (Ref 42)

C51000, and ones that have been cold-rolled more than 50% exhibit high-quality blanked edges at die clearances between 3 and 12% of the strip thickness. Small additions of lead to brass and other copper alloys decreases their tendency to form burrs. The lead additions also reduce the shear-to-break ratio in blanking operations, although formability, as measured in most forming operations, suffers somewhat in leaded alloys (Ref 38).

Press Bending and Forming

Many copper-base connectors, terminals, and spring-like components are fabricated by simple bending operations. Bend formability is usually expressed as the minimum bend radius, R, for a given strip thickness, t, for example, as the ratio, R/t, which can be formed without cracks appearing on the outer (tension) surface. Bend formability is primarily limited by the ability of the material to distribute strain, a property that manifests itself in the necking strain. This depends on the composition and temper of the alloy. Cold work (for example, harder temper) reduces the necking strain, although composition and other strengthening mechanisms are also influential.

Copper connector alloys frequently contain alloy additions to increase their work-hardening rate. The zinc in brasses is the most common example. Precipitation strengthening is also important. Its advantage is that strip can be formed in the solution-annealed condition then aged to the desired strength level (Ref 38).

Directionality and Texture. Bend formability varies with orientation relative to the direction of rolling (Fig. 4.59). All cold-rolled materials exhibit such directionality. Its extent varies from alloy to alloy, but it always increases with increasing cold reduction. Bend directionality results from the alignment of impurities and the generation of crystallographic textures during rolling. Crystallographic texture describes the preferential alignment of specific planes in the crystal structure of the metal during rolling. It is defined in terms of the Miller indices of the dominant crystallographic planes and directions. Connector alloys such as C26000 (Cartridge Brass), which have low stacking fault energies, develop strong $\{110\}$, $\langle 112 \rangle$ textures during rolling. Such alloys can exhibit bend directionality after as little as 30% reduction by cold rolling. Dilute copper alloys and copper-nickel alloys do not develop well-defined rolling textures and exhibit little directionality even after as much as 70% reduction.

In general, alloys that are strengthened by cold work and solute effects, permit sharper bends in the direction of rolling ("good way") than transverse to it ("bad way"). Figure 4.60 shows the springback behavior of copper alloys as a function of temper, sheet thickness, and bend radius. On the other hand, excessive cold working, as by rolling to increase temper, decreases the capability of the alloy to be formed further. Thus, Table 4.16(a, b) gives the maximum level of strength allowable to make particular minimum-radius bends successfully. The situation is more complex in the case of precipitation-hardened alloys, some of which are strengthened by a combination of precipitation and cold work. In such cases, bend anisotropy is strongly process dependent.

Figure 4.61 shows how bend directionality influences the way a part is laid out. Layout is considerably more restricted when the part is made from an alloy with strong directionality (for example, phosphor bronze, C51000, in the spring temper) than with less directional alloys such as C68800 or C72500. Therefore, in addition to other factors, alloy selection affects cost by influencing the amount of scrap losses.

Coining

As its name implies, coining involves the compression of metal between two dies. The metal fills and takes the form of the die cavities. The most familiar application of the process is the minting of coins, many of which are made from copper alloys. Electrical and electronic connectors and leadframe leads are also often coined (Ref 38).

Ease of coining is determined by the strength and work-hardening rate of the material. Copper,

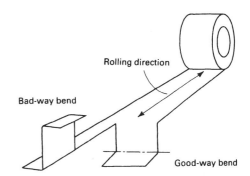

Fig. 4.59 Bend formability of copper alloys as a function of rolling direction. Bends with the axis transverse to the rolling direction are termed good way bends; bends with the axis parallel to the rolling direction are bad way bends

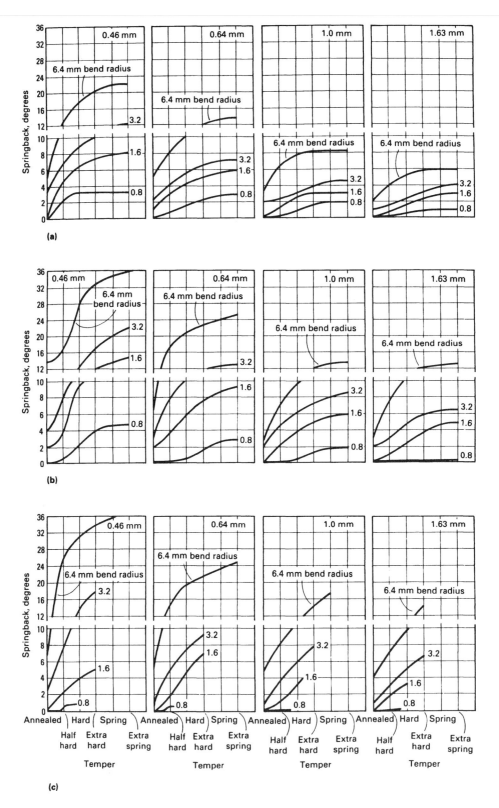

Fig. 4.60 Springback behavior of copper alloys as a function of temper, sheet thickness, and bend radius (90° bends). (a) Alloy C21000. (b) Alloy C26000. (c) Alloy C35300

low-zinc brasses, low-alloy nickel silvers, and copper-nickels, all of which exhibit low work-hardening rates, have good coinability.

Embossing and Swaging

Embossing and swaging, which are compressive operations closely related to coining, are also frequently used in cold-forming of copper alloys. Embossing entails impressing letters, numerals, or designs into a surface by displacing metal to either side. Swaging is the radial compression of a cylindrical object, in which diametrically opposed dies strike the workpiece repeatedly at a frequency of several hundred blows per minute while indexing circumferentially between each impact. The principles of alloy selection described for coining apply equally to embossing and swaging, although embossing can be performed on any copper alloy if special attention is given to tooling alloy temper. Swaging is used to produce electrical contacts and gas-welding and -cutting torch nozzles.

Cupping and Single-Step Drawing

Cupping forms a closed-bottomed shell in a flat blank by pressing it between dies. Drawing is similar, with the exception that the product is deeper and has a higher draw ratio, l/D. Drawability is described by the limiting draw ratio (LDR), the deepest single draw ratio a material can withstand without failure at the punch radius. Also important is the so-called "r" value, the ratio of true width strain to true thickness strain in the region of uniform uniaxial elongation during a tensile test. It measures resistance to thinning and correlates with deep-drawing performance. Copper alloys with high "r" values permit the largest LDRs. Among the copper alloys, these include phosphor bronze C52100, followed by the brasses (in order of decreasing zinc content) and copper.

Multiple-Step Deep Drawing

Copper alloys with lower work-hardening rates can be redrawn and ironed more times without

Table 4.16(a) Maximum allowable strengths to make minimum bend radii in good way bends in selected copper alloys

| | Tensile strength, MPa (ksi) | | |
| | At sheet thickness 0.25 mm, bend radius 0.4 mm | At sheet thickness 0.50 mm, bend radius 0.8 mm | At sheet thickness 0.76 mm, bend radius 1.2 mm |
UNS number			
C11000	372 (54)	352 (51)	352 (51)
C17200(a)	896 (130)	896 (130)	896 (130)
C17500(a)	...	...	724 (105)
C15100	428 (62)	400 (58)	400 (58)
C19400	538 (78)	510 (74)	496 (72)
C19500	614 (89)	572 (83)	572 (83)
C19700	538 (78)	510 (74)	496 (72)
C23000	593 (86)	593 (86)	593 (86)
C26000	662 (96)	662 (96)	662 (96)
C35300	641 (93)	572 (83)	572 (83)
C41100	517 (75)	496 (72)	496 (72)
C42500	621 (90)	621 (90)	621 (90)
C50500	490 (71)	469 (68)	469 (68)
C51000	710 (103)	662 (96)	648 (94)
C52100	765 (111)	745 (108)	731 (106)
C63800	827 (120)	807 (117)	793 (115)
C65400	745 (108)	731 (106)	731 (106)
C66600	669 (97)	655 (95)	641 (93)
C68800	786 (114)	744 (108)	745 (108)
C70250(a)	690 (100)	655 (95)	...
C70600	524 (76)	496 (72)	496 (72)
C72400(a)	793 (115)	690 (100)	621 (90)
C72500	572 (83)	517 (75)	517 (75)
C73500	579 (84)	579 (84)	579 (84)
C74000	648 (94)	600 (87)	586 (85)
C75200	579 (84)	579 (84)	579 (84)
C77000	807 (117)	571 (109)	717 (104)

Good way refers to the orientation of the bend with respect to the sheet or strip rolling direction. Tensile strengths of 965 and 1103 MPa (140 and 160 ksi) are available in 0.25 and 0.5 mm (0.010 and 0.02 in.) thicknesses with specially supplied mill tempers. (a) Mill hardened to strength shown, then formed.

Table 4.16(b) Maximum allowable strengths to make minimum bend radii in bad way bends in selected copper alloys

| | Tensile strength, MPa (ksi) | | |
| | At sheet thickness 0.25 mm, bend radius 0.4 mm | At sheet thickness 0.50 mm, bend radius 0.8 mm | At sheet thickness 0.76 mm, bend radius 1.2 mm |
UNS number			
C11000	365 (53)	331 (48)	345 (50)
C17200(a)	896 (130)	896 (130)	896 (130)
C17500(a)	...	...	724 (105)
C15100	407 (59)	400 (58)	400 (58)
C19400	517 (75)	496 (72)	490 (71)
C19500	592 (86)	572 (83)	558 (81)
C19700	517 (75)	496 (72)	490 (71)
C23000	572 (83)	552 (80)	538 (78)
C26000	627 (91)	524 (76)	524 (76)
C35300	496 (72)	483 (70)	469 (68)
C41100	468 (68)	448 (65)	434 (63)
C42500	552 (80)	475 (69)	462 (67)
C50500	490 (71)	468 (68)	469 (68)
C51000	621 (90)	572 (83)	538 (78)
C52100	614 (89)	558 (81)	552 (80)
C63800	724 (105)	696 (101)	696 (101)
C65400	627 (91)	627 (91)	627 (91)
C66600	613 (89)	586 (85)	579 (84)
C68800	786 (114)	745 (108)	731 (106)
C70250(a)	552 (80)	517 (75)	...
C70600	489 (71)	483 (70)	483 (70)
C72400(a)	793 (115)	690 (100)	621 (90)
C72500	531 (77)	504 (73)	503 (73)
C73500	525 (76)	518 (75)	517 (75)
C74000	593 (86)	565 (82)	552 (80)
C75200	558 (81)	558 (81)	558 (81)
C77000	758 (110)	696 (101)	676 (98)

Bad way refers to the orientation of the bend with respect to the sheet or strip rolling direction. Tensile strengths of 965 and 1103 MPa (140 and 160 ksi) are available in 0.25 and 0.5 mm (0.010 and 0.02 in.) thicknesses with specially supplied mill tempers. (a) Mill hardened to strength shown, then formed.

intermediate annealing. Ironing occurs when the clearance between punch and die is smaller than the thickness of the drawn product. The curves shown in Fig. 4.62 suggest that ETP copper C11000 should have better redrawing and ironing characteristics and require lower press forces than copper alloys containing zinc, tin, or silicon.

Drawing Characteristics of Copper Alloys. The higher zinc brasses, such as C24000 (low brass), C26000 (cartridge brass), and C26200 (high brass), have strengths and high ductilities comparable to those of low-carbon steel. They are

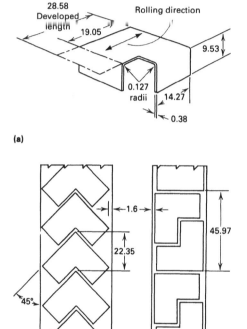

(a)

(b)

Layout A   Layout B

Fig. 4.61 Impact of bend anisotropy on part layout. (a) Hypothetical part, which has equal-radius bends at 90° orientations in the plane of the strip. Selection of the appropriate copper strip alloy for this application depends on the material strength and the bend properties in the relevant orientations. (b) Potential nesting of blanks for the part shown in (a). Layout A is required for directional alloys such as C51000 and results in 38% scrap; a nondirectional alloy such as C68800 would allow the more efficient layout B, with 23% scrap. Dimensions given in millimeters (1 in. = 25.4 mm).

outstanding materials for deep drawing and stretch forming. The phosphor bronzes represent another families of copper alloys with good deep-drawing and stretch-forming properties. Among these, phosphor bronze A (C51000) offers an excellent combination of high strength and high ductility. It is used to form deep-drawn, thin-wall shells that are corrugated to produce bellows.

Drawability of nickel-silvers (C73500–C79800) is similar to that of the high-zinc brasses, although the former have somewhat higher work-hardening rates and require more intermediate annealing. When fully annealed, low-aluminum bronze, C63800, also exhibits good deep drawability. Tin brass alloys, such as C40500, C41100, C42200, and C42500, respond well to drawing and redrawing operations. In this regard, C40500 and C41100 are similar to the high-copper brasses, and C42200 and C42500 are similar to C24000. Starting in the solution annealed temper, beryllium copper can be drawn as much as 80% reduction (C17200) before re-annealing. Subsequent heat treatment produces tensile strengths ranging from 1275 to 1380 MPa (185 to 200 ksi).

For alloys lacking drawability data, forming characteristics can be estimated on the basis of composition and mechanical properties and by comparison with standard alloys whose drawability is known. If drawing is to be combined with machining, it is advantageous to used a leaded alloy.

Grain Size Effects. For a given alloy and sheet thickness, ductility generally increases, and strength decreases, with increasing grain size; however, when the grain size is so large that there are only a few grains through the strip thickness, both ductility and strength decrease. Table 4.17 lists recommended grain sizes for annealed strip for drawing and stretch-forming operations. One rule of thumb suggests that for light-gage stock, the best balance between strength, ductility, and surface appearance can be struck with grain sizes approximately one-third the thickness of the drawn cup. Figure 4.63 indicates that peak elongation for 0.15 mm (0.006 in.) thick cartridge brass occurs at an average grain size of 0.020 mm (0.0008 in.). For 0.40 mm (0.0157 in.) brass strip, average grain sizes ranging between 0.038 and 0.061 mm (0.0015 and 0.0024 in.) provide maximum drawability. For 0.81 mm (0.032 in.) material, optimum performance occurs with average grain sizes between 0.060 and 0.090 mm (0.0024 and 0.0035 in.) (Ref 38).

Grain size also affects surface finish. Coarse-grained metal, when deep drawn or stretch-formed, produces rough surfaces resembling an orange peel. Such a surface is costly to polish, therefore, highly polished parts require fine-grained metals. A classic example of this situation is the one-piece brass or bronze doorknob shown in Fig. 4.64. Doorknob shapes are difficult to produce on draw presses. They are usually produced in transfer presses, and the process can include 15 to 20 operations with one intermediate anneal or partial anneal. The 0.76 mm (0.030 in.) strip (C36000 or C22000) used to produce the doorknob shown in Fig. 4.64 calls for grain sizes between 0.030 and 0.035 mm (0.008 and 0.0014 in.) or 0.015 and 0.030 mm (0.0006 and 0.0012 in.) to avoid surface roughening.

Roll Forming

Contour roll forming is an automated high-speed production process that produces tubular, box, angular, and folded parts of in a variety of profiles. The process is applicable to copper alloys, and the material characteristics that determine suitability are the same as those that govern bend and stretch formability. Complicated shapes and profiles incorporating sharp bends require annealed tempers.

Stretch Forming

The stretch formability of copper alloys correlates with the total elongation measured in a tension test, and annealed alloys that show high work-hardening rates offer the best stretch-forming characteristics. Cold rolling to increase strip temper (strength) significantly reduces stretch formability. The data shown in Fig. 4.65 suggest that high-tin and high-zinc alloys offer the most favorable combinations of strength and stretch formability (Ref 38).

Hammer Forming

Hammer forming is used to make artistic or decorative articles from flat sheets. Before working, the sheet is heated to cherry red and allowed to cool slowly. Forming is completed using chasing tools, hammers, mallets, and planishers (Ref 51).

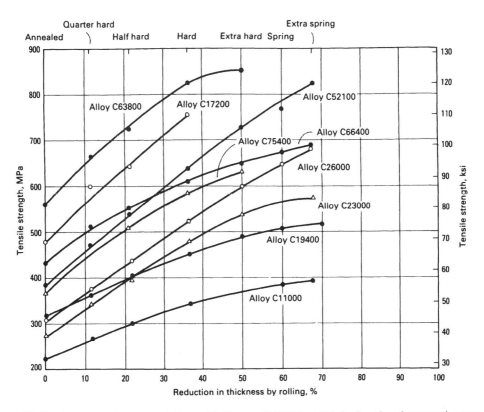

Fig. 4.62 Redrawing characteristics of 1.0 mm (0.040 in.) thick sheets of several copper alloys. Curves of lower slope indicate a lower rate of work hardening and, therefore, a higher capacity for redrawing

Electromagnetic Forming

Electromagnetic forming, also known as magnetic pulse forming, is a process for forming metal by the direct application of an intense, transient magnetic field. The workpiece is formed without mechanical contact by passing a pulse of electric current through a forming coil. Electromagnetic forming can be used on copper and some brasses because of their high electrical conductivity and excellent formability.

Electrical connections are made by electromagnetically swaging a copper band onto the end of a stranded electrical conductor wire before insertion into a brass terminal. The process yields joints with optimum conductivity, high mechanical strength, and long life under severe service conditions.

Spinning

Spinning is a method of forming sheet or tube into seamless circular shapes by force the metal against a rapidly rotating die with a suitable tool. Copper alloys with high "r" values, high tensile elongation, and low work-hardening rates exhibit the highest spinability.

Tough pitch copper, C11000, can be spun easily, usually without the need for intermediate annealing. Brasses are also readily spun, although the

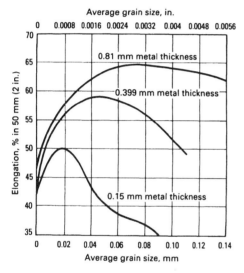

Fig. 4.63 Elongation versus grain size for alloy C26000 sheets of various thicknesses

Fig. 4.64 Doorknob fabricated by deep drawing

Table 4.17 Grain size ranges for selected forming operations

Type of operation and surface characteristics	Average grain size	
	mm	in.
Shallow forming or stamping. Parts will have good strength and very smooth surface. Also used for very thin metal	0.005–0.015	0.0002–0.0006
Stamping and shallow-draw parts. Parts will have high strength and smooth surface. General use for metal thinner than 0.25 mm (0.010 in.)	0.010–0.025	0.0004–0.001
Shallow-drawn parts, stampings, and deep-drawn parts that require buffable surfaces. General use for thickness under 0.3 mm (0.012 in.)	0.015–0.030	0.0006–0.0012
This grain size range includes the largest average grain that will produce parts essentially free of orange peel. Therefore, it is used for all types of drawn parts produced from brass up to 0.8 mm (0.032 in.) thick	0.020–0.035	0.0008–0.0014
Begins to show some roughening of the surface when severely stretched. Good deep-drawing quality in 0.40 to 0.64 mm (0.015 to 0.025 in.) thickness range	0.010–0.040	0.0004–0.0016
Drawn parts from 0.4 to 0.64 mm (0.015 to 0.025 in.) thick brass requiring relatively good surface, or stamped parts requiring no polishing or buffing	0.030–0.050	0.0012–0.002
Commonly used for general applications for the deep and shallow drawing of parts from brass in 0.5 to 1.0 mm (0.020 to 0.040 in.) thicknesses. Moderate orange peel may develop on drawn surfaces	0.040–0.060	0.0016–0.0024
Large average grain sizes are used for the deep drawing of difficult shapes or deep-drawing parts for gages 1.0 mm (0.040 in.) and thicker. Drawn parts will have rough surfaces with orange peel except where smoothed by ironing	0.050–0.119	0.002–0.0047

higher zinc brasses sometimes require intermediate annealing. Tin brasses containing at least 87% copper require higher spinning pressure and more frequent annealing than brasses. Nickel silvers that contain at least 65% copper are also well suited for spinning, as are copper-nickels. Phosphor bronzes, aluminum bronzes, and silicon bronzes present some difficulties, but they can be spun into shallow shapes under favorable conditions. Copper alloys that are difficult to spin include Muntz metal (C28000), nickel silvers containing 55% copper or less, beryllium coppers, alloys containing more than about 0.5% lead, Naval Brass, C46400, and other multiphase alloys. Annealed tempers are almost always used.

Alloys can be heated before spinning to reduce the required force, spin thicker material, and permit more severe deformation. The forming characteristics of Muntz metal, extra high-leaded brass, and Naval Brass are also improved at elevated temperature. Special care must be taken to avoid the unintentional heating of the workpiece when spinning brasses that contain more than 0.5% lead and more than 64% copper.

Forming Considerations for Electrical Springs

Electrical contacts sometimes rely on stepped or tapered beam profiles to optimize deflection or contact forces. These profiles can be obtained by coining, although this reduces formability (Fig. 4.66). The use of pre-contoured strip avoids this problem. Contour is produced by profile milling, skiving, or by the fabricating structures from several strip layers using longitudinal electron beam (EB) welding. Electron beam welding can also join dissimilar metals, for example, ductile C19500 and high-strength, mill-hardened C17200. Examples of such copper-alloy strip forms are shown in Fig. 4.67.

Springback causes a formed part to have a geometry different from that of the tooling. The degree of springback varies directly with the radius of the bend and is strongly related to the tensile properties of the material. The springback performance of beryllium coppers, which are used for high-performance electrical and electronic connectors, is shown in Fig. 4.68.

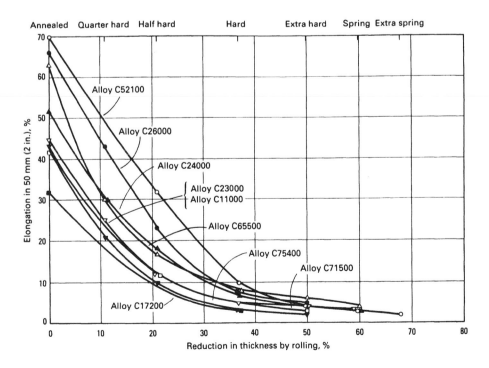

Fig. 4.65 Stretch-forming characteristics of 1.0 mm (0.040 in.) thick copper alloys. Elongation values for a given percentage of cold reduction indicate the remaining capacity for stretch forming in a single operation

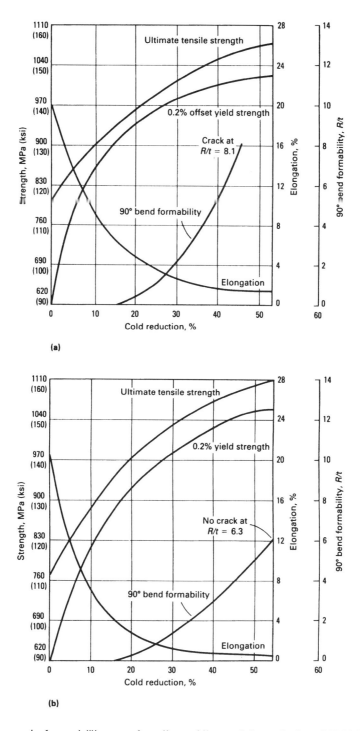

Fig. 4.66 Change in formability as a function of the coining of alloy C17200 in longitudinal (a) and transverse (b) directions. The effect of coining is simulated by cold reduction. Original strip thickness in both cases was 0.41 mm (0.016 in.). Bend formability is measured as the ratio of bend radius, *R*, to strip thickness, *t*

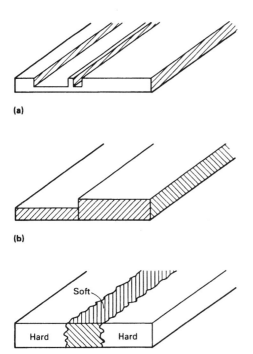

Fig. 4.67 Special treatment of copper alloy strip for optimized combinations of formability and spring characteristics. (a) Profile milled strip. (b) Dissimilar thicknesses longitudinally welded; this method can also be used to join dissimilar alloys. (c) Localized heat treatment (electron beam softening)

Forming Limit Analysis

Forming limit analysis provides the means to assess sheet metal formability over a wide range of forming conditions. The amount of deformation occurring during sheet forming, that is, major and minor strains, e_1 and e_2, respectively are measured using fiducial markings printed or etched onto strip surfaces before fabrication. The analysis makes use of empirically determined curves, which characterize individual materials. Forming limit diagrams (FLD) define the ability of the material to distribute localized strain. Limiting dome height (LDH) curves describe the overall ductility needed to form the material. These define the biaxial strain or deformation limits beyond which failure will occur during sheet metal forming. Regions beneath the curves are "safe," for example, fracture will not occur if minor and major strains (in the case of the FLD) do not exceed the values delineated by the curves.

Forming limit and limiting dome height curves for 13 copper alloys are shown in Fig. 4.69 and 4.70. The alloys referred to are described in Table 4.18. These data show that in annealed tempers, high-copper and copper-zinc alloys exhibit the highest formabilities, followed closely by alloys UNS C72500, C51000, and C74300; these materials in turn are slightly more formable than alloys C19400, C70600, and C75200. Increasing the temper by cold rolling decreases formability. The LDH data essentially follow the trend shown in FLD behavior (Ref 36).

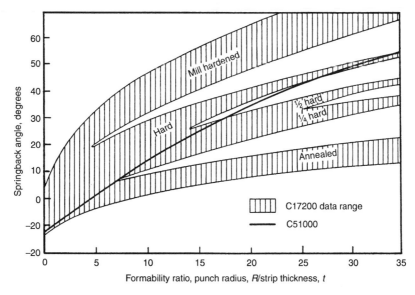

Fig. 4.68 Angular springback of heat-treatable and mill-hardened tempers of beryllium-copper C17200 and phosphor bronze C51000 strip (90° V-block plane-strain bends)

Tube and Pipe Fabrication

Copper tubular products are produced from shells extruded or pierced from copper billets. Tube is also produced by seam welding roll-formed strip and by continuous casting.

Extrusion. Most copper-base tube shells are extruded by forcing a heated billet over a piercing mandrel fixed within a die orifice. The clearance between mandrel and die determines the wall thickness of the extruded tube shell. The process is virtually identical to that described for rod except

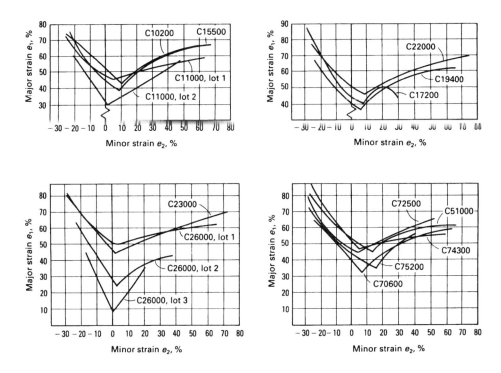

Fig. 4.69 Forming limit curves (FLCs) for selected copper alloys. FLCs reveal local ductility forming

Table 4.18 Copper and copper alloys evaluated using forming limit analysis

UNS number	Common name	Condition	Thickness, mm (in.)	Grain size, mm (in.)	Ultimate tensile strength, MPa (ksi)
C10200	Oxygen-free copper	Annealed	0.66 (0.026)	0.014 (0.0006)	234 (34)
C11000, lot 1(a)	ETP copper	Annealed	0.74 (0.029)	0. 016 (0.00063)	224 (32.5)
C11000, lot 2(b)	ETP copper	Half hard	0.69 (0.027)	...	268 (38.8)
C15500	Silver copper	Annealed	0.71 (0.028)	0.009 (0.00035)	288 (41.8)
C17200	Beryllium copper	Annealed	0.25 (0.010)	0.019 (0.00075)	491 (71.2)
C19400(c, d)	HSM copper	Annealed	0.69 (0.027)	...	319 (46.3)
C22000(d)	Commercial bronze	Annealed	0.69 (0.027)	0.017 (0.00067)	234 (34)
C23000(e)	Red brass	Annealed	0.69 (0.027)	0.024 (0.00094)	293 (42.5)
C26000, lot 1(f)	Cartridge brass	Annealed	0.64 (0.025)	0.025 (0.00098)	345 (50)
C26000, lot 2(e, g)	Cartridge brass	Half hard	0.69 (0.027)	...	407 (59)
C26000, lot 3	Cartridge brass	Full hard	0.51 (0.020)	...	531 (77.0)
C51000	Phosphor bronze A	Annealed	0.69 (0.027)	0.014 (0.0006)	374 (54.3)
C70600	Copper nickel, 10%	Annealed	0.81 (0.032)	0.016 (0.00063)	361 (52.4)
C72500	Copper-nickel-tin alloy	Annealed	0.69 (0.027)	0.023 (0.0009)	356 (51.6)
C74300	Nickel silver	Annealed	0.69 (0.027)	0.035 (0.0014)	387 (56.1)
C75200	Nickel silver	Annealed	0.69 (0.027)	0.020 (0.0008)	405 (58.7)

(a) LDH curves are medians based on 0.69, 0.74, and 0.79 mm thickness data. (b) 20% tensile elongation. LDH curves are medians based on 0.64, 0.69, and 0.79 mm thickness data. (c) 29% tensile elongation. (d) LDH curves are medians based on 0.69 and 0.74 mm thickness data. (e) LDH curves are medians based on 0.69, 0.79, and 0.81 mm thickness data. (f) LDH curves are medians based on 0.66 and 0.69 mm thickness data. (g) 28% tensile elongation

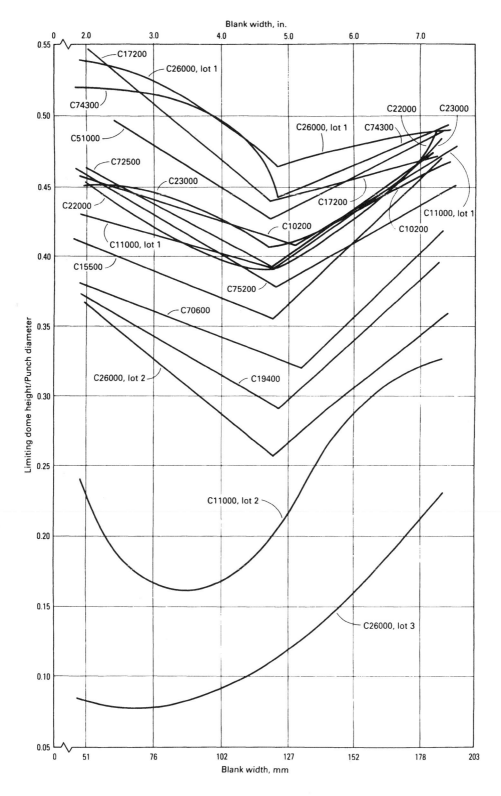

Fig. 4.70 Limiting dome height (LDH) curves for copper alloys. LDH curves illustrate the overall ductility of the coppers and copper alloys evaluated.

that tube extrusion requires somewhat higher pressures. Long tubular billets of, for example, OF copper are extruded over a fixed mandrel in an arrangement such as those illustrated in Fig. 4.71. Hot and cold isostatic extrusion processes have also been applied to copper-base materials, but conventional extrusion remains the dominant commercial practice (Ref 52).

The variation of extrusion pressures with alloy composition is not straightforward. Rather, it depends on the flow stress of the alloy at the extrusion temperature. Free-cutting brass, C36000, (61.5%Cu-3%Pb-35.5%Zn) requires a relatively low pressure, whereas cartridge brass C26000 (70%Cu-30%Zn) and admiralty brass C44300 (71.5%Cu-1%Sn-27.5%Zn-0.06%As) require the highest pressure of all of the brasses. Most of the coppers require an extrusion pressure intermediate between those for C26000 and C36000. Copper-nickel C71500 (70%Cu-30%Ni) requires very high pressure.

Cold drawing to size is performed on draw blocks for coppers and on draw benches for brasses and other alloys. The metal is cold worked as its diameter is reduced. Wall thickness is reduced concurrently by drawing over a plug or mandrel fixed in the die orifice. Because the metal work hardens, tubes may require intermediate annealing when drawing to small sizes; however, this is not always necessary when dealing with coppers (Ref 53, 54).

Tube reducing is an alternative to tube drawing. Here, grooved semicircular dies are rolled or rocked back and forth along the tube and a tapered mandrel inside the tube controls its inside diameter and wall thickness. The process yields tube with very accurate dimensions and better concentricity than can be achieved by tube drawing. Reduced tubes are used in air-conditioning equipment. See Chapter 3.

Tube reducing can be used for all alloys that can be drawn on draw benches. Slight changes in die design and operating conditions may be required to accommodate different alloys. Very small diameter tube is sometimes produced by block- or bench-drawing reduced tube (Ref 53, 54).

Rotary Piercing. Seamless copper and brass tubes can also be produced by rotary piercing. The operation is performed on Mannesmann mills, in which one end of a heated cylindrical billet is fed between rotating, horizontally oriented work rolls that are inclined at an angle to the axis of the billet (Fig. 4.72). The billet is simultaneously rotated and driven forward toward the piercing plug, which is held in position between the rolls. Pressure around the periphery of the billet opens up tensile cracks at the center of the billet just ahead of the piercing plug. The piercing widens the resulting hole, establishes the wall thickness and smoothes the inner diameter of the tube. Coppers and alpha brasses can be pierced, provided the lead content is held to less than 0.01%. Duplex

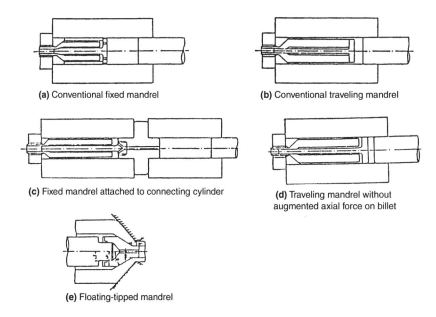

(a) Conventional fixed mandrel

(b) Conventional traveling mandrel

(c) Fixed mandrel attached to connecting cylinder

(d) Traveling mandrel without augmented axial force on billet

(e) Floating-tipped mandrel

Fig. 4.71 Various tool arrangements for tube extrusion

brasses can tolerate higher levels of lead without adversely affecting their ability to be pierced (Ref 53).

Each alloy has a characteristic piercing temperature range. Below this range, the central hole does not open properly under the applied peripheral forces. Overheating may lead to cracked surfaces. Suggested piercing temperatures for various alloys are given in Table 4.19 (Ref 54).

Surface enhanced tubes, sometimes called integrally finned tubes, have special inner and/or outer surface profiles. The profiles enhance heat transfer between the tube and external media, raising the overall heat transfer coefficient by up to 100% or more. External fins are obtained by roll forming the outer surface of a plain tube. This process results in an increase in strength of the finned sections, where as the plain ends remain soft. Internal grooving is obtained by extrusion using an internal die (Ref 55).

Wolverine Rotary Extrusion Process. Tube surfaces can also be profiled by rotary extrusion, a process developed by the Wolverine Tube Company. This process, performed cold, yields tubes between 12.7 and 38 mm (0.5 and 1.5 in.) containing from four to nine external helical fins per inch, depending on design. The fins are extruded from the surface of the tube by roll pressure. Rotary-extruded tube is manufactured in copper, copper-base alloys, and aluminum. Bimetallic tube consisting of a copper liner with a finned aluminum sheath are also produced.

Welded copper tube is made by drawing strip through a conical die such that it assumes a circular cross-section. Edges are induction- or seam-welded as they are forced together, forming the tube. Most copper alloys with conductivities lower than 30% IACS can be seam welded. Copper and high-conductivity copper alloys can also be seam welded, which requires more skill. After welding, tube blanks are annealed and drawn to finished sizes (Ref 53, 56).

Continuous and Semicontinuous Casting. Both solid and hollow tube blanks (or shells, if hollow) are usually produced by continuous or semicontinuous casting, which yields shells that are significantly better in quality than those produced by static casting. The continuous casting process (Fig. 4.73), has been in use since 1967 (Ref 57).

Semicontinuous casting of tube shells limits the length of the product by the size of the casting machine. Casting is performed vertically; shell lengths ranging from 3 to 6 m (9.8 to 19.7 ft) are typical. Diameters range between 170 and 310 mm (6.7 and 12.2 in.). Solidification takes place in a 125 to 455 mm (4.9 to 17.9 in.) long water cooled mold at approximately 200 mm/min (7.9 in./min). As with continuous casting, tube shells are cast over a water cooled copper mandrel. Figure 4.74 shows a longitudinal section of a semicontinuous casting mold (Ref 57, 58).

Outokumpu Process. A relatively new method of producing copper plumbing tube preforms is the Outokumpu "cast and roll" process. Here, tube shells with wall thicknesses between 20 and 30 mm (0.79 and 1.2 in.) and outside diameters of 87 mm (3.4 in.) are continuously cast in lengths of 12.5 m (41 ft) and weights up to 500 kg (1100 lb). The shells are worked over a mandrel in a planetary rolling mill, where cross-sections are reduced by as much as 95%. This severe deformation is sufficient to create hot-working conditions, avoiding the need for annealing. After rolling, tubes are further reduced by conventional cold drawing (Ref 59).

Quality Control Test Methods. Standard ASTM test methods used to determine the properties and chemical compositions of copper tube are listed in Table 4.20. These tests permit the nearly

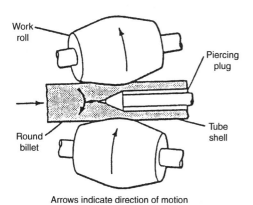

Work roll

Piercing plug

Round billet

Tube shell

Arrows indicate direction of motion

Fig. 4.72 Schematic diagram of metal piercing rolling

Table 4.19 Piercing temperatures for selected copper alloys

UNS number	Piercing temperature	
	°C	°F
C11000	815–870	1500–1600
C12200	815–870	1500–1600
C22000	815–870	1500–1600
C23000	815–870	1500–1600
C26000	760–790	1400–1450
C28000	705–760	1300–1400
C46400	730–790	1350–1450

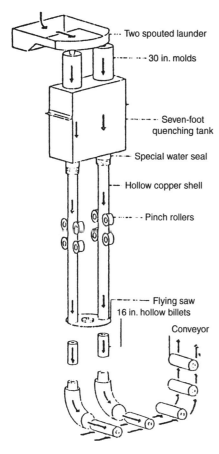

Fig. 4.73 Ajax continuous casting machine. 1. Molten metal is poured from the new arc-furnace into the two-spouted launder. 2. The molds are water-cooled and equipped with water-cooled mandrels; their movement is a vertical oscillation through $3/16$ in. 3. Water is retained within the seven-foot quench tank by special seals developed by Halstead engineers. The temperature of the strand is reduced from 1150 °C to room temperature. 4. The tube shell strand is passed through pinch rolls, which support its entire weight. 5. Moving through a vertical distance of 16 in. the flying saw takes 45 s to cut through both strands simultaneously. A specially developed machine removes the copper chips produced. 6. After cutting off the billet lengths are lifted by a conveyor from the 45 ft pit to the casting floor level.

complete characterization of tube product. Grain size and microstructure are examined in a longitudinal section of the tube, for example, the orientation that gives a side view of the elongated or recrystallized grains. Tensile tests are normally conducted on full cross sections of the tube. If this is impractical, for example, with very large diameter tube, representative longitudinal samples may be taken instead.

Wire and Cable Production

Practical Aspects of Copper Wire Drawing

The success of copper wire drawing operations depends largely on the quality of the starting materials. These factors, and their effects on the drawing process, are described.

Oxygen Content. Tough pitch copper contains 0.02 to 0.06% oxygen. Most of this oxygen forms a cuprous oxide eutectic at the grain boundaries. It also forms oxides of other contained elements. At 0.05 to 0.06% oxygen, copper becomes brittle and this adversely affects drawability.

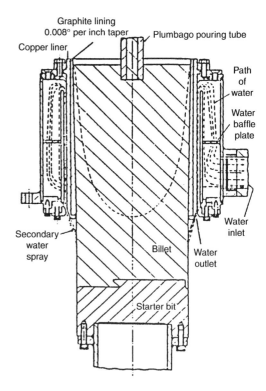

Fig. 4.74 Mold for semi-continuous casting of copper

Hydrogen Embrittlement. When tough pitch copper is heated in reducing atmospheres containing hydrogen, the gas can dissolve in the metal, where it combines with the grain boundary oxides to form water vapor. Molecular water is unable to diffuse into the metal. Instead, it forms microscopic voids in which extremely high pressures develop. The voids eventually lead to a brittle fracture phenomenon known as hydrogen embrittlement.

Impurity Segregation. Impurities tend to migrate to the last regions of metal to freeze. Once seen as rosette structures in wirebars, they now occur at the center of continuously cast wire rod. Excessive concentrations of such segregated impurities tend to create stringers or pockets of reduced ductility. This leads to checked wire or "crow's feet," as well as brittle wire breaks. Examples of these are shown in Fig. 4.80 and 4.81 (Ref 45).

Inclusions. Rods occasionally contain random foreign inclusions, which can come from a variety of sources. With the exception of improper drawing practice, they are reportedly the single most important cause of wire breaks. See Chapter 5 for a further discussion of this phenomenon (Ref 45).

Inclusions embedded in or near the surface (Fig. 4.75), usually originate from the rolling operation; those found below the surface can be traced to the casting process. Sources include drawing die particles—carbides or diamond—rolling mill roll particles, oxides, furnace wall refractories, slag, weld flash, pouring spout refractories, and casting wheel particles.

Voids (Fig. 4.76) are usually formed when the rod is cast, either through excess gas in the molten metal or by solidification shrinkage. When the stock is rolled into rod and drawn into wire, these voids become elongated. Breaks associated with them are referred to as piped wire breaks. This form of macroporosity can be overcome by not

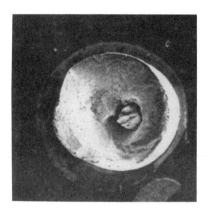

Fig. 4.75 SEM of iron oxide inclusion in continuous cast copper

Fig. 4.76 SEM of the cup usually associated with "cup and cone" breakage in wire, but also associated with "piped" wire breaks (Ref 43).

Table 4.20   Selected ASTM test methods used for the quality control of copper and copper alloy tube products

Test	ASTM designation	Name
Chemical analysis	E 55	Practice for sampling wrought nonferrous metals and alloys for the determination of chemical composition
	E 88	Practice for sampling nonferrous metals and alloys in cast form for the determination of chemical composition
	B 61	Specification for steam or valve castings
	B 442	Specification for tough-pitch chemically refined copper-refinery shapes
	B 62	Standard specification for composition bronze or ounce metal castings
	B 75	Standard specification for seamless copper tube
Grain size	E 112	Standard test methods for determining average grain size
Expansion (pin test)	B 153	Test method for expansion (pin test) of copper and copper alloy pipe and tubing
Mercurous nitrate	B 154	Method of mercurous nitrate test for copper and copper alloys
Tension	E 8	Standard test method for tension testing of metallic materials
Nondestructive test	E 243	Standard practice for electromagnetic (eddy-current) examination of copper and copper alloy tubes

casting copper with an excessive gas content or by providing sufficient cooling to eliminate the molten metal core in material leaving the casting stage of the continuous caster (Ref 45).

Slivers, shown in Fig. 4.77 and 4.78, occur primarily due to mechanical damage or to rolled-in overfills or fins that arise during the rolling operation. If the slivers are relatively free of oxides, the cause can usually be found in the drawing process, and may be due to rough capstans in the drawing machine, too shallow drafts when drawing soft wire, a lack of back relief in the dies, rough dies, improper die contours, or a lack of blend angles (Ref 45).

Fish-Mouth Breaks. Improper welding techniques cause breaks that have a characteristic fish-mouth appearance, Fig. 4.79 (Ref 45).

Flat and Rectangular Wire

Flat wire is a useful and convenient copper semi with numerous uses, mainly in electrical and electronics applications. Width-to-thickness ratios of 20 to 1 or less are the most popular, although ratios as high as 100 to 1 are now available. The tensile strength of flat wire is controlled by the degree of cold deformation. Wire can be softened by annealing.

Wire Drawing Mills

Drawing Machines. Wire drawing machines are devices that pull copper rod through a series of progressively finer dies. The two types are in use: stepped-cone and tandem machines. Figure 4.80 shows the difference in capstans used in these two types of machines (Ref 45).

Stepped-cone machines, or slip-type drawing machines, incorporate capstans of from two to six diameters—the stepped draw blocks—on each shaft. Stepped-cone machines are smaller and less expensive than tandem machines, but they offer limited flexibility in attainable reduction. In slip-type machines, a certain wire slip allowance is intentionally permitted to occur to compensate for changing wire speeds as the wire diameter decreases. These are the machines used in copper wire mills.

Tandem machines have one single-diameter capstan or block per die. Block speed is fixed by the drive gearing alone. The principal advantages of tandem machines are their higher speeds, less slip, improved capstan, and wire cooling and easier compensation for wear. On the other hand, these

Fig. 4.77 Sliver, showing characteristic "folded metal piece," is the result of overfire

Fig. 4.78 This sliver formation is apparently due to problems in mechanical drawing

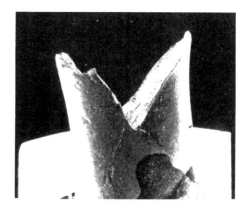

Fig. 4.79 Typical "fish-mouth" appearance of a failed wire weld with granular fracture surface structure

machines are larger and more expensive than stepped-cone machines.

Drafting

The reduction per pass is called drafting. When the reduction is largest in the first several dies (to

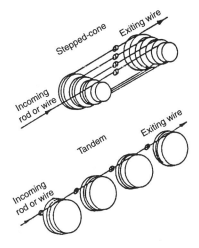

Fig. 4.80 Capstans used in wire-drawing machines. Stepped-cone: minimum floor space, high capstan maintenance, lower investment cost, slip determined by block diameters, gearing, and drafting. Tandem: maximum floor space, low capstan maintenance, higher invest-ment cost, slip determined by gearing and drafting, improved cooling, less wire damage, improved tension control, better wire-path alignment.

take advantage of the soft-annealed state of the rod) the reduction is known as tapered drafting. The equal amounts of reduction imposed in subsequent dies is called straight drafting.

It is the schedule of reduction used in the drawing of copper wire that governs the familiar American Wire Gage (AWG) sizes. The series was originally known as the Brown & Sharpe gage, after its inventor's firm. It is based on two defined wire diameters: size No. 0000 (often written 4/0) at 0.4600 in. (11.68 mm) and AWG No. 36 at 0.0050 in. (0.127 mm). The 38 intermediate steps correspond to an inverse geometric progression wherein the ratio of any diameter to the next smaller size is approximately 20.7% (Table 4.21) (Ref 60).

The Wire Drawing Process

The process of reducing the diameter of a wire by drawing it through a die involves the several factors described in Fig. 4.81 (Ref 45).

Raw Materials. The diameter of the rod and the size of the coil in part dictate the type of equipment and the drawing process. Common copper rod diameters and coil sizes are:

- 8.0 or 11.7 mm (0.315 in.) diameter rod in 113 kg (250 lb) coils
- 8.0 mm (0.315 in.) diameter rod in 272 kg (600 lb) coils
- 8.0 mm (0.315 in.) diameter rod in 2268 to 2722 kg (5000 to 7000 lb) coils

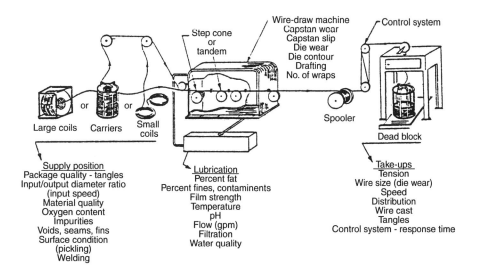

Fig. 4.81 Typical variables in a wire-drawing process

Large, continuous packages reduce the incidence of weld failures between succeeding coils and increase productivity, but they require larger and more expensive handling equipment, wider aisles, and heavier floor construction.

Welds. Threading wire drawing machines is time consuming, and a continuous supply of rod, spliced by welding, is preferred. The two common joining methods are resistance butt welding and cold pressure welding.

In resistance butt welding, the two wire ends are forced (butted) together under high pressure. A welding current is then turned on, and fusion heat is generated by electrical resistance of the wire itself. The wires ends are pre-formed with reduced diameters to force heating on a small cross-sectional area. Products between 0.127 to 31.75 mm (0.005 and 1.25 in.) diameter are successfully

Table 4.21   AWG standard wire sizes

AWG number	Diameter	
	in.	mm
4/0	0.4600	11.684
3/0	0.4096	10.404
2/0	0.3648	9.266
1/0	0.3249	8.251
1	0.2893	7.348
2	0.2576	6.554
3	0.22194	5.827
4	0.2043	5.189
5	0.1819	4.621
6	0.1620	4.115
7	0.1443	3.665
8	0.1285	3.264
9	0.1144	2.906
10	0.1019	2.588
11	0.0907	2.305
12	0.0808	2.053
13	0.0720	1.828
14	0.0641	1.628
15	0.0571	1.450
16	0.0508	1.291
17	0.04526	1.150
18	0.04030	1.024
19	0.03589	0.9116
20	0.03196	0.8118
21	0.02846	0.7229
22	0.02535	0.6438
23	0.02257	0.5733
24	0.02010	0.5106
25	0.01790	0.4547
26	0.01594	0.4049
27	0.01420	0.3606
28	0.01264	0.3211
29	0.01126	0.2859
30	0.01003	0.2546
31	0.00893	0.2258
32	0.00795	0.2019
33	0.00708	0.1800
34	0.00630	0.1601
35	0.00561	0.1426
36	0.00500	0.1270

welded by butt welding, although the process is not without drawbacks (Ref 45).

Cold pressure welding, or deformation welding, is a solid-state process that uses high pressure and plastic deformation to produce metallurgical bonds. It can be used to join almost all types of copper and is successfully applied in the production of magnet wire or finished wire for use in highly stressed applications. The process is conducted at room temperature. Wires to be joined are clamped in steel welding dies with a short length of wire projecting. The wire ends are butted together under considerable pressure, then upset several times. The flash formed is subsequently removed. Typical hydraulically powered (2 hp, 1.5 kW) butt welders have a maximum capacity of 4.75 mm (0.187 in.) diameter copper wire.

Previously cold-worked copper wire joined by cold pressure welding can be redrawn without annealing. On the other hand, cold pressure welding raises the hardness of soft annealed wire enough at the weld to cause a statistical increase in the number of rod breaks. As a result, cold pressure welding is not extensively applied to 7.9 and 9.5 mm (0.0625 and 0.375 in.) diameter rods, sizes that are normally drawn in the annealed state. For particularly clean materials such as OF coppers, the hardening effects are less pronounced, and cold pressure welds give satisfactory results.

Copper alloys containing beryllium, phosphorus, and zinc can be cold pressure welded, but copper alloys containing antigalling additions are more difficult to weld by this method (Ref 45).

Drawing Dies. Dies used for drawing copper and its alloys are made from diamond and tungsten carbide. Diamond dies are manufactured from a carefully selected and precisely oriented, high-quality natural or synthetic industrial diamond, supported in a corrosion-resistant metal case. Wires drawn with natural diamond dies range from 0.01 to 4.06 mm diameter. Synthetic diamond dies are polycrystalline. Polycrystalline dies have been used very successfully in copper wire drawing. In certain wire sizes, such dies outperform dies made from any other material (Ref 47).

Dies made from sintered tungsten carbide blended with cobalt powder were developed in Germany during the 1920s and at first, they were a substitute for natural diamond tools. They have excellent wear characteristics and are preferred for drawing copper wire of AWG 16 (1.291 mm, 0.05 in.) and larger. Carbide dies seldom fail by breaking but more often simply wear out. Despite its low strength in the annealed state, copper is very abrasive and tends to develop an abraded wear ring at the line where the wire first contacts the die.

Lubricants and Cooling Systems. The most common lubricants for drawing copper wire are soap and high-fat emulsions. Liquid products are easier to control and tend to run cleaner. Paste products are lower in price but run dirtier and require more maintenance. Water quality is very important. When using paste compounds, the alkalinity of the solution must be controlled between specific limits. If the pH drops too low, excessive amounts of copper soap are formed and the emulsion is said to run "dirty." A high pH favors foam formation and reduces lubricity. Bactericides, stabilizers, and other materials are added to ease removal of copper fines produced by the drawing operation (Ref 45).

Spoolers and Coilers. Wire drawing machines use basically four types of take-ups, or packaging units: spoolers, dead-block coilers, line-block coilers, and a combination of spool and a dead-block coiler known as a Bundpacker.

Spoolers are devices for filling reels with wire. They are used in conjunction with a wire drawing machine at high transfer rates. Spoolers are needed when the next machine only accepts spools, where subsequent wire feed rates exceed 225 m/min (750 ft/min) or for wires finer than about 0.40 mm (26 AWG). Otherwise, dead-block coilers are ordinarily used. In dead-block coilers, the block is motionless and a rotating flyer lays the wire, one strand at a time, on its surface.

Stranded Wire

Stranded copper wire and cable are made on machines known as bunchers or stranders. Conventional bunchers are used for stranding small diameter wires (34 to 10 AWG). Here, individual wires are paid off reels located alongside the equipment and are fed over flyer arms that rotate about the take-up reel to twist the wires. The rotational speed of the arm relative to the take-up speed controls the length of lay in the bunch. The individual wires in small flexible cables are usually 30 to 34 AWG, and there may be as many as 150 wires in each cable.

A tubular buncher mounts up to 18 wire-payoff reels. Wire is taken off each reel, threaded along a tubular barrel, and twisted together with other wires by a rotating action of the barrel. At the take-up end, the strand passes through a closing die to form the final bunch configuration.

For large diameter wire, supply reels in conventional stranders are fixed onto a rotating frame within the equipment and revolve about the axis of the finished conductor. There are two basic types of machines. In rigid-frame stranders, individual supply reels are mounted in such a way that each wire receives a full twist for every revolution of the strander. In planetary stranders, the wire receives no twist as the frame rotates (Ref 54).

Annealing

Although it is possible to cold work copper to a 99% reduction in area without annealing, copper wire usually is annealed after it has been reduced 90%. In some plants, electrical resistance is used to anneal the wire in less than a second as it exits from the drawing machines. Wire is also batch annealed, either in reels travelling through a controlled atmosphere furnace, or in batches in large bell-type furnaces. Annealing temperatures range from 400 to 600 °C (750 to 1100 °F), depending on wire diameter and reel weight (Ref 54).

Wire Coating

The four metallic coatings used on copper conductors for electrical applications are: lead, or lead alloy (80%Pb-20%Sn) designated as ASTM B 189; nickel designated as ASTM B 355; silver designated as ASTM B 298; and tin designated as ASTM B 33.

Coatings are used to improve the solderability of hookup wire. They also provide a barrier between the copper and insulating materials such as rubber, which react with copper and prevent oxidation of the copper during high-temperature service. Wires are coated by dipping in molten metal baths, by electroplating and by cladding. Electroplating has become the dominant process, especially because it can be done "on line" following the wire drawing operation (Ref 54).

Insulation and Jacketing

Polymeric Insulation. Whereas natural rubber was once the most common insulating material, it has largely been replaced by a variety of synthetic polymers, selected on the basis of service requirements. The most commonly used polymers are polyvinyl chloride (PVC), polyethylene, ethylene propylene rubber (EPR), silicone rubber, polytetrafluoroethylene (PTFE), and fluorinated ethylene propylene (FEP). Insulation is applied in extruders, which melt thermoplastic polymer and force the resulting paste through a die over the moving conductor (Ref 45).

Enamel Insulation. Fine magnet wire is commonly coated with a thin, flexible enamel film.

Enamel coatings are rated by temperature limits ranging from 105 to 220 °C (220 to 425 °F). The most commonly used enamels are based on polyvinyl acetals, polyesters, and epoxy resins. Enamel is applied by dipping or wiping, then cured in ovens. Several passes are sometimes necessary to build up the required thickness. Powdercoating, a relatively recent innovation, avoids solvents and is, therefore, more acceptable from hygienic and environmental standpoints. Electrostatic sprays, fluidized beds, and other nontraditional techniques are also used (Ref 53).

Paper and Oil Insulation. Oil-impregnated cellulose paper is used to insulate high-voltage cables for critical power-distribution applications. The paper is applied in tape form, helically around the conductors using special machines in which 6 to 12 paper-filled pads are held in a cage that rotates around the cable. Paper layers are wrapped alternately in opposite directions, free of twist. The wrapped cables are then pressure-impregnated with oil in such a way as to ensure that all air has been expelled from the wrappings (Ref 53).

Forming Bar, Tube, Shapes, and Wire

Bending

Bars are bent by draw bending, compression bending, roll bending, and stretch bending. Bending methods used for tube are similar, except that the tube usually requires internal support to restrain the tendency to buckle. Tube may also require lateral support to prevent flattening.

Wall thickness affects stress distribution in tubes during bending. A thick-walled tube will bend more readily to a small radius than will a thin-walled tube (Table 4.22).

Wire bending is performed on hand benders; kick presses; power presses equipped with dies; coiling devices; automatic forming and spring-coiling machines; and in special machines actuated by cams, air, or hydraulic cylinders (Ref 38).

Metallurgical Considerations. Annealed copper-alloy tubing is easily bent, and it has little springback. Copper and some brasses may not need to be annealed. Copper-nickel alloys are more difficult to bend and have greater springback. When copper alloys are annealed, as most of them are, oxides should be removed by pickling before the tube is bent to reduce friction and protect the tooling.

Draw bending, the most common bending method, requires the workpiece, usually a tube, to be clamped to a rotating form and drawn by the form against a pressure die (Fig. 4.82). Pressure dies are either fixed or movable along their longitudinal axes. Movable pressure dies move forward with the workpiece as it is bent (Ref 38).

In compression bending, the workpiece is clamped to a fixed form. A wiper shoe revolves around the form to bend the workpiece (Fig. 4.83). Compression bending is used to bend rolled or extruded shapes. It does not control the flow of metal as well as does draw bending (Ref 38).

In roll bending, the workpiece is bent as it passes between three or more parallel rolls. Roll

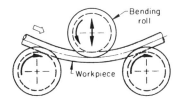

Fig. 4.83 Operating essentials in one method of three-roll bending

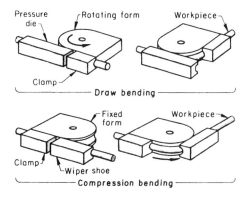

Fig. 4.82 Essential components and mechanics of draw bending and compression bending of bars and bar sections

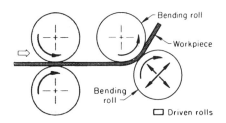

Fig. 4.84 Operating essentials in four-roll bending

benders may incorporate three (Fig. 4.83), or four (Fig. 4.84) rolls. Copper and copper-alloy rings, arcs, and helical coils are easily fabricated in a roll bender. It is difficult to control springback in a roll bender, and it may take several passes through the rolls to make the needed bend (Ref 38).

Stretch bending is used to generate large irregular curves. The workpiece is gripped at one end and stretched and bent simultaneously as it is forced around a form. This results in less springback than when the metal is simply bent. The tools, form blocks, or dies for stretch bending are simpler and less costly than conventional press tooling (Ref 38).

Hand and Power Bending. Bars are bent manually in fixtures, press brakes, and conventional mechanical and hydraulic presses, as well as in horizontal bending machines, rotary benders, and bending presses. Shapers can also be used to perform specific bending operations.

Hand-powered tooling is used for soft or light-gage material. Draw, stretch, and compression bending can be done manually. Roll bending is seldom done by hand. Manual bending tools are the same as those used on some power bending machines (Ref 38).

Press brake bending is especially useful in the production of copper and copper-alloy products in relatively small lots. Press brakes are versatile, as a variety of bend angles can be made with the same die (Fig. 4.85) (Ref 38).

Mechanical presses are generally reserved for mass production because the cost of expensive tooling must be amortized over large production lots. Figure 4.86 shows a round bar being bent into a U-bolt in a press (Ref 38).

Hydraulic presses are slower than mechanical presses, but they have the advantage of exerting full force over a long stroke. Therefore, deep bends can often be made on a hydraulic press that is much smaller than a mechanical press (Ref 38).

Tube Bending. Hand-powered machines used to bend copper tubes range up to 42 mm (1.65 in.) in capacity. They are small and light enough to be transported to the construction site. For larger diameters, ratchet action or geared machines should be used (Ref 61).

Copper tubes can usually be machine-bent without the need for internal support. Power benders equipped with mandrels support the sides of the tube to prevent it from distorting to an oval cross section. Light-gage tubes are easily bent, although some skills are needed. Hand bending of light-gage tubes is accomplished with the use of mandrels, usually in the form of flexible spiral springs. British standard BS 5431 applies to mandrels for copper tube in standard sizes from 10 to 22 mm (0.4 to 0.87 in.) diameter.

Low-melting-point alloys can be used in place of bending mandrels. Tubes are simply warmed and filled with liquid metal, which is then allowed

Table 4.22 Minimum practical inside radii for the cold drawn bending of annealed copper tube to 180 degrees

Tubing outside diameter		Minimum practical inside radius					
		Groove bending tools					
		With mandrel; ratio, <15 (best conditions)(a)		With mandrel or filler; ratio, <50 (normal conditions)(a)		Cylindrical bending block without mandrel; ratio, <30 (poor conditions)(a)	
mm	in.	mm	in.	mm	in.	mm	in.
3.2	1/8	1.6	1/16	6.4	1/4	13	1/2
6.4	1/4	3.2	1/8	7.9	5/16	25	1
9.5	3/8	4.8	3/16	9.5	3/8	50	2
12	1/2	6.4	1/4	11	7/16	75	3
16	5/8	7.9	5/16	14	9/16	102	4
19	3/4	11	7/16	17	11/16	152	6
22	7/8	13	1/2	19	3/4	203	8
25	1	14	6/16	22	7/8	254	10
32	1 1/4	17	11/16	25	1	381	15
38	1 1/2	21	13/16	29	1 1/8	508	20
44	1 3/4	24	15/16	32	1 1/4	686	27
50	2	27	1 1/16	35	1 3/8	889	35
64	2 1/2	35	1 3/8	41	1 5/8	...	...
75	3	41	1 5/8	48	1 7/8	...	...
89	3 1/2	48	1 7/8	54	2 1/8	...	...
102	4	54	2 1/8	60	2 3/8	...	...

(a) Ratio of outside diameter to wall thickness of tubing

to freeze. After the tube is bent, the assembly is dipped into a tank of boiling water to remove the fusible alloy.

Selecting the bending method for a particular application depends on the equipment available, the number of parts required, the size of the bar and/or the size and wall thickness of the tubing, the work metal, the bend radius, the number of bends in the workpiece, the accuracy required, and the amount of flattening that can be tolerated.

Successful bending depends in part on the type of lubricant used. Copper alloys are often simply lubricated with mineral oil. Severe bends in brass are readily performed with a creamy mixture of laundry soap and water. Excessive lubrication must be avoided because it causes wrinkling, and the cost to remove the excess can become significant (Ref 38).

Tube Forming

Copper and brass tubular sections are converted into a variety of products by means of press forming, contour roll forming, tube spinning, rotary swaging, hydraulic bulging, explosive forming, electromagnetic forming, and electrohydraulic forming.

Tube Bulging

Some capillary (soldered or brazed) connections require that the tube diameter be expanded to match the fitting. The process requires that the tube be heated with a suitable torch. Bulging can be performed in the field using special tools made for this purpose (Ref 61, 62).

Miscellaneous Tube Forming Operations

Nosing, a cold forming operation, is used to reduce the diameter of tube ends to match adjacent fittings. Tube shells are machined before nosing. Diameters can be reduced by as much as 30% by this operation.

Contour roll forming is used to make seam-welded tubing; it is also applied in the production of a variety of cross sections.

Tube spinning is used to alter the shape of tubing, usually to produce a tubular part with two or more wall thicknesses. Tube spinning often precedes some other forming operation.

Other methods used include rotary swaging, hydraulic bulging, explosive forming, and electromagnetic forming.

Wire Forming

All wire forming operations can be applied to copper metals. The metals are commonly formed into coils (as springs), flattened, stranded into cables, headed, swaged, welded, and threaded (Ref 54, 63).

Coiling and Spring Forming. When small quantities are needed, copper bobbins can be coiled in a lathe. An arbor is held in the chuck, and two wooden friction blocks are mounted on the cross slide. Numerous hand-operated devices are also used. Production coiling is done in single-purpose automatic spring coilers, in which a pair

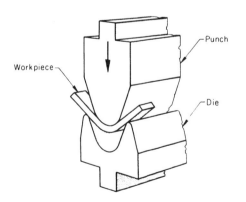

Fig. 4.85 Air bending of a bar in a press brake

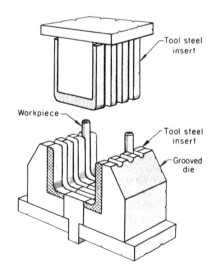

Fig. 4.86 Use of a grooved die in a mechanical press for bending a round bar into a U-bolt in one stroke

of feed rolls pushes a calculated length of straightened wire through restricting guides against a coiling point and around a fixed arbor, forming the coil.

Machining

Although much is known about machinability, the fundamental mechanisms are not thoroughly understood. In practical terms, however, machinability of particular copper alloys can be related to the type of chips they produce. Three basic chip types and, correspondingly, three degrees of machinability have been identified.

Alloy Types

Type I Alloys. This group contains the so-called free-cutting alloys, which are characterized by fine, frangible chips that break up easily during machining. Figure 4.87 shows the several shapes of turnings associated with each type of alloy. Many of the alloys contain lead, which occurs in globular form as shown in Fig. 4.88. This metallic lead promotes short-chip formation. Free-cutting brass (UNS C36000) and corresponding BS and DIN brasses are the best known examples of leaded type I materials. Other type I alloys contain tellurium, sulfur, or bismuth, which perform similar functions (Ref 64).

Because their chips readily free themselves from the tool or machine, type I alloys are well suited to rapid and automated machining. Rake angles, which are held to a minimum, break the chip almost immediately; minimum clearance angles provides greater support for the tool's cutting edge. Cutting fluids should be effective coolants. Water-based soluble oils are commonly used, although many operators prefer straight mineral oils.

Type II alloys are characterized by their tendency to form short, moderately brittle, and often curly chips. The metals are typically hard and strong, sometimes as a consequence of their complex microstructures, for example, duplex brasses. The intermittent shear process that governs the cutting operation in type II alloys raises the potential for tool chatter and poor surface quality, but these problems can be avoided by adjusting machining parameters appropriately. Chip breakers can be used to reduce the length of the coiled turnings. Power consumption varies with mechanical properties and work-hardening rate, and the shape of the turnings depends on the ductility of the metal. In part because of their relatively manageable chips, multiphase alloys are considered to have better machinabilities than ductile type III metals. The higher power consumption for the harder type II alloys is generally taken as being less important than tool wear or chip management, both of which are less favorable in type III alloys. Type II alloys are often processed on automatic screw machines, although production rates are considerably lower than those attainable with free-cutting alloys.

Type III alloys have uniform, single-phase microstructures. The group includes pure copper, high-copper alloys (except free-cutting types), low-zinc brasses, low-tin bronzes, and copper-nickels. Even highly alloyed single-phase alloys retain a considerable degree of ductility, and this is reflected in their long, stringy, and occasionally tightly curled turnings. Turnings are thicker than the feed rate because the copper is upset as it

Fig. 4.87 Broken chips typical of free-cutting brass (center) type I flanked by type II, short-chip turnings (right), and type III, long-chip turnings (left)

passes over the face of the tool. The accompanying cold work makes the chip hard and springy,

but it also consumes energy that, converted to heat in the chip and cutting tool, increases tool wear.

From the machinist's standpoint, however, the major problem with type III materials is that the stringy chips tend to snarl the cutting area. Type III alloys, while relatively difficult to machine in comparison with type I alloys, are not difficult to cut in an absolute sense. Their machining performance rather resembles mild or medium carbon steels and some of the non-free-cutting aluminum alloys. Type III copper alloys are routinely machined to close tolerances and fine surface finishes. Their low ranking is more a reflection of the unsuitability to high-speed automatic screw machines than it is of their inherent properties.

Machinability Ratings

Wrought Alloys. Machinability ratings for wrought copper alloys (Table 4.23), are semiquantitative comparisons with an arbitrary standard material, C36000 (free-cutting brass). The ratings are based in part on testing and partly on subjective considerations of machining speed, tool geometry, tool wear, attainable surface finishes, precision, and power requirements.

Free-cutting brass is customarily assigned a machinability rating of 100. Relatively easily machinable type I alloys rank above about 70; moderately machinable alloys rank between about 30 and 70, and difficult-to-cut alloys are given a rat-

Table 4.23   Machinability ratings of wrought copper and copper alloys

UNS No.	Alloy name	Machinability
Group 1 Free-cutting alloys		
C18700	Leaded copper	80
C14500	Tellurium copper	80
C33000	Low-leaded brass (tube)	60
C33500	Low-leaded brass	60
C34000	Medium-leaded brass	70
C33200	High-leaded brass (tube)	80
C34200	High-leaded brass	90
C35600	Extra-high-leaded brass	100
C36000	Free-cutting brass	100
C36500, C36800	Leaded Muntz metal	60
C37000	Free-cutting Muntz metal	70
C37700	Forging brass	80
C48500	Leaded naval brass	80
C38500	Architectural bronze	90
C54400	Free-cutting phosphor bronze	90
C70800	Leaded nickel silver, 10%	80
Group 2 Moderately machinable alloys		
C23000	Red brass, 85%	30
C24000	Low brass, 80%	30
C26000	Cartridge brass, 70%	30
C27000	Yellow brass	30
C28000	Muntz metal	40
C44300, C44500	Inhibited admiralty	30
C46400	Naval brass	30
C65100	Low-silicon bronze, (B)	30
C65500	High-silicon bronze, (A)	30
C67500	Manganese bronze, (A)	30
C53400	Leaded phosphor bronze, 5% (B)	50
C79200	Leaded nickel silver, 12%	50
C79400	Leaded nickel silver, 18%	50
C66100	Leaded silicon bronze, (D)	60
C63900	Aluminum silicon bronze	60
C68700	Aluminum brass	30
Group 3 Alloys difficult to machine		
C11000	Electrolytic tough-pitch copper	20
C12200	Phosphorus deoxidized copper	20
C10200	OF copper	20
C11300	Silver-bearing tough-pitch copper	20
C11400	Silver-bearing tough-pitch copper	20
C11000	Silver-bearing tough-pitch copper	20
C18200	Chromium copper	20
C21000	Gilding, 95%	20
C22000	Commercial bronze, 90%	20
C22600	Jewelry bronze, 87.5%	20
C51000	Phosphor bronze, 5% (A)	20
C52100	Phosphor bronze, 8% (C)	20
C52400	Phosphor bronze, 10% (D)	20
C61400	Aluminum bronze, (D)	20
C62800	Aluminum bronze, 10%	20
C70600	Copper nickel, 10%	20
C71500	Copper nickel, 30%	20
C74500	Nickel silver, 65-10	20
C75200	Nickel silver, 65-18	20
C75400	Nickel silver, 65-15	20
C75700	Nickel silver, 65-12	20
C77000	Nickel silver, 55-18	20
…	Beryllium copper (not heat treated)	20

(a) Approximately relative machinability rating (free-cutting brass = 100)

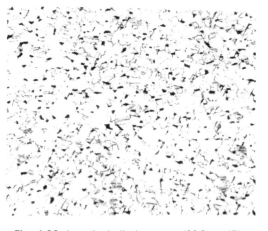

Fig. 4.88 Leaded tin-bronze (88Cu, 4Pb, 4Sn, 4Zn) strip, cold rolled and annealed. Globular particles of lead (black) and small, equiaxed, recrystallized grains of solid solution of tin and zinc in copper

ing of 20, mostly because of their long, stringy chips. There is some overlap due to unique machining characteristics of particular alloys.

Cast Alloys. Machinability ratings for copper casting alloys (Table 4.24) follow the same general pattern as for wrought alloys. Type I alloys are those containing sufficient lead to render them free-cutting. Type II alloys contain secondary phases that are harder or more brittle than the matrix. The silicon bronzes, several aluminum bronzes, and the high-tin alloys belong to this group. The microstructure of the manganese bronzes would qualify them as type II materials, but they produce a long spiral chip, smooth on both sides, which does not break. Some aluminum bronzes, on the other hand, produce long spiral chips that are rough on the underside and break readily, thus acting like type I alloys. The type III group is composed mainly of the high-strength manganese and aluminum bronzes with high iron or nickel contents.

Tool wear rates are one component of overall machinability ratings; to some machinists, they are the principal component. Approximate wear rates for wrought and cast-copper alloys are listed in Table 4.25.

Power requirements are often overlooked when dealing with high-strength alloys. A rough approximation of relative power required can be obtained by comparing the tensile strength of the alloy with that of C36000, because power and torque are roughly proportional to tensile strength (Ref 65).

Table 4.24 Machinability ratings of copper casting alloys

UNS No.	Alloy name	Machinability
Type I Free-cutting alloys(b)		
C83600	Leaded red brass	90
C83800	Leaded red brass	90
C84400	Leaded semi-red brass	90
C84800	Leaded semi-red brass	90
C94320	High-leaded tin bronze	90
C93700	High-leaded tin bronze	80
C93720	High-leaded tin bronze	80
C85200	Leaded yellow brass	80
C85310	Leaded yellow brass	80
C93400	High-leaded tin bronze	70
C93200	High-leaded tin bronze	70
C97300	Leaded nickel brass	70
Type II Short-chip alloys (moderately machinable)(b)		
C83500	Leaded tin bronze	60
C83520	Leaded tin bronze	60
C86500	Leaded high-strength manganese bronze	60
C63380	Silicon-aluminum bronze	50
C64200	Silicon-aluminum bronze	50
C90500	Tin bronze	50
C90300	Tin bronze	50
C95300	Aluminum bronze	35
C61800	High-strength manganese bronze	30
C67000	High-strength manganese bronze	30
...	Beryllium bronze	20 to 40
Type III Long-chip alloys (less easy to machine)(b)		
C86100	High-strength manganese bronze	20
C95200	Aluminum bronze	20
C95400	Aluminum bronze	20
C95500	Aluminum bronze	20

(a) Ratings are relative, based on free-cutting brass = 100. A material rated at 50 should be machined at roughly half the speed used for a material rated at 100 for equal tool life. (b) The alloy numbers refer to ASTM B 143, B 144, B 145, B 146, B 147, B 148, B 149, and B 198

Table 4.25 Tool wear rates for copper alloys

UNS No.	Approximate tool wear rating(a) number of tool dressings per basic production	
	Carbide tooling	HSS (standard high-speed steel) tooling
Type I Machinability rating group 70 to 100		
C36000	1	4
C35600	1	5
C18700	1½–2	6
C34500, C35300	1½	6
C14500	2	6
C14700	1½–2	6
C19100	2–3	7(b)
C31400	1½–2	5
C54400	3	8(b)
C79800	2–3	6
C33500	2	5
C34000, C35000	2	5
C48500	3	5
Type II Machinability rating group 30 to 60		
C63900	4	9(b)
C53400	4	9(b)
C79200	3	9(b)
C26000	3	6
C46400	4	8(b)
C65100	...	...
C67500	5	10
Type III Machinability rating group 20		
C10100, C10200	5	10(b)
C11000	3	6
C22000	3	6
C51000	4	8(b)
C63000	5	10(b)
C75200	...	...

Note: Approximate machinability rating based on C36000 (free-cutting brass) = 100. (a) The approximate tool wear rating can be best defined in the following example: If it is assumed that 100,000 units on any given shape and weight of C36000 (free-cutting brass) can be made per tool dressing, a similar piece made from C46400 (naval brass, uninhibited) using a carbide-tipped tool, it is expected to require four dressings for the same quantity; or the expected quantity of C46400 pieces per tool dressing = 100,000/4. Similarly, using standard high-speed steel tooling, an equivalent piece is expected to require eight dressings for the same quantity; or the expected quantity of C46400 pieces per tool dressing = 100,000/4. (b) Carbide-tipped tools or super high-speed steels were strongly recommended where practical when the tool wear rating is 7 or larger.

Turning and Form Cutting

Single Point Tools. Tool geometries for high-speed steel tools, carbide tools, and cutoff tools are given in Fig. 4.89(a and b) and 4.90 (Ref 65).

Form Tools. Recommended form tool geometries are given in Fig. 4.89(a, b). Free-cutting type materials require very little or no rake, but ductile or tough materials usually require a comparatively steep rake to prevent buildup on the face of the tool.

Speed and Feed. The cutting speeds and feed rates listed in Table 4.26 are typical for copper and copper alloys. They should be taken as guidelines

Table 4.26   Machining parameters (form and single-point tools, hollow and box mills) for copper alloys

Type of tool		Relief angles, degrees		Rake angles, degrees		Surface speed, ft/min	Roughing feed, thousandths of an inch per revolution	Finishing feed, thousandths of an inch per revolution
		Side	Front	Back	Side			
Type I: Alloy machinability rating 70 to 100	Standard high-speed steel	0–5	6	0–5	0–3	300–1000	2–15	2–3
	Carbide	4–6	4–6	0	2–6	500–1600	2–15	2–3
Type II: Alloy machinability rating 30 to 60	Standard high-speed steel	5–10	6–15	5–10	5–10	150–300	2–8	2–3
	Carbide	4–8	4–8	0–5	4–8	400–600	2–8	2–3
Type III: Alloy machinability rating 20	Standard high-speed steel	10–20	10–15	10–20	20–30	75–150	2–8	2–3
	Carbide	7–10	7–10	4–8	15–25	300–500	2–8	2–3

Type I Free-Cutting Alloys

Type II Short-Chip Alloys

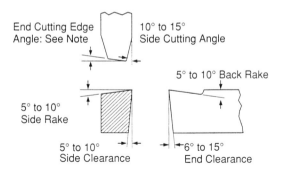

Type III Long-Chip Alloys

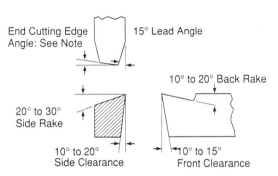

Note: A 19° to 5° cutting edge angle should prove satisfactory for most rough and finish turning operations. When the end cutting edge of a finishing tool is ground parallel with the axis, considerably heavier feeds may be employed on light finishing cuts. Tools should be ground and set so that the tool point is on center with the effective rake angles in correct relation to the center line of the work.

Fig. 4.89(a) Carbon and high-speed steel turning tools used for copper and copper alloys (Ref 65).

rather than rigid criteria. Good practice is to start at speeds and feeds about half-way between the stated limits, then adjust the cutting parameters until optimum performance is achieved. The values given are based on carbide tooling, with depths of cut between 1.15 and 3 mm (0.045 and 0.118 in.) for roughing and 0.4 and 0.75 mm (0.016 and 0.030 in.) for finishing.

Small diameter parts made from the alloys with machinability ratings between 70 and 100 are usually machined at the highest practical spindle speed, the data in Table 4.26 notwithstanding. Feeds are adjusted to suit such conditions as depth of cut, available power, coolants, and finish requirements.

Boring

Adherence to proper tool geometries is most important in boring to direct chip flow away from the cut surface and thereby achieve the smooth finished surfaces. Tools for boring copper and copper alloys are described in Table 4.27 and Fig. 4.90.

Chips should be flushed away by a copious supply of coolant.

Drilling

Where the volume of work does not call for use of special drills, standard carbon-steel or high-speed steel drills can be used to drill types I and II alloys. When ground as suggested in the drawings, they can also be used for type III alloys.

Special high-speed steel drills are frequently used with free-cutting brass, and flat and straight-flute drills with zero-degree rake angle are widely

Table 4.27 Boring with carbide tools

Classification	Back rake angle degrees	Side rake angle, degrees	Cutting speed, ft/per min
Type I: Alloys machinability rating 70–100	0	5	500–1000
Type II: Alloys machinability rating 30–60	0–5	5–10	400–600
Type III: Alloys machinability rating 20	5–10	15–20	200–500

Type I Free-Cutting Alloys

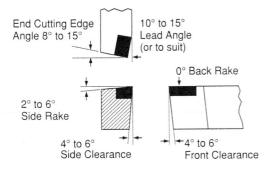

Type II Short-Chip Alloys

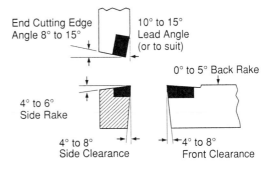

Type III Long-Chip Alloys

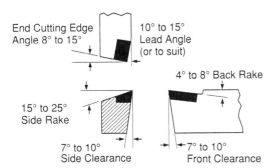

Note: Rake angles are based on the tool shank being set parallel with the center line of the work and with the tool point on center. Placing the tool point above or below center will change the effective rake angles appreciably, particularly on work of small diameter. On a set-up where the tool holder is not parallel with the center line, the rake angles should be ground so that when the tool is mounted, they are in correct relation.

Fig. 4.89(b) Carbide turning tools (Ref 65).

used for drilling Types I and II alloys. Standard drills have helix angles around 26°. Slow-spiral or brass drills have helix angles ranging from 10 to 22° as shown in Fig. 4.91. They also typically have wide, polished flutes and a thin web to facilitate chip clearance. Rake angles are between 0 and 30°. The point angle and lip clearance shown in the drawings for the various alloys can be used on all types of drills.

Hand feeding is frequently used for shop drilling, and feeds exceed those normally used for mild steel. For alloys that work-harden easily, such as annealed aluminum bronze, the drill should be cutting continuously to prevent glazing.

The drill should also be kept cutting, without interruption, as long as chips are being ejected. In holes that are deep in relation to drill diameter, the drill should be backed off occasionally for chip relief, especially when no lubricant is being used.

Many factors control the speed in drilling operations, for example, diameter of drill, wall thickness, depth of hole, but general recommendations are given in Table 4.28. The feed range suggested is for drills between 3 and 20 mm (0.125 and 0.75 in.) in diameter. Lighter feeds are used with smaller drills, for deep holes, and where it is necessary to maintain accuracy. Larger drills can take

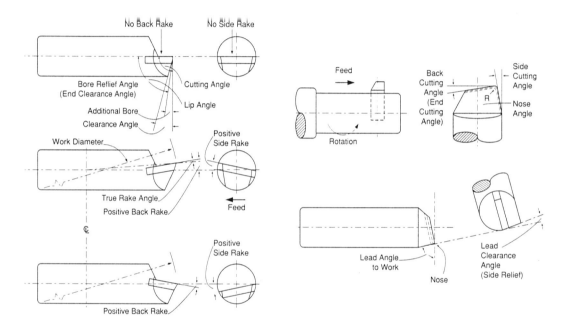

Fig. 4.90 Boring tools used for copper alloys (Ref 65)

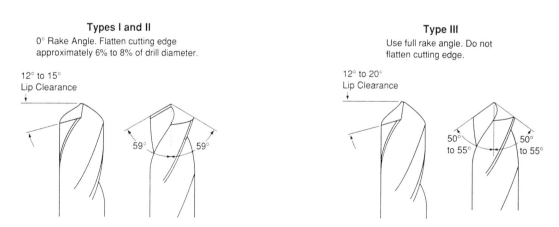

Fig. 4.91 Drill design types used for copper and copper alloys (Ref 65)

proportionately heavier feeds, especially the oil-tube type, where lubrication is supplied under pressure. If carbon-steel drills are used, speeds should be halved.

Milling

Practically all commercial types of milling cutters can be used to machine copper, brass, and bronze. Cutting angles and clearances should, however, be adjusted to suit the alloy and type of milling operation. As a general rule, the clearance behind the cutting edge should be sufficient to prevent a rubbing or burnishing action. Excessive vibrations and digging in are usually indications of too much rake or clearance, and sometimes of excessive speed.

Coarse-tooth spiral cutters with helix angles of 20 to 30 degrees and helical cutters with helix angles up to 53 degrees have a shearing action that tends to resist digging in, even with the free-cutting Type I alloys. With adequate rake and clearance, and with land width held to a minimum, these cutters produce fine finishes on all three groups of alloys, even at coarse feed and high speeds.

Staggered-tooth, side milling cutters with alternate spiral teeth are used for deep slotting operations, particularly on Type III alloys. Spiral-fluted end mills are fast cutting and produce a better finish than end mills with straight flutes.

Double angles on the back of the teeth, as shown in Fig. 4.92, are normally used for regrinding and give the cutting edges adequate clearance and strength. The clearance angle should be greater for small cutters than for large ones. The maximum clearance angles given are for cutters about 75 to 100 mm (3 to 4 in.) diameter.

The surface speeds recommended in Table 4.29 are based on the use of high-speed steel cutters with a suitable cutting fluid. For carbon-steel cutters, reduce the speed by about 50%. In many instances, these speeds may be increased several hundred percent, depending upon such variables as depth of cut, width of cutter, machine rigidity, desired finish, and rate of feed, which may vary from 0.15 to 6.0 or more meters per minute. When milling Type III alloys, some experimentation with clearance angles may be profitable (Ref 66, 67).

Reaming

Practically all standard types of hand and machine reamers can be used with copper materials. Straight-flute reamers with narrow lands and polished flutes are commonly used; but on some

Table 4.28 Recommended drilling parameters for copper metals

Classification of alloy	Speed, sm/min (sfm)	Feed, mm/rev (in./rev)
Type I: Free cutting	60–150 (200–500)	0.005–0.75 (0.0002–0.30)
Type II: Short chip	25–75 (75–250)	0.075–0.5 (0.003–0.020)
Type III: Long chip	15–40 (50–130)	0.075–0.5 (0.003–0.020)

Table 4.29 Milling data for copper metals

Alloy type	Milling speed, sm/min (sfm)
Type I alloys: Free-cutting	60–75 (200–250)
Type II alloys: Short-chip	45–60 (150–200)
Type III alloys: Long-chip	15–45 (50–150)

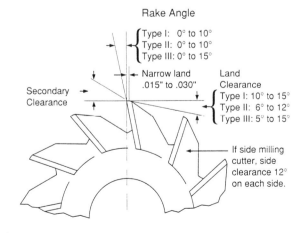
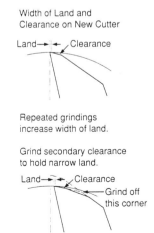

Fig. 4.92 Milling cutter used for copper and copper alloys (Ref 65)

types of work, they have a tendency to chatter. Standard spiral-flute reamers with a helix angle of between 7 and 12 degrees will overcome chatter and produce a smoother finish. Left-hand spiral and right-hand cut reamers gives excellent results either for straight or tapered holes in all three types of alloys. Depending on the diameter, length of holes, and wall thickness, high-speed steel reamers can be used over the range of speeds and feeds given in Table 4.30 (Ref 68).

Threading and Tapping

Thread chasers used with copper metals include circular, radial, and tangential types for self-opening die heads and tap chasers on collapsible taps. These are used for internal and external threading on all types of automatic turning machines and a variety of horizontal and vertical threading machines.

The rake angles and clearances indicated in Table 4.31 and Fig. 4.93 should be modified by the relation of the pitch to the diameter of the thread, the thread form, the thread fit needed, and other special considerations. Tools should be kept sharp and sharpening should be done with fine abrasives to prevent ragged or saw-toothed cutting edges. The milled type of chaser is used for alloys with a machinability rating of 70 to 100 and may be used for the other, although the hobbled type is some-

times used for tough alloys and tubular sections that have a thin wall.

Three-flute and four-flute taps are normally recommended for copper alloys under most conditions, with some preference for the four-flute taps. A greater number of flutes is advisable on sections with a thin wall and large diameter. On the other hand, the two-flute chip driver will give more chip clearance on coarse threads, where considerable metal is being removed, and will produce smoother threads.

It is important to select the proper coolant and the correct chaser or tap, depending on which copper alloy is being machined. The flow of the cutting compounds should be such that the taps or chasers are washed clean of chips after each operation. Building up of chips in the tools will not only produce rough threads but may also result in tap breakage. Alloys having a machinability rating of 20 often require a slight back taper on deep holes. Tools should be smooth, not too keen-edged and with no rake for alloys having a machinability rating between 70 and 100, but sharp and hooked for alloys having a machinability rating of 20 (Ref 65).

The cutting speeds suggested in Table 4.32 are for threads of moderate pitch; for coarse threads, speeds in the lower ranges should be used. On automatic screw machines working on alloys hav-

Table 4.30 Reaming data for copper metals

Alloy type	Speed, sm/min (sfm)	Feed, mm/rev (in./rev)
Group 1	<60 (<200)	0.2–1.0 (0.008–0.04)
Group 2	22–45 (72–150)	0.2–1.0 (0.008–0.04)
Group 3	20–30 (65–100)	0.2–1.0 (0.008–0.04)

Table 4.31 Tapping data for copper-base materials

Classification	Rake angle	Chamfer (2 or 3 threads)
Alloys having a machinability rating of 70–100	2–4°	10–30°
Alloys having a machinability rating of 30–60	5–8°	10–15°
Alloys having a machinability rating of 20	8–10°	10–15°

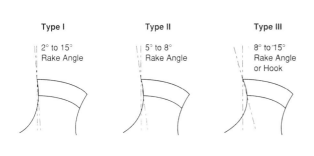

Type I — 2° to 15° Rake Angle

Type II — 5° to 8° Rake Angle

Type III — 8° to 15° Rake Angle or Hook

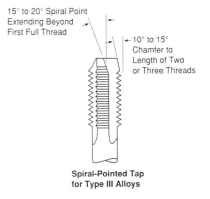

15° to 20° Spiral Point Extending Beyond First Full Thread

10° to 15° Chamfer to Length of Two or Three Threads

Spiral-Pointed Tap for Type III Alloys

Fig. 4.93 Threading tap design used for copper and copper alloys (Ref 65)

ing machinability ratings between 70 and 100, the maximum spindle speed is frequently used when threading work pieces that have small diameter or thread of fine pitch or when tapped holes are of relatively small size or with a fine pitch thread.

Sawing

Selection of the proper width of saw, number of teeth, pressure, feed, and speed is as important as the condition of the machine involved. Width of the hacksaw or bandsaw blade is usually governed by type of equipment. It is advisable that the blade be wide enough to withstand normal feeding pressures. Correct tension is vital. A loose blade will cause crooked cuts, buckling, and twisting as well as stripped teeth. Too great a tension will also cause snapping and throw too great a load on guides and the machine itself (Ref 65).

Feeding pressure should be determined by size and machinability of the alloy. On gravity- or hand-fed bandsaw machines, moderate feed is desirable on small sections or readily machined alloys. Increased pressure is recommended on heavy sections or alloys with low machinability ratings to reduce saw wear. Modern power hacksaw machines employ either mechanical or hydraulic feeds, permitting increased feed settings on small sections of soft alloys. Large sections and hard alloys require reduced feed settings.

Correct tooth specification is important to permit adequate chip clearance. Coarse teeth are desirable on soft or thick material and finer teeth are indicated on thin sections or hard alloys. Care must be taken to ensure correct set. If the set is worn to any great extent, crooked cuts and excessive heat with subsequent saw failure will result. For circular saw data, see Fig. 4.94 and Table 4.33. For band sawing of copper alloys, blades of 0.5 or 0.75 in. width are most commonly used and only for short radius cutting should narrow blades be used. The data in Table 4.34 will aid considerably in selecting a good combination of saw tooth and linear speed.

For power hacksaws, a good rule in selecting saw tooth pitch is to use fewer teeth for thick sections to provide for better chip clearance, and more teeth for thinner sections.

Grinding

Grinding copper alloys is not common, but in some applications grinding is the best means of

Table 4.32 Threading and tapping data for copper metals

Alloy type	Cutting speed, sm/min
Type I alloys	30–45
Type II alloys	15–27
Type III alloys	3–9

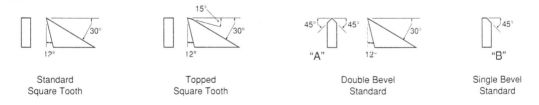

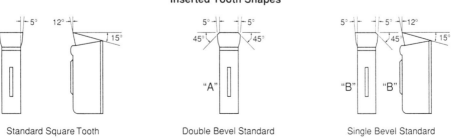

Fig. 4.94 Solid and insert tooth shapes for circular saws used for copper and copper alloys

producing the desired accuracy and finish. When finish grinding must supplement machining, one grinding operation can occasionally be used to replace both operations. Grinding speeds, feeds, and appropriate wheel types are given in Table 4.35.

Other Operations

Thread rolling is either performed in an automatic screw machine or as a separate operation in flat die rollers, depending on the shop equipment and the nature of the product. Commercial thread rolling on fine pitch threads can be done without excessive flaking or tool wear with the standard free-cutting brass; however, on coarse pitch threads, for example, 32 pitch and coarser, a rod with a light-drawn or light-annealed temper should be used. Due to the softer surface, the resistance to tool penetration is decreased with a

consequent increase in depth of penetration and filling up of the low places of the tool contour. For exceptionally coarse pitch work (16 pitch and coarser), the rod should be as soft as possible and still retain sufficient rigidity to withstand tool pressures. For precision threading, UNS C34000 (medium leaded brass), or UNS C34500 and C35300 (high leaded brasses) of suitable temper are preferred. When thread rolling is done as a separate operation in flat die rollers, the nonleaded alloys are most satisfactory and withstand the most severe rolling.

Knurling. There are two methods of producing knurls on brass parts, and differing results are produced with different tooling, alloys, and tempers. One method is by plowing from the end; the other is by butting or plunge-feeding from the side. Because leaded brass will stand only a limited

Table 4.33 Sawing with circular saws

Form and size	Machinability rating of alloys	Grade steel(a)	Tooth type(b)	Diam, in.	Thickness, in.	No. teeth	Hook angle degrees	Rim speed, fpm	Feed, ipm	Coolant (c)	Rockwell C hardness	Chromium plated 0.002 in. per side	Hollow ground taper, in.
Rod, up to 1½ in.	70–100	SHSS	SST	12	3/32	150–200	10–15	4000–8000	60	Grease stick	52–56	Yes	0.0035
	20 and 30–60	SHSS	TST	12	3/32	100–125	10–15	4000–5000	30	Grease stick	52–56	Yes	0.0035
	20, copper only	HSS	Ins. SST	12	1/8	75–100	5–10	2000–3000	20–30	Grease stick	60–62	No	0.0010
Rod 1½ to 4 in.	70–100	HSS	Ins. Alt. SST and BST "A"	16	3/16	60–64	10–15	1000	30–40	Compound	64–66	No	0.0035
	20 and 30–60	HSS	Ins. Alt. SST and BST "A"	16	3/16	60–64	10–15	1000	20–30	Compound	62–64	No	0.0010
	20, copper only	HSS	Ins. Alt. SST and BST "A"	16	1/4	60–64	5–10	750	10–20	Compound	62–64	No	0.0010
Rod, 4 to 8 in.	70–100	HSS	Ins. Alt. SST and BST "A"	18	1/4	60–80	10–15	600–750	20–30	Compound	64–66	No	0.0010
	20 and 30–60	HSS	Ins. Alt. SST and BST "A"	18	1/4	60–80	10–15	600–750	15–25	Compound	62–64	No	0.0010
	20, copper only	HSS	Ins. Alt. SST and BST "A"	18	1/4	60–80	5–10	500–650	10–15	Compound	62–64	No	0.0010

(a) SHSS is a trade designation for semi-high-speed saw in HSS for standard high-speed steel saw. (b) SST is standard square tooth. TST is topped square tooth. BST is bevel standard tooth. (c) Compound made up as water emulsion of soluble oil. Use 10% oil by volume and 1 lb heavy soap per 20 gal of compound.

Table 4.34 Band saw teeth and speeds

	Stock diameter of thickness, in.					
	1/16–1/4		1/4–1		Over 1	
Machinability rating of alloys	Teeth per in.	Velocity, ft/min	Teeth per in.	Velocity, ft/min	Teeth per in.	Velocity, ft/min
70–100	18	500–1000	10	350–450	4	250–350
30–60	18	250–350	10	230–250	4	200–230
20 (except copper)	18	250–300	10	200–250	4	150–200
Copper	18	800–1500	10	600–1000	4	300–600

amount of upsetting without flaking, it is difficult to produce an especially clean knurl from alloy C36000 by either of these methods. Because the lead limits the amount of cold work possible, a lower lead content is often used, which facilitates knurling but does not seriously affect machinability. That cautionary note notwithstanding, it should be mentioned that many brass rod suppliers, aware of this difficulty and motivated to reduce lead content for other reasons, now supply free-cutting brass with lead near the low end of the specified range.

Thus, for the majority of knurling applications, standard half-hard C36000 is entirely satisfactory. Using a front or turret knurling tool, a 70 to 80% straight or diamond knurl can be produced without excessive flaking or cracking. Where a butt or side knurling tool must be used, the knurl depth will be reduced to 40 to 60% depending on rod diameter. Because side pressures transmitted to the rod are appreciably higher with a side knurling tool, the depth of knurl is limited by rod size and sometimes a back rest is required.

For severe knurling, thread rolling, and other forming operations, alloys C34500 and C35300 may be more advantageous. Extremely severe operations may call for the use of alloy C34000. These alloys have lower lead contents than C36000. Their machinability, while very high in all cases, is slightly lower than that of free-cutting brass.

Roll lettering can be performed on standard screw machine rod. For sharp, clean letters, a somewhat softer temper is generally preferred. It is usually not necessary to change the alloy to accommodate roll lettering because the stresses imposed on the material are primarily compressive and the high-lead content does not adversely affect the letters. The temper used should be as soft as possible.

Spinning operations on rod products generally consist of forming a lip or flange on hollow screw machine product. Small flanges can be spun on standard half-hard screw machine rod. More severe deformations require softer tempers and/or alloys with reduced lead contents.

The greatest danger involved in spinning operations lies in the heat generated in the part by friction of the tool against the work. The combination of tensile stresses and high temperatures inherent in any spinning operation makes the use of a highly leaded alloy a poor risk. When this combination of stress and temperature becomes of sufficient magnitude, the leaded alloys display a brittle or "hot short" condition, resulting in a myriad of

fine cracks at the point of tool contact. In some instances, these cracks are so fine as to escape detection by the machine operator and do not become visible until the part is put into service.

In general, the alloys best suited for spinning operations are the non-leaded straight brasses and bronzes. Therefore, where a job requires a spinning operation, it is advisable to select an alloy with the lowest lead content possible which will still provide reasonable machinability for other operations involved.

Bending. In this case, the alloy's lead content is not so important a consideration as is temper. In general, half-hard C36000 will withstand a 120°

Table 4.35 Conditions for grinding copper alloys

Property	Value
Surface grinding	
Wheel classification	
Workpiece 20 to 70 R^B	C-46-K-V
Workpiece 60 to 100 R^B	A-46-K-V
Wheel speed	5000 to 6500 sfm
Work speed	100 sfm
Downfeed	
Rough	0.003 in./pass
Finish	0.0005 in./pass (max)
Crossfeed	$\frac{1}{3}$ wheel width/pass
Cylindrical grinding	
Wheel classification	
Workpiece 20 to 70 R^B	A-60-N-V
Workpiece 60 to 100 R^B	A-46-L-V
Wheel speed	5500 to 6500 sfm
Work speed	100 sfm
Infeed	
Rough	0.002 in./pass
Finish	0.0005 in./pass (max)
Traverse	
Rough	$\frac{1}{3}$ wheel width/work rev
Finish	$\frac{1}{6}$ wheel width/work rev
Centerless grinding	
Grinding wheel classification	A-60-L-V
Wheel speed	5000 to 6500 sfm
Work feed	50 ipm
Infeed	
Rough	0.005 in./pass
Finish	0.0015 in./pass (max)
Regulating wheel	
Angle	3°
Speed	30 rpm
Internal grinding	
Wheel classification	
Workpiece 20 to 70 R^B	A-46-J-V
Workpiece 60 to 100 R^B	A-60-L-V
Wheel speed	5000 to 6500 sfm
Work speed	100 to 200 sfm
Infeed	
Rough	0.002 in./pass
Finish	0.0002 in./pass (max)
Traverse	
Rough	$\frac{1}{3}$ wheel width/work rev
Finish	$\frac{1}{6}$ wheel width/work rev

cold bend around a radius equal to its radius or thickness. Provided that this radius is maintained and that no rough edges, nicks, or other imperfections are present, failure in this test is virtually unknown. However, as the radius of bend is decreased and the angular displacement increased, it becomes necessary to go to softer tempers.

Alloy composition does, of course, play its part in the ability of a material to withstand bending. For any given bend, harder tempers can be used in the higher copper, lower lead alloys. This fact is particularly applicable where a highly polished surface is desired on the outside radius of the bend. Use of soft, large-grained material may result in a rough "orange peel" surface, which is difficult and costly to polish. In such cases, it would be advantageous to select a fine grain size rod of C36000 or one of the lower leaded or higher copper alloys in a harder temper.

Staking is an operation wherein a relatively sharp pointed or edged tool is driven into the metal to form an ear or lip, which, in turn, secures an additional part. The presence of lead is detrimental in alloys used in this manner. The lead tends to produce a shearing action, rather than pure deformation, which may result in complete separation of the ear from the part. Where a leaded alloy is needed, it may be necessary to anneal the part before severe staking to ensure ample ductility. If machining is not a major consideration, the non-leaded alloys are recommended.

Flaring and Expanding. These assembly operations are generally accomplished by forcing a tapered punch into the part. Here again the choice of alloy and temper is largely dependent on the severity of the cold working performed. However, most flaring or expanding operations can be accomplished by annealing standard screw machine rod before flaring or using either UNS C34500, UNS C35300, or UNS C34000 in a light-annealed or quarter-hard temper.

Cutting Fluids and Lubricants

Copper alloys can be machined dry, but cutting fluids invariably increase the attainable speed and feed and therefore, productivity, improve surface finish, enhance accuracy, and lengthen tool life. Type I alloys are produced principally in rod form for use in automatic screw machines, and a lubricant is ordinarily used. Lubricants are also desirable when drilling Type II alloys, particularly for deep holes and where accuracy is necessary. A lubricant should always be used when drilling Type III alloys and coppers. The choice of cutting fluid depends on the type of machining operation, the workpiece material and surface finish, and tolerance requirements (Ref 69).

Soluble Oils. Water-soluble oils are the most widely used cutting fluids. They are the least expensive and are unexcelled in their ability to cool and to flush away chips. Soluble oils are nonflammable and nontoxic, and are safe to use with virtually all metals without fear of staining. The usual mixture is about 1 part oil to 20 parts water. Proportions are not critical, however, and in some cases 40 parts water may be used to dilute 1 part oil with no significant change in results. On the other hand, soluble-oil emulsions are far less effective than many other cutting fluids for promoting cutting action and preventing edge buildup. As smoothness and dimensional accuracy requirements increase, some oil or nonaqueous oil mixture will be needed.

Straight mineral oils are often used when soluble oils do not meet quality requirements, particularly when the work material is not free machining or when the finish specified exceeds the capability of soluble oil. Mineral oils with a viscosity of about 100 at 38 °C (100 °F) are most often used, but those with a viscosity of only about 40 (such as mineral seal oil) are used in many applications.

Blended cutting oils of various viscosities are readily available as proprietary compositions. Most of these are basically mineral oils, blended with sulfur compounds, animal fats, and other materials. They may be used as-purchased or cut back with mineral oils, depending on prior experience with similar jobs.

Although any straight oil is less effective than soluble oils (oil-water emulsions) for cooling and washing away chips, all oils (and particularly those containing sulfur compounds or other special additives) are more effective for improving cutting action. When chatter develops as the result of vibration or other causes, unacceptable surface finish and short tool life are inevitable. Cutting oils are more effective than soluble-oil emulsions for preventing chatter.

Special Oils and Mixtures. In applications that demand maximum performance of cutting fluids, high-viscosity thread-cutting oils or lard-oil mixtures are preferred. These special cutting fluids are especially effective for cutting threads that require smooth surfaces. Lard oil is one of the best fluids for promoting cutting action, but because of its high viscosity, it is impractical to use in high-production applications. Its most common use is in small lathes for toolroom of pilot-production applications.

Both thread-cutting oil and lard oil can be mixed with mineral oil to reduce viscosity to a practical level. These mixtures still retain some of the advantages of the undiluted oils. Neither of these special oils, however, is equal to a soluble-oil emulsion in ability to cool or to wash away chips.

Cutting Fluid Selection. Because of the many considerations that govern cutting operations, it is impracticable to make specific recommendations. The advice of reputable manufacturers of lubricants should be solicited to obtain the very best material for any specific operations. The following suggestions may serve as a guide.

Type I Alloys. For the free-cutting alloys included in this group, a light mineral or paraffin oil can be recommended for automatic and hand screw machines. It should be directed at the cutting edge of the tools. Unblended light mineral oil is particularly desirable when machining alloys of this group at high speed, light feed, and moderate depth of cut. An inexpensive cutting compound with good coolant properties consisting of 20 parts water to 1 part soluble oil may also be used.

Type II Alloys. The soluble coolant compounds may also be used for these alloys. However, a mineral oil base fluid fortified with 5 to 15% lard oil or a sulfurized fatty oil base thinned with a light mineral oil will likely prove to be more satisfactory as both lubricant and coolant. Higher concentrations of lard oil are used on those alloys, which produce tough, stringy chips.

Type III Alloys. A cutting fluid frequently used for these alloys is a mineral oil base with a 10 to 20% lard oil. For pure copper and high-nickel alloys, low-sulfur mineral lard oil compounds give excellent results.

When a sulfurized compound is used on any copper alloy, parts should be cleaned as soon as possible after machining to avoid discoloration. Tarnish can be removed by immersion in a 10% sodium cyanide solution, a bichromate dip, or a commercial bright dip.

Non-Mechanical Machining

Electrical discharge machining (EDM) is a method for producing holes, slots, or other cavities in materials by using an electric spark as the cutting "tool." The work is positively charged and the tool is the negative electrode (Fig. 4.95). The process is capable of cutting hard, refractory metals in any variety of shapes, and with exceedingly complex cuts, with very high precision (Ref 69).

The EDM method is often used to produce beryllium-copper molds and dies. The effectiveness of EDM is not dependent on the strength or hardness of the workpiece, and the beryllium copper can be cut in its age-hardened state without effect on strength and with no need for further heat treatment. This process is also used to drill small, burr-free holes and to make prototype quantities of contacts for the aerospace and electronic applications. The two major types of EDM are so-called conventional, or ram EDM, and travelling-wire (TW) EDM.

Electrochemical machining (ECM) is the controlled removal of metal by anodic dissolution in an electrolytic cell in which the workpiece is the anode and the tool the cathode. The electrolyte is pumped through the gap between the tool and the workpiece, while direct current is passed through the cell at a low voltage to dissolve metal from the workpiece at approximately 100% efficiency.

Electrochemical machining can be used to do work that would be difficult or impossible by mechanical machining. As with EDM, the work includes machining hard materials and odd-shaped, small, deep holes. The ECM process is used for operations as widely different as face milling, deburring, etching, and marking. Figure 4.96 shows a typical set-up for ECM (Ref 69).

Quality Control

Tool wear ratings in Table 4.25 are approximate. They are based on user and producer reports of shop experience and represent an overall average of a wide variety of types of product and operation. The values shown represent the comparative number of tool dressings required for a given quantity of production. Carbide-tipped tools require fewer dressings than standard high-speed steel tools. A generally outstanding characteristic of copper alloys as compared with other metals is their ability to be machined in long runs with a minimum of tool dressing.

Tool wear is, in general, affected by the factors discussed earlier, but other elements must be considered. Depth of cut is important. If 6.3 mm (0.25 in.) is being removed from a side in a turning operation, the tool cannot be expected to stand up as long as one removing only 0.8 mm (0.032 in.). More heat is generated in the first instance, and more stock is being removed. With deep cuts, it is advisable, where possible, to divide the operation between a roughing and a finishing tool.

Hardness of the cutting tool is another factor to be considered, especially where the alloy being cut is hard. Care must be taken in hardening high-speed tools to ensure the highest possible hardness without brittleness. In grinding, it is essential that the finish be excellent and that no burning has occurred to lower the hardness at the cutting edge. Rigid mounting of tools will add to their life. It is also important to set them either on center or slightly below center. However, tools for cutting tough alloys are generally set slightly above the center to compensate for springing. A tool set above center nullifies the front clearance angle. It, thereby, produces greater friction and tool wear and tends to burnish rather than to cut the material.

Finish is affected by practically every condition encountered in machining. Dull tools are the most common cause of poor finish. Better finishes, in many cases, can be obtained through the use of higher surface speeds, slower feeds, and minimum rake angles. However, the condition of the machine—play in the spindle, rigidity in the tool posts, and ways and the type of lubricant—affect the finish considerably. The rake and clearance of the tool are also important. When the front clearance is too small, a burnished or bumpy surface is formed, and increased heat, through friction, instantly reveals the source of trouble. If the finish cut is heavy, high rake and clearance angles cause hogging-in and chattering concurrent with tool vibration.

Accuracy is maintained by eliminating mechanical faults. In many cases, its loss can be attributed directly to too much heat generated through friction. Accuracy results from tool sharpness and chip control. If the tool is not cutting cleanly, the increased pressure between the work

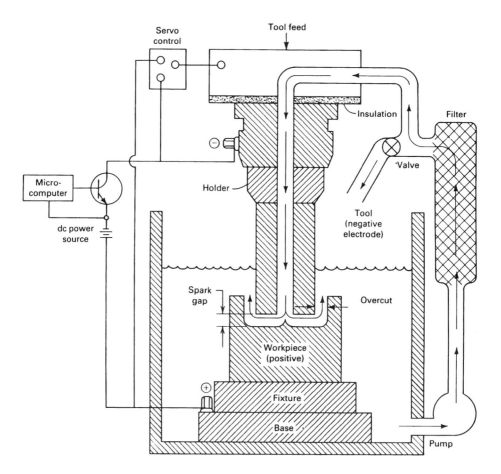

Fig. 4.95 Typical setup for electrical discharge machining. Negative (standard) polarity is shown. Positive (reverse) polarity is also extensively used

and the tool forces the work away from the tool and at the same time thrusts the carriage back. This condition normally accounts for oversized parts. Undersized dimensions are generally caused by excessive heat. Vibration in thin, long work pieces or in the tool or machine itself will lead to variation in dimensions. Lighter cuts will generally reduce this vibration. Also, rigidity of the workpiece permits better shape and accuracy in the finished part (Ref 65).

Heat Treatment

Only a few copper alloys derive their mechanical properties from heat treatment, but there are several thermal treatments applied to alloys that are not heat treated as such. Homogenizing anneals, for example, are used to soften alloys and render their compositions uniform throughout the microstructure; stress-relief anneals reduce residual stresses induced during cold working or

nonuniform solidification. Solution heat treatments or anneals drive alloying constituents into solid solution, from whence they can precipitate when needed, usually during aging treatments; martensitic transformations occur in some alloys when they are quenched from elevated temperatures. Ordering, or spinodal, decompositions have a strong influence on mechanical properties in certain copper-nickel-tin alloys; microduplex hardening is possible in some two-phase alloys.

Nomenclature and Temper Designations

Temper designations for coppers and copper alloys, therefore, require a broader-based classification scheme than one based solely on cold work. First, some alloys have much higher work-hardening rates than the coppers and brasses for which the scheme summarized in Table 4.36 was developed. Second, a temper-designation scheme based on cold reduction cannot be applied to product forms,

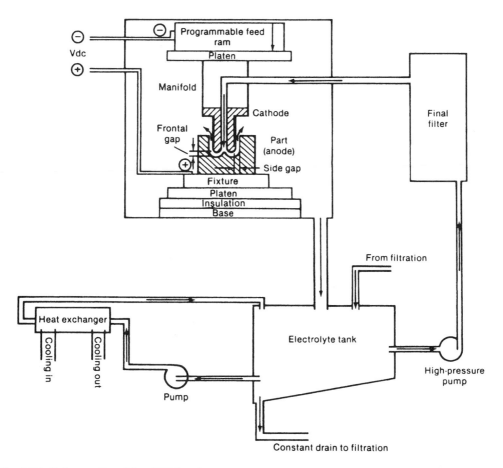

Fig. 4.96 Schematic of the ECM system

such as rod, tube, extrusions, and castings, or to heat-treatable copper alloys (Ref 53).

Similar problems exist for annealed tempers. With simple single-phase alloys, annealed tempers are generally specified on the basis of grain size of recrystallized material. This system cannot be applied to grain-size-stabilized alloys because there may be considerable variation in properties for material annealed at various temperatures up to several hundred degrees above the recrystallization temperature, even though the grain size remains quite stable. For these alloys, the terms "light annealed," "soft annealed," and "annealed to temper" are generally used instead of grain-size designations.

To clarify this complex situation, the American Society for Testing and Materials published Standard Recommended Practice ASTM B 601, "Temper Designation for Copper and Copper Alloys—Wrought and Cast," which is intended to be unambiguous and applicable to all product forms currently in widespread use. An alphanumeric code has been assigned to each of the standard descriptive temper designations (Table 4.37) (Ref 53).

Homogenizing and Preheating

Annealing at high temperatures for relatively long times reduces microsegregation (coring) in cast alloys, a process called homogenization. This process can be especially useful in products that are to be hot or cold worked. On the other hand, wrought brasses are seldom intentionally homogenized because they require repeated process annealing between reductions to finished size, and these annealing and cold-working cycles serve to homogenize the metal.

Diffusion and homogenization are slower and more difficult to accomplish in tin bronzes, silicon bronzes, and copper-nickels than in other copper alloys. Therefore, these alloys are usually subjected to prolonged homogenizing treatments before hot- or cold-working operations. With their wide solidification intervals, the high-tin-phosphor bronzes (above 8%Sn) commonly display high degrees of segregation. Although these alloys are sometimes hot worked, the usual practice is to roll them cold, making it necessary to first diffuse the brittle segregated tin phase. This serves to increase strength and ductility while decreasing brittleness. Excessive time at temperature must be avoided to avoid grain growth.

Annealing

Relation to Properties. The object of annealing is to prepare the metal for cold forming by creating an optimal combination of ductility and strength. Annealing also develops a sufficiently fine grain structure to avoid orange peeling and thereby simplify polishing. Annealing conditions are, therefore, always chosen with the type and degree of subsequent cold working in mind.

The relationship between grain size and mechanical properties required for cold forming varies considerably with the alloy and the amount and kind of cold work to be done. That notwithstanding, the recommended relationship between grain size, cold formability, and surface finish (Table 4.38) may be taken as a starting point for design planning. This table is particularly applicable to low-copper brasses such as UNS C26000 and C27000. Higher-copper alloys are more ductile and do not harden so rapidly with cold work. The 0.035 mm nominal grain size is generally acceptable for deep drawing such alloys. Temperatures commonly used for annealing cold-worked coppers and copper alloys are given in Table 4.39 (Ref 38).

Increasing the amount of cold work prior to annealing lowers the recrystallization temperature.

Table 4.36 Temper designations for wrought copper and brass based on degree of cold reduction

Nominal temper designation	Rolled sheet			Drawn wire		
	Increase in B & S gage numbers	Reduction in thickness and area, %	True strain(a)	Reduction in diameter, %	Reduction in area, %	True strain(a)
¼ hard	1	10.9	0.116	10.9	20.7	0.232
½ hard	2	20.7	0.232	20.7	37.1	0.463
¾ hard	3	29.4	0.347	29.4	50.1	0.694
Hard	4	37.1	0.463	37.1	60.5	0.926
Extra hard	6	50.1	0.696	50.1	75.1	1.39
Spring	8	60.5	0.928	60.5	84.4	1.86
Extra spring	10	68.6	1.16	68.6	90.2	2.32
Special spring	12	75.1	1.39	75.1	93.8	2.78
Super spring	14	80.3	1.62	80.3	96.1	3.25

(a) True strain equals $\ln A_0/A_1$ where A_0 is initial cross-sectional area and A is final area.

Table 4.37 ASTM B 601 temper designations for copper and copper alloys

Temper designation	Temper name or material condition	Temper designation	Temper name or material condition
Cold worked tempers		**Annealed tempers(a) (continued)**	
H00	$\frac{1}{8}$ hard	082	Annealed to temper—$\frac{1}{2}$ hard
H01	$\frac{1}{4}$ hard	**Annealed tempers(c)**	
H02	$\frac{1}{2}$ hard	OS050	Average grain size 0.005 mm
H03	$\frac{3}{4}$ hard	OS010	Average grain size 0.010 mm
H04	Hard	OS015	Average grain size 0.015 mm
H06	Extra hard	OS025	Average grain size 0.025 mm
H08	Spring	OS035	Average grain size 0.035 mm
H10	Extra spring	OS050	Average grain size 0.050 mm
H12	Special spring	OS070	Average grain size 0.070 mm
H13	Ultra spring	OS100	Average grain size 0.100 mm
H14	Super spring	OS120	Average grain size 0.120 mm
H50	Extruded and drawn	OS150	Average grain size 0.150 mm
H52	Pierced and drawn	OS200	Average grain size 0.200 mm
H55	Light drawn, light cold rolled	**Solution-treated temper**	
H58	Drawn general purpose	TB00	Solution heat treated
H60	Cold heading, forming	**Solution-treated and cold-worked tempers**	
H63	Rivet	TD00	TB00 cold worked to $\frac{1}{8}$ hard
H64	Screw	TD01	TB00 cold worked to $\frac{1}{4}$ hard
H66	Bolt	TD02	TB00 cold worked to $\frac{1}{2}$ hard
H70	Bending	TD03	TB00 cold worked to $\frac{3}{4}$ hard
H80	Hard drawn	TD04	TB00 cold worked to full hard
H85	Medium-hard drawn electrical wire	**Precipitation-hardened temper**	
H86	Hard-drawn electrical wire	TF00	TB00 and precipitation hardened
Cold worked and stress-relieved tempers		**Cold-worked and precipitation-hardened tempers**	
HR01	H01 and stress relieved	TH01	TD01 and precipitation hardened
HR02	H02 and stress relieved	TH02	TD02 and precipitation hardened
HR04	H04 and stress relieved	TH03	TD03 and precipitation hardened
HR06	H06 and stress relieved	TH04	TD04 and precipitation hardened
HR08	H08 and stress relieved	**Precipitation-hardened and cold-worked tempers**	
HR10	H10 and stress relieved	TL00	TF00 cold worked to $\frac{1}{8}$ hard
HR50	Drawn and stress relieved	TL01	TF00 cold worked to $\frac{1}{4}$ hard
Cold worked and order-strengthened tempers		TL02	TF00 cold worked to $\frac{1}{2}$ hard
HT04	H04 and order heat treated	TL04	TF00 cold worked to full hard
HT06	H06 and order heat treated	TL08	TF00 cold worked to spring
HT08	H08 and order heat treated	TL10	TF00 cold worked to extra spring
As manufactured tempers		TR01	TL01 and stress relieved
M01	As sand cast	TR02	TL02 and stress relieved
M02	As centrifugal cast	TR04	TL04 and stress relieved
M03	As plaster cast	**Mill-hardened tempers**	
M04	As pressure die cast	TM00	AM
M05	As permanent mold cast	TM01	$\frac{1}{4}$ HM
M06	As investment cast	TM02	$\frac{1}{2}$ HM
M07	As continuous cast	TM04	HM
M10	As hot forged and air cooled	TM06	XHM
M11	As hot forged and quenched	TM08	XHMS
M20	As hot rolled	**Quench-hardened tempers**	
M30	As hot extruded	TQ00	Quench hardened
M40	As hot pierced	TQ50	Quench hardened and temper annealed
M45	As hot pierced and rerolled	TQ75	Interrupted quench hardened
Annealed tempers(a)		**Tempers of welded tubing(d)**	
010	Cast and annealed(b)	WH00	Welded and drawn to $\frac{1}{8}$ hard
020	Hot forged and annealed	WH01	Welded and drawn to $\frac{1}{4}$ hard
025	Hot rolled and annealed	WM00	As welded from H00 strip
030	Hot extruded and annealed	WM01	As welded from H01 strip
040	Hot pierced and annealed	WM02	As welded from H02 strip
050	Light annealed	WM03	As welded from H03 strip
060	Soft annealed	WM04	As welded from H04 strip
061	Annealed	WM06	As welded from H06 strip
065	Drawing annealed	WM08	As welded from H08 strip
068	Deep drawing annealed	WM10	As welded from H10 strip
070	Deep soft annealed		
080	Annealed to temper—$\frac{1}{8}$ hard		
081	Annealed to temper—$\frac{1}{4}$ hard		

(continued)

Table 4.37   (continued)

Temper designation	Temper name or material condition	Temper designation	Temper name or material condition
Tempers of welded tubing(d) (continued)		Tempers of welded tubing(d) (continued)	
WM15	WM50 and stress relieved	WM50	As welded from O60 strip
WM20	WM00 and stress relieved	WO50	Welded and light annealed
WM21	WM01 and stress relieved	WR00	WM00, drawn and stress relieved
WM22	WM02 and stress relieved	WR01	WM01, drawn and stress relieved

(a) To produce specified mechanical properties. (b) Homogenization anneal. (c) To produce prescribed average grain size. (d) Tempers of fully finished tubing that has been drawn or annealed to produce specified mechanical properties or that has been annealed to produce a prescribed average grain size are commonly identified by the appropriate H, O, or OS temper designation.

The smaller the degree of prior deformation, the larger the grain size after annealing. For a fixed temperature and duration of annealing, the larger the original grain size before working is, and the larger the grain size after recrystallization will be.

Copper alloys are usually annealed at progressively lower temperatures, with intermediate cold reductions of at least 35% and as high as 50 to 60%, wherever practicable. The higher initial temperatures accelerate homogenization, and the resulting large grains permit more economical reduction during the early stages of mill reduction. Grain size should be decreased gradually in subsequent anneals until the final required grain size is approached. This practice helps maintain a uniform final grain size within a lot and from lot to lot.

Grain Size Stabilized Alloys. Several copper alloys have been developed in which the grain size is stabilized by the presence of a finely distributed second phase. Examples include copper-iron alloys, such as C19200, C19400, and C19500, and alumina-strengthened coppers, such as C15760 (Ref 53).

Stress Relief. Residual stresses induced by plastic deformation can be relieved mechanically, as by bending or roller levelling; thermally, by stress-relieving below the recrystallization temperature, or by a combination of heat and deformation. Stress relieving at high temperatures for a short time normally minimizes processing time and cost, even though the practice entails some sacrifice in mechanical properties. Using a lower temperature for a longer time provides complete stress relief with no decrease in mechanical properties, but it is more costly. Also, the hardness and strength of severely cold-worked alloys increases slightly when low stress-relieving temperatures are used.

Thermal stress-relieving promotes dimensional stability in cold formed parts and welded assemblies. Welded structures require stress-relieving

Table 4.38   Recommended grain sizes for forming operations and surface finish

Nominal grain size, mm	Typical uses
0.015	Simple forming operations
0.025	Shallow drawing, excellent polishing capability
0.035	Best combination of drawing and polishing capability
0.050	Deep drawing operations, fair polishing capability
0.070	Heavy drawing or thick gages, difficult to polish

temperatures 85 to 110 °C (150 to 200 °F) above those used for mill products.

Many copper alloys are susceptible to stress-corrosion cracking (SCC), once known as season cracking and now also referred to as environmentally assisted cracking, if they contain residual surface tensile stresses as a result of prior cold deformation. Stress-corrosion cracking can take place at room temperature in ammoniacal environments, certain amines, nitrites, and mercury and its compounds. Even higher copper alloys, such as aluminum bronzes and silicon bronzes, may crack under critical combinations of stress and specific corrodant. All copper alloys are susceptible to more rapid corrosion attack when stressed in tension. Thermal stress-relieving is frequently applied to finished parts to avoid SCC. Phosphor bronzes are relatively insensitive to SCC, and copper-nickels are essentially immune to the phenomenon. These alloys are, however, more susceptible to fire cracking than other copper-base materials.

Annealing for Specific Tensile Properties. Because mechanical properties change very rapidly with temperature in the vicinity of the recrystallization temperature, it is difficult to reliably obtain specific tensile strengths or hardnesses intermediate between heavily cold worked and fully recrystallized tempers. Batch annealing to intermediate tempers should, therefore, be specified only when accurate reproduction of mechanical properties is not critical. When batch annealing is the only available option, the best practice is to anneal completely and then generate the required properties by controlled cold working. Continuous strand annealing permits closer control of the thermal cycle, and this method, therefore, achieves more consistent (and frequently better) combinations of properties than is usual for batch annealing.

Age Hardening

A number of copper alloys can be strengthened by age hardening, also called precipitation hardening or precipitation strengthening. The process requires that a dissolved alloy species be more soluble at high temperatures than it is at low temperatures, a phenomenon known as retrograde solubility. Age hardening also requires that the approach to equilibrium, for example, the discharge of solute atoms from solid solution, be so sluggish that solute can be retained in its (dissolved) high-temperature state by simple quenching. This places the alloy in a metastable, supersaturated condition with respect to the solute element. The alloy is then reheated to an intermediate temperature sufficient to accelerate the precipitation of the solute. When done properly, this results in a fine, uniform dispersion of precipitate particles. It is these particles that strengthen the alloy, often considerably.

The most important age-hardening copper alloys are the beryllium coppers, chromium coppers, and zirconium copper.

Wrought beryllium coppers can develop a wide range of mechanical properties depending on heat treatment, the degree of cold work, and whether the cold work, is imparted before or after aging. Cast beryllium coppers are not cold worked, and mechanical properties depend solely on solution treating and aging conditions (Ref 53).

Solution Treating. Wrought beryllium copper mill products are supplied either in the solution-treated or solution-treated and cold-worked conditions. Solution-treated material can be fabricated directly into parts, as can cold-worked stock, although the latter requires greater care. Fabricators normally do not solution anneal beryllium coppers unless an anneal is needed to fulfill a special requirement, such as softening a semifinished product for additional forming, or to salvage incorrectly aged parts.

Solution treating must be carefully controlled to produce the desired grain size, dimensional tolerances, and mechanical properties. Surface oxidation must also be avoided. Temperature limits must be observed if optimal properties are to be obtained during subsequent aging. Overheating causes grain coarsening in wrought material; the

Table 4.39 Annealing temperatures for cold-worked copper metals

UNS No.	Alloy type	Annealing temperature, °C
Wrought coppers		
C11000	Electrolytic tough pitch	250–650
C12000	Deoxidized, low residual P	325–650
C12200	Deoxidized, high residual P	375 650
C10200, C14500, C18700	Oxygen-free; free-machining	425–650
C11300–C11600, C12700–C13000	Silver-bearing copper	400–475
Wrought copper alloys		
C21000, C22000	Gilding; commercial bronze	425–800
C22600, C26000, C60600	Jewelry bronze; cartridge brass; aluminum bronze, 5% Al	425–750
C23000	Red brass	425–725
C24000, C27000, C35300	Low brass; yellow brass; high-leaded brass	425–700
C31400–C35600, C37000	Leaded brasses; free-cutting Muntz metal	425–650
C28000, C36500–C38500, C44300–C48500, C66700, C67400, C67500, C68700	Muntz metal; leaded high-zinc brasses; admiralty metals; naval brasses; manganese bronzes; arsenical aluminum brass	425–600
C51000–C54400, C65100	Phosphor bronzes (except low Sn alloys); low silicon bronze B	475–675
C50500	Phosphor bronze, 1.25% E	475–650
C71500	Copper nickel, 30%	650–815
C70600, C74500–C78200	Copper nickel, 10%, nickel silvers	600–815
C65500	High silicon bronze A	475–700
C61300, C61400	Alpha aluminum bronze	815
C63800	Aluminum-silicon bronze	<650
C63000, C64200	Aluminum bronze	575–650
C17000–C17600	Beryllium coppers	775–1050(a)

(a) Solution-treating temperature

result is impaired formability. In castings, over-heating results in brittleness and the inability to respond fully to precipitation hardening. Too low solution temperatures result in insufficient solution of the beryllium-rich phase, which, in turn, results in lower hardness after aging.

In wrought beryllium coppers, solution annealing concurrently removes the effects of cold work, thereby permitting additional forming. Some grain growth invariably occurs because the solution-treating temperature is above the recrystallization temperature. To minimize such grain growth, time at temperature should be kept as short as possible, consistent with metallurgical requirements.

As-cast structures usually exhibit a large degree of microsegregation, and it is recommended that castings be homogenized for minimum of three hours.

Water quenching is the most common method for retaining the solution-annealed condition in both wrought and cast products. If the shape of the product makes it susceptible to cracking, oil or forced air quenching may be used instead. Slower cooling rates may, however, result in lower properties after precipitation.

Reducing furnace atmospheres are required for the solution annealing of beryllium coppers. Otherwise, a continuous and tenacious oxide surface layer will form on alloys with high-beryllium contents. Low-beryllium alloys form a loosely adhering scale, but these alloys are susceptible to internal oxidation.

The oxide layer on high-beryllium (high-strength, or "gold") alloys does not significantly affect the mechanical properties of the precipitation-hardened material, but it is abrasive and causes severe wear on tools and dies. In low-beryllium ("red") alloys, internal oxidation lowers strength by reducing the effective section thickness of the material. Oxide of both types of alloys can be removed by chemical or abrasive cleaning methods.

Aging. Cold-working solution annealed beryllium copper raises the strength attained after aging. The highest response to aging occurs in material that has been cold rolled to at least a TD04 (hard) temper. There is little advantage to work hardening beyond the TD04 temper because formability becomes undesirably poor and control of the aging treatment requires exceptional precision. Despite these difficulties, wire is occasionally drawn to higher levels of cold work prior to aging.

Beryllium coppers are aged at temperatures between 315 and 370 °C (600 and 700 °F). Aging may be conducted at any temperature within the normal aging range, although temperature influences the time required to reach maximum strength. High temperatures promote rapid aging, but they can reduce maximum attainable strengths compared with optimum treatments. Temperature should be controlled within ±5 °C (±10 °F).

The effect of grain size on properties is less significant for beryllium coppers than it is for solid-solution alloys such as brasses. The relatively high temperatures required for solution treating of beryllium coppers usually override the effects of cold work and time at temperature. Low solution-treating temperatures result in fine grain size, but if the temperature is too low to completely dissolve the beryllium-rich phase, response to aging is adversely affected and the benefits obtained from fine grain size are nullified. For this reason, grain sizes below about 0.015 mm are not practical for most beryllium-copper products. With normal commercial practice, grain size of solution treated material usually ranges from about 0.015 to 0.060 mm.

Copper-Nickel-Phosphorus Alloys. Alloys such as C19000, which contain about 1% nickel and 0.25% phosphorus, are used for a wide variety of small, high-strength products, such as springs, clips, electrical connectors, and fasteners. Alloy C19000 is solution-treated at 700 to 800 °C (1300 to 1450 °F). A reducing or neutral atmosphere should be used, especially when heating thin sections, to prevent internal oxidation. Water quenching is the preferred cooling method, although individually handled small parts may be air cooled. Such alloys are aged at 425 to 475 °C (800 to 900 °F) for 1 to 3 h. Material that must be softened between cold-working steps before aging can be satisfactorily annealed at temperatures as low as 620 °C (1150 °F). Rapid cooling from the annealing temperature is not necessary.

Chromium coppers, such as C18200, C18400, and C18500, are solution-treated at 950 to 1010 °C (1750 to 1850 °F) and rapidly quenched. Solution treating is usually done in molten salt, but controlled-atmosphere furnaces can also be used. Surface scaling and internal oxidation must be avoided. Solution-treated chromium copper is soft and ductile, and it can be cold worked like unalloyed copper. Chromium copper is aged between 400 and 500 °C (750 and 930 °F) (preferably 455 °C (850 °F)) for 4 or more hours to produce the desired mechanical and physical properties.

Zirconium copper, C15000 (99.8Cu-0.2Zr), is solution-treated at 900 to 925 °C (1650 to 1700 °F), then water quenched. Time at temperature should be minimized to limit grain growth and possible internal oxidation of zirconium. Because solution and diffusion of zirconium proceed rapidly at the solution-treating temperature, holding at temperature is not required.

High strength in zirconium copper depends primarily on cold work. Although aging results in some added strength, its chief effect is to increase electrical conductivity. Aging is conducted at 500 to 550 °C (930 to 1020 °F) for between 1 and 4 h. If the material has been cold worked following solution treating, the aging temperature may be reduced to 375 to 475 °C (700 to 900 °F). Maximum mechanical properties and resistance to softening are developed when all of the zirconium is dissolved during solution treating. If material containing 0.15% zirconium or more is heated above 975 °C (1790 °F), the Cu_3Zr phase will begin to melt. A slight amount of melting will not affect mechanical properties, but excessive melting degrades ductility.

Normally, when the solution temperature is increased above 900 °C (1600 °F), the aging temperature should also be increased. This practice maintains high electrical conductivity. The aging treatments listed in Table 4.40 produce the best combination of mechanical properties and electrical conductivity.

Aluminum Bronzes. The heat treatment of aluminum bronze depends strongly on composition.

Annealing and Ordering Treatments. Single-phase (alpha) aluminum bronzes that contain less than about 9.5% Al can only be strengthened by cold working and are not heat treated other than conventional annealing for softening at 425 to 760 °C (800 to 1400 °F). However, alloys in which the alpha phase is nearly saturated with a third alloying element can undergo an ordering reaction when they are highly cold worked and annealed at 150 to 400 °C (300 to 750 °F). Aluminum bronzes C61500 and C63800, as well as aluminum brass C69000 exhibit this behavior. Strengthening is attributed to short-range ordering of the solute atoms within the copper matrix, which greatly impedes the motion of dislocations through the crystals. The ordering anneal also acts as a stress-relieving treatment, and order-annealed alloys exhibit improved stress relaxation characteristics.

Transformation hardening (martensitic quenching) is observed in two-phase aluminum bronzes containing between 9.5 and 14% Al. At elevated temperatures, the structure fully transforms to beta. Rapid quenching produces a metastable, ordered, close-packed-hexagonal phase referred to as martensitic beta. Like martensite in steel, this phase is hard and brittle.

Heat-treatable aluminum bronzes are hardened by oil or water quenching from temperatures between 760 and 925 °C (1400 and 1700 °F). The alloys are then tempered between 425 and 650 °C (800 and 1200 °F), depending on composition and desired properties, to stabilize the structure and restore ductility and toughness. For example, tempering for 2 h at 595 to 650 °C (1100 to 1200 °F) causes reprecipitation of fine acicular alpha in the tempered beta-martensite structure, softening the structure somewhat and increasing ductility and toughness. Rapid cooling from the tempering temperature is advisable to avoid transformation of tempered beta to an embrittling eutectoid structure.

Nickel-aluminum bronzes respond to quench-and-temper treatments in a similar manner. An additional phase, ordered bcc kappa, generates structures resembling coarse pearlite in the alpha crystals in addition to stabilizing the quenched beta. Nickel-bearing alloys, such as C95500 and C63000, quench to a higher hardness than nickel-free aluminum bronzes. These alloys are more susceptible to quench cracking in heavy or complex sections, and oil quenching is usually advisable.

Cast two-phase aluminum bronzes can be normalized by heating to 815 °C (1500 °F), furnace cooling to about 550 °C (1020 °F), and cooling in air to room temperature. This treatment produces uniform hardness and improves machinability.

Process Control. Temperature variations of ±10 °C (±20 °F) during the heat treatment of aluminum bronzes do not materially affect final properties. Excessively high annealing temperatures increase grain size and thus decrease strength. For single-phase alloys, the critical annealing temperature is about 650 °C (1200 °F).

Protective atmospheres are not generally required when heat treating aluminum bronzes because these alloys form protective, aluminum oxide-rich surface films that retard massive oxidation.

Spinodal Alloys. Spinodal structures can be formed in alloy systems that exhibit a monotectoid miscibility gap. Spinodal structures are formed when the alloy is homogenized at a temperature

Table 4.40 Aging treatments for zirconium copper

Condition	Aging treatment
Solution treated at 900 °C (1600 °F)	3 h at 500 °C (930 °F)
Solution treated at 900 °C and cold worked	3 h at 400 °C (750 °F)
Solution treated at 975 °C (1790 °F)	3 h at 550 °C (1020 °F)
Solution treated at 975 °C and cold worked	3 h at 450 °C (840 °F)

above the monotectoid for a sufficient time such that only statistical variations in composition exist. If the alloy is then cooled rapidly to within the miscibility gap and held at that temperature, or slowly cooled through the gap, spinodal decomposition proceeds at a rate controlled by the diffusion rates of the two metals. Spinodal decomposition can function as a hardening mechanism, as in some copper-nickel-tin and copper-nickel-chromium alloys.

Microduplex hardening is applicable to alloys in which both of the two existing phases are stable or metastable to slightly elevated temperatures. The microduplexing process requires that the thermomechanical history of the alloy be closely controlled so that there is an optimum intermingling and dispersion of the two phases. The resultant structure is very fine (grain size usually less than 0.010 mm, 0.0004 in.). This produces a marked improvement in strength with practically no change in elongation.

Shape memory alloys (SMA) exhibit the ability to reverse the effects of plastic deformation by means of a reversible, thermally activated, martensitic transformation. Shape memory can be induced by first cooling the alloy from the beta range to below the temperature at which martensite forms, M_s. The martensite in low-aluminum SMAs is quite ductile and such alloys can be deformed readily. Upon reheating to a temperature above the M_s, the metal reverts to, "remembers," the shape the parent beta phase held before the martensite formed. In most known shape memory systems, thermally activated "memory" deformation takes place only upon heating. Copper-base SMAs are capable of reversible deformations on the order of 4 to 5%.

One of the several "training" processes used to induce reversible, or two-way memory involves subjecting the metal to a number of stress cycles at a temperature just above the beta-reversion temperature, the so-called A_f point. Another procedure requires thermal cycling through the martensitic transformation range, for example, between the M_s and M_f temperatures, under a constant stress.

Furnace Types and Methods of Heating

Although basic furnace design is similar for all copper alloys, consideration must be given to the annealing temperature range and method of cooling. Solid solution alloys that are not age-hardenable are usually annealed at temperatures below 760 °C (1400 °F) and may be cooled at any convenient rate. Heating and cooling rates for annealing single-phase alloys are likewise not critical; however, age-hardenable alloys are solution treated at temperatures up to 1038 °C (1900 °F) and require immediate water quenching. Complex alloys such as high-strength brasses (manganese bronzes) and aluminum bronzes may also require rapid cooling to suppress the formation of detrimental microstructural phases. Likewise, iron-containing alloys are sometimes rapidly cooled to retain iron in solution to suppress traces of ferromagnetism. The choice of furnace, therefore, depends in part on the types of alloys to be treated (Ref 70).

Batch-type atmosphere furnaces may be heated electrically, by oil, or by gas. When nonexplosive atmospheres are used, electrically-heated furnaces permit the introduction of the atmosphere directly into the work chamber. Furnaces heated by gas or oil and employing protective atmospheres sometimes use a muffle to contain the atmosphere and protect the work from the direct fire of the burners.

When protective atmospheres are used during annealing, the work must be cooled in the same atmosphere to near room temperature to prevent surface scale or discoloration. If some degree of surface oxidation and discoloration can be tolerated, direct-natural-gas-fired furnaces may be used. The products of combustion from the gas-air burners are controlled to yield reducing combustion products similar in composition to manufactured protective atmospheres. Traces of sulfur in carbonaceous fuels discolors copper metals, and parts annealed in combustion-gas reducing atmosphere require cleaning to restore their luster.

Continuous Atmosphere Furnaces. Versatile continuous controlled atmosphere furnaces are used to solution anneal a wide variety of products. Such furnaces commonly incorporate a vestibule to seal off the furnace chamber itself. In some instances, work can be preheated in the vestibule. A complementary chamber may be fitted to the exit end of the furnace. The cooling chamber may be either a long tunnel through which cool protective atmosphere is circulated or it may be a water (or other medium) quench zone supplied with a protective atmosphere.

Products such as stampings, machined shapes, castings, and small assemblies pass through the furnace on an endless belt or conveyor chain. Long sections, such as tubing, bar, and flat products, or heavy sections that permit stacking on trays may be conveyed on a roller hearth. In rolling mill and wire drawing operations, the product is uncoiled at the entrance of the furnace and pulled through the furnace by equipment at the exit end; thus, there are no moving parts within the furnace.

Other Furnaces and Heating Methods. Other heat treating furnaces include continuous furnaces with forced convection used for the annealing of copper and copper alloy strip and bell-type furnaces with bucket elevators for the vacuum annealing of drawn or rolled copper stock (Ref 70).

In field plumbing work, simple torches can be used to anneal hard copper tubes before bending. In this case, the deformation zone should be heated to dull red with a soft flame while constantly waving the burner. Forming should not be performed hot so as to avoid cracking, which sometimes occurs between 300 and 600 °C (570 and 1100 °F).

Salt baths can reduce total furnace time as much as 30%, compared with that required with atmosphere furnaces. Salt baths are particularly valuable when the age-hardening time is of short duration and precise control of time at the aging temperature is required.

Molten neutral salts can be used to anneal, stress relieve, solution-heat treat, and age copper alloys. The composition of the salt mixture depends on the temperature range required. For heating between 700 and 871 °C (1290 and 1600 °F), mixtures of sodium chloride and potassium chloride are commonly used. Various mixtures of barium chloride with chlorides of sodium and potassium are also used; these cover the wider temperature range from 593 to over 1093 °C, (1110 to over 2010 °F). The latter mixtures are compatible with each other. This is important because they are commonly used in multiple-furnace operations, when it is advantageous to preheat the work in one mixture at a low temperature and then transfer the work to a high-temperature bath. The least common of the neutral salts are mixtures of calcium chloride, sodium chloride, and barium chloride. Such mixtures have an operating temperature range between 538 and 871 °C (1000 and 1600 °F) but are usually employed between 538 and 649 °C (1000 and 1200 °F). Sodium chloride-carbonate mixtures are used between 593 and 927 °C (1100 and 1700 °F), primarily for annealing. For operating temperatures below 538 °C (1000 °F), the only practical mixtures are nitrate-nitrite salts.

Furnaces for Aging and Stress Relieving. Aging and stress-relieving operations require furnace equipment that can be controlled to within 2 °C (4 °F) throughout the work zone. Controlled-atmosphere or vacuum furnaces may be used, but these are not needed if parts can be cleaned after heat treatment.

Because of the need for close temperature control, forced-convection and salt bath furnaces are commonly used for aging and stress-relieving copper alloys. Commercially available nitrate-nitrite salt mixtures (40 to 50% sodium nitrate, remainder sodium or potassium nitrite) that melt at 143 °C (289 °F) are used for aging and stress relieving. Forced-convection furnaces are generally of the box, bell, or pit types, and are fitted with fans to circulate the atmosphere uniformly over the work. Forced-convection furnaces fired by gas or oil or operating with protective atmospheres normally incorporate a muffle to prevent the infiltration of air (Ref 70).

Protective Atmospheres

Above 700 °C (1290 °F). Selection of a protective atmosphere for heat-treating copper and copper alloys is influenced by the temperature used in the heat-treating process. For temperatures above 700 °C (1290 °F), an exothermic atmosphere is the least expensive protective atmosphere for use with copper alloys. Typically, the furnace air-gas ratio is adjusted to produce a combustion product containing 2 to 7% hydrogen for use in muffle furnaces operating between 700 and 1000 °C (1290 and 1832 °F). This atmosphere is used to solution-anneal alloys such as beryllium copper, chromium copper, zirconium copper, and copper-nickel-silicon.

Combustion gases are usually dried with a water-cooled surface cooler to maintain the water/hydrogen ratio at a reducing level throughout the heating and cooling cycle. It may be necessary to lower the dew point further by refrigerating the gas. If the furnace atmosphere is not sufficiently reducing, or if the muffle leaks air, internal oxidation of the alloying elements below the surface of the metal results. This so-called "subscale" formation can occur rapidly above 843 °C (1550 °F) if the atmosphere becomes oxidizing.

Dissociated ammonia is used primarily for annealing and brazing operations. The gas is very flammable, and can explode if air enters the furnace when it is hot or if the furnace is improperly purged before reaching its set temperature. Dissociated ammonia can be partially or completely burned with air to reduce its cost and flammability. The hydrogen content can be controlled within the range between 1 and 24%; the remainder of the gas then being nitrogen saturated with water vapor. Water must be removed to maintain a reducing atmosphere.

Hydrogen easily reduces copper oxide at elevated temperatures, and hydrogen atmospheres are recommended for bright annealing and brazing. Commercial-purity hydrogen contains about 0.2% oxygen, which should be removed because it can cause internal oxidation of reactive alloying elements.

Below 700 °C (1300 °F). Combustion gas (a lean exothermic atmosphere) is the most widely used protective atmosphere for annealing copper materials. Because of its low sulfur content, natural gas is the preferred fuel. The air/gas ratio is adjusted to produce a hydrogen content of 0.5 to 1%, and the combustion gas is dried before entering the furnace to prevent discoloration and staining of the metal by water vapor during the cooling cycle.

Steam is the cheapest atmosphere for protecting copper alloys during annealing. Although the annealed metal will not remain as bright as when heated in a clean flue gas atmosphere, it is satisfactory for many applications. For products such as tightly wound coils of strip, steam can be used during the heating cycle and combustion gas during cooling.

Inert gases, dissociated ammonia burned with air, and vacuum are more expensive than ordinary furnace atmospheres and are not commonly used for annealing copper alloys. Vacuum suffers an additional disadvantage in that heating and cooling are restricted to slow radiative heat transfer.

Hydrogen Embrittlement. When annealing copper that contains oxygen, hydrogen in the furnace atmosphere must be kept to a minimum to avoid embrittlement. For temperatures lower than about 400 °C (750 °F), the hydrogen content of the furnace atmosphere should approach zero.

Sulfur Stains. Sulfur, from carbonaceous fuels or residual lubricant films, will form red stain on yellow brass and reddish-brown stains on copper-rich alloys. It is, therefore, advisable to utilize low sulfur fuels such as natural gas, and to clean parts thoroughly before annealing.

Fire Cracking

Fire cracking occurs in some alloys when parts containing residual stresses are heated too rapidly. Leaded alloys are particularly susceptible to fire cracking. The remedy is to heat slowly until the stresses are relieved. Mechanical stress relief, such as by roller leveling or straightening, can aid considerably in preventing fire cracking. Hot short

materials such as leaded brasses should also be heated slowly and uniformly to avoid cracking due to excessive thermal stresses.

Sampling and Testing

Suitable annealing time/temperature ranges can be ascertained through the use of test samples, which are usually designed to represent expected temperature extremes in the furnace charge. For simple single-phase alloys, the best and most accurate measure of the extent of annealing in such samples is average grain size; however, grain size determination requires special equipment that may not always be available. In that case, Rockwell hardness can be used to approximate grain size. ASTM B 96 correlates Rockwell hardness with grain size for many copper alloys (Ref 71).

Joining Methods

Design and Preparation of Joints

Welded Joints. The most important properties controlling the weldability of copper and copper alloys are their high thermal conductivity and thermal expansion coefficient, their high gas solubility in the liquid state, and low gas solubility in the solid state. Weld quality also depends upon the type of copper, if oxygen-containing or OF, or alloy. The welding method used also has a significant effect.

Design. Joints for arc welding of copper and copper alloys do not differ greatly from those used for steel. Sections up to 3 mm (0.12 in.) thick can be joined using square groove welds without root openings. Thicker sections are ordinarily joined using either single or double-V-groove welds with root openings not more than 3 mm (0.12 in.) wide as shown in Fig. 4.97 (Ref 72).

Joint design and fixturing should be such as to minimize constraint and contraction stresses and to make allowance for the high coefficient of expansion of copper and copper alloys. This is necessary to prevent cracking, which may result from hot shortness of the base metal at temperatures close to and above the solidus line.

Backing strips or rings are used more extensively on copper alloys than on steel to avoid loss of molten metal, particularly on the highly fluid coppers and high-copper alloys. Backing strips and rings are usually made either of the alloy that is being welded, or of copper, carbon, or graphite.

Welding Position. Because of the high fluidity of copper and most copper alloys, the flat or horizontal position is used whenever possible. The horizontal position is also used for fillet welding of corners and T-joints. Vertical and overhead positions, as well as horizontally welded butt joints, are ordinarily restricted to gas tungsten-arc welding and gas metal-arc welding of aluminum bronzes, silicon bronzes, phosphor bronzes, and copper nickels. Small-diameter electrodes and filler wire and low welding currents are used for out-of-position welding.

Surface Condition. Grease and oxides should be removed from work surfaces before welding. Wire brushing or bright dipping are acceptable cleaning methods. Mill scale on aluminum bronzes and silicon bronzes should be removed for a distance of at least 12.7 mm (0.5 in.) from the weld zone, usually by mechanical means. Copper-nickels, in particular, can be embrittled by surface contaminants. Mill scale on copper-nickels must be removed by grinding or pickling because wire brushing is not effective.

Welding, Brazing, and Soldering Processes

Process Selection. Table 4.41 offers a qualitative overview of the copper alloys' suitability for joining methods. For resistance welding, the weldability of the work metals, along with their mechanical and physical properties, often determine which process can or should be used. For example, some copper-base metals can be spot-welded but not projection-welded because of low compressive strength of the projections at elevated temperature. When welding dissimilar metals, heat balance can influence the choice of process (Ref 73).

Size of the workpiece is also important. Spot and seam welds can be made in work metal as thin as 0.025 mm (0.001 in.), while spot welding of

metal as thick as 3.2 mm (0.125 in.) has been reported for copper alloys. Projection welding is best suited to work thicker than 0.5 mm (0.020 in.). The use of projection welding frequently can increase the quality of joints in high-conductivity alloys, because welding current can be concentrated where needed. Distortion and electrode pickup are minimized because the electrode contacts a large area of the work metal. Projection welding may be preferred when the components are self-locating, or when it can simplify fixturing or to improve dimensional accuracy.

The function of the joint should also be considered. Lap joints that must be liquid-tight are usually made most efficiently by seam welding. However, if a seam does not require leak tightness, spot welding may be preferred.

Resistance welding processes are classified according to the physical configuration of the joint, for example, butt, spot, seam, and projection or stud welding. All are applicable to copper metals, although not all copper metals can be joined by these methods. Butt resistance welding is illustrated schematically in Fig. 4.98; butt seam welding in Fig. 4.99; spot welding is described in Fig. 4.100, and projection welding is shown in Fig. 4.101. The lap seam welding of a pair of copper-nickel-manganese bellows is shown in Fig. 4.102 (Ref 74).

Pure copper has low electrical resistance and high thermal conductivity, and because of this, it requires welding currents that are two to three times as high as those needed for steel. Copper alloys have higher electrical resistivities than pure copper and are consequently somewhat easier to weld (Ref 72).

Spot welding is frequently used to join copper wire in electrical or electronic applications. Ca-

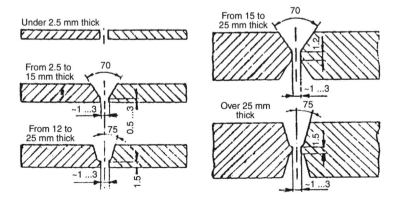

Fig. 4.97 Groove dimensions for copper alloy welding

Table 4.41 Welding and joining characteristics of wrought coppers and copper alloys

Material				Joining process										
	Designation			Soft solder	Braze	Oxyacet weld	GTAW	SMAW	Res. butt weld	Res. seam weld	Res. spot weld	Electronic beam weld	Cold pressure weld	Friction weld
Type	ISO	BSI	UNS											
Copper	Cu-ETP	C101	C11000	1	X	X	3	3	2	X	X	3	2	3
	Cu-FRHC	C102	C11020	1	X	X	3	3	2	X	X	3	2	3
	Cu-FRTP	C104	C12500	1	X	X	3	3	2	X	X	3	2	3
		C105	...	1	X	X	3	3	...	...	...	3	2	3
	Cu-DHP	C106	C12200	1	2	2	1	1	2	X	X	2	2	3
		C107	...	1	2	2	1	1	...	...	...	2	2	3
	Cu-OF	C103	C10200	1	3	X	X	X	2	X	X	1	2	3
	Cu-OFE	C110	C10100	1	3	X	X	X	2	...	X	1	2	3
Low alloyed coppers	CuS	C111	C14700	1	X	X	X	X	2	X	X	X	3	3
	CuTe	C109	C14520	1	X	X	X	X	3	X	X	X	3	3
	CuAg	...	...	1	X	3	2	2	3	3	3	2	2	3
	CuCd	C108	C16200	1	X	X	X	X	2	X	X	X	2	3
	CuBe1.7CoNi	CB101	C17200	2	X	X	3	3	2	2	2	X	X	3
	CuBe...	...	C18000	2	X	X	3	3	...	3	...	X	X	3
	CrCr1	CC101	C17500	2	X	X	3	3	3	2	X	X	X	3
	CuCr1Zr	CC102	C18100	2	X	X	3	3	2	X	X	X	X	3
	CuNi2Si	...	C64700	2	X	2	2	2	2	2	2	3	3	3
	CuNiP	...	C19000	2	X	2	2	2	2	3	3	X	3	3
	CuSi3Mn1	CS101	C65500	2	X	2	2	2	1	1	1	3	3	3
Phosphor bronzes	CuSn2	...	C50500	1	X	3	2	2	1	X	X	3	3	2
	CuSn4	PB101	C51100	1	X	3	2	2	1	3	2	3	3	2
	CuSn5	PB102	C51000	1	X	3	2	2	1	3	2	3	3	2
	CuSn6	PB103	...	1	X	3	2	2	1	1	1	3	3	2
	CuSn8	PB104	C52100	1	X	3	2	2	1	3	2	3	3	2
	CuSn10	...	C52400	1	X	X	3	3	1	3	2	3	3	2
Aluminum bronzes	CuAl5	CA101	C60800	2	X	X	2	2	2	2	2	3	X	3
	CuAl7	CA102	C61000	2	X	X	2	2	2	2	2	3	X	3
	CuAl7Fe3Sn	...	C61300	2	X	X	2	2	2	2	2	3	X	3
	CuAl12	...	C62400	2	X	X	2	2	3	2	3	3	X	3
	CuAl8	...	C61000	2	X	X	2	2	2	2	2	3	X	3
	CuAl8Fe3	CA106	C61400	2	X	X	2	2	2	2	2	3	X	3
	CuAl9Ni3Fe2	...	...	2	X	X	2	2	2	2	2	3	X	3
	CuAl9Mn2	...	C63200	2	X	X	2	2	2	2	2	3	X	3
	CuAl10Fe3	CA103	...	2	X	X	2	2	2	2	2	3	X	3
	CuAl9Ni6Fe3	CA105	C63000	2	X	X	2	2	2	2	2	3	X	3
	CuAl10Ni5Fe4	CA104	C63200	2	X	X	2	2	2	3	2	3	X	3
Cupro-nickels	CuNi5Fe1Mn	...	C70400	1	X	2	2	2	2	2	2	2	3	2
	CuNi10Fe1Mn	CN102	C70600	1	X	X	2	2	2	1	1	2	3	2
	CuNi20	CN104	C71000	1	X	3	2	2	1	1	1	2	3	2
	CuNi20Mn1Fe	...	C71000	1	X	3	2	2	1	1	1	2	3	2
	CuNi25	CN105	...	1	X	3	2	2	2	2	2	2	3	2
	CuNi30Mn1Fe	CN107	C71500	1	X	2	2	2	1	1	1	2	3	2
	CuNi44Mn1	...	...	1	X	...	...	...	2	2	2	2	3	2
Nickel silvers	CuNi10Zn27	NS103	...	1	X	2	3	3	2	2	2	X	3	3
	CuNi12Zn24	NS104	...	1	X	2	3	3	2	2	2	X	3	3
	CuNi15Zn21	NS105	...	1	X	2	3	3	2	2	2	X	3	3
	CuNi18Zn27	NS107	C77000	2	X	2	3	3	2	3	2	X	3	3
	CuNi18Zn20	NS106	C75200	1	X	2	3	3	2	3	2	X	3	3
	CuNi20Zn18	NS108	C73500	...	...	...	...	...	...	...	...	...	...	...
	CuNi25Zn18	NS109	C76390	...	...	...	...	...	...	...	...	...	...	...
	CuNi10Zn42Pb2	NS101	...	2	X	3	3	3	3	3	3	X	3	3
	CuNi14Zn44Pb2	NS102	...	...	...	...	...	...	...	...	...	...	...	...
	CuNi15Zn22Pb1	NA112	C78200	...	...	...	...	...	...	...	...	...	...	...
	CuNi18Zn19Pb1	NA113	...	2	X	3	3	3	3	3	3	X	3	3
Brasses	CuZn5	CZ125	C21000	1	X	2	2	2	2	X	X	X	3	2
	CuZn10	CZ101	C22000	1	X	2	2	2	2	X	X	X	3	2
	CuZn15	CZ102	C23000	1	X	2	2	2	2	X	3	X	3	2
	CuZn20	CZ103	C24000	1	X	2	2	2	2	X	3	X	3	2
	CuZn28	...	C26000	1	X	2	3	3	2	X	2	X	3	2
	CuZn30	CZ106	C26100	1	X	2	3	3	2	X	2	X	3	2
	CuZn33	CZ107	C27000	1	X	2	3	3	2	X	2	X	3	2
	CuZn37	CZ108	C27200	1	X	2	3	3	2	X	2	X	3	2
	CuZn40	CZ109	C28000	1	X	2	2	2	2	X	2	X	3	2

(continued)

Table 4.41 (continued)

| | Material | | | Joining process | | | | | | | | | | |
Type	ISO (Designation)	BSI	UNS	Soft solder	Braze	Oxyacet weld	GTAW	SMAW	Res. butt weld	Res. seam weld	Res. spot weld	Electronic beam weld	Cold pressure weld	Friction weld
Leaded brasses	CuZn9Pb2	...	C31400	1	X	X	X	X	3	X	X	X	3	2
	CuZn20Pb	CZ104	C24080	1	X	X	X	X	2	X	3	X	3	2
	CuZn34Pb1	CZ118	C35000	1	X	X	X	X	3	X	X	X	3	2
	CuZn36Pb2	CZ119, CZ131	C35330	1	X	X	X	X	3	X	X	X	3	2
	CuZn36Pb3	CZ124	C36000	1	X	X	X	X	3	X	X	X	3	2
	CuZn38Pb2	CZ128	C35300	1	X	X	X	X	3	X	X	X	3	2
	CuZn39Pb1	CZ129	C37000	1	X	X	X	X	3	X	X	X	3	2
	CuZn39Pb2	CZ120, CZ122	C37710	1	X	X	X	X	3	X	X	X	3	2
	CuZn39Pb4	CZ121, 4Pb	...	1	X	X	X	X	3	X	X	X	3	3
	CuZn40Pb	CZ123	...	1	X	3	3	3	3	X	X	X	3	3
	CuZn40Pb3	CZ1213Pb, CZ130	C38500	1	X	X	X	X	3	X	X	X	3	3
	CuZn43Pb1	...	...	2	X	X	X	X	3	X	X	X	3	3
High strength and special brasses	CuZn20Al2	CZ110	C68700	2	X	3	3	3	2	3	2	X	X	2
	CuZn28Sn1	CZ111	C44300	1	X	2	3	3	2	X	2	X	X	2
	CuZn30As	CZ105, CZ126	C26130	1	X	2	2	2	2	X	2	X	X	2
	CuZn36Pb2As	CZ132	...	1	X	X	X	X	3	X	X	X	X	2
	CuZn38Sn1	CZ112, CZ113	C46400	1	X	2	3	3	2	3	2	X	X	2
	CuZn39AlFeMn	CZ114	C67500	1	X	X	3	3	2	3	2	X	X	3
	CuZn39AlFeMn	CZ115	...	2	X	X	3	3	2	3	2	X	X	3
	CuZn17Al5FeMn	CZ116	C67000	2	X	X	3	3	1	1	1	X	X	3
	CuZn14AlNiSi	CZ127	...	2	X	X	3	3	X	X	X	X	X	3

Note: 1 = Excellent, 2 = Good, 3 = Fair, X = Not recommended

pacitive discharge power supplies are often used to avoid heat damage, either to the material itself or adjacent components. Electrodes are typically made from tungsten or molybdenum to preclude alloying. Electrode forces used to spot- and seam-weld copper metals are, in general, lower than those needed for welding low-carbon steel, but extremely low forces, which can cause electrode pickup and weak welds, should be avoided. The usual spacing is 5 to 7 spots per centimeter (12 to 18 spots per inch). If fewer than 5 spots per centimeter (12 spots per inch) are used, the spots may not overlap. On the other hand, spots that are too closely spaced can cause excessive hot working of the base metal.

Seam welding is difficult to perform on copper and many of the high copper alloys, but most low-conductivity copper alloys can be seam-welded readily using higher welding current and lower electrode force than those used for welding low-carbon steel. Projection welding is best suited to copper alloys with electrical conductivities lower than 30% IACS. To prevent collapse of the metal in the projection before the welding temperature is attained, coined projections are generally preferred to formed ones. Projections incorporate de-

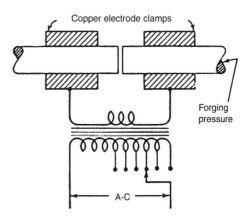

Fig. 4.98 Electric circuit for butt resistance welding

signs to limit expulsion of metal as shown in Fig. 4.103 (Ref 73).

Joints prepared for resistance welding should be free from dirt, scale, oil, drawing compound, or other foreign matter, and contact resistance should be uniform. Any effective cleaning method can be employed.

Gas tungsten arc welding (GTAW or TIG) is well suited to copper and copper alloys because its intense, high-temperature arc permits rapid welding speeds, which leave a narrow heat-affected zone (HAZ). A narrow HAZ is particularly beneficial when welding precipitation-hardened alloys in the form of products that cannot be re-heat treated after joining.

The GTAW of copper and copper alloys is most frequently used for sections up to about 3.2 mm (0.125 in.) thick, in which case square-edged joint

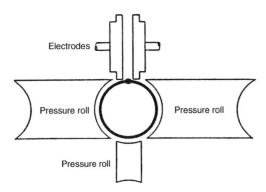

Fig. 4.99 Butt seam welding of pipe and tubing. The axes of the pressure roll and the electrodes are not necessarily in the same plane

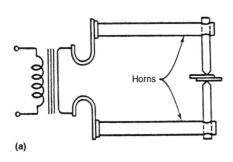

(a)

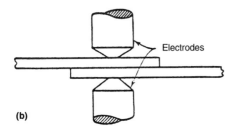

(b)

Fig. 4.100 (a) Electric circuit for spot welding. (b) Close-up of the electrodes

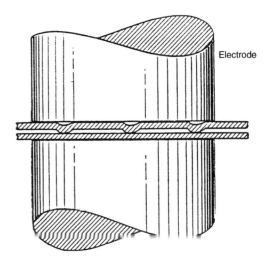

Fig. 4.101 Projection welding

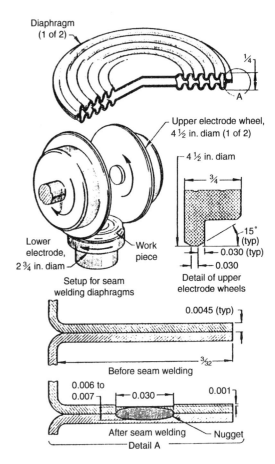

Fig. 4.102 Two 0.0045 in. thick diaphragms, made of a copper-nickel-manganese alloy, that were joined by resistance seam welding to produce a pressure-sensing capsule

preparations are adequate. Often, no filler metal is used in such cases, although filler metal is almost always used for thicker sections. Sections thicker than 12.7 mm (0.5 in.) are welded using GTAW only if gas metal-arc welding (GMAW or MIG) equipment is not available or if special conditions such as hot shortness of the base metal or adjacent heat-sensitive features make it necessary to limit the heat input.

Gas tungsten arc welding is performed using direct current, straight polarity (dcsp) to permit use of an electrode of minimum size for a given welding current and to provide maximum penetration. High-frequency alternating current is used with beryllium coppers and aluminum bronzes to prevent buildup of the tenacious oxide films that form on these metals.

Electrodes. Any standard tungsten or alloyed tungsten electrodes can be used. Except as noted for specific classes of copper alloys, thoriated tungsten (usually AWS EWTh-2) is preferred for its better performance, longer life, and greater resistance to contamination.

Filler metals most frequently used for GTAW are listed in Table 4.42. The filler metal composition is usually matched closely to the base-metal composition but, as discussed later in sections dealing with specific alloys, this is not always the case.

Suitability to Automation. While not as versatile as GMAW in this respect, GTAW, with or without filler metal, can be automated. Familiar applications in which this is done to advantage are the fabrication of large cylindrical vacuum arc remelt crucibles and the welding of heat exchanger tubes into tube sheets.

Gas metal-arc welding (GMAW or MIG) can be used to join all of the copper metals listed in Table 4.41. It is the preferred process for joining the aluminum bronzes, silicon bronzes, and copper-

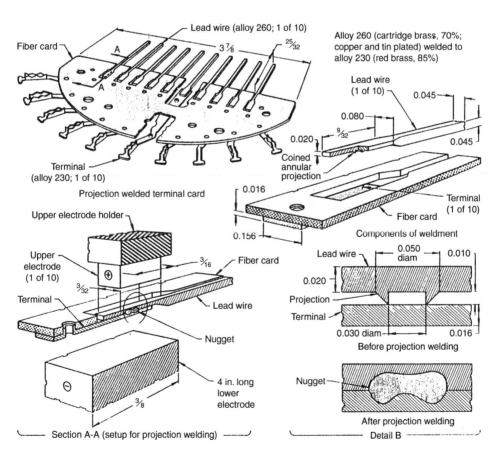

Fig. 4.103 Cartridge brass lead wires and red brass terminals that were joined by projection welding, instead of by spot welding, to prevent electrode sticking, using annular projections designed to limit expulsion of weld

Table 4.42 Filler metals for GTA welding of copper metals and alloys

Filler metal	AWS classification	Principal constituents(a)
Copper	RCu	98.0 min Cu + Ag, 1.0 Sn, 0.5 Mn, 0.50 Si, 0.15 P
Phosphor bronze	RCuSn-A	93.5 min Cu + Ag, 4.0–6.0 Sn, 0.10–0.35 P
Phosphor bronze	RCuSn-C(b)	7.0–9.0 Sn, 0.05–0.35 P, bal Cu + Ag
Aluminum bronze	RCuAl-A2	1.5 Fe, 9.0–11.0 Al, bal Cu + Ag
Aluminum bronze	RCuAl-B	3.0–4.25 Fe, 11.0–12.0 Al, bal Cu + Ag
Silicon bronze	RCuSi-A	94.0 min Cu + Ag, 2.8–4.0 Si, 1.5 Zn, 1.5 Sn, 1.5 Mn, 0.5 Fe
Copper nickel	RCuNi	1.00 Mn, 0.40–0.70 Fe, 29.0–32.0 Ni + Co, 0.20–0.50 Ti, bal Cu + Ag

Based on AWS A5.7 and A5.6. (a) Single percentages are maximum unless otherwise stated. Optional elements and impurities have been omitted. (b) ECuSn-C is classified by AWS as electrode wire for gas metal-arc welding, but it is used also as filler wire in gas tungsten-arc welding.

Table 4.43 Nominal GMAW conditions for butt welding copper and copper alloys

Weld types for butt joints	Work-metal thickness, in.	Root face, in.	Root opening, in.	Electrode	Electrode-wire diam., in.	Shield gas	Gas-flow rate, cfh	Current (dcrp), amp	Voltage, V	Travel speed, ipm	Number of passes	Preheat temperature, °F
Commercial coppers												
Square groove(a)	1/8	1/8	0	ECu	1/16	Argon	30	310	27	30	1	None
Square groove(b)	1/8	1/8	1–1/16	ECu	1/16	Argon(c)	30–35	325–350	28–33	...	1	None
Square groove	1/4	1/4	0	ECu	3/32	Argon	30	460	26	20	2	200
	1/4	1/4	0	ECu	3/32	Argon	30	500	27	20	1	200
75–90° single V-groove(b)	1/4	1/8	0–1/8	ECu	1/16	Argon(c)	30–35	400–425	32–36	...	2	400–500
	1/2	0–1/8	0–1/8	ECu	1/16	Argon(c)	30–35	425–450	35–40	...	4	800–900
90° single V-groove	3/8	3/16	0	ECu	3/32	Argon	30	500	27	14	(d)	400
	3/8	3/16	0	ECu	3/32	Argon	30	550	27	14	(d)	400
	1/2	1/4	0	ECu	3/32	Argon	30	540	27	12	(d)	400
	1/2	1/4	0	ECu	3/32	Argon	30	600	27	10	(d)	400
Alloy C17500 high conductivity beryllium copper												
90° single V-groove	1/4–1/2	1/32	...	C17500	0.045	A-He	30	200–240	...	...	3–4(e)	600
	3/4	1/32	...	C17500	0.045	A-He	30	200–240	...	...	6(e)	900
Alloys C17000 and C17200 high-strength beryllium copper												
90° single V-groove	1/4–1/2	1/32–1/16	...	C17000, C17200	0.045	A-He	45	175–200	...	...	3–4(f)	300–400(h)
30° double U-groove	3/4–1 1/2	1/16	...	C17000, C17200	1/16	A-He	60	325–350	...	...	10–20 (g, i)	300–400(h)
Low-zinc brasses												
Square groove(b)	1/8	1/8	0	ECuSi	1/16	Argon	30	275–285	25–28	...	1	None
	1/8	1/8	0	ECuSn-C	1/16	Helium	35	275–285	25–28	...	1	None
60° single V-groove(b)	3/8	0	1/8	ECuSi	1/16	Argon	30	275–285	25–28	...	2	None
	1/2	0	1/8	ECuSi	1/16	Argon	30	275–285	25–28	...	4	None
70° single V-groove(b)	3/8	0	1/8	ECuSn-C	1/16	Helium	35	275–285	25–28	...	2	500(j)
	1/2	0	1/8	ECuSn-C	1/16	Helium	35	275–285	25–28	...	4	500(j)
High-zinc brasses, tin brasses, special brasses, nickel silvers(k)												
Square groove (b)	1/8	1/8	0	ECuSn-C	1/16	Argon	30	275–285	25–28	...	1	None
70° single V-groove(b)	2/8	0	1/8	ECuSn-C	1/16	Argon	30	275–285	25–28	...	2	None
	1/2	0	1/8	ECuSn-C	1/16	Argon	30	275–285	25–28	...	4	None
Phosphor bronzes(l)												
90° single	3/8	0	1/8	ECuSn-A(m, n)	1/16	Helium	35	275–285	25–28	...	3–4(o)	200–300
V-groove(b)	1/2	0	1/8	ECuSn-A(m, n)	1/16	Helium	35	275–285	25–28	...	3–6(o)	350–400
Aluminum bronzes(p)												
Square groove(q)	1/8	1/8	0	ECuAl-A2	1/16	Argon	30	280–290	27–30	...	1	None
60–70° single V-groove	3/8	0	1/8	ECuAl-A2	1/16	Argon	30	280–290	27–30	...	2	None
	1/2	0	1/8	ECuAl-A2	1/16	Argon	30	280–290	27–30	...	3	Slight

(continued)

Table 4.43 (continued)

Weld types for butt joints	Work-metal thickness, in.	Root face, in.	Root opening, in.	Electrode	Electrode-wire diam., in.	Shield gas	Gas-flow rate, cfh	Current (dcrp), amp	Voltage, V	Travel speed, imp	Number of passes	Preheat temperature, °F
Silicon bronzes(r)												
Square groove	$\frac{1}{8}$	$\frac{1}{8}$	0	ECuSi	$\frac{1}{16}$	Argon	30	260–270	27–30	8 min	1	None
60° single	$\frac{3}{8}$	0	$\frac{1}{8}$	ECuSi	$\frac{1}{16}$	Argon	30	260–270	27–30	8 min	2	None
V-groove(b)	$\frac{1}{2}$	0	$\frac{1}{8}$	ECuSi	$\frac{1}{16}$	Argon	30	260–270	27–30	8 min	3	None
Copper nickels												
Square groove(b)	$\frac{1}{8}$	$\frac{1}{8}$	0	ECuNi	$\frac{1}{16}$	Argon	30	280	27–30	…	1	None
60–80° single	$\frac{3}{8}$	$0-\frac{1}{32}$	$\frac{1}{8}-\frac{1}{4}$	ECuNi	$\frac{1}{16}$	Argon	30	280	27–30	…	2	None
V-groove(b)	$\frac{1}{2}$	$0-\frac{1}{32}$	$\frac{1}{8}-\frac{1}{4}$	ECuNi	$\frac{1}{16}$	Argon	30	280	27–30	…	4	None
Commercial coppers to steel												
70–80° single V-groove	$\frac{3}{8}$	$\frac{1}{16}$	$\frac{1}{8}$	ERNi-3	$\frac{1}{16}$	Argon	60	375	29–31	…	4	800–1000
Copper nickel to steel												
70–80° single V-groove	$\frac{3}{8}$	$\frac{1}{16}$	$\frac{1}{8}$	ERNi-3	$\frac{1}{16}$	Argon	60	375	29–31	…	4	150 max
Aluminum bronze to steel												
60° single V-groove	$\frac{3}{8}$	0	$\frac{5}{16}-\frac{3}{8}$	ECuAl-A2	$\frac{1}{16}$	Argon	30	270–280	25–27	…	6	300–500
Silicon bronze to steel												
60° single V-groove	$\frac{3}{8}$	0	$\frac{5}{16}-\frac{3}{8}$	ECuAl-A2	$\frac{1}{16}$	Argon	30	270–280	28–30	…	6	150 max

The data in this table are intended to serve as starting points for the establishment of optimum joint design and conditions for welding of parts on which previous experience is lacking; they are subject to adjustment necessary to meet the requirements of individual applications. Thickness up to about 1 ½ in. are sometimes welded, by use of slightly higher current and lower travel speed than shown for a thickness of 1 ½ in. (a) Copper backing. (b) Grooved copper backing. (c) Or 75% argon, 25% helium. (d) Special welding sequence is used. (e) The final pass is made on the root side after back chipping. Grind after each pass. (f) The final pass is made the root side after back chipping. Wire brush after each pass. (g) Several passes are made on the face side, then several on the back side, until the weld is completed. Back chip the root pass before making the first pass on the back side. Wire brush after each pass. (h) Should not be overhead; as little preheat as possible should be used. (i) Welding conditions based on alloys C51000, C52100, and C52400; current is increased or speed decreased for alloy C50500. (j) Or ECuSn-C. (k) Hot peening between passes is recommended for maximum strength. (l) Slight preheat may be needed on heavy sections; interpass temperature should not exceed 600 °F. (m) With ⅛ by 1 in. aluminum bronze backing. (n) No preheat is used on any thickness; interpass temperature should not exceed 200 °F. (o) With ⅛ by 1 in. silicon bronze backing. (p) Steel should be well penetrated; and overlay is not usually needed. (q) Steel should be well penetrated; and overlay should be applied to avoid excessive dilution of the silicon bronze. (r) Except in welding silicon bronze to high-carbon or low-alloy steel, for which preheat temperature is 400 °F. Source: Ref 73

nickels in section thicknesses greater than about 3.2 mm (0.125 in.), and certainly beyond 12.7 mm (0.5 in.), where its high deposition rate is particularly advantageous. On the other hand, the GMAW high heat input and wide HAZ can be troublesome.

Nominal conditions for GMA are listed in Table 4.43. Direct current, reverse polarity is used exclusively. Argon is normally used for shielding. Helium or mixtures of argon and helium are used where hotter arcs are needed.

Square-groove joints are not ordinarily used for welding thicknesses greater than 3.2 mm (0.125 in.) except when welding coppers (Table 4.43). Otherwise, single V-grooves are used for workpiece thicknesses between 3.2 and 12.7 mm (0.125 and 0.5 in.). For material thicker than about 12.7 mm (0.5 in.), joints are usually prepared with double V- or double U-grooves.

Welding Position. Most GMAW is performed in the flat position, using spray transfer arcs. Fillet welds acceptable for many applications can be produced in the horizontal position. When it is necessary to weld in positions other than flat, GMAW is preferred to GTAW or shielded metal-arc welding. In the vertical and overhead positions, GMAW is usually restricted to the copper alloys, such as aluminum bronzes, silicon bronzes, and copper-nickels. Small-diameter electrode wire and low currents are preferred for such applications, and a globular or short-circuiting mode of transfer is ordinarily used.

Filler Metals. Compositions of electrode wires used for GMAW are given in Table 4.44.

Suitability to Automation. Gas metal-arc welding is somewhat more amenable to fully automatic and robotic welding than is GTAW. Semiautomatic GMAW welding is also widely applied.

Shielded Metal-Arc Welding (SMAW). When used with copper metals, the SMAW or "stick" welding process requires larger root openings, wider groove angles, more tack welds, higher preheat and interpass temperatures, and higher currents than are used with steel. A list of approved electrodes is given in Table 4.44. Note that similar compositions can be used for GMAW and SMAW.

The SMAW of copper metals is almost always restricted to flat-position welding. Out-of-position welding is usually limited to the joining of phosphor bronzes and copper-nickels. The process is not recommended for most alloys because of its detrimental effect on mechanical properties and the danger of unsound or dirty welds.

Coppers. Problems of porosity and low-weld strength due to oxygen content of the base metal and oxygen absorption during welding are more severe in joining coppers by SMAW than by gas-shielded processes. Shielded metal arc welding is, therefore, not advised for joining commercial coppers (Table 4.41).

Brass and Nickel Silvers. The low-zinc brasses, high-zinc brasses, tin brasses, special brasses, and nickel silvers listed in Table 4.41 are not generally recommended for welding by the SMAW process. However, some brasses are occasionally welded by this process when quality or strength requirements are not severe.

Phosphor-bronze electrodes such as AWS ECuSn-A and ECuSn-C have been used for welding the low-zinc brasses. The base metal is preheated and held at 204 to 260 °C (400 to 500 °F). Weld metal is applied in narrow and shallow stringer beads. The high-zinc copper alloys can be welded with AWS ECuAl-A2 (aluminum bronze) electrodes. Preheat and interpass temperatures should be between 260 and 371 °C (500 and 700 °F). The arc is held directly on the molten weld puddle rather than toward the base metal, and advanced slowly to minimize zinc volatilization.

Phosphor bronzes applicable to SMAW are listed in Table 4.41. Covered electrodes of AWS type ECuSn-A and ECuSn-C are used interchangeably. The phosphor bronzes flow sluggishly and must be preheated to 149 to 204 °C (300 to 400 °F), especially for thick sections. However, because of the hot shortness of these alloys, the interpass temperature must not be permitted to go above the preheat temperature. This is achieved by welding rapidly with light passes. For groove welding, the first two passes are made with a weaving technique. Width of the weave should not exceed two electrode diameters. The remaining passes are made without appreciable transverse weaving and with the use of narrow stringer beads. Careful attention to temperature and welding technique minimizes development of a coarse, brittle, low-strength dendritic structure. If coarse structures do develop, they can be broken up to some extent by hot peening after welding. For maximum ductility, the welded assembly should be post-heated to 482 °C (900 °F) and cooled rapidly. Joint grooves should be wide (80 to 90°) to achieve proper "washing" of the groove walls.

Aluminum Bronzes. Shielded metal-arc welding is readily performed on both wrought and cast aluminum bronzes. The aluminum oxides that form on the surface of these alloys are removed by the fluxing action of the electrode coatings. Except for thin sections, a 70 to 90° V-groove joint is used, usually with a backing strip of the same composition as the base metal. Deposition technique and bead thickness are not critical, because the weld metal has excellent hot strength and ductility.

Aluminum bronze AWS ECuAl-A2 and ECuAl-B electrodes are used for welding aluminum bronzes C61300 and C61400, which contain about 7% Al. Thick sections of these alloys may need preheating, usually to 204 °C (400 °F), and control of interpass temperature at 204 °C (400 °F). Depending on section thickness and overall mass, however, preheat and interpass temperature may vary between about 66 and 427 °C (150 and 800 °F).

Table 4.44 SMA and GMA welding electrodes used with copper and copper alloys

Type of electrode	AWS classification	Principal compositions(a)
Copper	ECu(b)	98.0 min Cu + Ag, 1.0 Sn, 0.5 Mn, 0.5 Si, 0.15 P
Silicon bronze	ECuSi	2.8–4.0 Si, 1.5 Sn, 1.5 Mn, 0.5 Fe, rem Cu + Ag
Phosphor bronze	ECuSn-A	4.0–6.0 Sn, 0.10–0.35 P, rem Cu + Ag
Phosphor bronze	ECuSn-C	7.0–9.0 Sn, 0.05–0.35 P, rem Cu + Ag
Aluminum bronze	ECuAl-Al(b)	6.0–9.0 Al, rem Cu + Ag
Aluminum bronze	ECuAl-A2	1.5 Fe, 9.0–11.0 Al, rem Cu + Ag
Aluminum bronze	ECuAl-B	3.0–4.25 Fe, 11.0–12.0 Al, rem Cu + Ag
Copper nickel	ECuNi	1.00 Mn, 0.6 Fe, 0.50 Si, 29.0 min Ni + Co, 0.6 Ti, rem Cu + Ag

Based on AWS A5.6; see current edition in this specification for complete compositions and qualifications. All electrodes listed are available in both bare and covered forms except ECu and ECuAl-Al, which are not available as covered electrodes for shielded metal-arc welding. (a) Single percentages are maximum unless otherwise stated. Optional elements and impurities have been omitted. For covered electrodes, the compositions are of the metal core. (b) Available only as bare electrodes for gas metal-arc welding

Weldments of the 7% aluminum bronzes need not be heat treated after welding.

Aluminum bronzes having an aluminum content higher than 7% are usually welded with electrodes that contain more aluminum than do AWS ECuAl-A2 and ECuAl-B. The higher aluminum-bronze electrodes are AWS ECuAl-C, ECuAl-D, and ECuAl-E, which are best known as surfacing electrodes (see AWS A5.13), and which have nominal aluminum contents of 12.5, 13.5, and 14.5%, respectively, and correspondingly increasing strength. In welding high-aluminum bronzes, thick sections may require preheating up to 621 °C (1150 °F), and fan cooling may be necessary to avoid cracking. Also, these alloys may require annealing at 621 °C (1150 °F), followed by fan cooling, for stress relief.

Silicon Bronzes. Shielded metal-arc welding of silicon bronzes is usually done with AWS ECuAl-A2 aluminum-bronze electrodes. Welding temperature is easily attained, because silicon bronzes have low thermal conductivities. However, because the alloys are hot short, the metal should not be preheated and interpass temperature should not exceed 93 °C (200 °F).

Groove dimensions are similar to those used for welding similar steel joints. Metal thicknesses up to 4 mm (0.156 in.) can be welded with square grooves; thicker sections, with a single V- or double V-groove of 60° included angle.

Properties of welds made in silicon bronzes by SMAW are usually substantially lower than those of welds made by the gas-shielded processes and may not meet code or design requirements for strength. Peening will reduce residual stress and minimize distortion.

Copper-Nickels. Both wrought and cast copper-nickels can be welded using the SMAW process. Having thermal conductivities close to that of low-carbon steel, these alloys behave like steel in most respects and are as readily welded. The 70-30 copper-nickel electrodes, AWS ECuNi (Tables 4.43 and 4.44), are used to weld alloys C70600 and C71500, usually with reverse polarity direct current.

Weld deposits ordinarily have a high center crown, and the slag is viscous when molten and adherent when cold. Therefore, special care is needed to ensure complete slag removal before complete solidification of the weld to prevent slag entrapment when cleaning between passes.

The copper-nickels can be welded in all positions with good results, although best results are obtained in flat-position welding. The SMAW process is preferred in applications where access to the joint is limited, as in the butt welding of copper nickel pipe.

Oxyacetylene (torch) welding has been applied to copper metals for many years, and while established practices are available for many alloys, the process is not generally recommended except for small, simple jobs. The process affords low heat input compared with arc welding and is, therefore, relatively slow. This is distinctly disadvantageous to many copper alloys.

Oxyacetylene welding is also relatively unprotective, and weld metal cleanliness must rely on fluxing agents, which prevent oxidation of the melt puddle. In addition, the fluxes dissolve high melting point oxides to prevent them from becoming entrapped in the molten metal. For copper and most copper alloys, a fluxing agent based on salts, such as sodium chloride, sodium fluoride, or sodium phosphate, are common. Aluminum bronzes require special fluoride mixtures to accommodate the refractory nature of aluminum oxide. Such fluxes are supplied as powder or paste.

Welding rods themselves may be flux coated or cored, for example, tubular and filled with granulated flux. Gas fluxing is also possible, one suitable system being 71.5% methyl borate and 28.5% acetone (Ref 72, 75, 76).

Due to the low heat input attending oxyacetylene welding and the relatively high thermal conductivities exhibited by copper alloys, preheating is generally necessary. The recommended preheat temperature range is 300 to 700 °C (572 to 1292 °F). For work more than 6 mm (0.25 in.) thick, simultaneous use of two burners is recommended. Work should be performed in the vertical downhand position as shown in Fig. 4.104 (Ref 72).

For copper, the seam may be hammered hot after each 100 mm (4 in.) advance to enhance tensile strength and formability. Hammering produces some work hardening and partial recrystallization (Ref 72).

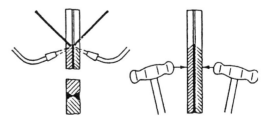

Fig. 4.104 Double-side welding and peening of vertical double v-seam (schematic)

Brazing and Soldering. By accepted definition, soldering is the joining process performed below 449 °C (840 °F), and brazing takes place above that temperature but below the melting point of the base metals. In actual practice for copper systems, most soldering is done at temperatures from about 177 to 288 °C (350 to 550 °F), and most brazing is done at temperatures ranging from 593 to 816 °C (1100 to 1500 °F). The choice between soldering and brazing generally depends on operating conditions. Solder joints are used where the service temperature does not exceed 121 °C (250 °F), and brazed joints can be used where greater strength is required, or where system temperatures are as high as 177 °C (350 °F) (Ref 77).

Brazing. Almost all copper alloys can be brazed, including those containing lead; however, special care should be taken when brazing leaded alloys to avoid overheating. Brazing temperatures must also be monitored carefully when joining heat-treatable alloys. Such alloys may require re-heat treatment after joining.

Joint Clearance. Clearance is the most important factor influencing the mechanical strength of brazed joints. Among other things, clearance establishes the capillary force required to fill the joint, and careful attention to clearance eliminates harmful voids in the joint area.

Typical diametral joint clearances for use with AWS filler metals commonly employed in brazing copper and copper alloys are given in Table 4.45. Note that the clearances given apply at room tem-perature and are applicable to brazing components of about the same mass made from the same cop-per or copper alloy. Adjustments may be required for brazing dissimilar metals to compensate for different coefficients of thermal expansion. DIN, ISO, and ASTM filler metals are listed in Tables 4.46 through 4.48.

Soldering. Proper flux practice is arguably the most important step toward ensuring acceptable solder behavior. Flux is supplied in liquid and paste forms. Formulations such as those given in ASTM B 813 ensure that soldering is simple to carry out, and that joints are leak tight and have satisfactory mechanical strength. In the United States, fluxes must be lead-free to the extent required by the 1906 Amendments to the Safe Drinking Water Act (Ref 78).

It is also important that fluxes not cause corrosion failures in service. This is generally ensured by flushing flux residues away after soldering. This characteristic is addressed in ASTM B 813. The U.S. Copper Development Association has also been especially concerned about the corrosiveness of fluxes and flux residues, and has issued a standard test method to measure this property. In Germany, flux requirements for the soldering of copper appear in work sheets GW2 and GW7 of the German Water and Gas Trade Association (DVGW) (Ref 78, 79).

Proper workmanship, particularly with respect to the correct amount of solder, is also important in the assembly of acceptable capillary joints. Good and bad situations are depicted in Fig.

Table 4.45 Nominal compositions, properties, and joint clearances for filler metals used with copper metals

| Filler metal metal, AWS | Composition, % | | | | | | | Solidus temperature, °F | Liquidus temperature, °F | Conductivity(a), %IACS | Typical diametral joint clearance (at room temperature), in. |
	Ag	Cu	P	Zn	Cd	Ni	Others				
RBCuZn-A	...	59.25	...	40	...	...	0.75 Sn	1630	1650	26	...
RBCuZn-D	...	48	...	42	...	10	...	1690	1715	...	0.002–0.005
BCuP-1	...	95	5	...	...	...	...	1310	1650	...	0.001–0.003
BCuP-2	...	92.75	7.25	...	...	...	...	1310	1460	...	0.001–0.003
BCuP-4	6	86.75	7.25	...	...	...	...	1190	1335	...	0.001–0.005
BCuP-5	15	80	5	...	...	...	...	1190	1475	10	0.002–0.005
	45	15	...	16	24	...	...	1125	1145	28	0.002–0.005
BAg-1	50	15.5	...	16.5	18	...	...	1160	1175	24	0.002–0.005
BAg-1(b)	35	26	...	21	18	...	...	1125	1295	29	0.002–0.005
BAg-2	50	15.5	...	15.5	16	3	...	1170	1270	18	0.002–0.005
BAg-3	45	30	...	25	...	...	...	1250	1370	19	0.002–0.005
BAg-5(b)	75	22	...	3	...	...	...	...	1365	...	0.002–0.005
BAg-8a	72	27.8	...	...	...	...	0.2 Li	1410	1410	89(c)	0.002–0.005
BAg-19	92.5	7.3	...	...	...	...	0.2 Li	1435	1635	88(c)	0.002–0.005
BAu-4	...	...	...	...	...	18	82 Au	1740	1740	6	0.002–0.005

(a) Ratio of electrical resistivity of International Annealed Copper Standard at 68 °F (20 °C) to the resistivity of the material at 68 °F (20 °C), expressed as a percentage and calculated on a volume basis. (b) Special filler metal used in brazing nickel-silver knife handles. (c) Conductivity of filler metal after volatilization of lithium in brazing. Source: Ref 73

4.105. If so much solder is added that the filler metal drips into the inside of the joint, the physical obstruction this produces could induce erosion-corrosion. Before leaded solders were outlawed for potable water systems, solder drips also constituted a potential source of lead contamination. The strength of the joint also diminishes as the layer of solder becomes thicker, as shown in Fig. 4.106 (Ref 80, 81).

The proper choice of solder compositions for a particular application is important because it is the solder that determines the strength, melting point, and properties of the joint. Figure 4.107 shows the melting characteristics of lead-free solders now required for potable water lines in many jurisdictions. Solder alloy compositions are listed in Tables 4.49 and 4.50. Rated internal working pressures over a range of service temperatures are listed in Table 4.51. The two brazing filler metals used most often for brazing copper and its alloys are BCuP (copper-phosphorus alloys) and BAg (silver alloys) (Ref 77, 82).

Table 4.46 Brazing alloys (DIN 8513) for copper and copper-alloys

DIN 8513	Abbreviation ISO 3677	ASTM	Average composition(a), wt %	Melting point, °C Solidus	Liquidus	Working temperature, °C	Basemetal applications	Shape of soldered joint	Solder feeding
L-CuZn40	BCu60Zn 890–900	...	60Cu, 0.2Si, bal Zn	890	900	900	Copper and copper alloys and melting "solidus" temperature above 900 °C	Square-butt joint, V-butt joint(b)	Placed or inserted
L-CuZn39Sn	BCu59Zn 870–890	RBcUzN-C	59Cu, 0.1Si, 1Sn, 0.6Mn, bal Zn	870	890	900	...	...	...
L-CuZn46	BCu54Zn 880–890	...	54Cu, bal Zn	880	890	890	Copper and copper alloys	Square-butt joint	Inserted
L-ZnCu42	BZn58Cu 835–845	...	42Cu, bal Zn	835	845	845	Preferentially nickel-silvers	...	...
L-CuP8	BCu92P 710	...	8P, bal Cu	710	710–750	710	Preferentially copper, rod brass	Square-butt joint	Placed or inserted
L-CuP7	BCu93P 710–820	BCuP-2	7.1P, bal Cu	710	820	720	Copper-zinc alloys and copper-tin alloys	...	...
L-CuP6	BCu94P 710–880	...	6.2P, bal Cu	710	880	730		...	...

(a) Compositional tolerances and allowed impurities, see DIN 8513. (b) For groove brazings without higher strength requirements. Source: Ref 31

Table 4.47 Silver-bearing brazing alloys (DIN 8513) containing less than 20% Ag

DIN 8513	Abbreviation ISO 3677	ASTM	Average composition(a), wt %	Melting range, °C Solidus	Liquidus	Working temperature, °C	Basemetal applications	Shape of based joint	Solder feeding
Copper-silver-zinc brazing rod or filler									
L-Ag12	BCu48ZnAg 800–830	...	12Ag, 48Cu, bal Zn	800	830	830	Copper and copper alloys	Square-butt joint	Placed and inserted
L-Ag5	BCu55ZnAg 820–870	...	5Ag, 65Cu, bal Zn	820	870	860	Copper and copper alloys	Square-butt joint, V-butt joint	Placed
Copper-silver-zinc-cadmium brazing rod or filler									
L-Ag12Cd	BCu50ZnAgCd 620–825	...	12Ag, 7Cd, 50Cu, bal Zn	620	825	800	Copper and copper alloys	Square-butt joint, V-butt joint	Inserted
Copper-silver-phosphor brazing rod or filler									
L-Ag15P	BCu80AgP 650–800	BCuP-5	15Ag, 5P, bal Cu	650	800	710	Copper; red brass	Square-butt joint	Placed and inserted
L-Ag5P	BCu89PAg 650–810	BCuP-3	5Ag, 6P, bal Cu	650	810	710	Copper-zinc alloys	Square-butt joint, V-butt joint	...
L-Ag2P	BCu92PAg 650–810	BCuP-6	2Ag, 6.2P, bal Cu	650	810	710	Copper-tin alloys	...	...

(a) Tolerances in composition, other alloy components and accepted impurity concentration after DIN 8513. Source: Ref 31

Mechanical Joining

Crimping, shown in Fig. 4.108, can be used both on tubular or flat parts, provided that the materials are sufficiently thin and ductile to tolerate the large localized deformation. An example of use of this method is copper roofing (Ref 48).

Adhesive Bonding. A number of adhesive formulations have been prepared to join copper plumbing tube used both for potable water and for waste, vent, and drain service. Several examples are given in Table 4.52. One epoxy adhesive, developed for the International Copper Research Association and sold under the name Copper-Bond (Marsh Laboratories, Pittsburgh, PA), is suitable for use with capillary spaces between 0.05 and 0.13 mm (0.002 and 0.005 in.) wide. Epoxies have been found to be safe for use in food and beverage contact applications by the U.S. Food and Drug Administration (Ref 83–85).

Table 4.48 Silver-bearing brazing alloys (DIN 8513) containing more than 20% Ag

	Abbreviation		Average composition (b), wt %	Melting range, °C		Working temperature, °C	Basemetal applications	Shape of based joint	Solder feeding
DIN 8513	ISO 3677	AWS(a)		Solidus	Liquidus				
Silver-copper-cadmium-zinc									
L-Ag50Cd	BAg50CdZnCu 620–640	BAg-1a	50Ag, 17Cd, 15Cu, bal Zn	620	640	640	Copper alloys	Square-butt joint	Placed and inserted
L-Ag45Cd	BAg45CdCuZn 620–635	...	45Ag, 20Cd, 17Cu, bal Zn	620	635	620	Copper alloys	...	...
L-Ag40Cd	BAg40CdCu 595–630	...	40Ag, 20Cd, 19Cu, bal Zn	595	630	610	Copper and copper alloys	...	...
Silver-copper-cadmium-zinc (continued)									
L-Ag34Cd	BAg30CuZn Cd610–680	...	34Ag, 20Cd, 22Cu, bal Zn	610	680	640	...	...	...
L-Ag30Cd	BAg34CuZnCd 600–690	BAg-2a	30Ag, 21Cd, 28Cu, bal Zn	600	690	680	...	...	...
L-Ag25Cd	BCu30ZnAgCd 605–720	...	25Ag, 17Cd, 30Cu, bal Zn	605	720	710	...	...	...
L-Ag20Cd	BCu40ZnAgCd 605–765	...	20Ag, 15Cd, 40Cu, bal Zn	605	765	750	...	Square-butt joint and V- butt joint	...
Silver-copper-zinc-tin									
L-Ag55Sn	BAg55ZnCuSn 620–660	BAg-7	55.5Ag, 21.5Cu, 3.5Sn, bal Zn	620	660	650	Copper alloys	Square-butt joint	Placed and inserted
L-Ag45Sn	BAg45ZnCuSn 640–680	...	45Ag, 27Cu, 35Sn, bal Zn	640	680	670	Copper and copper alloys	...	...
L-Ag44	BAg44CuZn 675–735	BAg-5	44Ag, 30Cu, bal Zn	675	735	730	...	...	...
L-Ag40Sn	BAg40CuZnSn 640–700	...	40Ag, 30Cu, 2Sn, bal Zn	640	700	690	...	...	...
L-Ag34Sn	BCu36AgZn 630–730	...	34Ag, 36Cu, 3Sn, bal Zn	630	730	710	...	...	...
L-Ag30Sn	BCu36ZnAgSn 650–750	...	30Ag, 36Cu, 2Sn, bal Zn	650	750	740	...	...	...
L-Ag30	BCu38ZnAg 680–765	BAg-20	30Ag, 38Cu, bal Zn	680	765	750	...	...	...
L-Ag25Sn	BCu40ZnAgSn 680–760	...	25Ag, 40Cu, bal Zn	680	760	750	...	...	...
L-Ag25	BCu41ZnAg 700–800	...	25Ag, 41Cu, bal Zn	700	800	780	...	...	...
L-Ag20	BCu44ZnAg 690–810	BAg-28	20Ag, 44Cu, bal Zn	690	810	810	...	Square-butt joint and V- butt joint	Placed and inserted
Zinc-free special brazing alloys									
L-Ag72	BAg72Cu 780	BAg-8a	72Ag, bal Cu	779	779	780	Copper and copper alloys	Square-butt joint	Inserted
Special brazing alloys									
L-Ag50CdNi	BAg50ZnCuCdNi	BAg-3	50Ag, 15.5Cu, 18Cd, 3Ni, bal Zn	645	690	660	Copper alloys	Square-butt joint	Placed and inserted

(a) After ANSI/AWS A5.6 and A5.7. (b) Tolerances in composition, other alloy components and accepted impurity concentration after DIN 8513. Source: Ref 31

Dissimilar Metal Joints

Metallurgical concerns regarding joints between different copper-base metals or between copper-base and other metals primarily deal with dilution effects at the interface. Dilution can reduce mechanical properties and can also affect corrosion resistance, either in copper metals or metals joined to them. These adverse effects can be avoided by giving due consideration to:

• Careful selection of the welding process
• The use of automatic welding
• Limiting heat input to the minimum required for the development of the fusion bond
• The application of an intermediate layer that is metallurgically compatible with both metals
• Stress relief of the metal and surface peening to reduce the extent of grain boundary penetration

Specific effects are not easy to predict, but reference to appropriate equilibrium diagrams may yield some guidance as to the appearance of undesirable phases. Dilution is often controlled by selecting a welding process, such as GTAW, with low heat input. Gas metal-arc welding is also beneficial if weld current is adjusted to produce droplet-type transfer. Friction welding offers another solution. Table 4.53 lists filler metals and joint characteristics as recommended under BS 2991 (Ref 83).

Copper to Steel. There is limited solubility of iron in copper and copper-rich alloys; therefore, a fusion weld between the two metals tends to leave iron as an insoluble, particulate dispersion in copper. This result has a minimum effect on the mechanical properties of the clad layer, but it reduces its corrosion resistance. Excessive dilution of copper or alloy into the steel substrate may cause hot cracking and, particularly, a reduction in fatigue properties in a structural material. Dilution must,

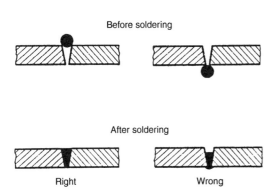

Fig. 4.105 Right and wrong placement of solder in V-shaped soldering slot

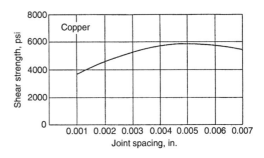

Fig. 4.106 Effect of joint spacing on the shear strength of copper soldered with 56:44 tin-lead alloy and zinc-chloride flux

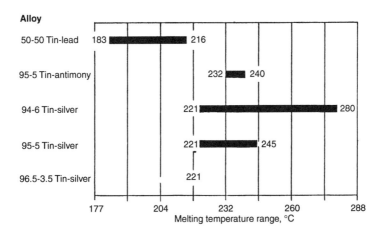

Fig. 4.107 Melting characteristics of alternative lead-free solders compared to 50-50 tin-lead solder

therefore, be minimized and at least two surfacing runs may be necessary to present a low iron or iron-free surface to the corrosive environment. The deposition of a nickel or nickel-rich alloy as a "buffer" or intermediate layer substantially reduces the risk in both directions. Figure 4.109 il-

lustrates a case of defective application of copper-weld cladding on steel (Ref 83).

Copper-Nickel to Steel. When welding copper-nickel to steel, it is generally possible to minimize dilution to a trouble-free level by weld process control. The higher the nickel content of the copper-

Table 4.49   Lead-tin and tin-lead solders for copper and selected copper alloys

DIN 1707	ISO 3677	ASTM(c)	Average composition(a), wt %	Solidus	Liquidus	F	I	K	T	Applications
	Abbreviation			Melting range, °C		Preferred soldering brasses(b)				
Antimonium-containing solders										
L-PbSn12Sb	BPb88Sn 250–295	...	12Sn; 0.45Sb; bal Pb	250	295	X	...	...	X	Radiator construction
L-PbSn20Sb	BPb80Sn 186–270	20B	20Sn; 0.70Sb; bal Pb	186	270	X	...	...	X	...
L-PbSn25Sb	BPb75Sn 186–260	25B	25Sn; 0.85Sb; bal Pb	186	260	X	...	...	X	Radiator construction, tin-lead bearing
L-PbSn40Sb	BPb60SnSb 186–225	40B	40Sn; 1.45Sb; bal Pb	186	205	X	...	X	X	Radiator construction
Low-percentage antimony solders										
L-PbSn8(Sb)	BPb92Sn 280–305	...	8Sn; 0.31Sb; bal Pb	280	305	X	X	...	X	Radiator construction, thermostats
L-PbSn33(Sb)	BPb67Sn 183–242	...	33Sn; 0.31Sb; bal Pb	183	242	X	...	...	...	Cable cover soldering
L-Sn50Pb(Sb)	BSn50Pb 183–215	50B	50Sn; 0.31Sb; bal Pb	183	215	X	...	X	X	Precision soldering and plumbers
L-Sn60Pb(Sb)	BSn60Pb 183–190	60B	60Sn; 0.31Sb; bal Pb	183	190	X	X	X	X	Electrical engineering, precision work soldering
Antimony-free solders										
L-PbSn40	BPb60Sn 183–235	40A	40Sn; bal Pb	183	235	X	...	X	X	Tableware
L-Sn50Pb	BSn50Pb 183–215	50A	50Sn; bal Pb	183	215	X	...	X	X	Electrical engineering
L-Sn60Pb	BSn60Pb 183–190	60A	60Sn; bal Pb	183	190	X	X	X	X	Electrical engineering, printed circuits
L-Sn63Pb	BSn63Pb 183	...	63Sn; bal Pb	183	183	X	X	X	X	Electrical engineering, electronic
Solders with copper addition										
L-Sn50PbCu	BSn60Pb 183–215	...	50Sn; 1.4Cu; bal Pb	183	215	...	...	X	...	Electrical engineering, electronic
L-Sn60PbCu	BSn60Pb 183–190	...	60Sn; 0.15Cu; bal Pb	183	190	...	...	...	X	Electrical engineering, electronic
L-Sn60PbCu2	BSn60Pb 183–190	...	60Sn; 1.8Cu; bal Pb	183	190	...	...	X	X	Printed circuits
Solder with silver addition										
L-Sn50PbAg	BSn50PbAg 178–210	...	50Sn; 3.5Ag; bal Pb	178	210	...	...	X	X	Electrical engineering
L-Sn60PbAg	BSn60PbAg 178–180	...	60Sn; 3.5Ag; bal Pb	178	180	...	X	X	X	Electronic
L-Sn63PbAg	BSn63Pb 178	...	63Sn; 1.4Ag; bal Pb	178	178	...	X	X	X	Printed circuits
Solders with phosphorus addition										
L-Sn50PbP	BSn50Pb 183–215	...	50Sn; bal Pb(d)	183	215	...	...	...	X	Electrical engineering, printed circuits,
L-Sn60PbP	BSn60Pb 183–190	...	60Sn; bal Pb(d)	183	190	...	...	...	X	principally for immersion soldering
L-Sn63PbP	BSn63Pb 183	...	63Sn; bal Pb(d)	183	183	...	...	...	X	Electrical engineering, printed circuits,
L-Sn60PbCuP	BSn60Pb 183–190	...	60Sn; 0.15Cu; bal Pb(d)	183	190	...	...	...	X	principally for immersion soldering

(a) Tolerance is tin contents ±0.5%. Other tolerances and accepted impurities after DIN 1707. (b) F = flame furnaces; I = Induction soldering; K = soldering iron; T = emersion soldering joining. (c) After ASTM B 32. (d) Phosphorus addition from 0.001 up to 0.004%

nickel, the smaller will be the penetration problem. It should not normally be necessary to provide an intermediate layer, but nickel-rich or nickel-copper alloys can be effective in cases where full weld control is not attainable (Ref 83).

Aluminum Bronzes to Steel. When welding aluminum bronzes to steel, it should be recognized that given adequate weld process control, these alloys, like copper-nickels, tolerate a reasonable amount of iron pick-up and penetrate steel less readily than pure copper. The simple binary filler metals such as those manufactured to BS 2901 (C13 or the more complex C20, Table 4.53) are

preferred. These alloys tolerate iron dilution without impairment of their corrosion resistance (Ref 83).

Copper to Aluminum. Both fusion and friction welding techniques are used to join copper to aluminum. Direct fusion welding produces brittle intermetallics, notably $CuAl_2$, which seriously impair the mechanical properties of the joint. A technique whereby a layer of silver brazing alloy is applied to the copper prior to fusion welding with a conventional aluminum-silicon filler metal has been described. The silver layer is about 0.75 to 1 mm (0.03–0.04 in.) thick and is metallurgi-

Table 4.50 Special solders (DIN 1707) for copper and copper alloys

Abbreviation DIN 1707	ISO 3677	ASTM(c)	Average composition (a), wt %	Melting range, °C Solidus	Liquidus	Preferred soldering processes(b) F	I	K	T	Applications
L-SnIn50	BSn50In 117–125	...	50Sn; bal In	117	125	X	...	X	...	Glass-metal soldering
L-SnPbCd18	BSn50PbCd 145	...	18Cd; 31.5Pb; bal Sn	145	145	X	X	X	X	Precision and cable soldering
L-SnCd20	BSn80Cd 180–195	...	20Cd; bal Sn	180	195	...	X	X	X	Electrical engineering
L-SnAg5	BSn95Ag 221–240	95TA	4Ag; bal Sn	221	240	X	X	X	X	Copper tube plumbing, electrical engineering
L-SnSb5	BSn95b 230–240	...	5Sb; bal Sn	230	240	X	...	X	X	Low-temperature industry
L-SnCu3	BSn97Cu 230–250	...	3Cu; bal Sn	230	250	X	X	X	X	Copper-tube plumbing, table ware
L-CdZnAg2	BCd82ZnAg 270–280	...	2Ag; 16Zn; bal Cd	270	280	X	...	X	...	Electrical engineering, electrical motors
L-CdZnAg5	BCd73ZnAg 270–310	...	5Ag; 22Zn; bal Cd	270	310	X	...	X	...	Electrical engineering, electrical motors
L-CdZnAg10	BCd68ZnAg 270–380	...	10Ag; 22Zn; bal Cd	270	380	X	...	X	...	Electrical engineering, electrical motors
L-PbAg3	BPb97Ag 304–305	...	3Ag; bal Pb	304	305	X	...	X	...	Electrical engineering, electrical motors
L-PbAg5	BPb95Ag 304–365	...	5Ag; bal Pb	304	365	X	...	X	...	High-service temperature
L-PbAg2Sn2	BPb96AgSn 304–310	...	2Ag; 2Sn; bal Pb	304	310	X	...	X	...	Electrical engineering, electrical motors
L-CdAg5	BCd95Ag 340–395	...	5Ag; bal Cd	340	395	X	...	X	...	High-service temperature

(a) Compositional tolerances and allowed impurities, see DIN 1707. (b) F = Reverberatory furnace, I = Induction furnace, T = immersion soldering, K = soldering iron. (c) After ASTM B 32

Table 4.51 Rated internal working pressure (psig) for soldered water tube joints

Solder used for joints	Service temperature	Tube size, types K, L, and M, in. 1/4 to 1	1/4 to 2	2 1/2 to 4	5 to 8	10 to 12
95-5 tin-antimony	100	500	400	300	270	150
	150	400	350	275	250	150
	200	300	250	200	180	140
	250	200	175	150	135	110
95-5 tin-silver	100	525	330	235	...	...
	150	365	245	235	...	...
	200	275	170	170	...	...
	250	200	120	120	...	...
50-50 tin-silver	100	200	175	150	130	100
	150	150	125	100	90	70
	200	100	90	75	70	50
	250	85	75	50	45	40

Source: Ref 82

cally compatible with both aluminum and copper. During welding, the weld pool should be established toward the aluminum side of the joint, by directing the arc away from the silver-brazed face. Fusion of the filler metal with the brazed layer is thus effected without excessive melting. For this reason, welding conditions should be as cool as possible and inter-run temperatures kept low to avoid over-dilution of the brazing alloy.

In properly made joints of this nature, the electrical and mechanical properties are good, and excellent ductility can be retained. Joints have been produced on a commercial scale with consistent quality; they have provided satisfactory operation under severe thermal cycling. It must be stressed, however, that in service environments where electrochemical corrosion can occur, there is likely to be severe attack on the aluminum and weld metal.

Copper to Copper-Nickels. Welding copper-nickel alloys to each other presents no undue problems. In dissimilar metal joints between copper-nickel and copper, it is common practice to use copper-nickel rather than the other copper-base filler metals to minimize the risk of porosity, and also to produce weld metal with strength compatible with the copper-nickel side of the joint. The large difference in thermal conductivity between copper and copper-nickel will require close attention to preheating the copper side of the joint to ensure that full fusion of the copper is attained. For this reason, the welding arc should be directed toward the copper. There will be dilution of the

Table 4.52 Epoxy adhesive formulations for joining copper tube

Adhesive components	Parts by weight, %
Epoxy Novolac adhesive	
Part A (resin)	
Dow chemical DEN 431	100
Tamms industry silica flour EP-2400	100
Part B (curing agent)	
Ciba-Geigy XU225	82
Tamms industry silica flour EP-2400	50
PPG chemicals Hi-Sil 233	22.5
Epoxy adhesive	
Part A (resin)	
Shell chemical epon 830	100
Tamms industry silica flour EP-2400	200
Part B (curing agent)	
3M Co. Cardolite NC 540	65.7
Union carbide aminoethylpiperazine	4.2
PPG chemical Hi-Sil 233	30
L-24 Epoxy adhesive (control)	
Dow chemical DEN 431	100
Tamms industry silica flour EP-2400	100
PPG chemicals Hi-Sil 233	15
3-Diethylaminopropylamine (DEAPA)	7

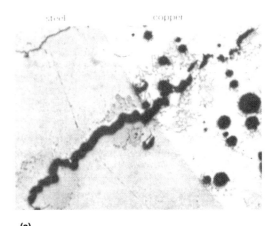

(a)

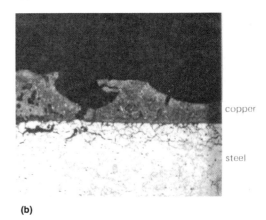

(b)

Fig. 4.109 Section through copper/steel weld clad interface showing what can go wrong without careful control. (a) Crack through interface and porosity in copper cladding. (b) Excessive penetration of copper into the steel substrate

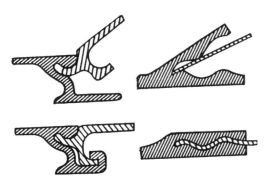

Fig. 4.108 Crimped joints applied for joining extrusions

copper-nickel weld metal by copper, and this must be taken into account when selecting the composition of the filler alloy in terms of maintaining matching corrosion resistance.

Copper to Aluminum Bronze. In welded joints of this nature, the aluminum bronze is often in cast form, the copper component wrought. Such joints present no undue problems providing the casting is sound. Filler metal should conform to compositions C13 and C20 as specified in BS 2901 (Table 4.53) (Ref 83).

Aluminum Bronze to Copper-Nickel. Such joints are satisfactorily made using filler wire matching the composition of the aluminum bronze.

Testing and Quality Control of Welded Joints

Inspection. Slag inclusions and porosity are the most common weld defects in copper-base weldments, and together with undercutting and incomplete fusion, they account for the majority of defects observed in these metals. The center-line cracking that occasionally occurs in single-phase alloys is probably the most dangerous defect.

Modern nondestructive examination methods, particularly computerized tomography (CT) and related techniques, provide the best resolution and the least ambiguity with regard to defect size, shape, and orientation. Providing weldments are accessible, these techniques can be used to identify and quantify all subsurface weld defects. Their major drawback is the high cost of the equipment required.

There are few situations in which conventional radiography could not be used successfully to inspect weldments in copper structures. On the other hand, radiography can be time-consuming and its resolution (minimum detectable defect size) depends on the dimension and configuration of the weld being examined. For example, a 1.6 mm ($1/16$ in.) diameter defect, detectable even to the novice radiographer in 12.7 mm ($1/2$ in.) welds, would be nearly invisible in a 150 mm (6 in.) thick joint. Radiography can detect slag inclusions and porosity; however, its ability to detect cracks varies strongly with the defect's orientation.

Dye penetrant (DP) examination, of which there are several variants, makes use of a colored or fluorescent powder, suspended in a suitable medium, which is sprayed or swabbed on the weld joint surface. The medium penetrates into surface cracks. After excess oil is wiped off and suitable developer is applied, any oozing from the cracks is visually detectable. Die penetrant tests are only sensitive to cracks that intersect the weld surface.

Ultrasonic (UT) inspection is also useful, although it is normally limited by the requirement that weld beads be ground flat. The UT method is fully applicable to copper metals. Unfortunately, the technique does not always pick up slag inclusions, and pattern interpretation becomes more difficult as the thickness of the material increases.

Powder Metallurgy (P/M)

Production of Copper and Copper Alloy Powder

Copper powder is produced by atomization, electrolysis, hydrometallurgy, and solid-state reduction; the first two methods are the most prevalent. Special applications, such as high-temperature superconducting ceramics, utilize powder made by aerosol processes and sol-gel, low-temperature

Table 4.53 Filler metals and joint characteristics for commonly encountered dissimilar metal combinations

Joint materials	Filler metals	Remarks
Copper/mild steel	Copper to BS 2901 C.7	Overlay steel with copper
Copper/alloy steels	Copper as above or nickel to BS 2901 NA32	Overlay steel with copper or with nickel
Cupro-nickels/steel	Cupro-nickels to BS 2901 C18	Overlay steel with cupro-nickel
Nickel-copper (Monel)/steel	Nickel-copper to BS 2901 NA33	Overlay steel with nickel-copper
Aluminum bronzes/steel	Aluminum bronze to BS 2901 C13, C20	Overlay steel with aluminum bronze
Manganese-aluminum-bronze/steel	Manganese-aluminum-bronze	Overlay steel with aluminum bronze
Silicon-bronze/steel	Aluminum bronze as above	Overlay steel with aluminum bronze
Brass/steel	Aluminum bronze as above	Overlay steel with aluminum bronze
Copper/aluminum	Aluminum/silicon to BS 2901 NG2	Overlay copper with silver-brazing alloy
Copper/cupro-nickel	Cupro-nickel as above	Direct weld
Copper/aluminum bronze and manganese-aluminum-bronze	Aluminum bronze with manganese-aluminum-bronze as above	Direct weld
Copper/brass	Aluminum bronze as above	Direct weld
Aluminum bronze/cupro-nickel	Aluminum bronze as above	Direct weld
Aluminum bronze/silicon bronze	Aluminum bronze as above	Direct weld

synthesis. Each method yields a powder having certain inherent characteristics (Ref 86–89).

Any copper alloy can be produced in powder form. Commercially available alloy powders include brasses ranging from 95%Cu-5%Zn to 60%Cu-40%Zn (and leaded versions of these alloys), nickel silvers, tin bronzes, aluminum bronzes and beryllium coppers. Most alloy powders are produced by atomization. Copper alloy products can also be pressed from blended powders, for example, mixtures of the several powders that, when blended, make up the desired composition. The alloy forms as the powder constituents interdiffuse during sintering.

Atomization is a relatively simple physical process. The metal is melted and allowed to flow through an orifice, where it is struck by a high-velocity stream of gas or liquid, usually water. This breaks the molten metal into droplets that solidify rapidly as powder particles. Atomization is followed by annealing in a reducing atmosphere to decompose surface oxides. Purity in most cases is better than 99%. Typical atomized particles are shown among the powders illustrated in Fig. 4.110 (Ref 86).

Electrolysis. Electrolytic copper powder is produced by a process similar to that used in electroplating except that plating conditions are deliberately altered to produce a loose, powdery deposit. After deposition, the powder is washed, annealed in a reducing atmosphere, milled, screened, classified, and blended to the desired particle size distribution. The starting material is pure cathode copper, and this level of purity can be carried over to the powder product. The powder is dendritic in shape. A wide range of powders having different apparent densities and high green strengths can be obtained by this method (Ref 86).

Hydrometallurgy. Hydrometallurgical processes are used to produce powder from cement copper, concentrates, or scrap copper. Copper is leached from these materials with sulfuric acid or ammoniacal solutions. The copper is precipitated from solution by reduction with hydrogen at 107 to 138 °C (225 to 280 °F), a partial pressure of hydrogen at 2794 MPa (400 psig) under a total pressure of 2930 MPa (425 psig). A thickening agent is added to minimize plating and control the particle size. The powder is pumped as a slurry to a centrifuge where it is separated from the liquid and washed. It is then dried in a reducing atmosphere, milled, classified, and blended. Purity averages higher than 99% Cu. The powder has fine particle sizes with relatively low apparent densi-

ties and, when pressed, high green strength (Ref 86).

Solid-State Reduction. In this method, copper oxides (including ordinary mill scale) are first ground to a controlled particle size, then reduced by carbon monoxide, hydrogen, or cracked natural gas at temperatures below the melting point of copper. Powder size and shape can be controlled within rather wide limits by varying the particle geometry of the oxides and the reducing temperatures, pressure, and flow of the gas. The resulting powder is milled, classified, and blended to the desired specifications. Product purity depends on the purity of the oxide. The powders produced by this method are normally porous and have high apparent densities. Pressed articles have high green strength. Particle shapes are irregular (Ref 86).

Blending and Mixing

Almost all powders must be blended and mixed thoroughly before forming to control sintering characteristics and obtain uniform products. Important mixing process variables include:

- Particle shape and size
- Specific gravities of the components
- Characteristics and condition of the particle surface

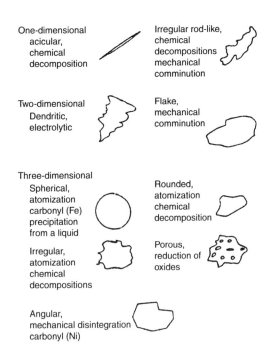

Fig. 4.110 Powder particle shapes

- Particle structure and plasticity
- The type of mixer employed

Powders are mixed in tumbling barrels, rotating drums, or double-ended cones fitted with baffle plates. Conventional cement mixers and edge runner mills have also been adapted to the purpose. Mixers not fitted with baffles contain balls or rods to ensure that alloy components coat one another. Organic lubricants may be added during mixing.

When blends of very hard and very soft powders are mixed, for example, diamond dust with bronze, an undesirable film of work-hardened metal usually forms around the soft particles. The presence of graphite or the addition of a volatile solvent reduces this tendency (Ref 90).

Compaction and Pressing

Consolidation and pressing are usually performed in closed dies, although roll compaction, isostatic compaction, extrusion, and forging are also practiced. Parts can also be slip cast. Uniform densification is favored by particle size distributions and pressing conditions that permit uniform packing in the die. Pure copper P/M parts require relatively low pressures (Table 4.54). An initial compacting pressure of 234 to 276 MPa (34 to 40 ksi) has been recommended for thin sections, although higher pressures can be used for heavier sections. The objective is to permit the escape of gases and water vapor formed by the internal reduction of oxides during sintering. Compacting pressures that are too high will prevent proper sintering in the center of the compact.

Sintering

Sintering can be performed at one of several stages in the production of P/M products. The sequence and nature of the operation is determined by the process employed. Sintering proceeds in three stages, during which particles bridge, recrystallize, and diffuse. Figure 4.111 shows the progress of densification in compacted copper powder as a function of time and temperature. Figure 4.112 shows the porosity and density of sintered bronze powder as a function of sintering temperature. The rate of sintering has a significant effect on properties. Sintering rate can be modified by physically or chemically treating the powder or compact or by incorporating reactive gases in the sintering atmosphere. The use of reducing gases to

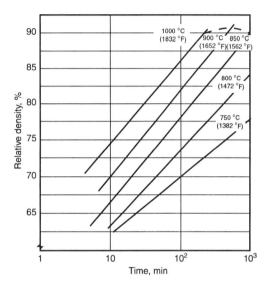

Fig. 4.111 Effect of sintering temperature and time on densification of copper powder compacts

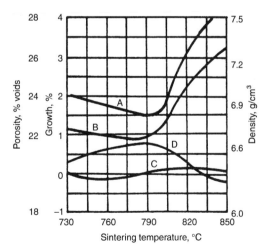

Fig. 4.112 Effect of sintering temperature on the physical properties of bronze, 90 parts copper, 10 parts tin, 2 parts graphite. (a) Porosity (% voids). (b) Axial growth. (c) Radial growth. (d) Density

Table 4.54 Typical compacting pressures and compression ratios for copper powder products

Typical P/M part	Compacting pressure		Compression ratio
	tons/in.^{2}	MPa	
Brass parts	30–50	414–690	2.4:1 to 2.6:1
Bronze bearings	15–20	207–276	2.5:1 to 2.7:1
Copper-graphite brushes	25–30	345–414	2.0:1 to 3.0:1
Copper parts	15–18	207–248	2.6:1 to 2.8:1

remove oxide films from copper powder grains inside compacts has been known for years. The practice improves densification, strength, and electrical conductivity of the compacts. Chemical treatments, such as with aqueous formic acid, are also beneficial as indicated in Fig. 4.113 (Ref 86).

In liquid-phase sintering, mixed powders are sintered at temperatures below the melting point of the high-melting constituent, but above that of the low-melting constituent. In the copper-tin system, for example, the tin melts and alloys with the copper to form a bronze. Figure 4.114 shows the structure of a typical sintered bronze part. In the iron-copper system, the copper melts and becomes saturated with iron. The copper-iron alloy then diffuses into the iron skeleton, causing it to expand. Pores remain at the sites vacated by the copper. In both the copper-tin and the iron-copper systems, growth or shrinkage of compacts can be modified by the addition of graphite. In the copper-tin-graphite system, sintering is inhibited by mechanical separation of the constituents and, as a result, there is less expansion. In the iron-copper-graphite system, the proportion of liquid phase is increased by formation of a ternary iron-copper-carbon eutectic, which restricts expansion (Ref 86).

Powder Forging

Powder forging is a process in which preforms made by conventional P/M techniques are cold or hot forged in closed dies. A hot-forging process developed in Japan utilizes brass powder pulverized from machining swarf or loosely sintered cut chips. In the so-called loose sintering process, powder is placed in a vessel that lends its shape to the preform. The process has been successfully applied to Cubralloy 11, an alloy similar to C95300 and C95400, aluminum-bronze casting alloys (Ref 86).

Sintering requirements for preforms differ somewhat from those for standard P/M parts, as dimensional considerations are of only secondary importance. Hot- or cold-working require different degrees of sintering, depending upon the nature of the forming operation and metal flow required. Use of higher sintering temperatures causes rounding of the porosity in the preform and increased contact between particles, thus making the preform more ductile and stronger. Higher temperatures also facilitate reduction of oxides in the preform.

Fabrication of Composite Materials

Composites are defined as multiconstituent materials formulated in which constituents generally retain their basic structure and remain as discrete phases. The oldest known examples among the copper metals are the leaded brasses, which are

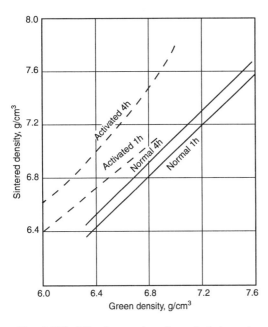

Fig. 4.113 Effect on density of sintered product of activating copper powder with aqueous formic acid before sintering

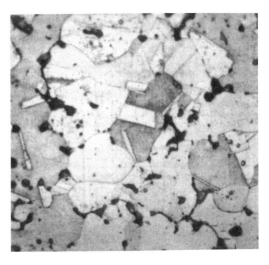

Fig. 4.114 Microstructure of sintered bronze (magnification: 200×)

not true alloys in that lead and copper phases are nearly insoluble. Each constituent, lead and the brass matrix, contributes to the desired property, in this case superior machinability.

Infiltration

Many metallic composites are formed by infiltrating one component (as a liquid) into interstices in another, usually a powder compact.

Tungsten Fiber Composites. Copper can be infiltrated into a bundle of continuous, unidirectional type 218CS tungsten filament at 1205 °C (2200 °F) in vacuum or hydrogen to form composites with accurately aligned unidirectional fiber orientation (Ref 54).

Graphite-copper composites are made by liquid-phase infiltration into or through graphite laminate, as shown schematically in Figure 4.115 (Ref 91).

Hot Isostatic Extrusion

The hot isostatic extrusion process, described earlier in this chapter, is also suitable for the extrusion of composite materials. Its large extrusion ratio strongly bonded composites and, because compaction, sintering, and extrusion are performed simultaneously, the process is relatively rapid (Ref 92).

Copper-Base Superconductor (SC) Composites

NbTi Superconductors. The best-known and commercially most important hot isostatically extruded product is superconducting cable, in which niobium-tin or niobium-titanium fibers are embedded in a matrix of OF copper. Figure 4.116 shows a schematic diagram of the processes used for NbSn and NbTi superconductors. The cable fabrication process depends primarily on the intended

application, which dictates the number and size of the superconducting filaments and the volume of stabilizer (copper) needed to protect the magnet (Ref 37).

Special attention must be given to grain size of the high purity copper, because coarse grains in as-cast and/or insufficiently deformed copper have been known to compromise the integrity of the superconducting composite during isostatic pressing and extrusion. All billet components must be free of oxides, inclusions, and other surface defects before assembly to avoid contamination of the copper and NbTi core once the composite is drawn to fine wire. In general, components are degreased, acid etched, rinsed in water, acetone, and a drying agent such as methanol. The parts are then blown dry and stored in a nitrogen atmosphere until they are assembled. Assembly takes place in a low-level clean-room environment.

Monofilamentary wires are made by co-extruding a large ingot of niobium-titanium alloy in a copper can. Billets are 200 mm (8 in.) in diameter, with a 120 mm (4.75 in.) diameter core. Extruded composite is drawn to final size, typically 0.5 to 1.0 mm (0.02 to 0.04 in.) in diameter. Intermediate heat treatments precipitate the α-phase of the titanium alloy. The volume ratio of copper to superconductor is approximately 1.8 to 1 but may vary depending on application.

Multifilament conductors incorporating less than 200 niobium-titanium filaments are made by inserting super conductor (SC) rods in close packed array of gun-drilled holes in a solid copper billet, then processing the billet to wire. Billet diameters are typically 250 to 280 mm (9.8 to 11 in.).

Conductors with more than 200 filaments are assembled using either the so-called kit method, the restacked monofilaments method, or the

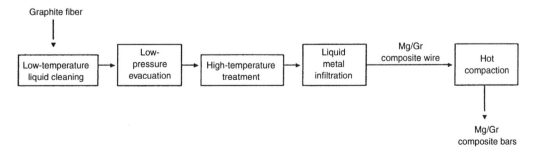

Fig. 4.115 Schematic representation of metal-graphite composite liquid infiltration process

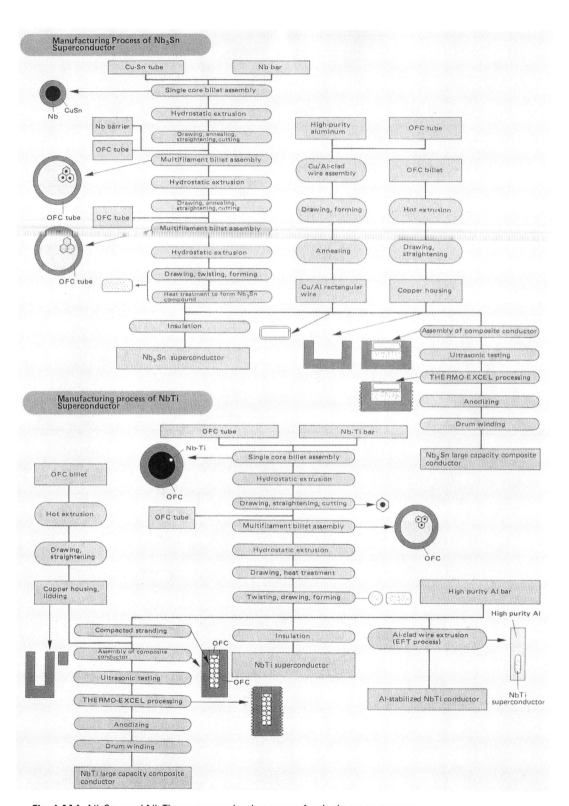

Fig 4.116 NbSn and NbTi superconductor manufacturing processes

restacked drilled billets method. In the first approach, also referred to as the CBA/Fermi kit method because it was used to manufacture the strand for the Fermilab Tevatron magnets, 3.18 to 7.11 mm (0.125 to 0.280 in.) diameter Nb-Ti alloy rods are inserted into round-bored hexagonal copper tubes. From several hundred to 2000 of these tubes are stacked in a hexagonal close packed array. A copper can with an outside diameter of approximately 250 mm (10 in.) is lowered over the assembled filaments, the ends sealed, and the billet processed to fine wire, as shown in Fig. 4.116 (Ref 87).

Diffusion barriers may be needed to prevent the formation of brittle copper-titanium intermetallic compounds (Fig. 4.117). Such barriers are made by wrapping one or more sheets of niobium foil around the niobium-titanium alloy ingot before encasing the assembly in a copper or copper-manganese can. Bonding occurs during extrusion, which is conducted at high temperatures and with high extrusion ratios.

The preparation of niobium-titanium monofilaments incorporating a diffusion barrier is shown schematically in Fig. 4.118. The initial niobium-titanium ingot is 146 mm (5.75 in.) in diameter by 610 mm (24 in.) long. Extruded monofilament is drawn to the proper diameter, then drawn through a hexagonal die, cut to length, and restacked into the desired multifilamentary array. Round monofilamentary subelements can also be used, although these result in a less uniform array. Billet diameters range from 178 to 203 mm (7 to 8 in.) for monofilaments, 203 to 203 mm (8 to 11 in.) for

drilled billets, and 250 to 356 mm (9.8 to 14 in.) for restacked monofilaments.

Hot isostatic pressing (HIP) is used to reduce void volume before extrusion. The void space in a drilled billet as assembled is relatively small, but a billet containing several thousand hexagonal subelements may contain 6% void space, and twice that much for round subelements. HIP, which is conducted over several hours in argon at 500 to 650 °C (930 to 1200 °F) and pressures between 103 and 206 MPa (15 and 30 ksi), is followed by low-temperature extrusion to add cold work.

Extrusion ratios typically range from 10 to 1 to 20 to 1. Insufficient reductions can lead to center burst, an internal tensile tear. To facilitate extrusion, billet and tooling are coated with MoS_2 or graphite lubricant. Extrusion is conducted relatively slowly (3 to 7 mm/s (0.12 to 0.28 in./s)) to ensure uniform temperatures over the length of the product. Rods are water quenched after extrusion to avoid grain coarsening.

A15-Type Superconductors. The second important type of low-temperature superconductors is based on A_3B intermetallic compounds typified by Nb_3Ge and Nb_3Sn. They are called A15 materials after their crystal structure. As might be expected, these compounds are exceedingly brittle and must be supported in use, a feat that is most often accomplished by encasing them in copper substrates.

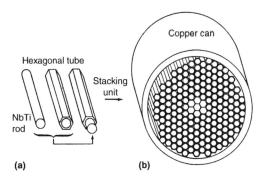

Fig. 4.117 Schematic of CBA/Fermi kit billet assembly. (a) Shows how NbTi rod is inserted into a copper tube with a hexagonal cross section and circular bore. (b) Individual stacking units arranged in a hexagonal close-packed (hcp) array in a 250 mm diam. copper can

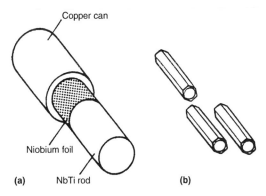

Fig. 4.118 Schematic illustration of diffusion barrier-clad monofilament assembly. (a) Diffusion barrier is obtained by wrapping an NbTi rod with niobium foil and placing foil-wrapped rod in a copper can. (b) Extruding and drawing of copper can assembly yields monofilamentary hexagonal rods with diffusion barrier

Because the intermetallics cannot be deformed, they must be formed in situ, for example, by reacting the constituent elements after the filament or ribbon shape has been established and often after the magnet coil has been formed. This is done (for Nb₃Sn, for example, one of the more popular compositions) by encasing wires or filaments of niobium in either tin or a tin-containing bronze, forming the magnet coil, then annealing the compact to drive the solid-state reaction that forms the intermetallic. The bronze gives up its tin by diffusion, leaving behind a matrix of dilute bronze or copper. The matrix then remains to serve as a thermal and electrical shunt, as described above.

For tape-type conductors, the SC intermetallic can be formed by vapor depositing bronze onto the surface of pure niobium ribbon. Bronze-Nb-Bronze sandwich structures can also be pack-rolled, formed, and reacted. Liquid Nb-Al-Ge alloys have also been quenched onto moving copper tape.

The manufacture of multifilamentary superconductors by the co-extrusion of niobium or vanadium rods in a tin-bronze matrix closely resembles the techniques described above for NbTi-type materials. In the so-called modified jelly-roll process, a sandwich structure of bronze sheet and niobium or vanadium expanded metal is co-wrapped, then compacted and drawn. Another process utilizes niobium tubes lined with tin bronze or copper over a tin core. Numerous other fabrication processes continue to be developed to accommodate the mechanical or metallurgical limitations of the particular SC materials or to satisfy the physical requirements of the finished magnet or conductor.

Electroplating

One of the idiosyncracies of copper and its alloys is that while their high natural corrosion resistance often obviates the need for protective electroplating, they generally accept such coatings very readily. Copper itself is also quite easy to plate onto other metals. Plating processes are addressed in numerous technical specifications and standards. Recommended practices published by ASTM are listed in Table 4.55 (Ref 93).

Surface Preparation

Among the more important precautions to be observed are careful cleaning and surface preparation of the base metal. Preparation normally includes precleaning with a solvent to remove most of the soil; intermediate cleaning with alkaline cleaners; electrocleaning to remove the last traces of solids and other contaminants that are especially adherent; and acid pickling and surface conditioning to remove light oxide films formed during previous cleaning processes. The last step also microetches the surface.

Table 4.55 Specifications and recommended practices for copper electroplating

Specification	Recommended practices
Copper plating	
AMS 2418	Copper plating
MIL-C-14550 (ord)	Copper plating
Copper plating in multiplate systems	
ASTM A 166	Electrodeposited coatings of nickel and chromium on steel
ASTM B 141	Electrodeposited coatings of nickel and chromium on copper and copper-base alloys
ASTM B 142	Electrodeposited coatings of nickel and chromium on zinc and zinc-base alloys
ASTM B 200	Electrodeposited coatings of lead on steel
ASM 2412	Plating, silver, copper strike, low bake
ASM 2413	Silver and rhodium plating
ASM 2420	Plating, aluminum for solderability, zincate process
ASM 2421	Plating, magnesium for solderability, zincate process
QQ-N-290	Nickel plating (electrodeposited)
Surface preparation	
ASTM A 380	Descaling and cleaning stainless steel surfaces
ASTM B 183	Preparation of low-carbon steel for electroplating
ASTM B 242	Preparation of high-carbon steel for electroplating
ASTM B 252	Preparation of zinc-base die castings for electroplating
ASTM B 253	Preparation of and electroplating on aluminum steel
ASTM B 254	Preparation of and electroplating on stainless steel
ASTM B 281	Preparation of copper and copper-base alloys for electroplating
ASTM B 319	Preparation of lead and lead alloys for electroplating
ASTM B 322	Cleaning metals before plating
MIL-HDBK-132 (ord)	U.S. Military Handbook, "Protective Finishes"

Copper Electroplating

Copper can be electrodeposited from numerous electrolytes. Typical compositions and operating conditions for alkaline copper plating baths are given in Tables 4.56 (cyanide) and 4.57 (pyrophosphate). Table 4.58 gives compositions and operating conditions used with acid baths.

Water purity is important for all plating operations. Iron in the water causes roughness in the deposits if the pH of the electrolyte is above 3.5, where iron can be precipitated. Chlorides, in concentrations greater than about 0.37 g/L (0.05 oz/gal), promote nodular deposits. Calcium, magnesium, and iron precipitate in the bath. Organic particles cause pitting of deposits.

When plating in concentrated sodium- or potassium-containing electrolytes, distilled or deionized water, or steam condensate should be used for solution make-up and replenishment. Most tap water, because of the relatively high content of calcium salts, will cause roughness of the deposit when added to these solutions. Softened water may be used, but the results will be less satisfactory. A slight precipitation is always noticeable when softened water is added to a cyanide electrolyte.

Known causes of roughness in copper deposits are drag-over from cleaners (which results in the formation of insoluble silicates in the electrolyte), poor anode corrosion, insoluble metallic sulfides due to sulfide impurities, organic matter in the water used for make-up and especially in rinse tanks, and insoluble carbonates due to calcium and magnesium in hard water.

Agitation during plating permits the use of higher current densities, which, in turn, makes possible more rapid deposition of copper. Permissible increases in current density vary for different baths. Cyanide and acid baths are agitated at the cathode, with air, or both. Pyrophosphate baths are air agitated (Ref 93).

When air agitation is used in periodic reverse plating, all air-line pipes should be covered with vinyl plastisol rack coating. The rack coating must be counterbored with a larger drill at the air holes to keep any swelling of the material from blocking the holes. Ultrasonic vibration does not materially improve the appearance of copper electroplates, but it can improve plating speed by permitting higher current densities without risk of burning the parts. The economics of ultrasonic vibration are questionable in view of the high efficiency of high speed baths at fairly high current densities (Ref 93).

Alkaline Plating Baths. Alkaline pyrophosphate baths are used primarily to produce heavy deposits. The alkaline cyanide baths can be better controlled to produce thin deposits of relatively uniform thickness on all surfaces. They have the best macro throwing power and are the most widely used.

Table 4.56 Typical compositions and operating conditions of cyanide-copper plating baths

| Constituent or condition | Dilute cyanide | Rochelle cyanide | | | High concentration | |
		Standard barrel	Low concentration(a)	High concentration(a)	Sodium cyanide(b)	Potassium cyanide(b)
Bath make-up, oz/gal						
Copper cyanide	3.0	6.0	3.5	6.0	16.0	8.0
Copper cyanide	4.4	9.0	4.6	7.0	18.0	12.5
Sodium carbonate	2.0	...	4.0	8.0	2.0	2.0
Sodium hydroxide	to pH	...	to pH	to pH	4.0	...
Rochelle salt	...	6.0–10.0	6.0	12.0	...	...
Potassium hydroxide	...	1.0–2.0	...	...	...	5.6
Bath analysis, oz/gal						
Copper	2.1	4.2	2.5	4.0	11.3	5.6
Free cyanide	1.0	2.4	0.8	2.0	0.5	1.0
Operating conditions						
Temperature, °C (°F)	32–43	54–60	54–71	79	77–82	77–82
Current density, amp/ft^2	(90–110)	(130–140)	(130–160)	(175)	(170–180)	(170–180)
Cathode efficiency	10–15	...	20–40	50	30–60	30–60
Voltage	30	...	50	70	100	100
pH colorimetric	6	6(c)	6	6	6	6
Anodes	12.0–12.6	...	12.0–12.6(d)	13	>13.0	>13.0
	Copper; steel	Copper	Copper	Copper	Copper	Copper

(a) Low concentration typical for strike; high concentration typical for plating. (b) Commonly employed with addition agents, as proprietary or patented processes. (d) For zinc-base die castings, maintain temperature at 60 to 70 °C (140 to 160 °F) and pH between 11.6 and 12.2. (c) At 6 V, the bath will draw approximately 2 amp/gal through the solution at 12 V, approximately 3 amp/gal

Pyrophosphate Bath. If pyrophosphate electrolytes are used, conventional cleaning cycles are satisfactory. A preliminary strike should be applied to steel, zinc-base die castings, magnesium, and aluminum. The strike solution may be a dilute cyanide copper, dilute pyrophosphate copper, or nickel. If a cyanide copper strike is used, adequate rinsing (or, preferably, a mild acid dip) following the strike is recommended before final pyrophosphate copper plating. These baths are used primarily to produce deposits ranging from 0.013 to 0.025 mm (0.0005 to 0.001 in.) in thickness. Operating conditions for one of the numerous patented pyrophosphate baths are given in Table 4.57 (Ref 93).

Cyanide Plating Baths. Although dilute cyanide baths (and the Rochelle cyanide baths) simultaneously act as cleaners during the plating operation, this does not eliminate the need for thorough cleaning before electroplating. High-concentration sodium cyanide and potassium cyanide electrolytes have virtually no surface-cleaning ability during plating because no hydrogen is evolved. In addition to thorough cleaning beforehand, parts must also first receive a 0.013 mm (0.0005 in.) copper strike in a dilute cyanide bath.

Dilute cyanide and Rochelle cyanide baths are used mainly to deposit a 0.0013 to 0.0025 mm (0.00005 to 0.0001 in.) strike coating of copper before copper plating or electrodepositing other metals. The high-concentration Rochelle cyanide bath can be used efficiently for plating up to about 0.00076 mm (0.0003 in.) thickness. Cyanide and Rochelle cyanide electrolytes may both be used for still-tank plating of racked work. With a modification in composition (Table 4.56), the Rochelle electrolyte may be used for barrel plating. Although the Rochelle cyanide can be used for periodic-reverse plating with good results, better results are produced by either current interruption or periodic-reverse plating in the high-concentration potassium cyanide bath. These baths exert a significant cleaning action on the part surfaces during the plating operation. Although plating baths should not intentionally be used for cleaning, the cleaning characteristics of these cyanide baths are advantageous in that difficult-to-clean parts can be given a copper strike in one of these baths with a high degree of success (Ref 93).

High concentration cyanide baths are characterized by relatively high operating temperature, high copper content, and rapid operation; deposition rates are three to five times faster than the rates for the dilute cyanide and Rochelle cyanide baths. Parts to be plated in the high-concentration electrolytes must be thoroughly cleaned. Otherwise, the plate will be of inferior quality and the bath will require frequent purification for removing organic contaminants.

The potassium complexes formed by combining potassium cyanide with copper cyanide are more soluble than those formed when sodium cyanide is used. This causes higher metal content and makes possible higher rates of deposition with the potassium complexes, sodium cyanide, high-concentration bath. The potassium bath has more operating flexibility than the sodium bath and is favored because it raises the burning point and permits higher current density without the use of robbers.

Current interruption is frequently used for operating high-concentration electrolytes to produce greater leveling and more nearly uniform distribution of copper on intricate shapes. Therefore, plating time and the amount of metal normally required for plating intricate shapes to a specified minimum thickness are reduced. With the use of current interruption, uniform thicknesses in the range of ±20% of specified thickness can be achieved.

With proprietary additives, high-concentration sodium and potassium cyanide baths (Table 4.56), are used to produce deposits of various degrees of brightness and leveling, in thicknesses ranging from 0.0076 to 0.051 mm (0.0003 to 0.002 in.). Thick deposits are ductile and bright. The high throwing power of the electrolyte produces adequate coverage of sufficient thickness in recessed areas under most plating conditions. These baths may, however, produce pitted deposits unless antipitting additives are used.

Before being plated in high-concentration baths, parts receive a strike coating of copper

Table 4.57 Concentration limits and operating conditions for a copper-pyrophosphate plating bath

Concentration limits(a), oz/gal	
Copper	3.0–4.0
Pyrophosphate	23–28
Ammonia	0.07–0.27
Weight ratio, P_2O_7 to copper	7.0–8.5
Nitrate	1–2
Operating conditions	
Temperature	100–140 °F
Current density	10–70 amp/ft^2
Cathode efficiency	95–100%
Voltage at tank	2–5 V
pH (electrometric)	8–8.5(b)
Anodes	Copper(c)

(a) Fluid ounces. (b) May be maintained with pyrophosphoric acid and potassium hydroxide. (c) Electrolytic or rolled electrolytic copper; not bagged about 0.0013 mm (0.00005 in.) thick from a dilute cyanide copper electrolyte.

Acid Plating Baths. Electrodeposition of copper from acid baths is used extensively for electroforming, electrorefining, manufacturing copper powder, and decorative electroplating. Acid copper plating baths contain copper in the bivalent form and are more tolerant of ionic impurities than alkaline baths, but have less macro throwing power and, consequently, poorer metal distribution. Acid baths, however, have excellent micro throwing power, which makes them effective in sealing porous die castings.

A cyanide copper or nickel strike must be applied to steel or zinc-alloy die castings before they are plated in acid copper. Assuming complete coverage, the strike may be as thin as 0.0013 mm (0.00005 in). A nickel strike will prevent deposition of copper by immersion, thus precluding peeling of the plate.

The copper sulfate bath (Table 4.58), is the most frequently used of the acid copper electrolytes and has its primary use in electroforming. It is used fairly extensively for applying copper as an undercoating for bright nickel-chromium, as on automotive bumpers, and plating cylinders used in rotogravure printing. It can produce heavy copper deposits to any thickness, and with additives, it produces a bright deposit with good leveling characteristics. Semi-bright deposits produced from sulfate baths are easily buffed (Ref 93).

The copper fluoborate bath (Table 4.58) is capable of high-speed plating and producing heavy, dense deposits to any required thickness. This bath is simple to prepare, stable, and easy to control; operating efficiency approaches 100%. Deposits are smooth and have excellent appearance. Deposits from the low-copper bath operated at 49 °C (120 °F) are soft and easily buffed to a high luster. The addition of molasses to either the high-copper or the low-copper bath operated at 49 °C (120 °F) results in deposits that are harder and stronger. Smooth coatings up to 0.51 mm (0.020 in.) thick can be obtained without addition agents. For thicknesses greater than 0.51 mm (0.020 in.), addition agents must be used to avoid excessive porosity.

Quality Control Parameters. Variations in processing during surface preparation or during plating have significant effects on the quality of the copper electrodeposit. Certain variations can adversely affect the adhesion of copper to the substrate metal, as well as brightness, adhesion, porosity, blistering, roughness, hardness, solderability, and leveling.

Brightness. Bright copper coatings can be obtained either by adding brighteners to the electrolyte or by buffing the electrodeposited copper coating. Coatings deposited from acid baths, especially from sulfate baths, are easily buffed to a high luster; electrodeposits from cyanide baths are more difficult to buff.

Buffing or electropolishing the work before plating it in an electrolyte not containing a brightener results in the deposition of a smooth and sometimes semibright coating. If an electrolyte containing a brightener is used, the luster of the coating will be enhanced. The high cost of labor is

Table 4.58 Typical compositions and operating conditions for acid copper plating baths

Constituent or condition	Copper sulfate	Copper fluoborate	
		Low copper	High copper
Bath make-up, oz/gal			
Copper sulfate, $CuSO_4 \cdot 5H_2O$	26–33	...	...
Sulfuric acid, H_2SO_4	4–10	...	...
Copper fluoborate, $Cu(BF_4)_2$	...	30	60
Fluoroboric acid, HBF_4	...	to pH	to pH
Bath analysis, oz/gal			
Copper	5.2–6.6	8	16
Sulfuric acid	4–10	...	...
Specific gravity at 80 °F, Be	...	21–22	37.5–39
Operating conditions			
Temperature, °F	70–120	80–170	80–170
Current density, amp/ft^2	20–100	75–125	125–350
Cathode efficiency, %	95–100	95–100	95–100
Voltage	6(a)	6	6–12
pH (colorimetric)	...	0.8–1.7	<0.6
Anodes	Copper	Copper	Copper

(a) Average; bath is generally used at less than 6 V, but higher voltage may be required.

a primary concern when buffing is considered. Plating from high-concentration cyanide baths with current interruption or periodic reversal of current also improves the luster of copper coatings.

Adhesion. The type of substrate surface and proper preparation of the surface before plating are important for good adhesion. In general, cast and other porous surfaces are less receptive to good-quality electrodeposited coatings than are wrought surfaces. The kind of material to be electroplated with copper is another important consideration. For magnesium-base or aluminum-base die castings, the zincate layer between the substrate and the copper deposit is a critical factor to control. For a properly activated stainless-steel surface, a controlling factor for assured copper adhesion is the speed with which the workpiece is immersed in the bath. Some brighteners, especially organic ones, adversely effect adhesion of later electrodeposited coatings. Adhesion of copper electrodeposited from acid baths can be assured only if a strike from a cyanide copper bath precedes acid copper plating.

Porosity. The degree of porosity in a copper coating depends upon the type of copper plating bath selected, the make-up and control of the electrolyte, the basis material to be plated, and the condition of the surface before plating. A porous surface has high surface area and, therefore, requires high current density for efficient plating. Impregnants, such as lead and tin, can be used to fill the pores of powder metallurgy parts before plating.

Blistering of copper electroplate, particularly when the plated work is subjected to heat, occurs mostly on zinc-base die castings or on parts made of magnesium or aluminum in any form. It is a result of poor casting quality or poor surface preparation, or both. Blistering of copper plate on zinc-base die castings plated in a Rochelle electrolyte and then subjected to heat can be reduced by lowering the pH of the Rochelle bath from the range of 12.0 to 12.6 to about 9.4. *Caution:* Operation at a low pH value may result in the release of hydrogen cyanide gas, which is highly poisonous. It is, therefore, imperative that the plating bath be thoroughly vented (Ref 93).

Blistering of copper-plated magnesium and aluminum, especially during later soldering or heating in service, is mainly caused by poor adhesion at the zincate-copper interface. Unfortunately, blistering often does not become visible until after a new additional electrodeposit has been applied and the coating has been subjected to heat. It is good practice to expose all copper-plated magnesium and aluminum parts to controlled heat representative of that to be encountered later. This exposure will cause blistering before deposition of additional metal coatings if there is poor adhesion at the interface.

Roughness in copper deposits is often caused either by foreign particles present in the bath due to faulty cleaning or by migration to the cathode of metallic copper of cuprous oxide particles that form at the anode. Such roughness is especially likely to occur with the sodium cyanide high-concentration electrolytes, and it can be prevented by the use of diaphragms or anode bags.

Hardness. Without the use of addition agents, cyanide electrolytes produce harder coatings than acid baths. However, the use of addition agents allows the hardness of copper deposits from any electrolyte to be increased. Hardness of the electrodeposit is in most cases associated with fine grain size. Hardness can also be increased by promoting a preferred crystal orientation in the absence of grain refinement. Changes in the copper sulfate or sulfuric acid concentration of acid baths have little effect on the hardness of copper plate.

Solderability. Good solderability requires oxide-free surfaces and coatings that are sufficiently thick and adherent. Direct soldering of electrodeposited copper is not unusual for parts subsequently used in hermetically sealed assemblies. Soldering is a routine operation for aluminum and magnesium electronic parts used in aerospace applications. A copper strike and copper plate frequently comprise the initial metal coating over the zincated surfaces of these parts. After this, electrodeposits of other metals are applied before soldering. A top coat of tin or cadmium plate that has been chromate conversion coated is a particularly effective means of producing good properties. A combination of solderability and corrosion resistance for parts exposed to the atmosphere will result.

Leveling has a significant effect on the appearance of the copper coating, as well as on the appearance of the final product when other metals are additionally plated over the copper. Often, the substrate metal does not have the degree of smoothness that is desired of the plated surface. Metal substrate surfaces can be mechanically or chemically worked to reduce surface roughness before electroplating. However, some copper electrolytes can produce strong leveling in the deposited coating, thus reducing costs related to elaborate prepolishing or other means of smoothing the surface. The high-concentration potassium cyanide electrolytes produce excellent leveling when certain inorganic agents are added and interrupted current is used during plating. Although somewhat less effective, high-concentration sodium cyanide baths, mixed sodium and potassium electrolytes, and Rochelle cyanide electrolytes also have good leveling characteristics with interrupted or periodically reversed current (Ref 93).

Electroforming

Industrial Electroforming

Electroforming is the process of forming products through the electrodeposition of metal in a mold or mandrel that is later removed. It generally differs from ordinary electroplating work in the thickness of the deposit. Electrolytic refining and electrowinning are not classified as electroforming operations. Electroforming is considered an art because its success relies upon the operator's skill and experience. Electroforming is widely used in the production of products requiring extremely faithful reproduction, as in molds for compact audio disks, phonograph records, holograms, lenses, and a huge variety of decorative items. Copper, nickel, and iron are the most commonly electroformed materials (Ref 94, 95).

Mandrel (Mold) Materials. With adequate handling, almost all solid materials can be used as mandrels. Mandrels can be reusable or single-use items. Single-use mandrels are stripped by melting or chemical dissolution; plastic mandrels are simply heat-softened. Copper is selected as a mandrel material when high conductivity is required and when the oxidation commonly associated with its surface can be tolerated, either by the electroforming process or the subsequent application. This requires some explanation: it is common practice in the electroforming industry to utilize a negative mold to form a positive product, and then to use the positive product to form another mold in a tree-like expansion of production facilities. The fine detail available with electroforming makes this possible. Copper can be used as both mold and product. On the other hand, electroformed copper lacks the hardness available with, for example, nickel- or iron-base electroforms, and it may not be suitable for products, such as compression molds, where natural abrasion may degrade its surface (Ref 95).

Plastic and other dielectric molds are rendered electrically conductive by applying suitable coatings. Among the most widely used are bronze powder contained in a binder, such as lacquer or varnish; dry graphite on a substrate, such as wax or oil cloth; graphite in a suitable binder (this can also be used as a parting agent on metallic mold surfaces); and chemically deposited metallic films. Of these, bronze is considered to give only fair results, as it has difficulty reproducing fine surface details, and graphite on wax or oilcloth gives excellent results.

The electroforming of copper is normally performed from sulfate or fluoborate baths. Baths must be capable of depositing relatively thick films efficiently, yet possess the high microthrowing power needed to reproduce fine details. Typical bath compositions and electroforming conditions are given in Table 4.59 (Ref 95).

Artistic Electroforming

Artistic electroforming begins by preparation of an expendable mold made from beeswax mixed 50% with 770-mesh graphite powder. An artist models the desired pattern on the mandrels, and they are fixed to a suitable substrate with an organic adhesive. Electrical resistivity is adjusted by dusting fine graphite powder onto the surface, which is slightly heated to facilitate adhesion (Ref 96).

Mandrels are next dipped into a 10 g/L sodium hydroxide solution containing proprietary additives for 5 min at room temperature; statically water-rinsed; dipped in sulfuric acid (10 vol%) for 1 min; again statically water-rinsed, and swilled in running water.

Copper is electroformed from a bath containing copper sulfate ($CuSO_4·H_2O$) of 150 g/L (20 oz/gal) (range: 120 to 180 g/L, 16 to 24 oz/gal) and sulfuric acid (H_2SO_4) of 75 g/L (10 oz/gal) (range: 50 to 100 g/L, 7 to 13 oz/gal) at room temperature. Its current density is initially plated at 50 A/m^2 in the still bath until the entire surface

Table 4.59 Commercial copper electroforming baths

Composition	Sulfate	Fluoborate
Copper sulfate, oz/gal	30–32	...
Copper fluoborate, oz/gal	...	30–60
Fluoboric acid, oz/gal	...	to pH 0.3–1.4
Sulfuric acid, oz/gal	6–10	...
Temperature, °C (°F)	27–43 (80–110)	27–49 (80–120)
Current density, A/ft^2	30–150	75–300
Baumé at 27 °C (80 °F)	...	29–31
Hardness, DPH	51–170	40–75
Tensile strength	241–290 (35–42)	117–220 (17–32)
Elongation (51 mm, or 2 in.), %	5–25	3–14
Internal stress (tensile), MPa (ksi)	0–34 (0–5)	...

is covered with copper. Later, the current density is raised to 324 A/m^2 with air agitation. Duration of plating is 80 h. Cast copper is bagged in polyester cloth. Lead sheet is used as an anode when CuSO$_4$ concentration increases to 180 g/L and above. The anode to cathode area is 1 to 1.The tank material is hard rubber-lined steel.

The electroform is smooth and ductile. It requires no post-plating treatment. The average thickness of copper in the electroform was 3 mm (0.12 in.).

Finishing

Abrasive Blast Cleaning

Abrasive cleaning is used to remove molding and core sand and investment material from the surfaces of copper-base castings. Selection of the proper kind and particle size of grit determines the type and color of the finish. Coarser grits clean faster but give a rougher finish (Ref 38).

Steel shot is used for general cleaning to remove sand and slight surface imperfections from the casting after mold shakeout. Sands are used to blend-in surface areas, to remove heat-treatment scale, and to produce a uniform surface texture. Graded bronze chips, together with the regular commercial abrasives, are used in some applications to impart a better color and finish. Abrasive blast cleaning is seldom used to produce decorative finishes on copper alloys. Dry abrasive cleaning of beryllium copper is usually confined to castings.

Wet blasting offers a means of cleaning previously blasted and machined surfaces without damaging the finished or threaded areas. Wet blasting produces satin finishes, whose roughness can be controlled. The process is ideal for removing oxide films acquired during brazing, soldering, welding, or heat treating, and for removing smudges or stains and finger marks. For example, bronze castings that have been machined and brazed are wet blasted with quartz (140 grit) for 0.5 to 5 min to remove braze discoloration and shop dirt. The parts are normally degreased before wet blasting, which precedes ultrasonic cleaning, final inspection, and assembly.

Surfaces cleaned by wet blasting have a uniform appearance, although color is not quite the same as the original grit-blasted surfaces because of the abrasive used. Because water is the carrier, cleaning action is gentle but effective.

Barrel Finishing (Burnishing)

Barrel finishing is best suited for stamped, formed, or machined parts. In this process, parts are tumbled or rolled in a barrel-shaped container together with an abrasive slurry. Castings with remnants of gates and parting lines, forgings with heavy scale, flash lines or die marks, and heavily burred, pitted, or dented parts are not well suited for barrel finishing.

Light burrs can sometimes be removed by a bright dip, after which tumbling may be used for radius blending, polishing, and burnishing. High thin burrs of soft alloys are likely to peen over. Also, barrel finishing of soft alloys at excessive speeds with insufficient amounts of solution can result in roughened and indented surfaces. Dry tumbling is generally restricted to small parts of simple shape and a maximum dimension less than 51 mm (2 in.).

Abrasives. Aluminum oxide, silicon carbide, limestone, and flint are the abrasives most often used in barrel finishing of copper and copper alloys. Combinations of these abrasives may be used for specific applications. For example, a blend of aluminum oxide and silicon carbide of mesh size 46 to 150 produces a reasonably fine matte surface on parts with heavy burrs; aluminum oxide has a cutting action, and silicon carbide has a planing action.

Compounds. Parts heavily coated with grease or oil or contaminated with dirt or chips should be degreased before barrel finishing, preferably in a separate barrel, dip tank, or degreaser. It is usually better practice to clean parts in the barrel when progressing from rough to finishing cycles and to bright-dip before burnishing. A dilute sulfuric-nitric acid bright dip should be used if plating follows burnishing; otherwise, a chromic-sulfuric dip is satisfactory.

Soft water and neutral compounds are preferred for barrel finishing copper and copper alloys. The use of liquid soap-free alkaline compounds for barrel finishing highly leaded free cutting brasses prevents the formation of lead soaps that impair the effectiveness of the operation.

A large selection of commercial products are available to the metal finisher. They are classified by function:

- Cleaning compounds are characterized by high detergency and strong buffering action, and are used to remove oils, greases, and residues.

- Descaling compounds are used to remove tarnish from copper alloys. Neutralizing cycles usually follow the use of these compounds.
- Grinding compounds are used with abrasive media for softening the water, saponifying oils, and keeping chips clean. The compounds inhibit tarnish and improve the color of parts.
- Abrasive compounds contain grits such as aluminum oxide, silicon carbide, emery, quartz sand, or pumice.

Surface Finishes. Although barrel finishing produces the final finish for many parts, it is used more extensively for cleaning before plating and

Table 4.60 Operating conditions for cleaning and pickling in water rolling barrels

Alkaline cleaner(a)	
Concentration of cleaner, oz/gal	2–3
Temperature of solution, °F	160
Speed of rotation, rpm	7–32
Time cycle(b), min	15–20
Pickling	
Concentration of H_2SO_4, %	1–2
Temperature of solution, °F	130
Speed of rotation, rpm	7–32
Time cycle(b), min	15–20

Barrel is made of type 304 stainless steel. (a) Cleaner may be sodium hydroxide or a proprietary compound. (b) Rinsed thoroughly in hot water after cycle is finished.

painting or for deburring and polishing before a final finish is applied.

Bright rolling (water rolling) in a barrel is an economical method for bulk finishing small brass products such as buckles and zippers. For successful water rolling, parts must be thoroughly cleaned beforehand. Cleaning and pickling conditions used before water rolling are given in Table 4.60 (Ref 93).

Vibratory Finishing

Vibratory finishing is particularly effective for deburring, forming radii, descaling, and removing flash from castings and forgings. It can also be used for burnishing. Media used for the vibratory finishing of copper and copper alloys are similar to those used for rotary tumbling. Dry media are used occasionally, but liquid is usually added to provide lubrication, suspend worn-off particles, and provide more gentle cleaning action.

Surface Activation

Preparation for Plating. Copper metal surfaces must be cleaned of oxides and chemically activated to ensure satisfactory plate adherence. Figure 4.119 shows a typical sequence of operations required for preparing buffed copper alloys for copper plating. These operations apply to lead-free and leaded materials as well as to soft-soldered

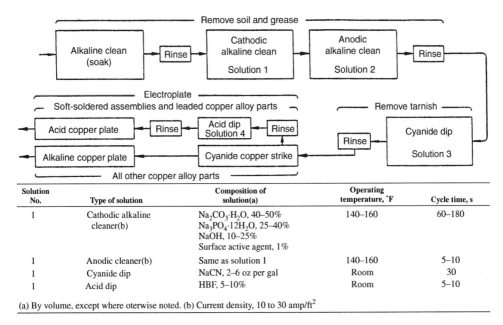

Solution No.	Type of solution	Composition of solution(a)	Operating temperature, °F	Cycle time, s
1	Cathodic alkaline cleaner(b)	$Na_2CO_3 \cdot H_2O$, 40–50% $Na_3PO_4 \cdot 12H_2O$, 25–40% NaOH, 10–25% Surface active agent, 1%	140–160	60–180
1	Anodic cleaner(b)	Same as solution 1	140–160	5–10
1	Cyanide dip	NaCN, 2–6 oz per gal	Room	30
1	Acid dip	HBF, 5–10%	Room	5–10

(a) By volume, except where oterwise noted. (b) Current density, 10 to 30 amp/ft^2

Fig. 4.119 Surface preparation required for the removal of buffing compounds and tarnish prior to plating of copper alloys (except beryllium copper)

assemblies, but they do not apply to beryllium coppers, which require the more aggressive preparation steps.

Oxides are removed by abrasive or chemical cleaning, as discussed earlier. Castings are always abrasively cleaned, but they do not necessarily require subsequent pickling or bright dipping to remove oxides. If acid cleaning is used, the castings must be thoroughly rinsed to prevent entrapped acid from spotting the plating. Stampings and drawn parts can be treated in the same manner as castings, although it may be cheaper to remove oxides by pickling. Screw-machine parts usually require no treatment for oxide removal before plating, and simple degreasing usually suffices.

Quality electroplating requires careful surface preparation procedures after removal. Electrolytic alkaline cleaning should be conducted anodically, or cathodically followed by anodically. Separate solutions are required. Anodic cleaning, particularly of brass, may cause slight tarnishing or etching if applied for more than a few seconds. Smut may form on the work if the cathodic cleaner is contaminated.

Beryllium Copper. Surface oxides are usually present because of the heat treatments these alloys require. Beryllia in these oxides makes them more difficult to remove. Immediately after pickling and rinsing, beryllium-copper articles should be dipped in a cyanide bath of the same pH as the subsequent cyanide strike-plating solution.

Powder Metal Compacts. The porosity of copper-base P/M parts makes them difficult to plate, and continuous electroplates can seldom be obtained unless the pores are sealed, as by buffing, rolling, burnishing, heat treating, and impregnation. These steps are taken when appearance is important and dimensional tolerances are not critical. In general, the methods applicable to porous cast products can be used for low-porosity powder metal parts, providing their density is greater than 95% of the theoretical value. The surface preparation and plating techniques varies for parts with somewhat lower porosity (density, 85 to 95% of theoretical). Sintered bronze bearings and filters with densities

as low as 40% of theoretical are seldom electroplated; however, when necessary, special plating procedures can be used to improve corrosion resistance.

Impregnation with copper, lead, tin, waxes, resins, or oils is also effective. Thermosetting polyester-styrene and silicone resins are suitable impregnating materials. When the polyester-styrene resin is used, the parts are degreased, baked to remove moisture, then impregnated under vacuum and pressure. An emulsion cleaner can be used to remove excess resin from blind holes and threads before curing the resin. Excess resin is removed by tumbling or polishing. When silicone resin is employed, the surfaces of the powder compact are coated with a thin water-repellent film. Then the compact is heated in air to 204 °C (400 °F), quenched in a solution of 4% silicone (type 200), and 96% perchlorethylene, and baked again for $\frac{1}{2}$ to 2 h. After pickling in acid, the part can be plated in the conventional manner, preferably in an alkaline bath.

To impregnate parts with oil, they should be heated in a vapor degreaser, then, while hot, immersed in cold oil. After cooling in the oil, the parts should be removed and drained of oil, then rinsed in cold solvent. Oil-impregnated parts can electroplated conventionally with copper. Bonding is satisfactory.

Chemical and Electrochemical Cleaning

The cleaning methods employed with copper alloys depends on the type of soil to be removed, the equipment available, and the degree of cleanness required. Oils or greases, such as tallow, lard oil, palm oil, and olive oil, can be removed by saponification in an alkaline solution. Dirt particles, abrasives, metal dust, and inert materials are removed by saponification emulsification. Mineral oils, such as kerosene and machine and lubricating oils, are more commonly removed by emulsion cleaning.

Table 4.61 Typical conditions for the solvent cleaning of copper-alloy parts

Part	Solvent cleaner	Temperature of solvent, °F	Immersion time, min	Soil removed
Dose cap(a)	Stoddard solvent or mineral spirits	Room to 120	2	Heavy drawing compound
Brass retainer ring(a)	Stoddard solvent or mineral spirits	Room to 120	2	Eyelet machine lubricant
Brass rods	Sawdust dampened with Stoddard solvent or mineral spirits	Room	5(b)	Mill lubrication

(a) Hand cleaning was necessary because parts were fragile. (b) Tumbled in barrel

Solvent and Vapor Degreasing. Solvent cleaning of copper alloys requires naphthas, such as Stoddard solvent, with flash points over 38 °C (100 °F). Solvents of this type are preferred to kerosene and to the naphthas used in paints because they leave less residue. Such naphthas are effective against light grease and oil.

Trichloroethylene with a boiling point of 87 °C (189 °F) and perchloroethylene with a boiling point of 121 °C (250 °F) are also useful. Chlorinated solvents are less of a fire hazard, but they can be more toxic than petroleum solvents. When used to clean brass before annealing, chlorinated solvents may cause staining. Nevertheless, stabilized trichloroethylene is an effective vapor degreasing agent because it does not attack copper alloys and because it has high solvency for the oils, greases, and related products used in the copper and brass industry. Perchloroethylene is also used, especially for removing high-melting pitches and waxes, for drying parts by vaporizing entrapped moisture, and for degreasing thin-gage materials.

Neither vapor degreasing nor solvent cleaning effectively removes inert materials and inorganic soils that are not generally soluble in chlorinated solvents. Solid particles entrapped in organic soil can be removed mechanically by the washing action of the solvent, especially if the cleaning medium is sprayed or agitated, as by boiling.

Straight-chain hydrocarbons are not effective for the complete removal of heavy grease, burned-on hydrocarbons, pigmented drawing compounds, and oils containing solid contaminants. Buffing compounds containing tallow, stearic acid, and metallic soaps require aromatic hydrocarbons such as toluol or xylol. Table 4.61 lists typical solvent cleaning operations used with copper alloys.

Emulsion and Alkaline Cleaning. Extremely heavy soils, such as machine oils, grease, and buffing compounds, are removed first with emulsion cleaners. Parts are then rinsed, and the remaining soil is removed by alkaline soak or electrolytic cleaning. Extreme caution must be exercised to avoid dragging emulsifiers through the rinses and into a plating solution, especially one of the acid type.

Emulsion cleaning requires 3 min or less in a mildly agitated solution. Spraying is helpful. The thin oil film remaining after cleaning acts as a temporary tarnish preventive; however, brass may be attacked by prolonged exposure to certain emulsifier films.

Alkaline cleaners for copper-base materials contain one or more of the compounds listed in Table 4.62. Solutions for soak cleaning usually contain 4 to 8 oz of cleaner per gallon of solution and are operated at 60 to 88 °C (140 to 190 °F). For every 5.5 °C (10 °F) rise in temperature above 60 °C (140 °F), the cleaning time is reduced by about 25% (Ref 93).

Some uninhibited alkaline cleaners may also cause a slight darkening or tarnishing of the work surface. If this appearance is objectionable, the darkening may be removed by dipping in dilute hydrochloric acid or in cyanide solution.

Ultrasonic Cleaning

Ultrasonic cleaning is used with copper-base materials when the size of particles remaining on the surface is less than 5 to 10 μm, and when the dirt and chips that cause smudge, as indicated by a white-cloth wipe test, are present. Ultrasonic cleaning utilizes alkaline cleaning solutions, solvents, vapor degreasing solutions, or acid pickling solutions to increase the rate of cleaning or to complete the removal of soil from areas not completely cleaned by soaking or spraying.

Alkaline Electrolytic Cleaning

Copper-base materials are electrolytically cleaned cathodically, followed by short-time anodic cleaning or soaking and further anodic cleaning. Positively charged particles deposited on the surface during cathodic cleaning cause smut, which leads to blistering and poor plate adhesion if not completely removed. When cleaning copper, reserve-current anodic cleaning cannot be used for more than a few seconds because the metal dissolves in alkaline cleaning solutions. Copper alloys will tarnish readily during exposure to oxygen released at the anode, although this problem can be minimized by the addition of inhibitors. On the other hand, the slight tarnish provides an indication that all soil has been removed.

Table 4.62 Alkaline cleaners for copper alloys

Compound	Amount, %	
	Soak cleaners	Electrolytic cleaners
Sodium hydroxide	10–20	10–15
Sodium polyphosphates	5–20	5–20
Sodium orthosilicate, sesquisilicate, metasilicate	30–50	30–50
Sodium carbonate, bicarbonate	0–25	0–25
Resin-type soaps	5–10	None
Organic emulsifiers, wetting agents, chelating agents	2–10	1–3

Pickling and Bright Dipping

General-Use Baths. Pickling in solutions containing 4 to 15 vol% sulfuric acid or 40 to 90% hydrochloric acid removes oxides formed on the surface of copper-base materials during mill processing. Pickling can be so effective that no additional surface preparation may be necessary to produce the uniform appearance required for further finishing. However, heavily scaled material may need a bright dip or color dip after pickling.

Except for bright annealed material, copper alloys must be pickled after each annealing treatment. They can be bright dipped to produce a natural surface color and luster suitable for electroplating. Bright dips for copper-base materials consist of sulfuric and nitric acids in widely varying proportions with a small amount of water and hydrochloric acid. After bright dipping and thorough rinsing in cold running water, stain or tarnish can be removed by dipping in a cyanide solution. When a semibright finish is satisfactory, a dichromate color dip is less expensive and more convenient to use than the conventional acid dip.

Baths for Specific Metals. To remove light oxide and/or lubricants, extruded or forged yellow brass is pickled in dilute sulfuric, or, less frequently, hydrochloric acid. Bright dipping completes the removal of all oxide. The brass is then given a color dip to produce a cartridge-brass color. Castings are seldom pickled, but if they must be, similar solutions can be used.

A typical acid treatment cycle for copper-base materials is given below. The cycle may be terminated after any water rinse if the desired finish and color have been obtained.

1. Pickling (Table 4.63)
2. Cold water rinse
3. Scale dip or bright dip (Tables 4.63 and 4.64)
4. Cold water rinse
5. Cold water rinse
6. Color dip (Table 4.65)
7. Cold water rinse
8. Cold water rinse
9. Hot water rinse
10. Air-blast dry

Aluminum Bronzes. Aluminum bronzes form a tough, adherent aluminum oxide film during hot fabrication. This film can be removed or loosened by immersion in the following alkaline solution: sodium hydroxide (10% by weight) and the remainder, water. The temperature of the solution is 77 °C (170 °F) and the immersion time is 2 to 6 min. After this treatment, aluminum bronze can be pickled in the same manner as other copper-base materials.

Table 4.63 Pickling solutions and temperatures for copper and copper-alloy products

Solution	Composition	Temperature, °C	Uses
Sulfuric acid	4–15 vol% H_2SO_4 (1.83 specific gravity), bal H_2O	Room to 60	Removal of black Cu oxide scale from brass forgings
Hydrochloric acid	40–90 vol% HCl (35% conc.), bal H_2O	Room	Removal of scale and tarnish from brass forgings; removal of oxide from Cu forgings
"Scale" dip A	40% conc. HNO_3, 30% conc. H_2SO_4, 0.5% conc. HCl, bal H_2O	Room	Used with a pickle and "bright" dip to give a bright, lustrous finish to Cu and Cu-alloy forgings
"Scale" dip B	50% conc. HNO_3	Room	Used with a pickle and "bright" dip to give a bright, lustrous finish to Cu and Cu-alloy forgings
"Bright" dip	25 vol% conc. HNO_3, 60 vol% conc. H_2SO_4, 0.2% conc. HCl, bal H_2O	Room	Used with a pickle and "scale" dip to give finish to Cu and Cu-alloy forgings

Table 4.64 Scale dip and bright dip conditions for copper-base materials

Constituent	Scale dip		Bright dip solution
	Solution A	Solution B	
Sulfuric acid (1.83 sp. gr.), vol%	0	25–35	50–60
Nitric acid (42° Be),vol %	50	35–50	15–25
Hydrochloric acid (20° Be), oz/gal	½	½	½
Water, vol%	50	35–40	bal
Temperature of solution	Room	Room	Room
Immersion time, s	15–60	15–60	5–45

Note: These solutions can remove 0.001 in. of metal in 5 s and should not be used when close dimensional tolerances must be maintained. Excess hydrochloric acid spots brass. Wood soot and activated charcoal are added to the solution to prevent this condition. These solutions remove scale that is not removed by sulfuric or hydrochloric acid solution. Lower concentrations of nitric acid and higher concentrations of sulfuric acid produce a bright lustrous finish.

Alloys containing silicon may form surface silicates during thermal processing. These can only be removed by hydrofluoric acid or proprietary fluorine-bearing compounds. If a dull brown-to-gray appearance is not objectionable, the material can be pickled in the conventional sulfuric acid solution to remove the copper oxides. If a brighter finish is required, one of the solutions in Table 4.66 should be used.

Nickel silvers and copper-nickels do not respond readily to pickling solutions used for brasses because nickel oxide has a limited solubility in sulfuric acid. Therefore, heavy scaling of these alloys should be avoided by annealing in a reducing atmosphere. For example, annealing 18% nickel silver in a rich reducing atmosphere results in a slight tarnish that is easily removed in the sulfuric acid pickle and dichromate solutions ordinarily used to descale brass.

Copper-30% nickel tubing can be annealed in a reducing (but not bright annealing) atmosphere to produce surfaces clean enough to avoid acid pickling. Nickel-silver (18%) wire is often pretreated in proprietary hot alkaline cleaning solutions followed by annealing in controlled atmospheres to produce a clean and bright surface. The wire is subsequently pickled in sulfuric acid and dichromate solution to remove zinc sweat, then pickled in 10 to15% sulfuric acid at 60 °C (140 °F).

Tarnish Removal. Tarnish is surface discoloration usually consisting of a thin film of oxide or sulfide. It is commonly removed by dipping in a solution of 28 to 227 g (1 to 8 oz) of sodium cyanide per gallon of water. Short-time immersions at room temperature usually suffice. Thorough rinsing is required, and the usual precautions regarding the handling of cyanide solutions must be observed.

Electrochemical Plating and Polishing

Immersion plating, sometimes called galvanic plating, occurs when metal ions are displaced from solution by a more active (less noble) metal. Only those metals more noble than copper can be displaced from solution by copper, therefore, this process is limited to metals such as gold and silver. Immersion-plated deposits are usually only between 5×10^{-5} and 5×10^{-4} mm (2 and 20 µin.) thick. Typical compositions and operating temperatures of solutions for immersion plating of gold and silver on copper are:

Gold Plating
- Potassium gold cyanide: 3.7 g/L (0.5 oz/gal)
- Potassium or sodium cyanide: 26g/L (3.5 oz/gal)
- Sodium carbonate: 30 g/L (4 oz/gal)
- Temperature: 60 to 82 °C (140 to 180 °F)

Silver Plating
- Silver cyanide: 7.5 g/L (1 oz/gal)
- Sodium or potassium cyanide: 15 g/L (2 oz/gal)
- Temperature: 18 to 38 °C (65 to 100 °F)

Electroplating onto Copper Metals

Gold, silver, rhodium, nickel, chromium, tin, and cadmium are electroplated onto copper and copper alloys for decorative purposes and to prevent tarnish or corrosion. Electroplated noble metal coatings are applied to copper electrical contacts in quality-critical electronic equipment.

Chromium is sometimes plated directly onto copper and copper alloys as a low-cost decorative coating and color match. Higher quality chromium plating requires a plated nickel undercoat to produce depth of color and resistance to corrosion, abrasion, and dezincification (Ref 97).

Decorative chromium deposits are characteristically thin and porous. On exposure to aggressive environments, galvanic action between the chromium and base metal can occur, accelerating corrosion of the underlying copper. If the substrate is a high-zinc brass, dezincification and eventual lifting and flaking of the deposit may result. Low-cost chromium deposits plated directly on the base metal are normally coated with clear lacquers or, for added durability, combina-

Table 4.65 Conditions for color dipping copper-base materials

Sodium dichromate, oz/gal	4–12
Sulfuric acid (1.83 sp. gr.), vol%	5–10
Water	bal
Immersion time, s	30

Note: To remove red copper oxide and to produce the uniform yellow color of cartridge brass. Imparts film that resists discoloration in storage or later working. Color dip should not be used if parts are to be plated or soldered subsequently.

Table 4.66 Pickling conditions for copper alloys containing silicon

Constituent or condition	Solution A	Solution B
Sulfuric acid (1.83 sp. gr.)	5–15%	40–50%
Hydrofluoric acid (52%)	½–5%	½–5%
Nitric acid (42° Be)	…	15–20%
Water	bal	bal
Temperature	Room	Room
Immersion time	½–10 min	5–45 s

Note: Percentages are by volume

tion coatings of silicone and acrylic thermosetting organic resins.

Decorative chromium deposits are normally 2.5×10^{-5} to 5×10^{-5} mm (1 to 2 μin.) thick. Chromium deposited directly onto the base metal is generally thicker, seldom less than 0.04 mm (0.0015 in.), and for many applications, a minimum of 0.05 mm (0.002 in.) is specified.

Hard (engineering) chromium deposits are used primarily to improve wear resistance and friction characteristics. Because most of the copper-base materials are soft and ductile, the basis metal lacks sufficient hardness to support thick deposits of hard chromium against heavy pressures without the risk of scoring or flaking.

Hard chromium is electroplated using solutions and conditions similar to those used for decorative chromium plating; however, the deposits are much thicker, frequently dull in appearance, and are generally deposited directly on the base metal. When exact dimensional tolerances are required, and to avoid buildup of deposit at corners, an excess of deposit is plated and the final dimension is attained by grinding or lapping.

Traditional chromium plating solutions consist of chromic acid and sulfuric acid or a suitable sulfate salt dissolved in water. The CrO_3 to SO_4 ratio should be approximately 100 to 1 to ensure good throwing power and economical deposition rates. Modern plating baths contain a number of proprietary additives designed to make the baths self-regulating. The subject is too complex to address in this volume. The reader is advised to consult any standard electroplating text (Ref 97).

Cadmium. Copper-base materials are easily cadmium plated using conventional plating baths. A typical bath has the following composition:

- Cadmium oxide: 22 to 34 g/L (3.0 to 4.5 oz/gal)
- Sodium cyanide: 86 to 112 g/L (11.5 to 15.0 oz/gal)
- Sodium hydroxide: 16 to 24 g/L (2.1 to 3.2 oz/gal)
- Brightening agent as required

The principal reason for cadmium-plating copper metals is to minimize galvanic corrosion between the copper alloys and other cadmium-plated metals in the same assembly. Cadmium-plated copper parts are used in aircraft, marine, and military applications involving atmospheric exposure. It is usually not necessary to apply an undercoat before the deposition of cadmium.

Gold is normally applied over electroplated nickel or silver, which serve as diffusion barriers.

Diffusion destroys the electrical and decorative properties of the deposit. The nickel barrier is normally a minimum of 0.0013 mm (0.00005 in.) thick; however, long-term elevated temperature exposure may require nickel barriers as thick as 0.127 mm (0.005 in.). Government specifications for electronic applications usually require a minimum thickness of 0.005 mm (0.0002 in.) of silver and 0.0013 mm (0.00005 in.) of gold.

Gold plated directly on copper requires a deposit of 0.0025 mm (0.0001 in.) or more to account for the fairly rapid diffusion between the two metals, especially at elevated temperatures. The fairly thick plates also help ensure total coverage and freedom from porosity, both necessary for corrosion protection. For best adhesion, parts should first be given a gold strike, as in 0.99 g/L (0.1 troy oz/gal) potassium gold cyanide (as gold) and 44.4 g/L (6 oz/gal) potassium cyanide. To avoid tarnishing, the struck part should be transferred immediately to the plating solution with the current flowing as the work enters the solution. An alternate procedure is to rinse the struck work, then acid dip, rinse, and plate.

Current density for racked parts varies from 21.5 to 108 A/m² (5 to 10 A/ft²), depending on the solution used. The rate of deposition also depends on the solution and varies from 100 mg/A · min (32.3 A/m² or 3 A/ft²) flowing for 14.4 min deposits 0.0025 mm (0.1 mil) of gold in most cyanide and neutral solutions to 33 mg/A · min in most acid solutions. For barrel plating, the limiting current density for all solutions is about one-third of that used in rack plating.

Cyanide solutions are the most efficient and have the best throwing power, whereas deposits from acid and neutral solutions are less porous and are not stained by incompletely removed solution. Ordinary, for example, unpatented hot gold cyanide solution has limited applications because it is extremely sensitive to impurities, deposits lack brightness, and gold metal anodes must be used.

Proprietary plating solutions contain 1 troy oz of gold per gallon (5.6 g/L), usually added as potassium gold cyanide, and proprietary brightening agents. Insoluble anodes are used with these solutions. The proprietary solutions are brightened or colored by co-deposited base metals, especially silver, nickel, cobalt and copper. When gold deposits of less than 3.3×10^{-4} mm (0.000013 in.) are employed to avoid the U.S. federal excise tax on plated items, the deposit should be protected by lacquer or chromate passivation.

Nickel is plated onto copper-base materials for decorative purposes and as an undercoating to increase the corrosion resistance of nickel-chromium coating systems. Because nickel deposits have a yellow cast, many nickel-plated parts are finished with electroplated chromium. Costume jewelry, lipstick cases, hardware for doors and windows, automotive hub caps, air valves, and plumbing fixtures are examples of copper-base parts regularly plated with nickel, either as a final coat or as an undercoat.

Nickel electrodeposits from some solutions reproduce to a marked degree the irregularities in the base metal surface, and the base metal may have to be polished and buffed before plating if a smooth uniform plated surface is desired. Leveling agents reduce the need for polishing and buffing.

Rhodium. Copper alloys plated with rhodium first receive an undercoat of nickel. Decorative rhodium plating is usually about 2.5×10^{-5} mm (1 µin.) thick. Deposits of 0.025 mm (0.001 in.) or more are used for functional purposes, but these heavier deposits do not have the brightness or color characteristics of the decorative finishes.

Silver plating is employed for decorative and functional purposes. The important functional applications include:

- High-skin conductivity for radio frequencies. A minimum of 0.0025 mm (0.0001 in.) of silver is required.
- Low resistance for electrical contacts. The thickness of silver ranges from very thin deposits such as those produced by immersion coating to electroplates 0.013 mm (0.0005 in.) thick.
- Antiseizing or antigalling surfaces, as for bearings.

Typical plating times for the deposition of 0.025 mm (0.001 in.) of silver as a function of current density are: 37.5 min for 108 A/m^2 (10 A/ft^2); 18 min for 215 A/m^2 (20 A/ft^2); 13 min for 322 A/m^2 (30 A/ft^2); and 9 min for 430 A/m2 (40 A/ft^2).

Tin. As a protective coating, tin is not necessarily anodic to copper alloys; therefore, a copper alloy may not always be fully cathodically protected under tin-plate coatings. Some corrosion products of tin are more noble than copper and can create the type of galvanic conditions that lead to pitting corrosion.

Tin-Copper Alloys. The 40 to 50% tin alloy known as speculum metal is occasionally used as a decorative coating because of its silver-like color.

Bronze-alloy coatings containing 10 to 20% tin can be produced to match the color of gold.

Tin-lead alloys can be plated on copper alloys in almost any proportion using fluoborate baths. An alloy of 7%Sn-93%Pb is used for corrosion resistance, especially against sulfuric and chromic acids. The 60%Sn-40%Pb eutectic alloy has excellent solderability and good electrical properties and corrosion resistance. This alloy can also be applied by hot dipping, but thickness control is difficult as it is with pure tin.

Other Metallic Coatings

Electroless plating involves the reduction of a metal salt to its metallic state by electrons supplied from the simultaneous oxidation of a reducing agent. Electroless plating solutions are designed so that the concentration of the reducing salts, metal salts, buffering salts, and the pH controls the rate of reduction of the metal salt and of the oxidation of the reducing agent.

On copper-base materials, the process is limited to the plating of nickel, tin, gold, and silver deposits. Solutions for plating these deposits are affected adversely by contaminants, such as cyanides, lead, zinc, manganese, and cadmium. Tin may render the nickel-plating solution inoperative. Therefore, tin-containing copper alloys must be plated with copper or gold before the final electroless nickel coating is applied.

Because copper is more noble than nickel, it will not act as a catalyst to start the deposition of nickel. This condition is alleviated by contacting the copper workpiece with an active metal such as iron or aluminum that causes the deposition of a thin nickel coating on the workpiece. The nickel coating is of sufficient thickness to continue the oxidation-reduction reaction for further nickel deposition after the activating metal is removed. Another procedure for starting the deposition of nickel is to make the part cathodic and apply a current to the part briefly as it is held in the electroless bath.

Electroless processes are used to plate copper with gold and silver for decorative purposes. Nickel, gold, and silver are frequently applied to copper electronic components to prevent tarnishing during subsequent processing and to aid in the soldering of semiconductors. Electroless tin has been used for copper tubes to prevent corrosion by carbonated waters.

Physical Vapor Deposition (PVD). Copper metal can be deposited onto metallic and non-metallic substrates using conventional vacuum coating techniques. The work or substrate is mounted in a vacuum chamber and positioned to expose the surfaces to be coated to the source of vapor. Vacuums at about 10^{-3} to 10^{-5} torr are required. Vapor is generated by heating metal or a metal compound to a temperature at which its vapor pressure appreciably exceeds the residual pressure in the chamber. Vacuum coatings are widely used in thin-film integrated electronic circuits. Vacuum-deposited copper, among other metals, is used in various thicknesses as a conductor, resistor, and capacitor (Ref 93).

Electrolytic Polishing

Electropolishing is accomplished by making the parts to be treated anodic in an acid solution. Application of direct current selectively removes more metal from high points, such as burrs, than from flats or depressions. Most commercial electropolishing solutions used with copper alloys are based on phosphoric acid.

Electropolishing for 4 to 6 min reduces surface roughness to between one-third and one-half the original root mean square (RMS) value. Chemical polishing is similar to electropolishing, although electropolished surfaces are usually smoother and brighter. Process selection is based on the following considerations:

- Complex part configuration renders mechanical polishing expensive and difficult.
- Entrapped buffing or polishing compounds. Chemically or electrochemically polished copper-base materials do not require additional cleaning operation other than rinsing.
- Parts racked for electropolishing may be electroplated in the same racks, saving handling costs.
- Electrolytically or chemically polished surfaces tarnish less readily than mechanically polished surfaces.
- Parts that might deform during mechanical polishing or tumbling will not be deformed by electrolytic or chemical polishing.
- Whereas mechanical polishing may result in a highly deformed skin that differs from the base metal and does not accept plating uniformly, electroplate on electropolished base metal has superior adhesion to that of mechanically finished surfaces.

Typical electropolished parts include fishing lures, plumbing fixtures, lamps and lighting fixtures, wire goods, brass appliance parts, jewelry, name plates, and bezels. Copper-plated parts, such as automobile tail pipes, die castings that have not been buffed, and high-altitude oxygen bottles, are effectively electropolished for improved finish. The thickness of copper plate should be in excess of 0.02 mm (0.0008 in.) before electropolishing.

The choice of electropolishing reagents and conditions depends strongly on the alloy in question. Current densities range from less than 1 to more than 250 A/dm^2, potentials from 1 to 50 V, and times from a few seconds to 40 min. Parts to be electroplated after electropolishing are anodically alkaline cleaned until the surface is light brown in color, rinsed, dipped in 5% sulfuric acid solution, double rinsed in water at room temperature, immersed in cyanide copper strike solution, and plated.

The quality of an electrolytically polished surface is affected by material composition, agitation, and ventilation. Most copper alloys are suitable except those containing appreciable amounts of lead. Materials successfully treated include copper, beryllium copper, single-phase bronzes, and single-phase unleaded brasses.

Organic Coatings

Tarnishing or discoloration of copper alloys may be retarded or, in many instances, delayed indefinitely by lacquering. For exterior service, a dry film thickness of 0.04 to 0.05 mm (0.0015 to 0.002 in.) is recommended. In less severe or indoor service, a dry film thickness of about 0.01 to 0.02 mm (0.0005 to 0.0007 in.) performs satisfactorily.

Air-Drying versus Thermosetting Lacquers. In general, the performance of thermosetting lacquers is superior to that of air-drying lacquers, and the use of the thermosetting types is preferred if ovens are available. True thermosets must be heated to 121 to 204 °C (250 to 400 °F) or higher for 5 to 60 min to cross-link the polymers present. However, the catalytic activity of copper facilitates complete curing of thermosetting lacquers at temperatures lower than those required with inert substrates. Many thermosetting lacquers discolor copper alloys severely when heated to temperatures recommended by their suppliers, but such discoloration can be minimized by curing at lower temperatures.

Resins in lacquers used with copper alloys include alkyds, acrylics, cellulosics (cellulose nitrates, ethyl cellulose, cellulose acetate, and cellulose acetate butyrates), epoxies, phenolics, polyesters, silicones, urea and melamine formaldehydes, vinyls, urethanes, and the polyvinyl fluoride film, Tedlar (Du Pont de Nemours, E.I., & Co., Inc., Wilmington, DE), applied by roll bonding (Ref 98).

Satisfactory short-term protection of an inexpensive product with a relatively short-life expectancy can be provided by an epoxy, cellulose nitrate, or alkyd-based lacquer. Satisfactory service at moderately elevated temperature can be obtained with a heat-resisting melamine. If economics permit, silicones formulated for high-temperature service are best. For resistance to degradation by weather, best results are obtained from high-quality acrylics. INCRALAC (International Copper Association, New York, NY), a clear coating based upon Acryloid B-44, a chelating agent (benzotriazole), and a leveling agent (silicone oil or epoxydized oil) belongs to this group, as do silicone coatings and melamine-modified alkyd resins (Ref 98).

Inorganic Coatings

Passivation refers to the process of forming a protective film on the metal. The blue-green patina developed during atmospheric exposure is a natural protective, or passive coating. A patina can be artificially produced or accelerated by a solution having the formulation: ammonium sulfate (2.7 kg, or 6 lb), copper sulfate (85 g, or 3 oz), ammonia (technical grade, 0.90 sp. gr.) (29.5 mL, or 1.34 fl oz), and water (24.6 L, or 6.5 gal), the total solution is 27.4 L or 7.25 gal.

A fine spray of the solution should be applied to a chemically clean surface, and the film should be permitted to dry before the part is sprayed a second time. Five or six repetitions of the spraying and drying sequence are required. The color will begin to develop in about 6 h and will at first be somewhat bluer than natural patina. A more attractive color will develop as the surface is exposed to natural weathering.

Small copper parts can be coated with an imitation patina by dipping in, or brushing with, a solution consisting of the following (proportions are given by weight): copper metal, 30; nitric acid (conc.), 60; acetic acid (6%), 600; ammonium chloride, 11; and ammonium hydroxide (tech.), 20. The copper is dissolved in nitric acid, and the remaining three constituents are added as soon as the action ceases. The solution should stand several days before use. Parts treated with this solution can be protected with linseed oil.

Coloring

Copper-base materials can be treated to produce a variety of colors, ranging from dark reds to black. The final color depends on base metal composition, solution composition, immersion time, and operator skill. Coloring is primarily an art, and practical experience is necessary to develop the skill required to produce uniform finishes consistently.

Copper alloys are colored chemically to enhance the appearance of a product, to provide an undercoating for subsequent organic finishes (as with brass), and to reduce light reflection in optical systems. Chemical coloring produces a thin layer of a compound on the surface of the base metal. This layer remains smooth, lustrous, or dull, depending on the condition of the substrate.

Coloring copper alloys is essentially a process for coloring copper, because zinc and tin compounds are colorless. However, these constituents and their concentrations greatly affect many of the chemical reactions and color tones of the coatings formed. A copper content of less than 85% is required to produce a good blue-black finish on brass by an ammoniacal copper sulfate or ammoniacal copper carbonate blackening (blue-dip) solution. Other solutions are more suitable for coloring high-copper alloys.

After machining and mechanical surface preparation, parts should be thoroughly cleaned to remove dirt, oil, grease, and oxide films. The cleaning and deoxidizing procedures should be selected so that the structure of the metal at the surface undergoes a minimum of undesirable change. Acid dipping or bright dipping (nitric-sulfuric acid solution) may be necessary to remove oxides and to activate the surface for chemical coloring. A certain amount of trial and error is usually required to establish the most suitable techniques for surface preparation (Ref 93).

The following formulations and conditions are commonly employed in commercial applications to produce the colors indicated.

Alloys Containing 85% or More Copper

Dark Red. Molten potassium nitrate 649 to 704 °C (1200 to 1300 °F); immersion time, up to 20 s; hot water quench. Parts must be lacquered (Ref 93).

Black. Solution A is liquid sulfur, 28 g (1 oz), or sulfurated potash, 57 g (2 oz); ammonium hydroxide (sp. gr. 0.89), 7.4 mL (0.25 oz); and water, 3.8 L (1 gal). The temperature of solution is at room temperature. It produces a dull black finish. Reddish-bronze to dark-brown finishes are obtainable by dry-scratch brushing with a fine wire or cloth wheel to the desired appearance.

Solution B is sulfurated potash, 0.25 oz per gal of water. Solution strength should be adjusted to blacken the part in approximately 1 min. Too rapid formation of coloring film can result in a nonadherent and brittle film.

Solution C is potassium sulfide, 0.5 to 1 oz per gal of water. Immersion time is up to 10 s. Lacquer protection is required.

Several proprietary processes are available for producing a satisfactory black finish. Alloys blackened by these materials include silicon bronzes, beryllium coppers, bronzes of all types, and brasses (leaded or unleaded) with zinc contents up to 35%.

Steel Black. Arsenious oxide (white arsenic), 113 g (4 oz); hydrochloric acid (sp. gr. 1.16), 238 mL (8 fl oz); and water, 3.8 L (1 gal). The temperature of solution is approximately 82 °C (180 °F). Immerse part in the solution until a uniform color is obtained. Scratch brush while wet, then dry and lacquer.

Black Anodizing. Sodium hydroxide, 454 g (16 oz) and water, 3.8 L (1 gal). The temperature is 82 to 99 °C (180 to 210 °F) with a current density of 21.5 to 108 A/m^2 (2 to 10 A/ft^2). The anode-to-cathode ratio is 1 to 1 with a voltage of 6 V. The cathodes are steel, carbon, or graphite. The anodizing time is 45 s to 3.75 min. The tank material is steel. Adequate ventilation is required. After anodizing, the parts are washed in hot and cold water, rinsed in hot water, dried, buffed lightly with a soft cloth wheel, and lacquered if desired.

Reddish Bronze to Dark Brown (Statuary Bronze). Sulfurated potash, 57 g (2 oz); sodium hydroxide, 85 g (3 oz); and water, 3.8 L (1 gal). The temperature of the solution is 77 °C (170 °F). Immersion time depends on final color desired. Parts are usually scratch brushed with a fine wire wheel. Lacquering is required.

Verde Antique. Solution A is copper nitrate, 113 g (4 oz); ammonium chloride, 113 g (4 oz); calcium chloride, 113 g (4 oz); and water 3.8 l (1 gal).

Solution B is acetic acid, 1.9 L (0.5 gal); ammonium chloride, 567 g (20 oz); sodium chloride, 198 g (7 oz); cream of tartar, 198 g (7 oz); copper acetate, 198 g (7 oz); and water, 1.9 L (0.5 gal).

Verde antique finishes are known also as patina. They are stippled on brass or copper plate that has usually been treated in a sulfide solution to produce a black base color, which results in a yel-

lowish color, and ammonium salts impart a bluish cast. Stippling can be repeated and, when the antique green color appears, immersion in boiling water will produce several different color effects. Other color effects are obtained by incorporating some dry colors such as light and dark chrome green, burnt and raw sienna, burnt and raw umber, ivory drop white and drop black, or Indian red. After coloring, the surface should be lacquered or waxed. A pleasing semiglossy appearance of the lacquered surface may be produced by brushing with paraffin, beeswax, or carnauba wax on a goats-hair brush rotated at about 750 rpm.

Light Brown. Barium sulfide, 14 g (0.5 oz); ammonium carbonate, 7 g (0.25 oz); and water, 3.8 L (1 gal). The temperature is at room temperature.

Brown. Potassium chlorate, 156 g (5.5 oz); nickel sulfate, 78 g (2.75 oz); copper sulfate, 680 g (24 oz); and water, 3.8 L (1 gal). The temperature of the solution is 91 to 100 °C (195 to 212 °F).

Alloys Containing Less Than 85% Copper

Black. In solution A, a suitable quantity of brass parts is placed in an oblique tumbling barrel made of stainless steel, and the parts are covered with warm water. Usually 11 to 19 L (3 to 5 gal) is sufficient. Copper sulfate (85 g, 3 oz) and sodium thiosulfate (170 g, 6 oz) are dissolved in warm water in a separate container and added to the contents of the barrel. The parts are tumbled for 15 to 30 min to obtain the desired black finish, after which the solution is drained from the barrel and the parts washed thoroughly with clean water. The parts are removed from the barrel, dried in sawdust, or with an air blast, and lacquered if desired (Ref 93).

Solution B is copper carbonate, 14 g (0.5 oz); ammonium hydroxide (sp. gr. 0.89), 133 g (4 oz); sodium carbonate, 7 g (0.25 oz); and water, 3.8 L (1 gal). The temperature of the solution is 88 to 93 °C (190 to 200 °F).

Statuary Bronze. Copper carbonate, 14 g (0.5 oz); ammonium hydroxide (sp. gr. 0.89), 118 mL (4 fl oz); sodium carbonate, 7 g ($\frac{1}{4}$ oz); and water, 3.8 L (1 gal). The temperature of the solution is 88 to 93 °C (190 to 200 °F).

Immerse the parts into the hot solution for about 10 s, rinse in cold water, and dip in solution of dilute sulfuric acid. Rinse in hot and then cold water, clean with soft cloth or sawdust, and coat with clear lacquer.

Blue Black. Copper carbonate, 454 g (1 lb); ammonium hydroxide (sp. gr. 0.89), 0.9 L (1 qt);

and water, 2.8 L (3 qt). The temperature of the solution is 55 to 79 °C (130 to 175 °F). Excess copper carbonate should be present. The proper color should be obtained in 1 min.

Brown. Solution A is copper sulfate, 113 g (4 oz); potassium chlorate, 227 g (8 oz); water, 3.8 L (1 gal).

Solution B is liquid sulfur, 28 g (1 oz) or sulfurated potash, 57 g (2 oz); and water, 3.8 L (1 gal).

Immerse the parts in Solution A for approximately 1 min, then without rinsing, immerse in Solution B for a short time, and rinse in cold water. The dipping operation should be repeated in both solutions until the desired color is obtained. Finally, rinse the work in hot water, dry in hot sawdust or with an air blast, scratch brush with fine wire wheel and lacquer.

Old English Finish (Light Brown). Solution A is liquid sulfur, 14 g (0.5 oz) or sulfurated potash, 28 g (1 oz), and water, 3.8 L (1 gal).

Solution B is copper sulfate, 57 g (2 oz), and water, 3.8 L (1 gal).

Immerse the parts in Solution A and, without rinsing, immerse in Solution B; then rinse in cold water. Repeat the operations until a light color is produced. For a uniform finish scratch brush and then repeat the dipping operations, first in Solution A and then in Solution B, until the desired color is obtained. Finally, rinse the parts thoroughly in cold and hot water, dry in sawdust, scratch brush on a fine wire wheel, and lacquer.

Antique Green on Brass. Nickel ammonium sulfate, 227 g (8 oz); sodium thiosulfate, 227 g (8 oz); and water, 3.8 L (1 gal). The temperature of the solution is 71 °C (160 °F).

Hardware Green on Brass. Ferric nitrate, 28 g (1 oz); sodium thiosulfate, 170 g (6 oz); and water, 3.8 L (1 gal). The temperature of the solution is 71 °C (160 °F).

Brown on Brass or Copper. Potassium chlorate, 156 g (5.5 oz); nickel sulfate, 78 g (2.75 oz), copper sulfate, 680 g (24 oz); and water, 3.8 L (1 gal). The temperature of the solution is 96 to 100 °C (195 to 212 °F)

Light Brown on Brass or Copper. Barium sulfide, 14 g (0.5 oz); ammonium carbonate, 7 g (0.25 oz); and water 3.8 L (1 gal). The color is made more clear by wet scratch brushing and re-dipping.

Post-Treatment. Many chemical films must be scratch brushed to remove excess or loose deposits. In addition, contrast in colors may be obtained by relieving by scratch brushing with a slurry of fine pumice; hand rubbing with an abrasive paste; and barrel finishing or buffing to remove some or all of the colored film from the highlights.

Clear lacquers are usually necessary for adequate life and service of chemical films used as decorative finishes outdoors. Finishes for exposure indoors are often used without additional protection to the conversion coating.

Mechanical Finishing

Polishing and Buffing. Copper-alloy parts are polished after scale removal and dressing or rough cutting, but before final finishing operations. Rough castings normally require two polishing operations before buffing; forgings and stampings require one polishing operation prior to buffing. Pipe, tubing, and some stampings can be buffed without previous polishing. Buffing is not required when a brushed or satin finish is desired as the final finish (Ref 93).

Because copper-base materials are softer than steel, fewer stages of successively finer polishing are required to achieve a uniformly fine surface finish. For many parts, especially those having machined surfaces or otherwise, essentially free of defects, a single-stage polishing operation employing 180- to 220-grit abrasive on a lubricated belt or setup wheel may be all that is required before buffing. Poor quality surfaces require preliminary rough polishing, which is accomplished on a dry belt or wheel with 80- to 120-grit abrasive. Surfaces of intermediate quality may be given a first-stage polishing with 120- to 160-grit abrasive, either dry or lubricated (Ref 93).

Belt polishing is generally advantageous for high-production finishing except when special shapes are processed. These are best handled by contoured faces of setup polishing wheels.

Buffing of copper and copper alloys is usually performed on standard sectional cloth wheels operating at moderate speeds of 1200 to 1800 rpm. Typical wheel speeds (surface feet per minute) for various finishes are as follows:

- 3000 to 5500 for dull finish using 120- to 200-grit aluminum oxide
- 4000 to 6000 for satin finish
- 5500 to 7000 for cutting and coloring
- 7000 to 8000 for high luster, using tripoli, lime, and silica with no free grease binder

When it is necessary to "mush" a buff to the contour of a complicated part, buffing speeds range between 200 and 1000 rpm (Ref 93).

When polishing and buffing precede electroplating, essentially neutral compounds that are free of sulfur must be used, to avoid staining in the plating operation. Excessively high temperatures must not be developed in the work during polishing and buffing, because this may cause difficulties in subsequent cleaning and plating operations. When flawless chromium-plated surfaces are required, it is necessary both to buff and color buff the polished copper-alloy surfaces before plating. Chromium reproduces all imperfections in the underlying plating or base metal and, being hard and having a high melting point, is more resistant to flow and thus less readily buffed by normal methods. Although not flawless, a good chromium-plated surface can be obtained without the color buff operation, that is, by polishing and cut-down buffing only (Ref 99).

Scratch brushing is used to produce a contrasting surface adjacent to a bright reflective surface (although some parts are scratch finished all over); to produce an uneven surface for better paint adherence; during final finishing, to remove metal from parts with intricate recesses that are inaccessible to polishing and buffing wheels, and in some applications, to remove impacted soil and buffing compounds from previous finishing operations, before subsequent finishing. Scratching media used on copper alloys include:

- Wire wheels: Used on copper or brass grill work to clean intricate recesses, holes, or ribbed areas, and to produce a decorative non-continuous scratch pattern on ornamental parts such as vases and lamps
- Emery cloth or paper: Common medium used to produce a series of linear or circular parallel lines on flat objects with no sudden changes in contour. This type of decorative finish is applied to fireplace accessories
- Polishing wheels: Loaded with greaseless compounds to produce scratch-brush patterns.

Used on decorative items such as jewelry and building paneling
- Soft tampico and manila brushes: Used to remove soil from scrollwork and embossed areas on ornate tableware serving sets and jewelry before final processing

Table 4.67 lists the media and sequence of operations for scratch-brush finishing several copper-alloy products. Although scratch-brush finishing is useful for producing eye-appealing finishes and as a mechanical means for preparing surfaces for subsequent processing, certain hazards must be recognized. For example, extreme care and control are required when the part being worked contains patterns with sharp corners or embossments, because sharpness of detail may be destroyed. Also, in salvage or rework operations, it is difficult and sometimes impossible to blend the original brush pattern into a repaired area from which a defect has been removed by grinding.

Cladding

Metal laminate composites characterized as clad metals usually consist of a thin foil or sheet bonded to a heavier substrate. The foil characteristically offers special properties (more corrosion and/or abrasion resistance), and it is usually more expensive than the substrate, whose major function is support. Examples of clad products include copper-nickel-clad carbon steel ship hulls, copper-nickel-clad copper coins, and copper-clad stainless steel for roofing and other uses (Ref 99, 100).

The most common fabrication methods include electrocladding, roll bonding, coextrusion, weld plating, explosive welding, and brazing. Laser and

Table 4.67   Media and suggested sequence of operations for scratch brushing copper-alloy parts

Part	Finish desired	Abrasive	Type of wheel	Size of wheel	Speed, rpm	Comments
Black fuse body, yellow brass	Dull, smooth, black	None	Tampico	7 in. diam by 3 in. thick, 5 rows wide	1200	Clean brush often by running pumice stone across face of wheel
Silver-plated red-brass lipstick case	Semi-bright	Solution of soap bark and cream of tartar	Nickel-silver wire, 0.004 in. diam	6 in. diam by 3 in. thick, 6 rows wide	850	Lacquer after scratch brushing
Black-on-bronze bookends, highlights relieved	Black background, colored copper highlights (sulfurated potash)	Pumice in water	Cloth, sewed sections	7 in diam by ½ in. wide	850	Lacquer after scratch brushing
Silver-plated lipstick cap	Satin	Greaseless rouge (proprietary)	Loose cloth wheel	6 in. diam by 2 in. wide	1800	Lacquer after finishing
Nickel-plated refrigerator panels or stove parts	Satin	Greaseless compound	Loose cloth wheel	12 in. diam by 2 to 20 in. wide	1800	Chromium plate after finishing

ultrasonic welding systems have also been investigated (Ref 99).

Electrocladding involves electrodepositing a uniform coating on a metallic substrate. Methods include electroplating and sputtering. Rochelle cyanide copper solution A is recommended for plating copper onto stainless steel. Deposition requires a current density higher than 10 A/m² (0.93 A/ft²). High pH, high temperature, and high current densities cause the copper layer to separate from the stainless-steel surface. Electrode effi-

ciencies are less than 20% for plating times less than 30 min and vary from 40 to 70% for times longer than 1 h (Ref 93).

Roll Bonding. Roll-bonded sheets rely on the metallurgical bond formed when two metals are hot rolled together. Cladding thickness typically vary between 2.5 and 20% of the total laminate thickness (Ref 99, 100).

Coextrusion. Metals to be clad by coextrusion are assembled in an extrusion slug, as shown in Fig. 4.120, which schematically depicts the coex-

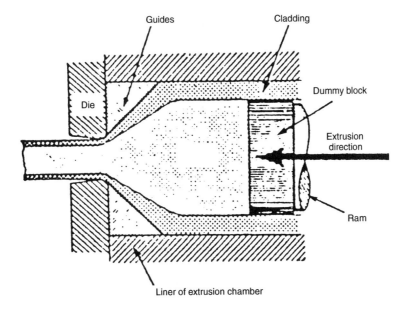

Fig. 4.120 Scheme of coextrusion process

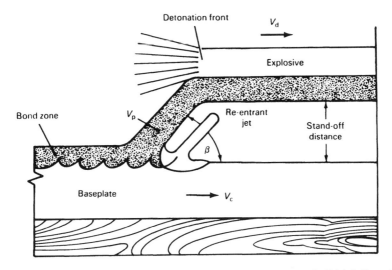

Fig. 4.121 Scheme of explosive-welding setup with parallel standoff. V_d is the detonation velocity. V_c is the velocity of the collision point.

trusion of cable. Here, the cable or core to be clad is fed through a central mandrel into a chamber where the heated cladding is extruded over it through the die opening. Copper-clad aluminum wire is made by this process (Ref 100).

Explosive Bonding. Two dissimilar metals can be explosively bonded using the parallel-plate standoff technique schematically illustrated in Fig. 4.121. A rubber or plastic buffer layer serves to protect the flyer plate from the explosive that, when detonated at one end of the flyer plate, causes it to strike the stationary plate with an impact pressure much greater than the yield strengths of the metals to be joined. If the velocity of the collision point is less than the velocity of sound in the two metals, a metal jet is formed at the lower surface of the flyer plate, which scours the interfacial surfaces. The explosive pressure then bonds them together (Ref 100).

Explosive welding is especially well suited for cladding tubing, tube sheets, and wear surfaces in machinery, engine components, and nuclear-reactor and chemical-process equipment. Figure 4.122 illustrates two methods for cladding tubing. The arrangement for explosively cladding the inside of a tube is shown in Fig. 4.122(a), and Fig. 4.122(b) depicts the method used to implode cladding onto the outside surface of a tube. Table 4.68 lists sev-

eral combinations of copper-base metals that have been successfully bonded by this method. (Ref 101)

Weld Overlay. Cost and low rates of deposition preclude the use of weld overlaying as a method to clad large surface areas. The process is known, however, and copper-nickel coatings, in particular, are applied to steel substrates to provide corrosion resistance, often against seawater. Although 90/10 copper-nickel, flux-covered electrodes, and filler wire are available, it is more common for weld overlaying to be undertaken with 70/30 electrodes and/or filler wire. A barrier layer of 65/35 nickel-copper minimizes the effects of iron dilution and improves interface ductility.

Where copper-nickel coatings are required more for biofouling prevention than for corrosion resistance, such as concrete surfacing, metal-sprayed coating may be adequate (Ref 99).

Brazing. Clads can be formed by brazing two components of a laminate together. The system is heated to a temperature between the melting point of the brazing alloy and that of the laminae to be bonded. The brazing alloy may be in the form of foil or wire and used with or without flux. When the brazing alloy forms a low-melting eutectic between the clad and substrate layers, the process is known as eutectic brazing. This method has been used to bond titanium to steel using a copper-silver eutectic alloy. The process is conducted without flux under high vacuum (Ref 99).

Semi-Rigid Linings. The semi-rigid lining of steel fabrications such as water boxes with copper-nickel sheet is a well-established practice. It is usually less expensive than constructing the fabrication from pre-clad plate. One popular technique is to attach the copper-nickel to the substrate with a series of GMAW rosette or plug welds.

Adhesive-Bonded Linings. Adhesive-bonded, wrap-round systems have been used to protect offshore platform legs and cross members from biofouling. The wrapped clad consists of a thin sheet of 90/10 copper-nickel. If necessary, the copper-nickel sheet can be adhesively bonded to a flexible

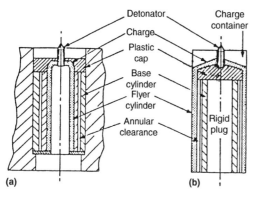

Fig. 4.122 Setup for cladding cylinders. (a) Explosive system. (b) Implosive system

Table 4.68 Copper-base metals successfully clad by explosive bonding

	Molybdenum	Gold alloy	Tantalum	Titanium and Ti alloys	Bronze	Cupro-nickel	Brass	Copper
Copper	X	X	X	...	—	...	...	X
Brass	...	...	...	X	...	...	X	...
Cupro-nickel	...	...	...	...	...	X	...	...
Bronze	...	...	...	...	X	...	...	...

Note: X denotes feasible conditions. — means bonding of that combination has not been attempted. It does not mean those metals cannot be explosion bonded.

rubber, which insulates the copper-nickel from the steel to prevent galvanic corrosion.

REFERENCES

1. "Casting of Copper Alloys" ("Le Fonderie Des Alliages Cuivre"), Publication IMNG 003, Centre D'Information Des Métaux Non Ferreux (ASBL), Brussels, 1983
2. K.S.S. Vasan, Industrial Furnaces for Non-Ferrous Melting, *Melting and Casting of Copper and its Alloys Seminar Proc.*, (Calcutta, India) Indian Copper Information Centre and Small Industries Service Institute, Nov. 4, 1968
3. A.K. Biswas and W.G. Davenport, *Extractive Metallurgy of Copper*, 3rd ed., Pergamon Press, Tarrytown, NY, 1994, p 437
4. A.K. Biswas and W.G. Davenport, *Extractive Metallurgy of Copper*, 3rd ed., Pergamon Press, Tarrytown, NY, 1994, p 9
5. *Modern Refractory Practice*, 2nd ed., Harbison-Walker Refractories Co., Pittsburgh, PA, 1938
6. *Modern Refractory Practice*, 3rd ed., Harbison-Walker Refractories Co., Pittsburgh, PA, 1950
7. Revista Mineria Chilena, Vol 12 (No. 131), May 1992
8. K. Dies, Copper and Copper Alloys in Technology (Kupfer und Kupferlegierungen in der Technik), Springer-Verlag, Berlin, 1967
9. K.R. Van Horn, Ed., *Aluminum*, Metals Park, OH, 1968
10. *Refractories*, North America Refractories Co., Cleveland, OH, Sept. 1990, p 1–135
11. "Refractory Bricks for an Ascro Furnace" ("Ladrillos Refractories Horno ASARCO"), Colada Continua Chilena S.A., personal communications, 24 Dec. 1991
12. "Catalog of Refractory Products" ("Catálog de Productos Refractarios"), Lota Green, Santiago, Chile, 1990
13. "Product Catalog" (Catálogo de Productos"), Refractorios Chilenos S.A., Santiago, Chile, 1990
14. Roskill Reports on Metals & Minerals, *The Economics of Copper 1988*, 4th ed., Roskill Information Services Ltd., London, 1988
15. Copper (*Kupfer*), Deutsches Kupfer-Institut E.V., Berlin, Germany, 1982
16. P. Klare, *Copper-Oxygen-Hydrogen* (*Kupfer-Sauertoff-Wasserstoff*), The German Mining and Metallurgical Society (der Gesellschaft Deutscher Metallütten und Bergleute e.V.), Frankfurt, Germany, 1962
17. P.S. Ramaswany, Melting and Casting of Copper and Copper and Copper Alloys for Processing, *Wrought Copper and Copper Alloy Products Seminar Proc.* (Calcutta, India), Indian Copper Information Centre and Small Industries Service Institute, Sept. 19, 1969
18. Developments in Process Control and Instrumentation, *Metallurgia*, (No. 3), 1984, p 89–92
19. E. Brunhuber, *Copper Alloy Castings* (*Guss aus Kupferlegierungen*), Fachverlag Schiele & Schon GmbH, Berlin, Germany, 1986
20. M.C. Flemings, *Solidification Processing*, McGraw-Hill Book Co., New York, NY, 1974
21. W.H. Salmon, E.N. Simons, and E.G. Gardner, *Foundry Practice*, Sir Isaac Pitman & Sons Ltd., London, 1959
22. D.C. Ekey and W.P. Winter, *Introduction to Foundry Technology*, McGraw-Hill Book Co. Inc., New York, NY, 1958
23. M. Kleinau, *Centrifugal and Continuous Casting of Heavy Metals* (*Schwermetall-Schleuder und Strangguss Technische und Wirtschaftliche Möglochkeiten*), Deutsches Kupfer-Institut, Bestellnummer s. 165, Heft 9, 11 und 12, Berlin, 1981
24. L.W. Collins, Nonferrous Wire Rod, Vol 1, *Nonferrous Wire Handbook*, The Wire Association International Inc., Guilford, CT, 1977
25. A.M. Wagner, Continuous Casting and Rolling of Copper by the Properzi Method, Metallurgical Society Conferences (Pittsburgh, PA), Nov. 29–Dec. 1, 1965, *Pyrometallurgical Processes in Nonferrous Metallurgy*, Vol 39, Gordon and Breach, Science Publishers, New York-London-Paris, 1966
26. A.K. Biswas and W.G. Davenport, Continuous Casting and Rolling of Copper by the Properzi Method, Metallurgical Society Conferences (Pittsburgh, PA), Nov. 29–Dec. 1, 1965, *Pyrometallurgical Processes in Nonferrous Metallurgy*, Vol 39, Gordon and Breach, Science Publishers, New York-London-Paris, 1966, p 439
27. A.K. Biswas and W.G. Davenport, *Extractive Metallurgy of Copper*, 1st ed., Pergamon Press, Terrytown, NY, 1976, p 348
28. G. Raimanns, "Description of Existing Industrial Scale Continuous Casting Systems" ("Descripción de los Sistemas de Colada Continua Existentes en Escala Industrial"), Informe Interno CODELCO, GTEC-SPG-I-159/80, Santiago, Chile, 1980
29. M. Rantanen, Upward Continuous Casting Technique in the Production of Nonferrous Wires, *Wire J.*, Vol 67 (No. 3), 1980, p 102–104
30. Outokumpu OY, Technical Export Division, *Upcast Method for Producing Copper Wire Rod*, Outokumpu OY, Espoo-Finland, 1990
31. F. Ullmann, Encylopedia of Industrial Chemistry, Enciclopedia de Química Industrial, Editorial Gustavo Gili S.A., 2nd ed., Barcelona, España
32. "Continuous Casting Machines," Rautomead Ltd., Dundee, Scotland
33. *Elements of Rolling Practice*, 2nd ed., The United Steel Co., Ltd., Sheffield, 1949
34. B. Avitzur, *Metal Forming Processes and Analysis*, McGraw-Hill Book Co., New York, NY, 1968
35. E.C. Larke, *The Rolling of Strip, Sheet, and Plate*, Chapman and Hall Ltd., London, 1963
36. C. Carmichael, *Kents' Mechanical Engineers' Handbook*, 12th ed., Wiley Handbook Series, John Wiley & Sons, New York, NY, 1953
37. P. Devroey, "On the Forging of Brass" (Ö Forjamiento a Quente Dos Latoes"), Technical Bulletin No. 97 CEB302.200.320.4-204.11, Centro

Brasileiro de Informação do Cobre, Sao Paulo, Brazil, Dec. 1974

38. Forming and Forging, Vol 14, *Metals Handbook,* 9th ed., American Society for Metals, Materials Park, OH, 1988

39. Copper Brass Bronze, *Forgings,* CDA No. 705/5 application data sheet, Copper Development Association Inc., Greenwich, CT

40. K. Laue and H. Stenger, Extrusion: Processes, Machinery, Tooling, ASM International, Metals Park, OH, 1981

41. Hitachi Hollow Conductors, Hitachi Cable Ltd., Tokyo, 1986

42. *Copper for Electric Generators,* Outokumpo Oy, Espoo, Finland

43. D.W. Rowell, A New Process for Extruding Large Tubes, *Wire Ind,* Nov. 1976, p 903–906

44. K. Laue and H. Stenger, *Extrusion Processes, Machinery, Tooling,* ASM International, Metals Park, OH, 1981

45. O.J. Tassi; *Bare Wire Nonferrous Wire Handbook,* Vol 2, The Wire Association International Inc., Guilford, CT, 1981

46. *Conform Babcock Wire Equipment Limited Extrusion Technology,* Babcock, Ltd., England

47. *Conform Copper Welding Wire Recovery,* Metal Box P.L.C., England, 1986

48. S. Kalpakjian, *Manufacturing Processes for Engineering Materials,* Addison-Wesley Publishers, London, 1985

49. E. Schlowag and H. Mueller, *Cold Impact Extrusion of Copper Cases: an Example of a Technological Development,* Vol 13 (No. 3), Umformtechnik, 1979, p 31–34

50. R.J. Dower, The High-Speed Extrusion of Some Common Metals, *Journal of the Institute of Metals,* Vol 95 (No. 6), 1967, p 1–7

51. G.A. Banner, Art Metal Work, *Brass World and Platers' Guide,* Vol 21 (No. 6), p 187–190

52. T. Matsushita, Y. Yamaguchi, M. Noguchi, and M. Nishihara, Hydrostatic Extrusion of Tubes, Kobe Steel *CIRP Ann.,* Vol 23 (No. 1), 1974, p 75

53. Properties and Selection: Nonferrous Alloys and Pure Metals, Vol 2, *Metals Handbook,* 9th ed., American Society for Metals, Metals Park, OH, 1979

54. Properties and Selection: Nonferrous Alloys and Special-Purpose Materials, Vol 2, *Metals Handbook,* 10th ed., ASM International, Materials Park, OH 1990

55. *GEWA Finned Tubes Type D Safety Tubes PYRAMID/GEWA-D,* Catalog HE9-3GewaD, Wieland-Werk AG, Ulm

56. Copper and Copper Alloys, *1989 Annual Book of ASTM Standards, Nonferrous Metal Products,* Vol 02.01, Society for Testing and Materials, Philadelphia, PA, 1989

57. Economical Production of Copper Tube, *Metals,* Vol 2 (No. 14), 1967, p 52–53

58. P.T. Gilbert, E.B. Lockyer, and D.R. Warner, The Manufacture and use of Copper and Copper Alloy

Tubs, *Wrought Copper Materials Symp.,* Copper Information Centre, Bombay, Dec. 8–9, 1972

59. *Cast and Roll,* Outokumpu Engineering, Oy Espoo, Finland, 1988

60. Alambres y Cables Eléctricos, Cobre Cerrillos S.A., Santiago, Chile, 1990

61. H. Blaschke and K. Rustenbach, Craftsmanlike Copper Tube Installation (Die fachgerechte Kupferrohr-Installation), DKI Sonderdruck, Vol 158 (No. 7–9), Berlin, 1979

62. *Copper Tube in Domestic Water Services,* CDA UK, TN33, Copper Development Association, Potters Bar, Herts, England, March, 1988

63. Properties and Selection of Metals, Vol 1, *Metals Handbook,* 8th ed., American Society for Metals, Metals Park, OH, 1961

64. Atlas of Microstructures of Industrial Alloys, Vol 7, *Metals Handbook,* 8th ed., American Society for Metals, Metals Park, OH, 1972

65. *Machining Rod Handbook,* Copper/Brass/Bronze, No. 702/9, Copper Development Association Inc., New York

66. *Machining Copper and its Alloys,* Publication TN3, Copper Development Association, Potters Bar, Herts, England, Dec. 1980

67. *Copper Rod Alloys for Machined Products,* No. 7142/1699, Copper Development Association Inc., New York, NY

68. *Wrought Beryllium Copper,* Brush Wellman Inc., Cleveland, OH, 1984

69. Machining, Vol 3, *Metals Handbook,* 8th ed., American Society for Metals, Metals Park, OH, 1967

70. F. Reverchon, *Heat Treatment of Copper and its Alloys,* (Traitements Thermiques du Cuivre et de ses Alliages), Centre d'Information du Cuivre Laitons et Alliages, No. 106, Paris, in Extrait de la Revue Cuivre, Laitons, Alliages, No. 106 and 112, Riegel Editeur

71. Copper and Copper Alloys, *1991 Annual Book of ASTM Standards, Nonferrous Metal Products,* Vol 02.01, American Society for Testing and Materials, Philadelphia, PA, 1991

72. L. Dorn, *Welding of Copper and Copper Alloys (Schweiben von Kupfer und Kupferlegierungen),* Schweißtechnik (DDR), 1990, p 170–172

73. Welding and Brazing, Vol 6, *Metals Handbook,* American Society for Metals, Metals Park, OH, 1971

74. H. Udin, E.R. Funk, and J. Wulff, *Welding for Engineers,* John Wiley & Sons Inc., New York, NY, 1954

75. *Welding of Copper* (Scheweißen von Kupfer), Deutsches Kupfer-Institut, Bestellnummer I.11, Berlin, Germany

76. Welding, Brazing, and Soldering, Vol 6, *Metals Handbook,* 9th ed., American Society for Metals, Materials Park, OH, 1983

77. Copper Brass Bonze, *Soldering and Brazing Copper Tube,* Application Data Sheet, Copper Development Association Inc., Greenwich, CT

78. A. Cohen, *Solder Flux Evaluation and Listing Program,* Copper Development Association Inc., Greenwich, CT, March, 1991

79. "Copper as a Material for Water Tubing" ("Kupfer Als Werkstoff für Wasserleitungen"), Deutsches Kupfer-Institut, Berlin, 1986

80. Copper Brass Bronze, *Mechanical Properties of Soldered Copper,* CDA No. 163/0, Application Data Sheet, Copper Development Association Inc., New York, NY

81. "Soldering of Copper and Copper Alloys" ("Löten von Kupfer und Kupfer-Legierungen"), Deutsches Kupfer-Institut, Bestellnummer I.003, Berlin, 1986

82. Copper Brass Bronze, *Lead-Free Solders for Drinking Water Plumbing Systems,* Application Data Sheet No. 406/7, Copper Development Association, Greenwich, CT

83. Joining of Copper and Copper Alloys, TN 25, Copper Development Association, Potters Bar, Herts, England, Dec. 1980

84. A.R. Bunk, M.A. Roe, and M. Luttinger, Epoxies for Residential Copper Plumbing, *Adhes. Age,* Dec. 1984

85. Copper-Bond Epoxy Adhesive, Marsh Laboratories, Pittsburgh, PA, 1986

86. Copper Brass Bronze, *Powder Metals Properties and Applications,* Technical Report No. 129/6, Copper Development Association Inc., New York, NY

87. *Superconductors,* Hitachi Cable Ltd., Tokyo, 1989

88. H. Murakami, S. Yaegashi, J. Nishino, Y. Shiohara, and S. Tanaka, Preparation of Y-Ba-Cu Powder by Sol-Gel Method at Low Temperature, *Science & Technology in Japan,* Vol 9, 1990, p 59–61

89. J.J. Dunkley and D.F. Berry, Production and Application of Copper and Alloy Powders, *Performance Materials*

90. J.F.C. Morden, Powder Metallurgy-VII, *Metal Industry,* Feb. 8, 1957

91. E.G. Kendall, *The Development of High-Modulus Graphite Reinforced Lead, Zinc and Copper Composites,* p 291–300

92. R.W. Evans and G.M. McColvin, Hot-Forged Copper Powder Compacts, Powder Metall., Vol 19 (No. 4), 1976, p 202–209

93. Heat Treating, Cleaning and Finishing, Vol 2, *Metals Handbook,* 8th ed., American Society for Metals, Metals Park, OH, 1964

94. A.K. Graham, *Engineering Manual of Electrolytic Lubricants (Manual de Ingeniería de los Lubricantes Electolíticos),* Reinhold Publishing Hand, México, 1967

95. C.M. Rodia, Electroforming, *Metal Finishing Guidebook and Directory,* Metals and Plastics Publications Inc., Hackensack, NJ, 1989

96. G. Govindarajan and S. Ramamurthi, Electroforming of Copper for Artistic Patters, *J. Electrochem.,* Vol 5 (No. 8), 1989, p 593–595

97. W. McMullen, Chromium Plating, *Metal Finishing Guidebook and Directory,* Metals and Plastics Publications Inc., Hackensack, NJ, 1989

98. Copper Brass Bronze, *Clear Organic Finishes for Copper and Copper Alloys,* CDA No. 161/0, Application Data Sheet, Copper Development Association Inc., New York, NY

99. B.B. Morton, *The Cladding of Steel with Copper-Nickel for Marine Industry Applications,* International Copper Research Association Inc., Personal Communication, 1990

100. L.J. Broutman and R.H. Krock, Ed., *Composite Materials,* Vol 4, *Metallic Matrix Composites,* K.G. Kreider, Ed., Academic Press, New York, NY, 1974

101. T. Astoddart, *Engineering Production,* Vol 2, 1971, p 25–28

5

Applications

Introduction

Whether in the pure state, in alloys, or in chemical compounds, copper's physical, mechanical, and chemical properties find its applications in a very large variety of products. This chapter describes common applications for the most important copper products.

W.C. Butterman of the U.S. Bureau of Mines accounted for the widespread application of copper in the following way (Ref 1):

"... At least four physicochemical properties, singly or in combination, account for the widespread use of copper and its alloys. Foremost among these is electrical conductivity. Among the electrically conductive metals, copper is second in conductivity to silver; among the metals that are available in quantities sufficient for large-scale use, its only close competition is aluminum, which it exceeds in conductivity by more than one-half. It has been said that the ready availability of copper made the age of electricity.

"Next in importance is thermal conductivity. Again, copper is second to silver in conductivity, and aluminum, which has about half [sic] the thermal conductivity of copper, is its only close competitor. Copper and its alloys are widely used as the materials of construction for the heat exchangers used in motor vehicles, refrigeration units, water desalination plants, and numerous other industrial applications.

"The chemical stability [i.e., corrosion resistance] of copper, conferred in most environments by a rapidly formed surface oxide film, is important in both of the foregoing classes of end use and is the principal reason that copper and its alloys are used extensively for tubing and valves in systems carrying potable water or other aqueous fluids. It is also used for roofing, siding, gutter and downspouts.

"A fourth important property of copper is workability. This refers mainly to plastic work and machining. Annealed copper and copper alloys can easily be rolled or drawn. With suitable heat treatments and alloying additions, favorable machinability is as well obtained. In an extended fashion the term workability may also be used to refer to giving desired color, texture or luster. Color is changed through alloying. Special texture and luster are brought out through chemical etching or abrasive procedures."

As shown in Table 1.9, this distribution of functional uses of copper are reflected in the sale of copper products in the marketplace.

The beneficial aspects of the electrical conductivity of copper became increasingly appreciated during the last quarter of the 20th century, when energy savings and related environmental issues gained wide public recognition. Its high conductivity enabled the metal to recover electric applications previously lost to aluminum.

Copper usage is governed by its performance and cost-effectiveness compared with alternative materials. Electrical conductivity is especially important, because wire accounts for more than 50% of copper consumption worldwide. The percentage is nearer 70% in the United States, when all copper alloys used in electrical products are included. One major exception is overhead transmission cable, where aluminum is used almost exclusively to reduce weight. However, portions of the power transmission market have begun to revert to copper, as the economic and environmental benefits of undergrounding become more widely recognized (Ref 2).

The high thermal conductivity of copper finds applications in many types of heat exchangers. Automobile and truck radiators account for most of the volume, but large quantities of copper are also used in industrial heat exchangers, steam condensers, waste heat boilers, space heaters, and air conditioner evaporators and condensers.

Superconducting cables represent an application in which both the high electrical and thermal conductivities of copper are essential. This application currently accounts for a very small fraction of total copper consumption, but advances in high temperature superconductors may make this a significant use for copper in the future.

Other technically valuable properties of copper include:

- *Resistance to corrosion and marine fouling.* These properties are primarily important to the use of copper in shipbuilding, offshore structures, seacoast power plants, and other marine applications, including desalination equipment.
- *Favorable mechanical properties.* Copper and brass offer high malleability and formability, while copper alloys are available in a surprisingly wide range of strengths. Copper-base metals are unique in that they combine these useful mechanical properties with high conductivity and corrosion resistance.
- *An attractive appearance.* The range of colors displayed by copper alloys encourages the use of copper where technical properties and aesthetic appeal are important.

The electrical and thermal properties of copper make the metal essential to the development of basic industrial infrastructures (see Chapter 1), and copper consumption tends to mirror industrialization and industrial expansion. The following sections describe the applications of copper in important market sectors.

For statistical purposes, the applications of copper are divided into market sectors. This text will follow the market sector distribution system used in the United States. Table 5.1 indicates the distribution of usage of copper and copper alloys in several market sectors.

Building Construction

In industrialized countries, most copper used in building construction is consumed in the form of plumbing goods: tubing, taps, valves, tanks, connectors, and fittings. Copper tube and fittings are widely used for hot and cold domestic water and central heating systems. Copper tube is also used to connect gas service intakes and home heating systems and appliances. Because of its conductivity, and the consequent energy efficiency obtained compared to that of substitute materials, copper tube is the predominant material for air conditioner condensers and evaporators in both domestic and commercial units. There is also a growing trend to specify copper for fire sprinkler systems in both commercial and residential buildings.

The remaining copper used in buildings is in the form of electrical wiring, roofing, gutters, downspouts, façades, and architectural and builders' hardware. Of these, wiring is the most important. Plumbing tube is made from copper, whereas valves, faucets, and some fittings are made from brass. Roofing and related applications are made from copper, and hinges, locks, fasteners, and window and door hardware are made from brass or bronze.

The expanding use of electrical and electronic equipment in all buildings has increased the need for copper wire for power, communications, and signal transmission. In office buildings, the increased use of electrical and electronic equipment obviously requires more copper. In addition, however, it has also created a demand for larger space which, in turn, calls for larger capacity ventilation and air conditioning systems, all of which further

Table 5.1 U.S. copper and copper alloy consumption by major market 1991

Markets	Total, %
Building construction	41.3
Electrical/electronic products	24.4
Industrial machinery/equipment	13.1
Transportation equipment	11.4
Consumer/general products	9.8

Source: Copper Development Association, Inc.

increase the demand for copper. The same trend has been detected in residential structures (Ref 3).

Architectural Features

Copper sheet is light in weight, easy to work and join, attractive, and extremely durable. It resists attack by air and moisture. Bronzes are also used in façades and as door and window mullions. Most bronzes are easy to form by rolling, extrusion, or casting. Bronzes offer a variety of colors and finishes, and, like copper, they are durable beyond reasonable concern.

Copper metals have been used in architecture for thousands of years, and many early examples have survived to the present. Among the best known examples are the bronze doors on the cathedral in Aachen, Germany, which date to about A.D. 800. Perhaps even better known are the two bronze Baptistery doors in the Cathedral of Florence, the first of which was completed by Ghiberti in A.D. 1423 and the less well known south door, which was designed and cast by Andrea Pisanos about 93 years earlier. As mentioned in Chapter 1, among the oldest examples of copper in European architecture are the bronze ties and

clamps found in the Pantheon in Rome, which also date from the second century A.D. These and countless other architectural artifacts have survived despite significant changes in the atmosphere's corrosivity.

The range of colors exhibited by brasses and bronzes is very similar, both in the natural and oxidized states. This may explain why architects commonly refer to all of the copper alloys they use as *bronzes*, whether the alloys are true bronzes or not. Among the most common copper-base architectural alloys are commercial bronze, C22000, and Muntz metal, C28000, (both brasses), red brass, C23000, architectural bronze, C38500, and silicon bronze, C65500. Architects frequently refer to these alloys as yellow bronzes. Similarly, nickel silvers such as C74500 and C77000 are commonly called white bronzes. Statuary bronzes refer to the range of naturally weathered or chemically oxidized brown-to-black surfaces produced on bronze-casting alloys. So-called green bronzes allude to the various shades of natural and artificial patina (Ref 4–10).

Table 5.2 summarizes the most important architectural end-use applications for copper, brass, and bronze. Some of these are discussed below.

Table 5.2 Typical architectural uses for common copper alloys

Product	Coppers		Brasses, architectural bronze					Silicon bronze, nickel silver		
	C11000	C12200	C22000	C23000	C26000	C28000	C38500	C65500	C74500	C79600
Art works	•	...	•	•	•	•	•	•	•	•
Bank equipment	•	•	•	•	...	•	•	...	•	•
Builders' hardware	...	...	•	•	•	...	•	...	•	•
Curtain walls, storefronts	...	...	•	•	...	•	•	...	•	•
Ecclesiastical equipment	•	...	•	•	•	•	...	...	...	...
Educational equipment	•	•	•	•	•	•	•	...	...	...
Elevators, escalators	...	...	•	•	...	•	•	...	•	•
Entrance doors and frames	•	•	•	...	•	•	•	•	•	•
Fire-fighting devices	...	...	•	•	...	...	...	...	...	...
Flag poles	...	...	•	•	...	...	...	...	...	...
Floor tile	•	•	•	•	...	...	...	...	...	...
Food service equipment	•	•	...	•	...	•	...	•	...	...
Furniture	•	...	•	•	•	...	...	...	•	•
Grilles, screens	•	...	•	•	•	...	...	...	•	•
Gutters, downspouts	•	•	...	...	...	...	...	...	...	...
Identifying devices	•	•	•	•	•	•	•	...	...	...
Lighting fixtures	•	•	•	...	•	...	•	...	...	...
Louvers	•	•	•	•	...	...	...	...	...	...
Mansards	•	•	•	•	...	•	...	...	...	...
Plumbing fixtures	...	...	•	•	•	...	...	...	...	...
Postal specialities	•	•	•	•	•	•	...	...	...	...
Railings	...	•	•	•	•	•	•	...	•	•
Roofing, flashing	•	•	...	...	...	...	...	...	...	...
Skylights	•	•	...	...	...	•	...	...	...	...
Solar collector panels	•	•	...	...	...	...	...	...	...	...
Sun control devices	•	•	•	•	...	...	...	...	...	...
Toilet and bath accessories	...	...	•	•	•	...	...	...	...	•
Wall coverings	•	•	•	•	•	...	...	...	...	...
Wall panels	•	•	•	•	...	•	...	...	...	...
Wall tile	•	•	•	•	•	...	...	...	...	...
Weatherstripping	...	...	•	•	...	...	...	...	...	...
Windows	•	•	•	•	...	•	•	•	•	...

Roofing, Guttering, and Flashing

Roofing materials should offer long life, an aesthetically pleasing appearance, easy and economical application, and little or no need for maintenance. Copper roofing combines these qualities better than any other material, and as a result, it has found worldwide acceptance on both old and new structures (Ref 11, 12).

The most important technical features of copper roofing are its resistance to corrosion and its good thermal stability. Most metal roofing failures are caused by material deterioration and thermal fatigue (repeated flexure under cyclical temperature variations), but such failures will not occur in copper when the metal is properly installed. Moreover, copper will not creep when fixed vertically, as in façades. Copper also expands and contracts less than most other roofing metals—40% less than lead or zinc for a given temperature range. This accounts for the extensive use of copper as a roofing material in regions that experience wide temperature swings. Finally, copper roofing is relatively light in weight and does not require heavy timber or steel-and-timber support structures (Ref 13–16).

The materials required in the installation of a copper roof include:

- *Copper sheet and strip.* Table 5.3 lists the common sizes for copper roofing materials. Electrolytic tough pitch copper sheet (UNS C11000) in 16 oz weight, the most common roofing material, is specified under ASTM Standard B370 and is supplied as cold-rolled, soft, and lead-coated sheets and coils. Cold-rolled sheet is used for flat surfaces, and annealed material is used where good malleability is required. Lead-coated copper (plumbum), used for severely corrosive atmospheres, is equivalent to copper in workability, ease of joining, and strength. Copper roofing is always laid over a suitable felt backing to cushion contact with the substrate, dampen rain and hail noise, and avoid galvanic contact with dissimilar metal fasteners. Technological developments during the 1980s made thinner copper sheet available. The lighter gages reduce both the weight and cost of copper roofing systems. Rigidity is provided by redesigned support systems (Ref 6, 11, 17).
- *Copper nails.* Copper roofing requires nails made from copper or a copper alloy, usually brass. Nails should not be shorter than 25 mm (1 in.) under the head, not less than 2.6 mm (0.1 in.) thick, and weigh not less than 1.5 kg (3.3 lb) per 1000. The heads should be flat with a diameter of not less than 6 mm (0.25 in.); shanks should be barbed throughout their length (Ref 18).
- *Copper screws.* Screws for clips and other components should be made from brass. Aluminum or steel fasteners, if used, must be insulated from the copper to avoid galvanic corrosion (Ref 19).
- *Substrates* should be constructed as continuous decking made from moisture-resistant chipboard or plywood. Concrete and other materials are also suitable. Copper battens may be used in place of decking (Ref 11, 20).

Types of Joints. There are two kinds of seams used in copper construction, rigid and loose, and one kind of joint used to permit expansion and contraction. Most seams are of the expansion type and, therefore, expansion joints are used to accommodate movement between building components. Figure 5.1 shows the details of a typical standing seam joint used in roofing applications, and Fig. 5.2 shows the details of typical ridge and batten seams. Figure 5.3 shows the details of two types of ridge caps. These are but a few of many types of seams used in copper roofing. The type used depends on the slope and nature of the roof and the architectural effect desired (Ref 21).

A system known as "long strip copper roofing" has been used in the United Kingdom and Europe, particularly Switzerland, for many years. Its main advantage is reduced installation cost by the elimination of cross welts on sloping roofs and drips on flat roofs. The system utilizes long continuous lengths of one-eighth to half-hard temper copper strip. The rigidity of the strips, combined with special fastening clips, accommodates movement

Table 5.3 Recommended sizes and thicknesses of copper sheet for traditional roofing

Thickness, mm (in.)	Bay width, mm (in.)		Standard width of sheet to form bay, mm (in.)	Length of each sheet, m (ft)
	Standing seam	Batten roll		
0.45 (0.018)	525 (20.7)	500 (19.7)	600 (23.6)	1.8 (5.9)
0.60 (0.024)	525 (20.7)	500 (19.7)	600 (23.6)	1.8 (5.9)
0.70 (0.028)	675 (26.6)	650 (25.6)	750 (29.5)	1.8 (5.9)

resulting from thermal expansion. Attachment details at ridges and eaves are somewhat different from those used with traditional roofing (Ref 11).

Preformed copper panels in tile or shingle form are available commercially in several countries. Profiled panels having the appearance of individually scalloped tiles is also used. The panels interlock with concealed fasteners designed to take up thermal movement. They are suitable for roofing with pitches greater than 15°. The panels are pressed from 0.40 mm sheet and can be laid on flat or curved surfaces.

Tools and Equipment for Roofing. The fabrication and installation of copper roofing was once very labor-intensive, requiring much field fabrication. Today, portable roofing pan formers and power seamers have greatly simplified the process. Pans can be produced in continuous lengths,

which in some cases can be paid off directly onto the roof (Ref 11, 22).

Patina: Natural and Artificial Coloring of Copper. The natural weathering of copper to the characteristic blue-green or gray-green of patina is a basic copper sulfate similar to the naturally occurring copper mineral, brochantite. The patina forms by reaction of copper with airborne sulfurous and sulfuric acids. Once formed, the mineral surface layer essentially halts further corrosion (Ref 23–25).

As copper weathers, it initially passes from salmon pink through a series of russet browns, which signal the formation of copper oxide films. Cuprous and cupric sulfide conversion films then develop within the oxide. Their colors range from chocolate brown to black, and as they develop, the exposed metal surface darkens appreciably. Continued weathering results in the conversion of the sulfide films to the blue-green basic copper sulfate patina. Some patinas contain varying amounts of carbonates and chlorides. In seacoast locations, basic chloride salts may form an essential part of the patina film (Ref 26, 27).

Natural patinas form within 5 to 7 years in marine and industrial atmospheres. In cleaner rural atmospheres, the process may require from 10 to 20 years. In arid environments lacking sufficient

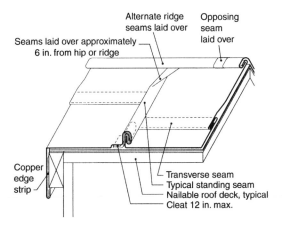

Fig. 5.1 Details of a typical standing seam roofing joint (Ref 11)

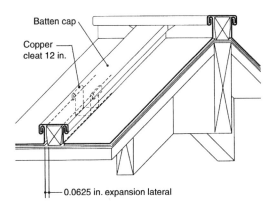

Fig. 5.2 Details of a typical ridge and batten seam (Ref 11)

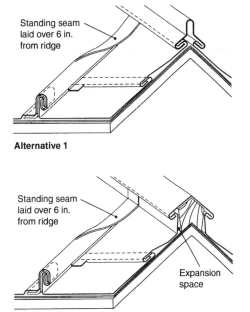

Fig. 5.3 Details of alternative ridge caps (Ref 11)

moisture to carry the chemical conversion process to completion, the basic sulfate patina may never form. Exposed horizontal surfaces develop a patina more rapidly than sloping surfaces which, in turn, patinate more rapidly than vertical surfaces. The critical variable is apparently the dwell time of moisture on exposed surfaces.

Patina films and their precursors are only 0.02 to 0.03 mm thick. They are strongly adherent but have relatively low abrasion resistance. Neither the intermediate oxide nor the sulfide films are particularly corrosion resistant in media other than the atmosphere in which they form. The basic sulfate patina is, however, highly resistant to all forms of atmospheric corrosion, and once it has had an opportunity to form completely, it significantly increases the underlying durability and service life of the metal (Ref 28).

While natural weathering yields the most durable installation, it may be aesthetically desirable to preserve the intermediate rust or chocolate brown shades, or hasten patina formation through chemical means. Both options are possible, although there are advantages and disadvantages to the processes involved.

Most of the commercial chemical treatments contain a pigmented material, sometimes powdered brochantite, to produce the desired color. Other coatings react with the copper to form a colored surface. Because of the chance for non-uniform coloring, and the cost of the process, some architects consider it impractical to artificially patinate large surfaces. Nevertheless, many large roofs have been artificially patinated to provide an immediate patina in a location where a patina will ultimately develop or to provide a patina where the atmosphere is too dry for a natural patina to develop. Tokyo, Japan is an example of the former and Santiago, Chile, an example of the latter. While none of the presently known treatments produce the natural basic sulfate patina, they do provide surrogate coatings until the natural patina forms or substitute for a natural patina where one cannot form.

To accelerate natural patina formation, the copper is first oxidized to a brown color by washing with a dilute solution of hydrochloric and perchloric acids. This can be followed by a sulfidation treatment to create a grey-to-black color. Depending on local atmospheric conditions and the specific pretreatment used, the natural blue-green patina can then begin to form within months, although it may require several years to fully develop. To ensure uniformity, prepatination should be performed before installation. Several proprietary processes have been developed to prepatinate copper sheets in the factory. Such sheets are commercially available in Europe, Japan, and the United States. One process has also been developed in the United States, Incrapatine (International Copper Research Association (INCRA), New York, NY), that duplicates the composition of the natural patina (Ref 29).

Light coatings of linseed oil, paraffin, and even crude oil can be used to provide a moisture barrier layer to retard subsequent reactions. The protective coatings must be reapplied at periodic intervals, but surfaces can be preserved in desired shades of brown or black. Roofing and flashing is usually oiled, and waxing is reserved for more visible architectural trim.

The natural color of copper can be preserved with clear organic coatings. The 4.5 acre domed copper roof of the Mexico City sports palace (erected for the 1968 Olympics) was preserved in this manner using an air-drying acrylic lacquer known as Incralac (International Copper Association, New York, NY). The lacquer contains benzotriazole to inhibit underfilm corrosion and an ultraviolet absorber to limit decomposition of the polymer. Despite the highly polluted Mexico City atmosphere, the dome's coating lasted nearly 15 years before tarnishing marred the beauty of the natural copper. A similar treatment was applied to the dome of the Arizona State Capital building in 1981. In this case, the bright surface has been preserved by periodic cleaning, polishing, and relacquering.

Flashing is used to prevent water intrusion through walls and wall penetrations. Concealed or through-wall flashing is used between outer and inner walls, in chimney construction, and where vertical and horizontal building elements meet. Visible flashing, sometimes called roof flashing, is used at caps, bases, valleys, windows, and doors. Visible flashing is much thicker than the through-wall type, up to 20 oz weight in copper- or lead-coated copper. Through-wall flashing is usually factory-made and can be as thin as 1 oz copper sheet, sometimes sandwiched between paper or fabric (Ref 30).

Gutters and Downspouts. Gutters are conduits, but built-in gutters and leaders can also function as architectural features, such as fascia. Figure 5.4 shows the details of a typical copper gutter installation. Copper scuppers eliminate the need for leaders. They are easy to maintain and add interesting sculptural elements to the building (Ref 31).

Thermal Expansion. Copper has a higher coefficient of thermal expansion than most building materials (metals excluded), and if sufficient allowance is not made in building design, the metal will buckle or tear. Thermal expansion coefficients increase with zinc content; high-zinc brasses, such as C26000, C28000, and C38500, expand as much as 18% more than copper itself. The thermal expansion coefficients of the nickel silvers are about 10% smaller than that of copper (Ref 32).

Architectural Hardware

Copper alloys are used extensively in the manufacture of door and cabinet hardware. The alloys' range of warm colors and finishes are aesthetically pleasing. Alloys commonly used are gilding, C21000; commercial bronze, C22000; and jewelry bronze, C22600. The metals readily lend themselves to a variety of manufacturing processes:

- Large parts, such as entrance handles, may be cast or forged.
- Curved brass latch bolts are produced by powder metallurgy to accommodate complex shapes.
- Doorknobs are made from flat strip that is either drawn, drawn and formed, or hydro-formed, depending on shape.
- Rosettes and other doorknob components are stamped from flat brass strip; heavier items may be forged.
- Lock cylinders are machined from brass rod.

- Latch faces and striker plates are stamped from flat brass strip.
- The housing and stock ring of deadbolt locks, as well as the deadbolts themselves, are machined from brass bar stock.
- Brass and bronze hardware is readily polished to a mirror finish. Natural or antiqued finishes are usually coated with a tarnish-retarding lacquer. Chromium plating can be applied for decorative reasons.

Building Wire

Building wire comprises all electric conductors within or on buildings and other structures, including mobile homes, recreational vehicles and boats, and industrial substations. The designation also includes conductors that connect to the supply of electricity and installations of other outside conductors on the premises at potentials less than 600 V (Ref 34).

Copper has almost always been the material of choice for building wire. Attempts to replace copper with aluminum were curtailed in most countries when it was found that aluminum connections gradually loosened due to a slow relaxation process known as creep. Insulating aluminum oxide, formed on loose connections, produced enough resistive heat to initiate electrical fires. Because copper oxides are semiconductors, this problem does not occur even in the presence of a loose connection. While improved spring-loaded contacts have largely alleviated this problem, alu-

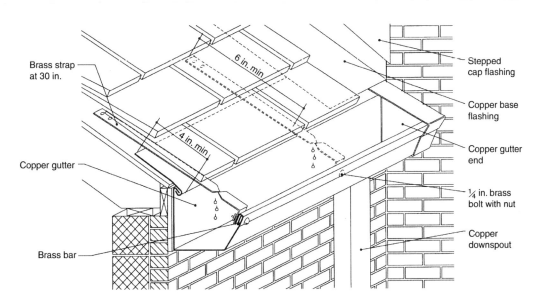

Fig. 5.4 Details of a typical copper gutter installation (Ref 11)

minum building wire has never regained the market share it once enjoyed. While in 1974 aluminum had 31% of the building wire market, by 1991, it had but 8%. The present market for aluminum building wire is essentially confined to the larger gage sizes used in supply circuits. For branch-circuit sizes, virtually no aluminum is used today (Ref 34).

Building wire design, installation, and/or hardware standards are found in safety codes such as those published by the American National Standards Institute, the American Society for Testing and Materials, DIN/VDE (Deutsches Institut für Normung/Verband Deutscher Electrotechniker), and the NCh/NSEG (Norma Chilena/Norma Segtel, Superintendencia de Servicios Eléctricos y Gas). In the United States, the National Electrical Code (NEC) serves as a model for many state and local codes. The national code bodies of several

countries are listed in Chapter 3 under "Cables" (Ref 33, 35).

Building wire sizes generally range from No. 14 AWG (American Wire Gage) through 740 kcmil (kcmil is 1000 circular mils; 740 kcmil denotes wire having a cross-sectional area of about 0.58 in.^{2}, or 375 mm^2). Table 5.4 lists cross-sectional areas corresponding to some commonly encountered AWG round wire sizes. Wire sizes as small as No. 22 AWG are currently used for signal and communications functions (although there is a trend toward smaller gage numbers); larger gages are used for power transmission. Solid wire is used from No. 14 AWG through No. 10 AWG; larger sizes are stranded to provide flexibility. Table 5.5 shows conductor sizes required for air conditioning systems, and Table 5.6 shows minimum conductor sizes for ground wires used in raceways and other equipment.

Cable, whether standard or of proprietary design, denotes wire used for special and/or multiple purposes. For example, a single multifunctional cable for a residential system may carry power, telephone, video, and control/communications signals. Like electrical wire, cable is normally insulated with a rubber or polymeric compound, as discussed in Chapter 3. Examples of common copper electrical cables are shown in Figure 5.5 (Ref 33).

Grounding. Since the 1950s, house wiring circuits in most countries have included a copper ground wire at each outlet. Safety codes, such as the U.S. NEC, spell out the installation of grounding circuits in detail, including the specification of standard color coding for grounding conductors. Grounding systems have been traditionally designed to protect people and equipment and normally carry no current. The growth in computer use has made grounding more important than ever. Computer logic circuits operate at <5 V DC; therefore, a low, uniform ground potential is important when one computer talks to another. In

Table 5.4 American wire gage (AWG) sizes

AWG No.	Area, mm^2	Area, in.^{2}
16	1.31	0.00289
14	2.08	0.00322
12	3.31	0.00513
10	5.26	0.00815
8	8.37	0.00130
6	13.3	0.0206
4	21.2	0.0328
3	26.7	0.0413
2	33.6	0.0520
1	42.4	0.0657
1/0	53.5	0.0829
2/0	67.4	0.104
3/0	85.0	0.132
4/0	107.2	0.166
250 kcmil	126.7	0.196
300	152.0	0.235
350	177.3	0.275
400	202.7	0.314
500	253.0	0.392
600	304.0	0.471
750	380.0	0.589
1000	506.7	0.785

Source: Ref 37

Table 5.5 Copper conductors for AC systems

Size of largest service entrance conductor or equivalent area for parallel conductors		Size of copper grounding electrode
Copper	Aluminum or copper-clad aluminum	
No. 2 AWG or similar	1/0 or smaller	No. 8 AWG
No. 1 AWG or 1/0	2/0 or 3/0	No. 6 AWG
2/0 or 3/0	4/0 or 250 kcmil	No 4 AWG
Over 3/0 through 350 kcmil	Over 250 kcmil through 500 kcmil	No. 2 AWG
Over 350 kcmil through 600 kcmil	Over 500 kcmil through 900 kcmil	1/0
Over 600 kcmil through 1100 kcmil	Over 900 kcmil through 1750 kcmil	2/0
Over 1100 kcmil	Over 17,750 kcmil	3/0

Source: Ref 103

Fig. 5.5 Typical copper cables as used in construction applications (Ref 37)

installations containing a large number of computers, a special low-resistance, low-impedance grounding system is required. Minimum sizes for copper grounding electrodes and conductors are listed in Table 5.5 (Ref 33, 36).

Single Family Dwellings. Residential electric power is supplied from underground or overhead conductors through service laterals and service drop conductors, respectively. These may be copper or aluminum. The service cable is connected through the electric meter to the service entrance equipment or the electrical panel (Fig. 5.6) and from there to the individual house wiring circuits. Interior wiring, that is, wiring down stream of the service entrance, is almost always copper. As a general rule, residential circuits with 8 to 10 outlets are rated at 15 A and are wired with either No. 14 or No. 12 AWG copper. Twenty-ampere circuits, which can handle up to 13 outlets, are wired with No. 12 AWG copper (Ref 33, 37).

A survey conducted by the Copper Development Association, Inc., found that U.S. residential buildings use an average of 52 kg (115 lb) of con-

ductors per 100 m² (1000 ft²) of dwelling space. Almost 90% of the conductors are copper, the balance being aluminum, which is primarily used in the service entrance cable. The study found that No. 14 and No. 12 AWG represent 92% of the lineal feet of residential wiring, comprising 73% of the (copper) wiring's weight (31 kg per 100 m², or 68.5 lb per 1000 ft²). Subsequent studies have shown an increase in copper wiring in residential usage as home sizes increase and the number of electrical amenities increase. There is a trend toward 200 amp service, for which No. 12 AWG serves as branch circuit wiring.

In low-cost housing, the wiring system may only amount to electric power distribution and door bell wiring. More expensive homes often include built-in telephones and intercoms, cable TV, antenna networks, thermostat controls, electric door opener systems, wiring for personal computers, and for closed-circuit security imaging devices.

Multifamily Houses. Electrical systems in modern multifamily housing are normally designed around main load centers, for example,

Table 5.6 Minimum size copper grounding conductors for raceways and equipment

Rating or setting of automatic overcurrent device in circuit ahead of equipment or raceway, not exceed, amperes	Minimum copper conductor size
15	No. 14 AWG
20	No. 12 AWG
30, 40, or 60	No. 10 AWG
100	No. 8 AWG
200	No. 6 AWG
300	No. 4 AWG
400	No. 3 AWG
500	No. 2 AWG
600	No. 1 AWG
800	1/0
1000	2/0
1200	3/0
1600	4/0
2000	250 kcmil
2500	350 kcmil
3000	400 kcmil
4000	500 kcmil
5000	700 kcmil
6000	800 kcmil

Source: Ref 37

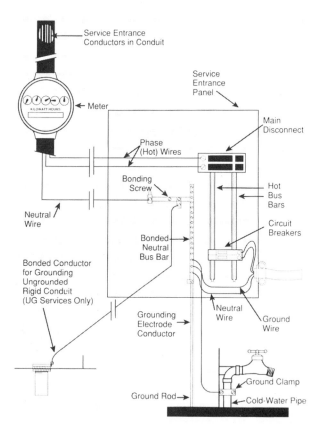

Fig. 5.6 Typical residential service entrance panel devices

where three-phase-plus-neutral power enters through a step-down transformer. Diesel-powered emergency generators and a battery-powered emergency light circuit may also be included. Telephone entrance panels connect to individual apartments, and there may be common TV, and radio antennas, cable TV, and other circuits, including security alarms.

Intelligent Buildings. A number of so-called "intelligent buildings" in which lighting, heating/air conditioning, and communications are controlled through computers and microprocessors have been demonstrated throughout the world. Copper finds extensive use in such structures in the form of specialized proprietary cables.

One example of an intelligent building (there are several commercial designs) is the Smart House system developed by the National Association of Home Builders, which contains hybrid branch cables, communications cable, and applications cable. Hybrid branch circuit cables contain wire of several different sizes and types. The cable illustrated in Fig. 5.7 is Smart House version of conventional type NM building wire. It contains three No. 12 AWG or No. 14 AWG wires for 120 V/15- and 20-amp AC power, as well as six No. 24 AWG conductors for control and communications. The conductors are in ribbon form, folded and enclosed in an outer jacket so that the cable can be installed behind walls, similar to the practice with conventional type NM wiring (Ref 38–42).

Smart House communications cable (Fig. 5.8) connects the communications entry box to distribution interface units, which serve as splitter/combiners, and from there to outlets, or "convenience centers." Included in the cable are two coaxial conductor cables (RG-59) for upstream and downstream distribution of audio and video signals and high-speed data. The cable also contains four twisted pair conductors (No. 24 AWG) for telephone service (Ref 43).

Smart House applications cable (Fig. 5.9) contains two No. 18 AWG wires for a 12 V DC uninterruptible power supply and six No. 24 AWG wires for control and communications. The applications cable is installed in a bus that branches from the system controller (the "brains" of the intelligent house) to applications' outlets in the home. It also is used for low voltage between switches and lights.

The hybrid branch cable alone contains 36 to 42 kg (79 to 92 lb) of copper per 100 m² (1000 ft²) of dwelling space, an increase of 16 to 35% over conventional housing, assuming the same length of cables for both types of construction. Commu-

nications and applications cables require additional copper. In all, the Smart House would likely utilize at least 30 to 40% more copper wire than homes wired using traditional practices. Use of copper-alloy strip would also be greater to accommodate the larger number of displacement-type connectors.

Plumbing

Copper tube is the standard plumbing material for potable water and heating systems in Canada, the United Kingdom, the United States, and parts of Europe. In Japan, however, it is mainly used for hot-water systems. In the United States, plumbing

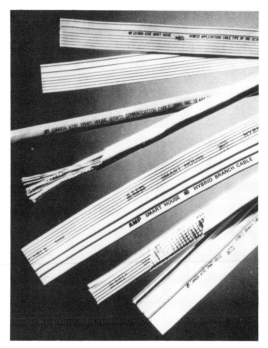

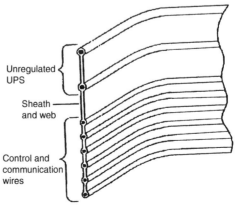

Fig. 5.7 Smart House hybrid cables

Unregulated UPS

Sheath and web

Control and communication wires

tube is made from phosphorus-deoxidized high-residual phosphorus (DHP) copper, C12200. Similar grades are used elsewhere. Plumbing tube is sold in several wall thicknesses: types K (heaviest), L(standard), and M (lightest) represent wall thickness gages used in plumbing systems. Type DWV (drain, waste, and ventilate) is used for drainage application. All four types are available in 20 ft (6.096 m) lengths in the drawn (hard) temper. Annealed (soft) temper tube is available in coils in 60 ft (18.29 m) and 100 ft (30.48 m) coils and straight lengths of types K and L are also available. Coils are frequently used as the underground service line between the water main in the street and the water meter in the home for the purpose of avoiding an intermediate joint. While straight lengths are generally used within the home in the United States, coils are frequently used in the United Kingdom and Europe. All copper tube produced to commodity or commercial tube classifications has a minimum chemical composition of 99.9% copper (including silver). Fittings for copper tube (water and drainage) are made to the American National Standards Institute (ANSI)/ASME standards. Wrought and cast copper and copper-alloy pressure fittings are available in all standard tube sizes and in a wide variety of types to cover needs for plumbing, heating, air conditioning, and fire sprinkler systems. Table 5.7 shows the allowable pressure ratings for copper tube as a function of temperature. These ratings are only a small fraction of the ultimate tensile, or burst, strength of copper. However, in practical systems, allowance must be made for joint strength. See Chapter 3 for a description of tubular products.

The advantages cited to justify the choice of copper in plumbing and heating systems include:

- *Experience.* Copper has been used in plumbing systems for more than 65 years.
- *Versatility.* Copper products are used for virtually all types of plumbing goods.
- *Durability.* Copper plumbing tube usually outlasts the buildings in which it is installed.
- *Health and safety.* Copper has been used to make drinking vessels, pipes, and water storage containers for thousands of years. It has been used by primitive cultures to purify water for drinking. Because copper is biostatic, it inhibits the growth of bacterial and viral organisms in water systems.
- *Cost-effectiveness.* The price of copper, in constant currency, has been stable since at least the 1940s. The ease of handling, forming, and joining of copper reduce installation time and cost.
- *Recyclability.* Nearly half of all copper currently used has been recycled from service in previous use.
- *Heat, corrosion, pressure, and fire resistance.* These attributes are cited as the reasons for the expanding use of copper in fire sprinkler systems.
- *Impermeability.* Ground-borne pollutants such as gasoline, organic solvents, fuel oil, and pesticides cannot permeate copper tube.
- *Professional endorsement.* Copper is the material of choice among plumbers.
- *Worldwide industry support.* Plumbing tube products are manufactured according to established standards. The industry provides training to installers and invests in technical support.

Potable Water Systems. Water tube is usually joined using capillary fittings and soldering. Cap-

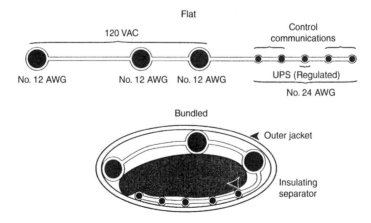

Fig. 5.8 Smart House 120 V AC/data communication 20 A cable

illary joints and brazing are used where greater strength is required and is required by some national codes for some or all tube sizes. Mechanical joints involving flared tube ends are frequently used for underground tubing joints where the use of heat is impractical and for joints that may have to be disconnected from time to time. Whereas 50-50 lead-tin solder was once the most prevalent solder used, the use of solders containing lead for potable water systems is now discouraged; some

authorities now forbid its use entirely. Solder-free joining with epoxy and other glues may gain prominence in the future, but lead-free solders are more common. Commercial grades suitable for plumbing use include: 95Sn-5Sb (wt%) (ASTM B 65, Grade S65), 95Sn-5Ag (wt%) (ASTM B 65, Grade Sn95), 94Sn-6Ag (wt%), and 96.5Sn-3.5Ag (wt%) (ASTM B 32, Grade Sn96). These solders have melting points 50 to 100 °F above that of lead-tin solder, but their fluidity facilitates joint completion (Ref 44).

The service line connecting the water main to the user frequently contains a water meter made of cast brass as illustrated in Fig. 5.10.

Drainage Systems. Copper DWV (drain, waste, ventilate) tubes claim several advantages for use as soil stacks, vent stacks, and soil, waste, and vent branches. DWV tube is furnished in 6 m (20 ft) lengths, reducing the number of joints and simplifying assembly. The entire vent stack for an average residence can be cut from a single 6 m length. Also, a 7.6 cm (3 in.) copper stack has an 8.6 cm ($3\frac{3}{8}$ in.) outside diameter at the fitting; it can, therefore, be installed in a 5 by 10 cm (2 by 4 in.) stud partition. Equivalent steel and plastic pipes have larger outside diameters and are more difficult to fit into restricted spaces. DWV tube is also light enough to be hand carried.

Thermal Expansion of DWV Systems. Copper expands about 37% more than iron or steel pipe but considerably less than plastic. A copper tube stack may move approximately 450 μm/°C (0.001

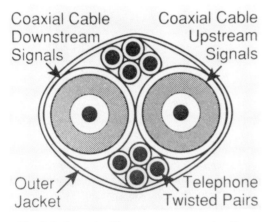

Fig. 5.9 Smart House communication cable

Fig. 5.10 Typical cast brass domestic water meter

in./°F) per floor. Thermal movement can be controlled by anchoring the stack. Anchoring at every eighth floor suffices for a maximum temperature rise of 28 °C (50 °F), while anchoring at every fourth floor accommodates a 56 °C (100 °F) temperature rise.

Allowance for Hydrostatic Testing of DWV Systems. Other than during hydrostatic testing, copper drainage systems do not ordinarily operate under pressure.

Corrosion. Certain water conditions, including those listed in Table 5.8, can lead to corrosive attack of copper water tube. In most cases, waters that are aggressive to metals are treatable and water utilities are encouraged to provide such treatment. Five particularly aggressive conditions include:

- *Hard well waters,* which cause pitting. Aggressive, hard well waters characteristically contain high total dissolved solids (tds), including sulfates and chlorides. They typically have a pH between 7.3 and 7.8, a carbon dioxide content (CO_2) greater than 10 ppm, and some dissolved oxygen. They can be identified by chemical analysis and can be treated to raise the pH and combine or eliminate dissolved CO_2. Sometimes simple aeration by spraying or sparging is sufficient to render them harmless.
- *Soft, acidic waters,* that inhibit formation of the protective film that normally forms inside the copper tube. Soft acidic waters cause staining of fixtures, a form of so-called "blue water." Raising the pH to 7 or higher usually solves the problem. Simple treatments include flowing the water through a bed of marble chips or injecting sodium carbonate (Na_2CO_3).
- *Excessive water flow velocity or turbulence* in the tube produce "erosion-corrosion or impingement" attack. Velocity effects can be aggravated if the water is also chemically aggressive or if it contains entrained solids (silt). To avoid this problem, flow rate should not exceed 1.5 m/s (4.9 ft/s).
- *Shoddy workmanship,* such as solder and flux residue left inside the tube. Dribbles of solidified flux left in the tube lead to pitting due to the continuing corrosive action of the flux, and erosion-corrosion, caused by disturbance of the smooth water flow around the obstruction.
- *An aggressive soil composition,* which deteriorates buried tube. Most natural soils do not attack copper, but any buried metal pipe laid in cinders is subject to attack by the acid generated when sulfur compounds in the cinders combine with water. The presence of chlorides and an acidic pH aggravates the situation. Under such circumstances, the tube should be isolated from the environment with a wrapping of insulating tape, a plastic coating or an asphaltic paint (Ref 45).

Table 5.7   Pressure ratings for copper tube joints

Maximum service temperature		Tube size		Pressure rating	
°F	°C	in.	mm	psi	kPa
Joints made with 50Sn-50Pb solder(a)					
100	38	1¼–2	32–51	95	655
		3–4	76–102	80	552
		5–8	127–203	70	482
150	66	1¼–2	32–51	70	482
		3–4	76–102	55	379
		5–8	127–203	45	310
200	93	1¼–2	32–51	50	345
		3–4	76–102	40	276
		5–8	127–203	35	241
Joints made with 95Sn-5Sb solder(a)					
100	38	1¼–2	32–51	220	1,517
		3–4	76–102	165	1,138
		5–8	127–203	120	827
150	66	1¼–2	32–51	195	1,344
		3–4	76–102	150	10,348
		5–8	127–203	100	690
200	93	1¼–2	32–51	140	965
		3–4	76–102	110	758
		5–8	127–203	80	552

(a) Solder alloys are described in ASTM B 32. Note: Metric numbers are conversions of U.S. customary units and are not necessarily standards expressed in SI units. Source: Ref 62

Copper drainage tube rarely corrodes except when misused or when improperly installed. Insufficient slope can permit corrosive solutions to lie in the tube and attack it. Hydrogen sulfide gas that vents back into the drainage system can also attack the tube. Additional information regarding copper water tube corrosion can be found in Chapter 2 (Ref 46).

Heating and Ventilation. As with plumbing systems, copper heating tube offers light weight, a choice of tempers, long-term reliability, and ease of joining, bending, and handling. Copper is used in radiator-type heating systems, forced air convection systems, and hydronic systems that use floors or ceilings as radiating panels. In steam-heating systems, copper return lines reduce the trap and valve maintenance.

Hard-drawn tube is recommended when rigidity and appearance are important. Softer annealed tube is better suited for panel heating, snow melting, and short runs to radiators, convectors, and the like. Soft, flexible tube also reduces the number of fittings required.

Radiators and Other Heat Exchangers. Cast iron or pressed steel radiators are often plumbed with copper tube. In North America, stand-alone radiators have largely given way to baseboard heating units that use tube-and-fin assemblies (copper tube, aluminum fins) mounted in a metal casement. Forced-air convection heating systems, more common in commercial buildings, also utilize heat exchangers fabricated from copper tubing and aluminum fins. In England, mini/micro-bore space heating systems are common in small- to medium-sized buildings. Such systems utilize small-bore (6, 8, or 10 mm) copper pipework (Ref 47).

Radiant Surface Heating. Radiant-surface, hydronic, or panel heating systems circulate warm water through tubes embedded in floors and/or ceilings. Copper tube is preferred because of its corrosion resistance, formability, and ease of installation. Tubes are plastic-coated to protect against mechanical or chemical damage and to allow for longitudinal thermal expansion. Types K and L tube are normally used for radiant heating and snow melting systems. Type L (soft temper) is used for coils or loops and type M (drawn) is used in straight lengths. Type M is recommended for hot-water and low-pressure steam systems in sizes up to 3 cm (1.2 in.). Type DWV is used for sizes larger than 3 cm. Condensate return lines should be made from type L tube.

Hydronic snow melting systems for driveways and sidewalks circulate a 50-50 solution of water and ethylene glycol through embedded copper tubes. As with floor-heating systems, soft-temper copper tube is suitable for both sinusoidal and grid-type coils, and hard-temper tube is best for larger grid coils and for mains.

Boilers. Copper and copper-alloy boiler tubes may be integrally finned or fabricated as fin-and-tube assemblies. Integrally finned tubes (see Chapter 3) contain helical fins extruded from the tube by a rolling process. Fin-and-tube assemblies are made of copper tubes that have been hydrauli-

Table 5.8 Effect of water conditions on copper corrosion

Water conditions	Corrosive effect
pH	
When pH is raised to give the water a slightly positive calcium carbonate saturation index, corrosion problems with copper are unlikely.	General attack on copper as a function of pH
CO_2 content and water mineral hardness	
Soft water, for example, water with less than 25 ppm mineral content, with 15 ppm or more, under stagnant conditions	Green staining and metallic taste due to dissolution of copper. Treatments include aeration to reduce CO_2 concentration, elevation of pH to neutralize CO_2 and blending in hard water to lay down a protective film.
Bicarbonate:sulfate ratio <1	Pitting attack
Well water at pH 7.5 plus >5 ppm $CO_2 + O_2$ and intermittent use	Pitting attack
pH < 7.8 plus > 25 ppm free CO_2 >17 ppm sulfate and sulfate:chloride ratio 3:1	Pitting attack
pH between 7.1 and 7.8 plus >100 ppm magnesium sulfate	Pitting attack
<4.2 ppm potassium	Pitting attack
>25 ppm silicate	Pitting attack
<25 ppm nitrate	Pitting attack
Soft water at pH < 6.5 with 10 to 50 ppm CO_2, low in chloride, sulfate and nitrate ions, in hot water systems	Pitting attack
Hard water	No dezincification of duplex brass fittings

Source: Ref 46

cally expanded into a stack of regularly spaced copper or aluminum fins. Boiler-type heat exchanger tubes are tinned to protect against corrosive flue gases (Ref 36).

Valves. Cast brasses and bronzes are preferred for water circulation and distribution valves. The alloys offer high strength, pressure tightness, corrosion resistance, and low cost, and they require little maintenance.

Sand-cast, pressure-retaining copper-alloy components for U.S. systems comply with American Society of Mechanical Engineers standards ASME SB61, "steam or valve bronze castings (alloy C92200);" "SB62," "Composition Bronze or Ounce Metal (alloy C83600) Castings;" B16.15, "Threaded Fittings," and B16.18, "Solder Joint Fittings," among others. In the United Kingdom, drainage taps for hot and cold water installations and heating systems must meet the requirements of BS 2879; diaphragm-type float-operated valves must comply with BS 1212: Part 2, and globe, stop, check, and gate valves must satisfy BS 5154 (Ref 48).

Quiescent potable waters high in dissolved oxygen and CO_2 corrode brasses by a mechanism known as "dezincification" or "parting corrosion." Susceptibility apparently increases with zinc content, and the beta phase constituent of duplex brasses is particularly vulnerable. Localized, or "plug-type" dezincification is encountered in warm, neutral, or slightly alkaline waters with high-salt contents; uniform or layer-type dezincification occurs in slightly acidic water with low-salt contents, even at room temperature. Dezincification can be avoided by using so-called inhibited brasses, whose effective ingredients may include tin, arsenic, antimony, and/or phosphorus (Ref 49).

Domestic Hot Water. The preferred tube materials for domestic hot water systems are C12200 and phosphorus-deoxidized non-arsenical copper, C10600. Some units utilize the primary space heating boiler as a source of domestic hot water. These generally involve immersion heating coils installed in the top of the boiler. For stand-alone units, copper sheathing is used to surround the heating element. Conversely, hot water for space heating can be provided by the domestic hot water unit. Here, the water is heated as it passes through a copper heat exchanger contained within a copper tank of heated water in the primary circuit. Instantaneous water heaters route water through a copper gas or electrically fired heat exchanger immediately ahead of the tap.

Solar energy systems are practical for space heating, hot water, and/or swimming pools, depending on location. Solar space cooling may become economically viable in the future. Copper flat-plate solar collectors generally utilize a water-propylene glycol solution as the heat transfer medium. The absorber plate with its attached copper-tube array can be enclosed with a modular collector panel, glass covers, and insulation, or it may be incorporated into the roof. An example is shown in Fig. 5.11. Solar water-heating systems require pressurized storage tanks. Such tanks may be lined with a copper alloy for corrosion protection. Space-heating systems are normally not pressurized, and storage containers may be made from a variety of structural materials (Ref 50, 51).

Air Conditioning and Refrigeration Systems

Air conditioning and refrigeration systems have become large and growing consumers of copper mill products. Here, both smooth bore and inner-grooved tubes are used. Copper is preferred for its formability, corrosion resistance, and ease of installation and maintenance; the metal is compatible with fluorocarbon refrigerants but cannot be used in ammonia-based systems. Copper is utilized for heat-transfer surfaces, tubing, motor windings, electrical controls, and valves and fittings. Further information regarding these uses can be found in Chapter 3 (Ref 51–53).

Gas Distribution Systems

Natural and Liquified Petroleum Gas (LPG). In the United States, the governing standards for the use of copper in gas distribution systems can be found in the National Fuel Gas Code, published by the National Fire Protection Association as NFPA 51 and accepted as the American National Standards Institute standard ANSI Z223.1. The standards apply to natural gas and LPG fuels, as well as manufactured gas and some gas mixtures. Copper tube called out in these standards must conform to type G, ASTM B 837, or to the "Specification for Seamless Copper Tube for Air Conditioning and Refrigeration Field Service" (ASTM B 280). Copper is unaffected by the normal constituents of natural and LPG fuels, but copper is proscribed under NFPA 54 if such gases contain more than 0.3 grains of hydrogen sulfide per 100 standard cubic feet (0.7 mg/L). In the presence of moisture, hydrosulfuric acid (formed from H_2S gas) can be corrosive to copper. In Canada, the appropriate code is CAN/CGA-B149.1, Natural Gas Installation Code, prepared and issued by the Canadian Gas Association (Ref 54, 55).

Home gas distribution systems in the United States must conform to ASTM B 837, type G seamless copper tube. Sizes and wall thicknesses are shown in Table 5.9. LPG distribution systems for one-family dwellings use type G copper tube at gage pressures up to 140 kPa (21.7 psi). Connections may be soldered, subject to approval by fire safety authorities (Ref 56–58).

Conventional gas distribution systems make use of ordinary screw- or compression-type fittings and connectors. Greater flexibility is offered by so-called gas-carcassing systems, which consist of plastic-coated copper gas tube fitted with plastic termination units containing brass terminal fittings. Terminal fittings, similar in appearance and size to an electrical outlet, can be positioned as

required. Gas appliances are connected to these outlets by quick-release fittings.

Manufactured Gas. In the United States, copper gas tube for manufactured gas distribution should conform to ASTM B 88, types K or L. The relevant Italian quality standard is UNI 6507, "Seamless Copper Tubes for Distribution of Fluids"; the German standard for gas and water distribution networks is DIN 1786 ("Seamless Copper Tubes for Installation"); the British standard is BS 2871: Part 197, but BS 1306:1975 is also applicable in the some cases. The latter is equivalent to ASTM B 302-88 and ISO 274:1975 (Ref 59–61).

Gas distribution tubing manufactured according to Italian standard UNI 9034 may be made from brass, bronze, steel, or plastics, and fittings for systems operating at pressures less than or equal to 5; bar may be of copper, brass, or bronze in

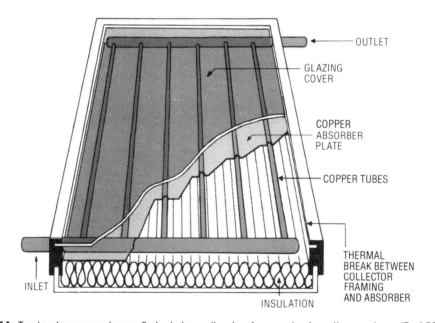

Fig. 5.11 Typical copper-base flat plate collector for a solar heating system (Ref 50)

Table 5.9   Sizes and wall thicknesses of type G tube for gas distribution

Standard size, in.	Actual outside diameter		Average outside diameter tolerances(a)				Wall thickness and tolerances				Theoretical mass	
			Annealed		Drawn		Wall thickness		Tolerance(b)			
	in.	mm	in.	mm	in.	mm	in.	mm	in.	mm	lb/ft	kg/m
3/8	0.375	9.52	0.002	0.051	0.001	0.025	0.030	0.762	0.003	0.076	0.126	0.187
1/2	0.500	12.7	0.0025	0.064	0.001	0.025	0.035	0.889	0.004	0.01	0.198	0.146
5/8	0.625	15.9	0.0025	0.064	0.001	0.025	0.040	1.02	0.004	0.01	0.285	0.424
3/4	0.750	19.1	0.0025	0.064	0.001	0.025	0.042	1.07	0.004	0.01	0.362	0.539
7/8	0.875	22.3	0.003	0.076	0.001	0.025	0.045	1.14	0.004	0.01	0.455	0.677
1 1/8	1.125	29	...	...	0.0015	0.038	0.050	1.27	0.005	0.013	0.655	0.975

Source: Ref 58. (a) The average outside diameter of a tube is the average of the maximum and minimum outside diameter, as determined at any on cross section of the tube. (b) Maximum permissible deviation at any one point

accordance with standard UNI-CIG 9034. Joints may be soldered, brazed, welded, screwed, or flanged. DIN tube sizes applicable to various types of joints are listed in Table 5.10. The corresponding Italian standard is UNI 8050/1 (Ref 59, 60).

Displacement meters common to domestic gas distribution systems can be cast in brass or bronze. Figure 5.12 shows an installation of such meters.

Medical Gas. United States Safety codes require the use of types K and L copper meeting ASTM B 819 for medical vacuum, oxygen, and anesthetizing gases. The ASTM standard contains special cleanliness requirements as a safety measure because pressurized oxygen can spontaneously ignite oils and/or drawing lubricants remaining inside tubes after processing. The National Fire Protection Association standard NFPA 99C, "Gas and Vacuum Systems," spells out installation requirements for such systems. Canadian Standard Association (CSA) Standard Z305.1, non-flammable medical gas piping systems, applies in Canada.

Fire Sprinklers

Copper tube and fittings for fire sprinkler systems have become increasingly popular in North America and now account for as much as 5% of a market dominated by steel. System design, installation, and maintenance are standardized in National Fire Protection Association Standard NFPA 13D, standard for the installation of sprinkler systems in one- and two-family dwellings and mobile homes. Applicable materials specifications are listed in Table 5.11. Copper tube conforming to ASTM B 88 types K, L, or M may be used. Joints are either soldered using 95Sn-5% Sb filler metal or brazed, depending on the size and type of system involved. The principal advantages of copper over conventional steel sprinkler systems are corrosion resistance, for longer service life (Chapter 2), and easier and somewhat less costly installation, particularly for retrofit situations. Dimensions of types K, L, and M tube can be found in Table 3.24 (Ref 33, 45, 62–65).

Electrical and Electronic Applications

Roughly one-half of all copper mined is used in the form of electrical wire and cable. This large area of application includes conductors for power generation, transmission and distribution, and for communications, electronic circuitry, data processing, entertainment, instrumentation, motors, and other electrical equipment.

Power Distribution

Overhead electrical transmission lines are predominantly made from a composite aluminum/steel cable. Aluminum saves weight, which reduces the number of transmission poles or pylons, but it does so at the expense of efficiency, because it has only between 54 and 66% the electrical conductivity of copper. Aluminum also has a larger coefficient of thermal expansion than copper. Copper is still used for overhead transmission lines in a few countries, and in some regions its market share is as high as 30%. While aluminum undoubtedly will remain the material of choice for overhead lines,

Table 5.10 Dimensions of copper gas and water tube according to DIN 1786

Outside diameter, mm	Wall thickness, mm
Capillary brazed or soldered joints	
12	1
15	1
18	1
28	1.5
35	1.5
42	1.5
54	2
64	2
76	2
22.9	2
102	2.5
Butt-welded, screws, or flanged joints	
133	3
159	3
219	3
267	3

Source: Ref 60

Table 5.11 Specifications for copper-base fire-sprinkler system components

Materials or components (title of standard)	Standard number
Seamless copper tube	ASTM B 75
Seamless copper water tube (types K, L, and M)	ASTM B 88
General requirements for wrought seamless copper and copper-alloy tube	ASTM B 251
Copper drainage tube (DWV)	ASTM B 306
Cast copper-alloy solder joint pressure fittings	ANSI/ASME B16.18
Wrought copper and copper alloy	ANSI/ASME B16.22
Cast copper-alloy pipe flanges and flanged fittings	ANSI/ASME B16.24
Brazing filler metal (classification BCuP-3 or BCuP-4)	ANSI/AWS A 5.8
Solder metal (95–5 tin-antimony, alloy grade Sb5) (50–50 tin-lead, alloy grade Sn50)	ASTM B 32

Source: Ref 62, 65

there is a growing trend in many countries to convert medium- and low-voltage overhead systems to underground cables.

Copper is preferred for underground transmission lines operating at high and ultrahigh voltages (64 to 400 kV). The predominance of copper underground stems from its higher volumetric electrical and thermal conductivities compared with aluminum. These properties act to conserve space, minimize power loss, and maintain lower cable temperatures. Some aluminum is used for medium-voltage underground transmission cables, but even these cables typically contain as much as 50% copper in the form of neutrals and protective screens. Copper continues to dominate low-voltage lines located under water and in mines, as well as for electric railroads, hoists, and other outdoor services.

Bus Bars

Bus bars are robust conductors that function as electrical manifolds to distribute power from a single source to several users. One or more bus bars may, for example, serve an array of machinery or equipment. Reasons cited for the preference of copper over aluminum include:

- High electrical and thermal conductivity for greater energy efficiency
- Good mechanical strength in tension, compression and shear, and a high resistance to fatigue failure
- Freedom from high-resistivity surface films, a property needed to make permanently efficient joints
- Ease of fabrication
- High-corrosion resistance
- Resistance to mechanical and electrical damage, including the ability to withstand the effects of flash-overs
- Reasonable first cost and high scrap value

Bus bar copper should be of the highest purity to ensure high-electrical conductivity. OF Copper C10200, with an electrical conductivity of 101 to 102% IACS (100% IACS is the International Annealed Copper Standard for conductivity, 0.580 Megmho/cm at 20 °C) fills this requirement. Silver-bearing OF coppers also have conductivities of at least 100% IACS and are used where resistance to annealing (softening) at moderately elevated temperatures is required. Most bus bars take the form of long, narrow plates with a rectangular cross-section. Channels, angles, rounds, tubular sections, and special shapes are also used.

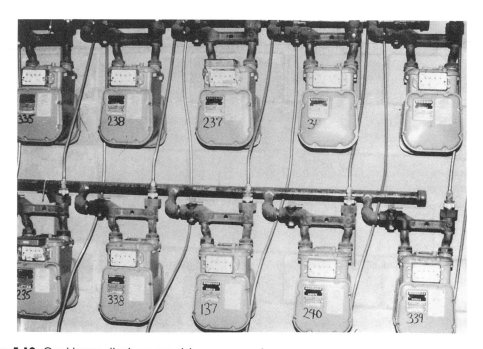

Fig. 5.12 Cast brass displacement-type gas meters

The four common bus bar types are:

1. Totally enclosed, with compound or oil insulation
2. Air insulated, high and low voltage
3. Air insulated, extra high voltage, for indoor installation
4. Air insulated, extra high voltage, for outdoor installation

Each type has its own electrical and mechanical characteristics, and this influences the form of copper used. Type 2 bus bars usually must have a high current capacity (ampacity) and sufficient mechanical strength to withstand the electromagnetic forces resulting from short circuits. Mechanical considerations govern design in bus bar types 3 and 4. For high voltages, a minimum number of insulators is desirable, and ampacities are usually small. Bars must be widely spaced to reduce corona discharge and minimize electromagnetic forces. Copper rods or tubes are usually preferred for high-voltage bus bars, particularly for indoor installations. In outdoor installations, the additional weight of ice and snow must be taken into consideration, and flexible cable or tubular conductors give better service than rigid constructions.

Ampacity is usually limited by the maximum temperature at which the bus can operate. Ampacity is calculated by taking into account the dissipation of I^2R heat through radiation and natural convection. Direct current ampacities differ from AC ampacities because the AC skin effect increases the bus bar's surface temperature relative to the interior.

Electrical Machinery and Equipment

Both copper and aluminum are widely used in electric motors and some generators; however, the higher electrical efficiency of copper (least I^2R losses) and lower life-cycle cost make it the preferable material for high-duty cycle installations (Ref 66).

Generators. The trend in modern generators is to operate at ever higher temperatures. This has prompted a trend toward the use of OF copper, C10200, for field bars and magnet wire in place of the deoxidized copper formerly used. The major advantage cited is the high electrical conductivity of OF copper, as much as 2% higher than that of deoxidized copper. Oxygen-free copper is far more conductive, more malleable, and more readily solderable than aluminum (Ref 67).

Motors. Copper is used in coil windings, bearings, collectors, brushes, and connectors. As in the case of generators, high-conductivity copper is increasingly replacing aluminum in high-duty cycle electric motors. So-called high-efficiency motors, which feature very low I^2R losses and several design and manufacturing improvements, are more economical on a life-cycle basis than even conventional copper-wound motors. Their use is encouraged by power companies eager to reduce system loads. In some countries, installation of high-efficiency motors is mandated by energy conservation laws. Table 5.12 shows a comparison of motor efficiency, copper contained, and payback time for typical 100 hp, Type D AC motors (Ref 34).

Transformer windings are normally made from ETP copper, C11000. Aluminum competes with copper where weight and first cost are decisive design elements, for example, in portable and pole-mounted transformers. While copper-wound transformers may cost more initially than those with aluminum windings, the lower I^2R losses of copper invariably result in lower life-cycle costs. One line of 25 kVA pole-top distribution transformers could vary from 14 lb of aluminum in the primary winding for the least efficient model to 58 lb of copper for the highest level of efficiency. The purchase decision is generally a function of the loss valuations expressed in dollars per kilowatt. For a large power transformer, the difference in the amount of copper used can be even more startling, again depending on the value placed on I^2R losses. The most efficient unit contains 132% more copper than the basic unit—19,000 versus 44,000 lbs (Ref 34).

Welding Equipment. Welding machines use copper in control circuitry, transformer windings, and in the cable to the welding gun or electrode holder. Welding machines produce AC and/or DC welding power, supplied by generators, alternators, transformers, transformer-rectifier combinations, or three-phase rectifiers. Generator-type machines are driven by electric motors or internal combustion engines. Transformer- and transformer-

Table 5.12   Pay-back time and copper content of high-efficiency motors

Motor designation	Efficiency, %	Copper content, lb	Pay-back, years
Standard	93.0	73.5	...
Efficient	94.5	113	2.62
High-efficient	95.4	124	2.09

Note: Basis: 100 hp, type D, AC motors

rectifier-type machines rely on line current as the primary power source. Welding machines utilize copper for internal conductors, control circuitry and transformer and motor windings.

Welding cables constitute a specialized but important copper market. Copper is favored over aluminum because its higher conductivity reduces line losses and permits the use of lighter, more flexible cables for a given ampacity. Table 5.13 lists cable sizes recommended for various lengths of leads and welding conditions (Ref 36, 68–70).

GMAW Contact Tips. In GMAW (MIG) welding, current is transmitted to the welding wire through short, small-bore copper tubes known as contact tips or welding tips. In the United States, more than 25 million of such tips are consumed each year. Tips are made from ETP copper, C11000, and occasionally free-cutting tellurium copper, C14500. Heavy-duty tips for automatic and robotic welders are made from chromium-zirconium coppers and copper-tungsten powder metallurgical composites.

Batteries. The press to develop efficient electric vehicles is driving research on high-performance batteries, among which lithium-based systems are leading contenders. It is projected that a practical electric vehicle based on lithium batteries will be available early in the 21st century. It is expected that electric vehicles will require sizeable quantities of copper for motor windings, bus bars, and wiring harnesses. Because of the premium placed on efficiency, copper will likely be favored over aluminum for these applications. Widespread application will likely occur first in regions affected by high levels of air pollution. In the United States, these regions include portions of California and ten eastern states (Ref 71–73).

Electrical Contacts and Connectors

Because of their high-electrical conductivity, availability and reasonable cost, copper and copper alloys are used almost universally for electrical and electronic contacts and connectors. Copper alloys are used where specific physical and mechanical properties are required (Ref 74).

Contacts. The four general types of electrical contacts are classified as: make-break, plug, sliding, and fixed. Their particular electrical and mechanical requirements, in addition to cost considerations, determines which copper or copper alloy should be used. Specific contact bearing materials are described in Table 5.14, which lists the contacts by type of service, required properties, and the type and form of contact material. A cross reference between ISO, DIN, and UNS designations is given in Table 5.15 (Ref 51).

Switches and Circuit Breakers. Copper is generally used for all internal conducting parts of small air-break circuit breakers. The contact points themselves may be made from or faced with more durable materials such as silver-cadmium oxide, silver-tungsten, or silver-graphite. Anti-weld contacts in transformers are often made from copper alloyed with 2 to 8% silver and up to 1.5% cadmium. Some high-current, low-voltage air-type circuit breakers employ sacrificial arcing contacts made from copper or silver alloys containing 3.75 to 10% copper (Ref 74–76).

Figures 5.13 and 5.14 illustrate how copper-base materials are used in typical large circuit

Table 5.13 Recommended sizes for welding cables

Weld current, A	AWG cable size for cable lengths listed, m (ft)					
	18 (60)	31 (100)	46 (150)	61 (200)	91 (300)	122 (400)
Manual (low-duty cycle)						
100	4	4	4	2	1	1/0
150	2	2	2	1	2/0	3/0
200	2	2	1	1/0	3/0	4/0
250	2	2	1/0	2/0		
300	1	1	2/0	3/0		
350	1/0	1/0	3/0	4/0		
400	1/0	1/0	3/0			
450	2/0	2/0	4/0			
500	2/0	2/0	4/0			
Automatic (high-duty cycle)						
400	4/0	4/0				
800	4/0 (2)	4/0 (2)				
1200	4/0 (3)	4/0 (3)				
1600	4/0 (4)	4/0 (4)				

(a) Ref 69

breakers. The breaker contains a movable copper rod, which makes contact against fixed copper blades, or contact fingers. The contacts points in this case are tipped with arc-resisting materials such as copper-tungsten composites containing between 20 and 70% copper. Optimum welding resistance under very high contact pressures is obtained with a 60Cu–40W composite (Ref 72).

Switch contacts can be made from ordinary copper, C11000, although they are more frequently made from inexpensive alloys such as cartridge brass, C26000. The industry considers copper, brass, and nickel silver, C76200, as first generation contact materials. Better combinations of strength and conductivity are provided by copper-iron alloys such as C19400, an aluminum brass, C68800, and beryllium copper, C17200. Better still are zirconium copper, C15100; a copper-iron-phosphorus alloy, C19700, and nickel copper, C70250. Third generation alloys will exhibit long-term stability at temperatures between 150 and 200 °C (300 and 390 °F), electrical conductivities higher than 50% IACS, and yield strengths in excess of 480 MPa (70 ksi) (Ref 74, 76–78).

Sliding contacts such as contact brushes and slip rings must resist wear at relative velocities that can exceed 50 m/s (164 ft/s). Copper-graphite composites containing relatively little graphite are used for slip rings in low-voltage DC machines operating at very high-current densities. Higher concentrations of graphite are used for conditions having lower current densities and those requiring better cooling conditions. Commutators in high performance motors are made from C11300, a copper alloy containing 0.12% Ag. High-conductivity OF copper (C10200), cold-worked to a hardness between 80 and 85 HV and an ultimate tensile strength of 275 to 310 MPa (40 to 45 ksi), is also used (Ref 74).

Plug connectors are essentially tribological (sliding wear) systems in which one workpiece moves in relation to the other. The most important functional characteristic of a plug-in connector is the ability to withstand numerous connect-disconnect cycles without a substantial increase in resistance. This calls for good resistance to oxidation-corrosion and wear. Other requirements for plug connectors include good spring qualities and resistance to fatigue, creep, and stress-relaxation. Plug connectors are often made from copper-tin alloys (phosphor bronzes), with contacts points plated with tin, nickel, silver, or gold. Other general-use connector alloys include C19500, C71000, C72500, and C75200 (Ref 79, 80). Figure 5.15

Table 5.14 U.K. and European copper strip and wire alloys for electrical and electronic applications

Alloy	Significant properties	Applications
CuFe2P	Good electrical and thermal conductivity	Leadframes; automotive fuse supports
CuSn0, 15	Very high electrical and thermal conductivity	Leadframes for semiconductors
CuZn0, 5	Higher mechanical strength than pure copper	
CuSn2, CuSn4, CuSn5, CuSn6, CuSn8	Low to moderate electrical conductivity (conductivity decreases with increasing Sn contents). Higher mechanical strength and good springiness (both increase with increasing tin content). Anneal resistance up to 200 °C (392 °F), good flexibility, readily plated, welded, clad, and soldered	Plug connectors, female plug supports, relay-springs, alligator connectors for radio, TV, and video, conducting alligator connectors for neon light supports
CuNi12Zn24	Good springiness, moderate electrical conductivity	Relay springs
CuNi18Zn20	Good springiness. Anneal resistant up to 200 °C (392 °F)	
CuNi18Zn27	Moderate flexibility. Good tarnish resistance (increases with increasing nickel and copper contents). Good plateability and solderability, adequate cladability	
CuNi9Sn2	Similar to Cn-Ni alloys, but somewhat lower springiness, higher electrical conductivity	Relay springs, conducting springs, switching devices
CuSn3Zn9	Higher conductivity and higher mechanical strength than $CnSn_6$ and $CnSn_8$	Automotive electrical systems, plug connectors
CuZn15, CuZn30, CuZn36, CuZn37	Moderate electrical conductivity (decreases with increasing zinc contents). Mechanical strength inferior to $CnSn_6$, $CnSn_8$, and Cn-Ni-Zn alloys. (Good cold-forming properties, good plateability and solderability, anneal resistant to 210 °C (410 °F)	Safety clips in plug connectors, relay springs, $CnZn_{15}$ also used for leadframes
CuZn23Al3Co	Lower conductivity than the copper-zinc alloys. High-mechanical strength with springiness similar to precipitation-hardened alloys. Anneal resistant up to 200 °C (392 °F)	Contact springs in switches, plug connectors, relays, alligator conductors, sliding contacts
CuBe1.7, CuBe2	Lower conductivity. Highest mechanical strength, good cold formability, anneal resistant up to 300 °C (572 °F)	Plug connectors

Source: Ref 74, 77

shows an assortment of electrical connectors using copper alloys as contacts.

Electronic Equipment

Electronic equipment constitutes a relatively small (in terms of tonnage consumed) but commercially important application for copper. The stringent demands placed on modern electronic devices have strengthened the role of copper in this market (Ref 49, 76). Figure 5.16 shows an assortment of copper-alloy lead frames and electrical contacts.

Printed Circuit Boards. Copper foil is the basic conductor in printed circuit boards (PCBs). The foil is produced either by rolling or electroplating, the latter being more common. Rolled foil is used where the printed circuit must flex, as in telephone headsets; plated foil is used in rigid boards. Rolled foil can be produced as thin as 0.5 mils whereas electroplated foil can be produced as thin as 5 microns (by plating on 3 mil aluminum and chemically stripping away the aluminum).

PCBs are made by the so-called print and etch process. A copper-clad resin laminate is coated with the circuit pattern in a protective "resist" ma-

Table 5.15   Cross reference of copper alloys for electrical spring contacts

ISO designation	DIN No.	DIN standard	UNS No.
CuSn2	2.1010	…	C50500
CuSn4	2.1016	17662	C51100
CuSn5	…	…	C51000
CuSn6	2.1020	17662	C51900
CuSn8	2.1030	17662	C52100
CuSn3Zn9	…	…	C42500
CuZn15	2.0240	17660	C23000
CuZn30	2.0265	17660	C26000
CuZn36	2.0335	17660	C27000
CuZn37	2.0321	17660	C27200
…	…	…	C27400
CuZn23Al3Co	…	…	C68800
CuNi9Sn2	2.0875	17664	C72500
CuNi12Zn24	2.0730	17663	C75700
CuNi18Zn20	2.0740	17663	C75900
CuNi18Zn27	2.0742	17663	C77000
CuFe2P	2.1310	17666	C19400
CuBe1.7	2.1245	17666	C17000
CuBe2	2.1247	17666	C17200
CuCo2Be	2.1285	17666	C17500
CuNi2Be	2.0850	17666	C17510

Source: Ref 77

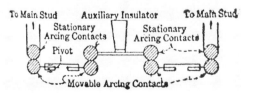

Fig. 5.13 Allis-Chalmers circuit-breaker contacts four breaks in series

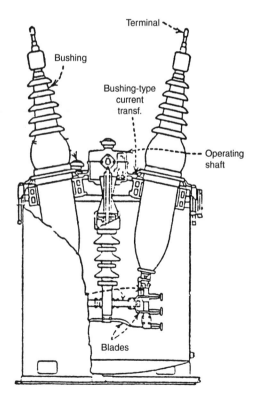

Fig. 5.14 Type R-72, 110 kV, six break, single-pole, oil circuit breaker

terial. The unprotected copper areas are then chemically etched to remove the unwanted copper from the laminate. The resist is then removed, leaving a copper replica of the desired pattern on the laminate. For reverse prints, the uncovered copper areas are electroplated with a protective metal, usually tin, tin-lead solder, gold, or tin-nickel alloy. After plating, the organic resist is removed and the plated metal serves as the resist in the subsequent copper-etching operation. Copper foils can also be incorporated into the resin laminate to improve heat dissipation in boards with high-component densities. The foils add mechanical and thermal stability, thereby reducing the incidence of thermally or vibration-induced fatigue failures.

Electronic Connectors. About two-thirds of the copper-base products used in electronic components are found in connectors. Copper is favored for this application because of its high conductivity; low-contact resistance; good spring properties; high-stress relaxation resistance; adequate strength, and excellent formability and plateability. The dominant electronic connector alloys are C26000, C50500, C51000, and C72500. Other connector materials include C11000, C15500, C17200, C17410, C17500, C19400, C19500, C68800, and C76200 (Ref 44, 78, 80).

Alloys with high-stress relaxation resistance are favored for contacts subjected to elevated temperatures, including the thermal spikes applied during burn-in. Highly stress-relaxation resistant alloys include C17200 and other beryllium coppers, a nickel copper, C70250, C51000, C52100, C72500, C15100, and a Cu-Ni-Sn spinodal alloy.

Connectors are normally stamped from strip, typically ranging from 0.25 to 0.6 mm (0.01 to 0.025 in.) thick. Heavy-duty, high-reliability products such as cylindrical military-type connectors are machined from rod or wire. Again, contact points are usually plated with oxidation-resistant metals. Tin and tin-lead contacts can be used in less-demanding applications, but noble metal contacts are more common. Gold-plated contacts require an underlying nickel barrier to prevent diffusion of gold into the copper substrate.

Beryllium coppers are used almost exclusively for the female sockets in cylindrical military connectors; male pins are made from brasses. Phosphor bronzes, alloy C72500, or other alloys are used in commercial products. Larger contacts can be stamped from strip, but screw-machined or cold-headed rod and wire dominate otherwise.

Insulation-displacement connectors were developed in the 1970s to permit efficient terminations in flat (ribbon) cable. Electrical contact is made by

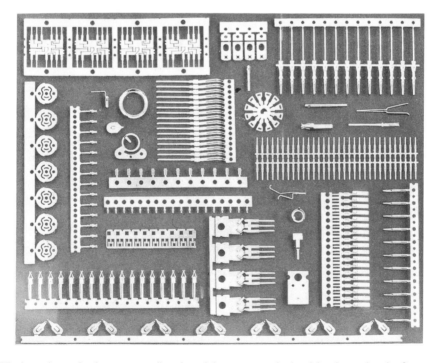

Fig. 5.15 Assortment of copper alloy lead frames and electrical contacts. Source: Copper Development Association, Inc.

piercing the cable insulation with prongs on the connector. In female insulation-displacement connectors, springiness must be maintained at the piercing end to hold the (typically four) points in contact with the wire and at the socket end. The preferred materials are phosphor bronzes, beryllium coppers, and UNS alloys C68800 and C72500. The male halves are less critical and can be made from brasses or other copper alloys.

Leadframes are the mounting media for integrated circuits (ICs) and similar devices. For many years, leadframes were stamped from nickel and 40 wt% Ni (alloy 42) strip. While these alloys are still used, copper-base materials now dominate in plastic dual-in-line (P-DIP) packages (Ref 72, 80, 81).

Copper alloys are increasingly favored for leadframes because they have high-thermal conductivities; their thermal expansion coefficients match those of the plastic encapsulating material better than iron-nickel alloys; they have high fatigue strength to withstand multiple insertions, and they are readily formable. Alloy C19400, whose conductivity is 25 times higher than that of alloy 42, is the leading choice worldwide. Alloy C19500 is used where higher strength is needed, and alloy C15100 is used for applications demanding high

conductivity. Strip for P-DIP applications is typically 0.25 mm (0.010 in.) thick. It can be as thin as 0.15 mm (0.006 in.) in other devices (Ref 80).

Communications

Before World War II, copper cable was the traditional communications medium for both long- and short-range communications. Copper cable usage has been affected over the years by changes in communications technology. The major factors involved include the development of microwave transmission and satellite communication; improved amplification systems, which reduce the gage of wire needed for a given distance; the reduction in wire gage (from 24-gage to 26-gage, which reduces the amount of copper required by one-half); multiplexing, which permits the transmission of several signals over a single twisted pair of conductors; electronic switching, which replaced copper-intensive mechanical equipment with solid-state devices using considerably less copper, semicolor, and optical fiber systems, which use copper only in interface devices. In addition, the rapid growth of "wireless" portable phone systems, particularly in developing countries where a telephonic infrastructure is not fully

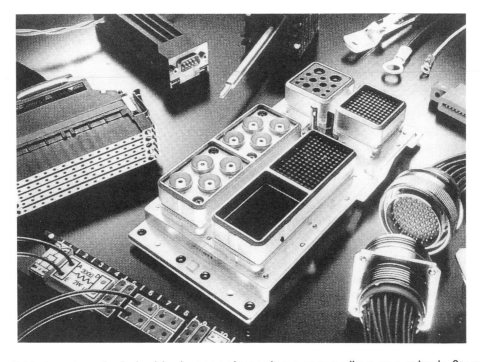

Fig. 5.16 Assortment of electrical connectors using copper alloys as contacts. Source: Copper Development Association, Inc.

developed, has undoubtedly impacted on the use of copper in telephonic communications (Ref 82).

Copper Conductors and Optical Fiber Systems. Forecasts of a rapid displacement of copper from the telecommunications market have not been not borne out. Rather, there has been a slow erosion of copper usage, which has been supported by the increase in the number of circuits (for Fax machines and computers) and by a reversion to 24-gage wire. Copper usage in communications did drop from its all-time high of 847,000 tonnes in 1979 in the combined Japanese, United States, and Western European markets to 648,300 tonnes in 1982, but it then rebounded to 722,000 tonnes by 1991. In the United States, a high was reached in 1979 with 388,000 mT of copper consumed in telecommunications cable. In 1994, this had dropped to 159,000 mT. Nonetheless, in the United States in 1992, 17 times as many miles of copper cable were installed as optical fiber cable. American data for the 1990s suggested that copper wiring systems were more economical than optical fiber for clusters of less than 24 residences, for distribution loops in urban areas with high user densities and short interconnect distances, and in the feeder plant for stable, low-growth areas. According to estimates by the American Telephone & Telegraph Company, the investment in new and replacement copper was more than $3 billion in 1992 (Ref 83).

A number of technological improvements have been advanced to strengthen the position of copper in communications. These include moisture-resistant cable terminals and improved connectors and closures for easier and more reliable installation. Multiplexing technology, such as high digital rate subscriber line (HDSL), which extends the capacity per twisted wire pair to the extent that systems that would have required 14,500 kg (32,000 lb) of copper using conventional technology could be downsized to only 590 kg (1300 lb), is another factor, as are band-widening systems that expand the message-carrying capacity of twisted wire pairs to that of optical fiber over distances of several miles. Probably the most significant technological development in this field has been in the field of digital signal processing (DSP), which has led to stunning increases in the amount of digital data that can be transmitted to residences over existing twisted pair telecommunications cable. While video transmission and telephone twisted pair cables are both ubiquitous in most developed countries, until recently they have never met. Asymmetrical digital subscriber line (ADSL) systems now make this possible over relatively short distances; namely, from the central telephone exchange to the residence or business office. The ADSL technologies are a family of digital twisted pair systems that are capable of delivering up to four independent, switchable VCR-quality video-on-demand streams, together with conventional telephone and electronic data transmission, over at least 3750 m (12,000 ft) of standard, 24-gage twisted pair cable. Thus, through ADSL technology, video, telephonic, and data transmission would be carried over long distances by optical fiber cable (a function not now served by copper) and into the place of usage by existing copper twisted pair cable (Ref 84–89).

Currently, the main use of copper cable in telecommunications systems is in the "outside plant"—the wire between the central station and the user. This loop consists of the feeder loop—from the local office to a distribution point near the user, and the distribution link—that part of the outside plant that is closest to the customer. In many regions, for commercial customers, fiber optic cable is rapidly moving into the feeder loop and is competing for copper in the distribution link. Copper still holds the major share of the market in residential service. For the commercial customer, T-1 lines and HDSL systems and the newly developed ADSL systems provide competition between copper and fiber optic systems. For the residential customer, a combination of optical fiber and coaxial cable known as hybrid fiber coax (HFC) seems to be in the future (Ref 90).

Superconductor Applications

Copper has been used for many years in the fabrication of filamentary superconductors (SC) based on Nb_3Sn, Nb-Ti alloys and other commercial superconducting materials. Copper is used as the matrix in which the superconducting filaments are imbedded and serves as a stabilizer. In its stabilizing role, copper provides for heat transfer and for bridging electrical conductivity. High field strength magnets for magnetic resonant medical imaging (MRI) equipment constitute the largest commercial application. The second largest use of superconducting devices is for magnets for use in the study of particle physics and the development of atomic fusion. Most other applications, such as in transportation and power transmission, are still under development although some prototypes have been built (Ref 91, 92).

Since the discovery of the so-called high-T_c metal-oxide SC compounds in 1986, there has been speculation regarding the replacement of copper-matrix superconductors by ceramics. In fact, many oxide superconductors contain copper ($YBa_2Cu_3O_{7-x}$ is the best-known example). However, it is possible that copper may be also utilized as a supportive and thermally conductive matrix for the oxide superconductors.

Industrial Machinery and Equipment

Copper and copper alloys are widely used in industrial equipment, generally because they offer a particularly beneficial combination of electrical and/or thermal conductivity, corrosion resistance, strength, and wear resistance.

Heat Exchangers and Condensers

Copper alloys have traditionally been the preferred materials for marine and fresh water condensers and heat exchangers. The most commonly used alloys are the copper-nickels, aluminum bronzes, aluminum brasses, and tin brasses. Alloys are chosen on the basis of their corrosion resistance (in the expected operating environment) and their ability to withstand high-velocity fluid flow (erosion-corrosion resistance). Equally important is the ability of copper alloys to inhibit the growth of marine organisms (biofouling) which, if unchecked, reduce flow rates, promote corrosion, degrade heat transfer, and necessitate costly maintenance.

Condenser tubes are one of a power plant's most critical elements, especially in plants that use saline or brackish water for cooling. Condenser tubes dare not leak, because intrusion of oxygen and salt water to the boiler rapidly leads to corrosion and hydrogen embrittlement of boiler tubes and other components. Even fresh-water cooled plants are sensitive to this problem because the evaporative cooling towers often used in such installations can concentrate dilute salts by factors of from 2 to 8. Condenser tube failures generally result from stress-corrosion cracking (ammonia attack), pitting, parting corrosion, or erosion-corrosion due to entrapped foreign matter or faulty tube-rolling (Ref 93).

Hydraulic shear forces in rapidly flowing media can damage protective oxide films, a process which eventually leads to erosion-corrosion. Condenser tube materials are rated by how well they withstand such hydraulic forces. Ratings are expressed in terms of shear stress or by the maximum or critical flow velocities the materials can tolerate without damage. The critical shear stress

for aluminum brass is 19 Pa (2.8 ksi); for 90Cu-10Ni it is 43 Pa (6.2 ksi), and for 70Cu-30Ni it is 48 Pa, (7.0 ksi). Alloy C72200, a chromium-iron-manganese-modified 80Cu-15Ni, exhibits a critical sheer stress of 297 Pa (43 ksi), well above that encountered in normal condenser service. The critical coolant velocity for this alloy is 6 m/s (19.7 ft/s), almost twice that of alloy C70600. Alloy C71500 can be used at flow velocities between 3.5 and 4 m/s (11.5 and 13 ft/s).

Salt Water Cooling. The three most important copper alloys for saltwater cooled condensers (in order of increasing corrosion resistance) are arsenical aluminum brass, C68700, 90Cu-10Ni, C70600, and 70Cu-30Ni, C71500. Table 5.16 lists the performance of three copper alloys in polluted seawater; ratings are based on the percent of tubes failed per 10,000 service hours.

Fresh Water Cooling. Thin-walled condensers cooled with recirculated fresh water are often tubed with 90Cu-10Ni, C70600. This alloy withstands evaporatively concentrated corrosive species, aeration, low pH (required for scale control), and high biomass contents (bacteria, mussels, and slime). C70600 tolerates such waters at temperatures up to 90 °C at velocities up to 3.9 m/s (12.8 ft/s).

Alloy Selection. The aqueous corrosion resistance of copper alloys is based on the formation of adherent and protective layers of insoluble corrosion products. Such layers cannot form on ordinary copper-zinc brasses in seawater, therefore special brasses, copper-nickels, and aluminum bronzes are better suited for such service (Ref 94).

Special brasses are alloys of the CuZn30 type (similar to C21630, nominally 70Cu-30Zn) modified with 0.03% As to inhibit dezincification. An uninhibited 70-30 brass such as C26000 can be used in river water with salt contents as high as 2 g/L unless the pH is acidic, but tin brasses such as C44300 (CuZn29Sn1) give better service. Better still is aluminum brass C68700 (CuZn22Al2),

Table 5.16 Comparative reliability of alloys C70600, C71500, and C68700 in polluted seawater

UNS No.	Median failure rate	Probability of reaching X years life with no more than Y % tube failures	
		X = 20 Y = 30 %	X = 30 Y = 10 %
C70600 (Cu-10% Ni)	0.05	85	85
C715000 (Cu-30% Ni)	0.06	81	81
C68700 (Al brass, arsenical)	0.33	52	57

Source: Ref 44

which was actually developed for seawater service. Handbook data indicate that this alloy should be limited to flow velocities between 0.8 and 2 m/s (2.6 and 6.6 ft/s), but the limit can be extended to 2.5 m/s (8.2 ft/s) in very clean water.

All brasses are susceptible to stress-corrosion cracking (SCC) in the presence of ammonia, amines, nitrites, and mercury and its compounds when the alloys are under residual or applied tensile stress. The organic amines morpholine and hydrazine used as oxygen scavengers in modern power plants break down in service to yield ammonia and, therefore, constitute a potential threat to copper condenser tubing. Experience suggests, however, that the oxygen and ammonia contents in the main body of well-monitored condensers are quite low, and corrosion is not a serious problem. Corrosion can be more severe in the air removal section of the units, and in general, during shutdown and startup. If ammonia contamination is high, it is better to specify SCC-resistant alloys such as copper-nickels.

Copper-nickel alloys (also called cupronickel) are single-phase binary alloys containing small amounts of additional alloying elements to enhance particular properties (iron improves erosion-corrosion resistance, niobium improves weldability). Uses for copper-nickels C70600, 71500, and C72200 are described above. Other important alloys in this family include C70610 (CuNeFe1Mn); the Cr/Mn/Zr/Ti-modified 70-30 alloy C71900, and the 70Cu-30Ni-2Fe-2Mn alloy, C71640. Like other copper-nickels, these alloys are not susceptible to stress-corrosion cracking, and unlike brasses, their corrosion behavior is not sensitive to changes in pH. Especially good resistance to high-velocity impingement attack has been demonstrated by alloy C71900 (Ref 45).

Alloy C71640 is intended for use in seawater. It has excellent corrosion resistance; however, high-flow velocities may induce pitting even in this 30%-Ni alloy. Very low seawater velocities (less than 1.5 m/s) and interrupted flow conditions can also be troublesome because stagnation may give rise to the formation of deposits, which invite crevice corrosion. The iron-bearing 10% copper-nickels are less affected by low-flow velocities and, therefore, present the best cost/benefits compromise for saline heat-exchanger applications.

Alloy C70600 has performed successfully in absorption type air conditioning units charged with concentrated lithium-bromide solutions and in air, hydrogen, and oil coolers in which the cooling water was contaminated with ammonia. It also gives good service in dormitory shell- and tube-type heating plants; dishwashers and laundries; softened waters containing high concentrations of oxygen and CO_2; seawater-cooled compressor inter- and aftercoolers, and air ejector sections of power plant condensers. In the latter case, the copper alloy successfully replaced stainless steel that had failed by chloride-induced SCC.

Food and Beverage Processing Equipment

While the use of copper to process food and beverages is not extensive, the metal is approved for use, and even preferred, in certain industries.

Copper is an essential trace element (see Chapter 6), and the presence of small concentrations of copper in foodstuffs is a nutritional necessity. Limitations placed on the copper content of potable waters, as well as concerns regarding the contact of copper with foods and beverages, are based more on the degree to which traces of copper-corrosion products impart a metallic taste than they are on concerns over toxicity. As a general rule, contact between copper and acidic foodstuffs or beverages (as, for example, carbonated beverages) should be avoided (Ref 76, 95).

Copper and copper alloys can safely be used in contact with potable noncarbonated water, and brass valves and fittings have long been used with, for example, tea and coffee. Copper tube is used in the refining of beet and cane sugars and has historically been considered the best material in which to brew beer. Cast nickel silvers, particularly C97600, are used for fittings and pump and valve components in the dairy industry (Ref 76, 96).

Bearings

Figure 5.17 illustrates the elements of a typical cast-bronze sleeve bearing. Sleeve bearings are also called journal bearings and/or radial bearings, because loading is oriented radially to the axis of the journal. Bronzes are considered to be the most versatile class of sleeve-bearing materials. The many grades available offer a broad range of properties to suit individual-bearing operating conditions.

Sleeve-Bearing Design. Copper products for sleeve bearings are divided into two classes—those that are machined from tubes and those that are bent from strips. Good bearing design requires a thorough understanding of the anticipated service conditions. The many factors involved determine in which of the three basic modes the bearing will operate.

- *Hydrodynamic operation*, in which the load is supported on a film of lubricant that is hydrodynamically pressurized by the relative motion between the bearing and journal
- *Mixed-film operation*, in which the load is supported partially by the lubricant and partially by contact between the bearing and journal surfaces. This is probably the most common of the three types of operation
- *Boundary lubrication*, in which the load is completely supported by the bearing and journal elements, which are separated only by a thin "boundary" layer of lubricant

Materials Selection. The principal cast-bronze bearing alloys, their bearing characteristics and typical uses are listed in Table 5.17. Compositions of these and other UNS-listed cast alloys can be found in Table 3.5; physical and mechanical properties are given in Tables 2.3 and 2.25, respectively. Material selection is an important design consideration for all sleeve bearings; however, particular attention should be given to the choice of materials for bearings that operate in the

mixed-film or boundary modes (Ref 92, 97–99, 100).

The sleeve bearing designer's job has been simplified considerably by the bearing selection tables and, more recently, by computerized design programs such as that published by the Copper Development Association Inc. This program facilitates the design of bronze-sleeve bearings for hydrodynamic, mixed-film, and boundary operation (Ref 100, 101).

Transportation

Automobiles and trucks account for the largest share of copper usage in the transportation sector. Trains, ships, and aircraft, in that order, make up the balance. Copper is mostly used for electrical products, followed by heat transfer devices such as radiators and oil coolers, and bronze sleeve bearings.

Countless fittings, fasteners, and other screw machine products made from leaded free-cutting brass, C36000, the most widely used of all copper alloys, are used in transport applications. In many

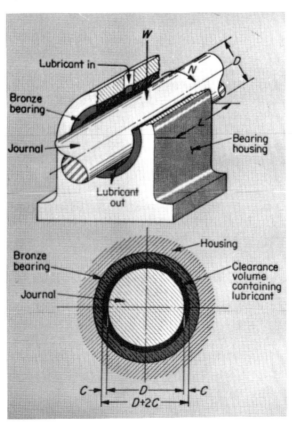

Fig. 5.17 Typical bronze sleeve bearing

Table 5.17 Characteristics and applications of copper-base bearing alloys

Alloy family UNS No.	Maximum unit load, MPa	Embedability conformability	Shaft hardness	Speed	Characteristics, typical uses
Tin bronzes					
C90300	...	M-L	M-H	M-H	Good pounding resistance; high loads at slow-to-moderate speeds; high thermal conductivity. Cool operation; low frictional coefficients against steel; high wear resistance, good seawater corrosion resistance; work well with greases; can function under boundary lubrication, but should be used with reliable lubricant supply; shafts should be hardened to 300 to 400 HB. Applications: Piston pin bushings, valve guides, rolling mill bearings, and rocker shaft bearings for internal combustion engines
C90500	4000	M	M	M	
C90700	4000	M-L	M-H	M-H	
Leaded-tin bronzes					
C92200	...	...	...	...	Similar to tin bronzes but more readily machinable
C92300	3500	...	...	...	
C92700	4000	M	M	M	
High-leaded tin bronzes					
C93200	4000	M-H	M-L	M-L	Slightly lower in strength and ductility than tin bronzes; ideal for moderate loads and moderate-to-high speeds; impact loads should be avoided; high thermal conductivity, wear resistance and corrosion resistance; very high anti-seizing protection in the event of lubricant failure; tolerate dirty lubricants; softer alloys accommodate some misalignment, can be run against unhardened shafts. Applications: Most popular alloys; best all-around choices for grease-lubricated and hydrodynamic bearings; broadly used in automotive, off-highway, agricultural, railroad, mining, and industrial equipment. C93200 is considered the workhorse alloy. C93600 has improved pounding resistance, better corrosion resistance and machinability; C93700 has similar properties and resists acidic and mineral waters, paper-mill sulfite liquors.
C93400	...	...	...	...	
C93500	3000	M-H	M-L	M-L	
C93700	4000	H-M	L-M	M-H	
C93800	3000+	H	L	L	
C94100	...	...	L	L	
C94300	...	VH	L	L	
High-strength brasses (manganese bronzes)					
C86300	...	L	H	H	
C86400	...	...	...	...	
Aluminum bronzes					
C95300	...	...	H	...	Higher strength and fatigue resistance than all alloys except beryllium copper; compressive strength of C95400 at 260 °C (500 °F) equals that of tin bronze at room temperature; resist shock and pounding well; very high corrosion resistance but can be difficult to machine; HT alloys heat-treated; require clean, reliable lubrication, fine shaft finishes and shaft hardness > 500 HB. Applications: heavy-duty bushings for power shovels, off-highway equipment, roll-neck bearings, hydroelectric turbine wear rings, thrust bearings, machine tool slides, boring bar guide bearings
C95300HT	...	...	H	...	
C95400	4500+	L	H	H	
C95400HT	...	...	H	...	
C95500	...	L	H	H	
C95500HT	...	...	H	...	
C95520	...	...	H	...	
C95800	...	...	H	...	
Silicon brass					
C87500	...	L	H	...	Moderate strength for medium loads and high speeds; good castability; relatively good machinability; require hard shafts and clean, reliable lubrication. Applications: small motors, appliances
Beryllium copper					
C82500	...	L	H	...	Ultrahigh strength and fatigue resistance, exceptional thermal conductivity; require hard, precisely aligned shafts and clean, reliable lubrication; used in heat-treated condition. Applications: Aircraft bearings (wrought alloys)
Leaded coppers					
C98200	...	H	L	...	High conductivity, low to moderate strength, depending on composition; somewhat better fatigue resistance than babbitts; usually supported by steel backing rings; may require lead-tin overlay for corrosion resistance. Applications: automotive main and connecting rod bearings
C98400	...	H	L	...	
C98600	...	H	L	...	
C98800	...	H	L	...	
C98820	...	H	L	...	
C98840	...	H	L	...	

Source: Ref 98

cases, brass screw machine products can be made at lower finished cost than they can when made from leaded free-cutting steels, the principal competitor to brass in the screw machine product sector. This is possible for three largely unappreciated reasons:

- Production rates with brass are significantly higher than those attainable with leaded steel
- Brass turnings, which are valuable, are recycled economically, whereas leaded steel turnings, which are virtually worthless, are difficult to dispose of.
- Unlike steel, brass ordinarily does not require protective electroplating to resist corrosion

Motor Vehicles

The average 1990s-vintage North American-made automobile contains approximately 23 kg (50 lb) of copper. About 14 kg (30 lb) of copper can be found in the average foreign-made automobile sold in North America (Ref 102).

Traditionally, copper usage was distributed about equally among electrical systems (motors, generator/alternators, wiring harnesses), heat transfer systems (radiators, oil coolers, heater cores, and air-conditioning heat exchangers), and mechanical components such as bearings and shifter forks. Beginning in the late 1980s, electrical uses steadily increased while heat transfer applications were gradually taken over by aluminum. By 1990, 79% of the total copper usage in an average North American automobile was in the electrical system and less than 10% was in heat transfer systems. However, despite downsizing and a general reduction in the weight of automotive components, the large increase in the number and complexity of electrical systems has actually led to an increase in copper usage per vehicle.

Automotive Radiators and Heaters. For most of the 20th century, copper and brass were the materials of choice for radiators and heaters. Aluminum, aggressively marketed on claims of lighter weight, lower cost, and higher reliability, gradually replaced copper and brass after 1978, the year Volkswagen introduced a car equipped with an aluminum radiator. The majority of car automobile radiators for new cars are now made from aluminum, although truck, bus, heavy vehicle, and most aftermarket radiators continue to be made from copper and brass (Ref 103).

Modern radiator design requirements have become increasingly demanding. Radiators operate at higher temperatures and pressures than early units, yet they are made from thinner materials. Operating stresses are, therefore, higher. In addition, there is increased potential for corrosion because of the growing use of de-icing road salts and the corrosive constituents (SO_3, H_2S, NO_x) found in catalytically processed automotive exhaust.

Because they were not designed to accommodate high-operating stresses, failure in conventional copper/brass radiator systems normally resulted from solder corrosion and fatigue, with 70% of the failures occurring between the fourth and ninth year of the car's life. On the other hand, 64.8% of copper/brass radiators are economically repaired, 26.5% are re-cored, and only 8.7% have to be replaced. Aluminum radiators, which fail primarily for the same reasons that plague copper/brass radiators, are difficult to repair and are usually scrapped and replaced (Ref 104, 105).

Competition from aluminum has prompted a substantial redesign of the copper/brass radiator. New designs, principally originating in the United States and Japan, offer improved materials, more efficient use of traditional materials, effective anticorrosion coatings, stronger supports, and improved fabrication and assembly methods. Significant improvements are listed in Table 5.18. Radiator weight has been reduced without sacrificing performance or reliability. Better joining methods, improved corrosion resistance, and better long-term fatigue performance have enabled modern copper-brass radiators to meet the automotive industries' long-sought 10-year life criterion (Ref 106).

Modern copper-brass radiator cores are as light or lighter than aluminum cores, primarily because copper's higher thermal conductivity permits the use of thinner fins. Thus, radiator cores made from 32 μm (1250 μin.) copper fins are lighter than brazed aluminum cores having the same fin density. "Splitter-fin" cores using 25 μm (0.001 in.) copper strip and a high fin density have a higher weight efficiency than conventionally brazed aluminum radiators (Ref 107–109).

Radiator tube weight can be reduced by replacing conventional induction-welded longitudinal lock seam joints with laser-welded butt joints. An additional advantage of laser welding is that it permits as much as a five-fold reduction in the tube wall thickness. Figure 5.18 shows the details of some of the design improvements listed in Table 5.18. Figure 5.19 shows a copper and brass automotive radiator core built to incorporate these design improvements.

Corrosion-resistant fin materials and barrier coatings are common in Japanese radiators but are seldom seen in European and North American units. One Japanese copper-base fin stock contains 0.003% Pb and 0.03% Sn. Its thermal conductivity is 96% that of pure copper, but its corrosion rate in salt air and salt spray environments is reportedly reduced approximately 50% compared with standard copper fins. Other Japanese solutions to fin corrosion include a zinc-coated plus diffusion-alloyed copper strip and high-quality lacquers to coat all exposed surfaces of the core.

Automotive Wiring. Whereas cars once typically had only three electric motors (for the starter, windshield wiper, and heater/ventilator blower), modern vehicles contain up to 70 motors for various safety, comfort and/or convenience features, many of which are now standard equipment. These motors, along with their wiring harnesses and connectors, add significantly to the modern vehicle's copper content. This is illustrated in Fig. 5.20, which shows a portion of the wiring network of a modern automobile. The wiring harness of a third-generation (1990s era) Volkswagen, for example, contains 26 branch circuit modules, each with six to eight contact plugs. In all, a VW Golf's circuitry comprises 800 contacting pairs and 500 m (1640 ft) of wire. Accessory wiring in the four-wheel-drive Golf accounts for about 11.25 kg (25 lb) per vehicle, Table 5.19 (Ref 110, 111).

New Automotive Applications. The trend toward so-called "smart" vehicles has increased copper consumption by 40% for devices such as antilock-brake systems (ABS), burglar alarms, gyroscopes, collision-avoidance systems, and navigation computers. The most significant non-electrical development is the use of corrosion-resistant alloy C70600 brake line tube as standard equipment on several European and British cars (Ref 112).

Other growing automotive uses for copper include the copper found in aluminum-alloy engine

Table 5.18 Recent improvements in copper/brass automotive radiators

Development/improvement	Reduce material cost or weight	Reduce fabrication cost	Improve reliability
Materials			
Fins			
Corrosion-resistant alloy and/or effective coatings	•	...	•
Reduced fin thickness	•	...	...
Tubes			
Corrosion-resistant brass	...	...	•
Butt-welded tube seams	•	•	•
Reduced tube wall thickness	•	...	...
Tanks			
Molded plastic replacing brass	•	...	...
Core geometry			
Fins			
Fin density ≥ 472 fins/m	•	...	...
Fin entire tube length with water-side soldering	...	...	•
Tubes			
Louver pitch 0.8 to 1.0 mm	•	...	...
One tube row or multiple one-row elements	•	•	...
Mechanical design			
Supports			
Improved side support strap	...	...	•
Straps not joined to header	...	...	•
Thermal expansion joint in side strap	...	...	•
Use of tie rods or wires	...	...	•
Fins			
Fin entire length or use comb insert	...	...	•
Tubes			
Expand all tubes before soldering	...	...	•
Deep drawn collar for tube holes in header	...	...	•
Joining, fabrication			
Fins			
Tin fin tips only	•	...	...
Tubes			
Laser-welded tube seams	•	•	•
Improved tube/header soldering process	...	...	•
Stronger tube/header solder alloy	...	•	•

Source: Ref 107–109

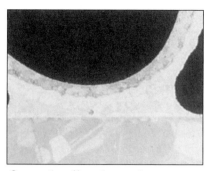

Cross section of brazed copper fin
to brass tube wall

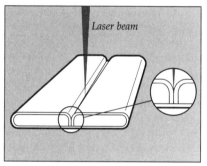

Laser-welded thin brass "twin" tube

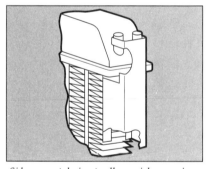

Side support design to allow axial expansion

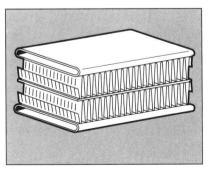

Compact core design

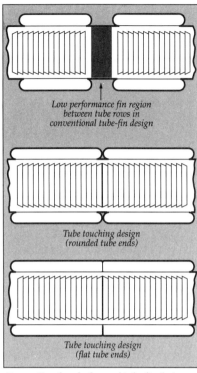

Conventional and advanced tube-fin designs

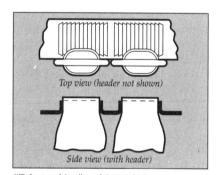

"Tube touching" multi-row design

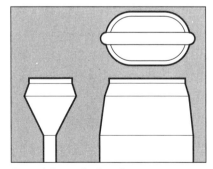

Tapered ob-round tube end

Fig. 5.18 Elements of an advanced technology copper/brass radiator. Source: International Copper Association, Ltd.

blocks, radiators and wheels, the addition of several percent copper to some high-strength low-alloy (HSLA) automotive steels, and the use of copper in asbestos-free brake linings.

Marine Applications

Copper alloys are very commonly used in marine heat exchangers and condensers, and in seawater piping, pumps, valves, sea chests, storage vessels, and sleeve bearings. There has been some replacement of copper by titanium in primary heat exchangers due to the rising sulfide pollution levels in commercial harbors.

Industrial marine applications usually call for copper-nickels or aluminum bronzes. Manganese bronzes and silicon bronzes are also widely used, mainly in pipes, fittings, pump, and valve components and other high-strength mechanical products. Small pleasure and racing craft utilize propellers cast from manganese bronzes C86100 or C86800. The manganese bronzes combine high-corrosion resistance with exceptional mechanical properties; some reach tensile strengths greater than 690 MPa (100 ksi). Small boat propellers are also available in cast silicon brass, C87500. Large propellers are cast in nickel-aluminum bronze, C95800.

Traditional antifouling paints are compounded with copper oxide, an effective biostatic agent. The insoluble paints are inhospitable to marine organisms. Copper-base systems are considered environmentally safer than, for example, tributyl tin-based coatings, which continuously release toxins into the surrounding water. Small boats which are infrequently used, are more vulnerable to marine fouling than commercial vessels. For light craft, adhesively bonded copper-nickel foil and copper powder-filled epoxy resins offer permanent remedies to biofouling. The relatively high initial costs are offset by significantly reduced hull maintenance.

Castings for Ships and Offshore Platforms. The high corrosion resistance, strength, and low life-cycle cost of cast-copper alloys make them economical choices for pipes, flanges and fittings, pumps and valve components, bearings, and other mechanical devices. In piping, the antifouling characteristics of the alloys reduce maintenance costs considerably compared with galvanized steel, fiberglass-reinforced plastic, or rubber-lined products, all of which may be cheaper initially. Also, because copper-alloy pipes need not be oversized to allow for biofouling, they can be smaller and lighter than those made from other materials.

Aluminum Bronzes. With their favorable combination of strength and corrosion resistance, cast aluminum bronzes have earned a reputation for reliable service in marine environments. Aluminum bronzes offer a broad range of mechanical properties; some can be heat treated to strengths higher than those of high-strength hull-plate

Table 5.19 Copper usage in a four-wheel-drive 1990s VW Golf

Component	Copper content	
	kg	lb
Central electrical system	0.6	1.32
Contacts	0.85	1.9
Conductors	4.2	9.3
Generator	1.0	2.2
Starter motor	0.8	1.8
Radiator ventilators	0.17	0.37
Fresh air ventilators	0.12	0.26
Ignition coil	0.14	0.31
Fuel pump	0.2	0.44
Air conditioner clutch	0.5	1.1
Power window motors	0.6	1.32
Windshield wiper motor	0.15	0.33
Windshield washer pump motor	0.12	0.26
Headlight aiming motor	0.09	0.2
Headlight washer pump	0.16	0.35
ABS system pump	0.2	0.44
Central door lock actuators	0.2	0.44
Rear view mirror heater	0.1	0.22
Mirror positioning motor	0.15	0.33
Magnetic circuit breakers	0.4	0.88
Digital command module	0.2	0.44
ABS command module	0.2	0.44
Centralized door lock command module	0.1	0.22
Total	**11.25**	**24.87**

Source: Ref 107, 112

Fig. 5.19 An automotive radiator core using advanced technology design and manufacturing concepts. Source: International Copper Association, Ltd.

steels. Aluminium bronzes are commonly used for shafts and mechanical components in pumps and valves for seawater cooling systems. Their anti-fouling properties are particularly valuable in safety-related systems such as fire-fighting equipment. Being non-sparking, aluminum-bronze hand tools can safely be used on tankers and oil and gas platforms. Because the aluminum bronzes are non-magnetic when properly processed, the alloys are also used for chains, slings, and other mechanical components in mine sweepers. Alloy C95800, a cast nickel-aluminum bronze containing small amounts of iron and manganese, has become the standard material for large propellers and the hubs and gearboxes used with them. A large nickel-aluminum bronze propeller is shown in Fig. 5.21 (Ref 113–115).

Copper-Alloy Marine Surfaces. The copper industry has sponsored extensive research and development directed at the use of copper-nickel hull materials for commercial vessels and for off-shore structures. Copper-nickel was chosen because it offers the best combination of erosion-corrosion resistance and antifouling properties. The idea of cladding boats with copper metals is hardly new; copper plates were commonly used in the 17th century to protect wooden hulls against wood-boring ship worms. Because early copper-clad craft were also free from the fronds of algae that formed on wooden hulls, they were significantly faster than unclad vessels. It is claimed that the speed of Admiral Nelson's copper-sheathed ships contributed to the British victory at Trafalgar. In the case of off-shore structures the pres-

Fig. 5.20 A portion of the wiring harness for a modern automobile. Source: Copper Development Association, Inc.

ence of fouling necessitates the use of a heavy structure to resist the wave action on the increased cross section presented by the fouled surface. The use of copper-nickel cladding enables the reduction in the amount of steel required in the submerged portion of the structure (Ref 111, 116–123).

Copper-Nickel Hulls. Ship hulls can be fabricated from solid copper-nickel plates, from metallurgically bonded copper-nickel clad steel (cladded), or from steel hulls overwhich has been welded sheets of copper-nickel alloy (sheathed). Steel and light-gage copper-nickel plates can be metallurgically bonded by hot rolling or explosive bonding to produce a composite structure. Solid copper-nickel plate is slightly stronger than hot-rolled hull steel and is economically competitive in thicknesses up to 10 mm (0.4 in.); roll-bonded material, up to 35 mm (1.4 in.); and weld overlay at total hull thicknesses greater than 100 mm (4 in.). Cladding thickness is at least 1.5 mm (0.06 in.) and normally between 2 and 3 mm (0.8 and 1.2 in.). Clad plate is commercially available in thicknesses of 5 mm (0.196 in.) and heavier in lengths to 13 m (43 ft) and widths to 2.5 m (8.2 ft). Plate supplied to ASTM B162 is guaranteed to exhibit a minimum shear strength of 137 MPa (20 ksi) between cladding and substrate (Ref 111, 117–120).

Copper-Nickel-Clad Vessels. Among the first commercial copper-hulled vessels to be commissioned in recent years was the *Copper Mariner*, a 20 m (67 ft) shrimp boat operating in highly biofouling waters off Nicaragua. The ship's hull was fabricated from plates of alloy C70600. The boat did not require drydocking for hull maintenance during its first 10 years of service. The *Copper Mariner* was followed by a small fleet of copper-clad shrimp boats, all of which were still in service in the mid-1990s. These vessels feature roll-bonded C70600-clad steel hull plates, which apparently perform as well as the earlier all-copper-nickel design. Several copper-nickel clad Italian fire boats, several Finnish ice breakers, and one large commercial vessel have been placed in service.

Copper-Nickel Sheathing. U.S. Maritime Administration studies have shown that for large cargo ships, tankers, and naval vessels, thin plates of copper-nickel fitted over conventional steel hulls offer the same advantages as solid or clad copper-alloy plate at significantly lower cost. The studies also demonstrated that hull sheathing is economically viable for new vessels and as retrofits on vessels less than 10 years old. Ships older than 10 years may not have enough remaining service life to amortize the investment required to retrofit the sheathing. An example of copper-nickel hull sheathing undergoing a field trial is illustrated in Fig. 5.22(a, b), which shows the 90,000 DWT, 271 m (889 ft) tanker *Arco Texas*. Along the side of the ship's hull can be seen four sets of three 3.2 mm ($\frac{1}{8}$ in.) thick panels of alloy C70600, each 91 by 305 cm (3 by 12 ft). The panels were attached by means of single-pass welds, and despite numerous impacts with moorings, the panels did not fail. Placing the plates at the light draft line exposed them to the most aggressive mechanical and corrosion conditions. In the case of the *Arco Texas*, these included repeated passages through the Panama Canal (Ref 124, 125).

Less than 0.0025 mm (0.0001 in.) of the surface of the copper-nickels' was lost to corrosion during the two-year test period, and there was no evidence of galvanic corrosion on the adjacent steel hull. The panels' mean roughness after exposure was 53 μm (2086 μin.) and that of the steel was 250 μm (9843 μin.). Research has shown that for surface finishes smoother than about 250 μm (9843 μin.), each 10 μm (394 μin.) increase in hull roughness is equivalent to a 1% power penalty. On this basis, had the hull of the *Arco Texas* been completely sheathed with copper-nickel, it would have been 19.7% more efficient (Ref 123–125).

Fig. 5.21 Large marine propeller cast in nickel-aluminum bronze. Source: Copper Development Association, Inc.

Consumer and General Products

The application of copper and copper alloys in consumer and general products covers an enormous variety of uses. The metals are favored because of their corrosion resistance, good formability, and bright colors. Highly malleable alloys, such as cartridge brass, C26000, yellow brass, C27000, forging brass, C37700, and architectural bronze, C38500, are widely used as stampings and forgings for mechanical and decorative products.

(a)

(b)

Fig. 5.22 (a) Copper-nickel test panels welded to the hull of a large tanker (b) Close-up of impact areas showing resistance of Cu-Ni panels to failure. Source: Copper Development Association, Inc.

Colors in Copper and Copper Alloys. Copper can be alloyed to produce colors ranging from red to silver. In brasses, color depends on zinc content, passing through the pinkish red of gilding, C21000, with 5% Zn; the red-yellow of jewelry bronze, C22000 (10% Zn); commercial bronze, C22600 (12.5% Zn); red brass, C23000, (15% Zn); and the increasingly lighter yellows of higher-zinc cartridge brass, C26000, (30% Zn); yellow brasses, C26800 to C27400, (34 to 37% Zn), and architectural bronze, C38500, (43% Zn).

Copper alloys containing nickel tend to have a pinkish tint, and the copper-nickels used for coinage closely resemble silver. Nickel silvers (or German silvers, which are alloys of copper, zinc, and nickel) also have a distinctly silver-like appearance. These alloys have good corrosion resistance and are commonly used for spectacle frames, camera hardware, and silver-plated flatware. In the latter case, the intent is to maintain a uniform appearance in areas where the softer silver plating wears away. Copper-tin alloys (phosphor bronzes) have a brownish color that becomes more distinct as a protective-surface tarnish layer forms. Beryllium coppers range from red to bright gold in color, depending on beryllium content.

Coinage

It has been estimated that coinage accounts for about 1% of copper consumption in the United States. Despite growing use of electronic and cheque transactions, coinage is becoming increasingly popular. Many countries have replaced low-denomination bank notes with coins, primarily because coins last longer (and therefore cost less) than paper notes. Among the more popular coinage metals are copper-nickels, brasses, nickel silvers, and copper itself.

Ordnance Items

Cartridge cases, bullets, primer caps, and rotating bands represent a relatively small and shrinking area of copper consumption. Military shell and cartridge casings, once predominantly made from alloy C26000 in calibers up to 105 mm, are now more often made from steel, aluminum, and, in some cases, consumable materials such as paper. Rotating bands on artillery projectiles, traditionally made from copper or gilding, C21000, can also be made from sintered iron and other materials. In accordance with international convention, small-caliber military bullets are clad with gilding, C21000, and cartridge cases for small-caliber military and sporting ammunition are still predominantly made from brass.

Medical Applications

While copper compounds are not commonly used in medical practice today, they have been used for a variety of applications since the beginnings of recorded history (Ref 126). This subject is discussed in more detail in Chapter 6. The pharmaceutical industry uses copper chemicals as a dietary supplement in baby food and in various vitamin and mineral preparations.

Copper in Agriculture

The earliest recorded use of copper in agriculture occurred in 1761, when it was discovered that soaking bean grains in a weak solution of copper sulfate inhibited later plant damage through seed-borne fungi. The greatest breakthrough for copper salts came in 1885 when the French scientist, Millardet became the first to recognize the fungicidal properties of copper. While looking for a cure for downy mildew disease, Millardet chanced upon the combination of copper sulfate, lime and water, which later become known as "Bordeaux mixture." Use of this mixture initiated the practice of protective crop spraying by this and other fungicidal agents. Copper chemicals are now commonly used in agriculture as fungicides, bactericides, herbicides, and micronutrients in soils. Copper compounds are also used to stimulate growth in chickens and swine (Ref 127–139).

As a Fungicide. A 0.5 to 1.0% Bordeaux mixture applied at 2- to 3-week intervals is normally sufficient to control most copper-susceptible fungi. Some 300 diseases have been found to be amenable to control by copper-based fungicides. In addition to a Bordeaux mixture, various mixtures of copper oxide, copper chloride, and copper oxychloride are used, sometimes as admixtures with organic compounds (Ref 139).

As a Micronutrient. Plants and animals show symptoms of disease both when deprived of or overexposed to micronutrients. Examples of copper deficiency include partially filled ears in wheat, interveinal chlorosis in maize, and exanthema in citrus (Ref 122, 133–136).

Diagnosis of copper deficiency is best performed by soil analysis for annual crops and by plant analysis for perennial deep-rooting crops. Measurement of total copper concentration in soils is of limited diagnostic value owing to the effects of variations in soil properties on the sorption and availability of copper to plants. Also, the presence of companion elements such as molybdenum and zinc can have a counter effect on copper nutrition. Nonetheless, soil analysis can be useful for identifying extremely deficient soils, such as those developed on coarse parent materials with inherently low-total copper contents, or those with very high levels as a result of either geochemical enrichment of the parent material or environmental contamination, where toxicity may be a problem (Ref 127, 139, 140).

In France, total copper concentrations in soil are interpreted as highly probable and cause deficiencies in crop growth at the level of 5 mg/kg. Potential crop deficiencies are at indicated concentrations up to and including 8 mg/kg. Likewise, in Scotland, 5 mg/kg of copper is considered to constitute a definitely deficient soil, and soils containing between 5 to 10 mg/kg are taken to be possibly deficient. In most other countries, concentrations above 10 mg/kg of copper in mineral soils and above 30 mg/kg in organic soils are normally regarded as adequate for most crops (Ref 127).

Copper deficiency in crops is corrected by incorporating copper compounds into the soil or applying them as sprays. Doses of copper vary between 0.7 and 23 kg/ha, the 300-fold range being due to differences in the method of application; sorption of copper by soil constituents, especially organic matter; the type of fertilizer used; and differences in the copper requirements of the crop (Ref 128).

Copper sulfate is the most common source of copper for soil application because it is widely available and relatively inexpensive. Its sulfate content also makes a useful nutritional contribution, although it is not adequate to satisfy crop requirements for sulfur. Finely ground slags containing copper and other metals in various forms are suitable for slightly acid to acid soils, which bring about the gradual solution of the copper. Such slags are popular on cultivated sandy podzols, especially where corn (maize) is grown. The slag's zinc content is also of value because corn has a high-zinc requirement.

Suggested maximum application rates range from 20 to 40 kg Cu/ha in total to a maximum of 44 kg/ha in one application. On the other hand, it has been demonstrated that cumulative applications of as much as 280 kg Cu/ha for 17 years and 523 kg/ha as copper sulfate for 24 years were not harmful to crops. Copper applications to soils generally need to be repeated at intervals between 5 and 18 years, depending on the soil, crop, and management. Soils with organic matter content require more frequent copper applications (Ref 127, 128).

Foliar (spray) applications are made in the form of organic chelates and copper oxychloride, often applied along with fungicides or herbicides. Repeated spraying over several years can provide sufficient copper concentration in surrounding soil to obviate the need for soil enrichment (Ref 127).

Copper as an Antifoulant

Copper chemicals and copper metal in powder- and flake-form have a wide spectrum of applications in the prevention of biofouling in both freshwater and seawater. Typical applications are in the painting of the submerged portion of vessels and other structures, such as piers, jetties, oil platforms, and water intake tunnels, exposed to water and in the dipping of nets, cages, and other equipment used in aquaculture. Copper chemicals, powders, and flakes are used in paint formulations and in coatings of both glass- and polymer base. They are traditionally applied by brush, roller or spray or by dipping. Copper-bearing formulations for this purpose are regulated in many countries as "pesticides."

Cuprous oxide (Cu_2O) is the most commonly used copper compound in antifouling applications. Formulations containing up to 70.4 wt% CuO are permitted in such applications. Other compounds used and their maximum permissible content are: copper thiocyanate (max. 50 wt%), copper naphthenate (max. 2.2 wt%), copper sulfide (30 wt%) and copper resinate (industry name for a reaction product of basic copper carbonate and rosin) (18.9 wt%). Copper sulfate is widely used in products for aquaculture. Here a maximum of 7.1 wt% is permitted.

Copper as a Preservative

Copper has a wide spectrum of effectiveness against the many biological agents of timber and fabric decay. It renders them resistant to insects and protects them from fungus attack. Three copper-based systems are used, generally marketed as proprietary formulations: copper naphthenate, chromated copper arsenate (CCA), and ammoniacal copper arsenate (ACA). These are used in both pressure and non-pressure treatment processes. Copper sulfate is the common starting chemical for these systems.

Copper in Water Treatment

Copper salts are commonly added to avoid algae growth in potable water tanks and swimming

Table 5.20 Industrial uses for selected copper compounds

Compound	Agriculture	Analytical chemistry	Antifouling paints	Catalysis	Cloud seeding	Corrosion inhibition	Electrolysis, electroplating	Electronics
Cu(I) acetate	...	...	...	•	...	...	ù	...
Cu(II) acetate	•	ù	...	•	...	...	...	...
Cu(I) acetylide	...	...	...	...	...	...	...	...
Cu(I) bromide	•	...	...	ù	...	...	...	...
Cu(II) bromide	...	...	...	•	...	...	...	ù
Cu(II) carbonate	•	...	ù	•	...	•	...	...
Cu(I) chloride	...	•	...	ù	...	...	...	•
Cu(II) chloride	•	ù	...	•	...	•	...	•
Cu(II) oxychloride	•	...	...	ù	...	...	•	...
Cu(II) chromate	•	...	...	ù	...	...	•	...
Cu(II) ferrate	...	...	...	•	...	...	...	ù
Cu(II) fluoroborate	...	...	...	•	...	...	ù	•
Cu(II) hydroxide	•	ù	...	...	...	...	•	...
Cu(I) iodide	...	•	...	ù	•	...	•	•
Cu(II) nitrate	•	...	...	...	...	ù	•	•
Cu(I) oxide	•	ù	•	•	•	•	...	•
Cu(II) oxide	•	ù	...	•	•	•	...	•
Cu(I) soaps	•	ù	...	•	•	...	...	•
Cu(II) sulfate	•	ù	...	...	...	•	•	•

Compound	Flame proofing, textiles	Flame proofing, other	Fuel oil treatment	Glass, ceramics	Food pharmaceuticals	Metallurgy	Mining and metals	Nylon
Cu(II) acetate	•	ù	...	...	•	...	...	•
Cu(I) bromide	...	...	...	...	...	•	...	ù
Cu(II) bromide	...	...	...	•	...	ù	...	...
Cu(II) carbonate	•	...	...	...	...	...	...	...
Cu(I) chloride	•	...	ù	•	...	•	•	•
Cu(II) chloride	•	ù	•	•	•	•	•	•
Cu(II) oxychloride	...	...	...	•	...	...	...	...
Cu(II) chromate	•	...	...	...	...	...	...	...
Cu(II) fluoroborate	•	...	...	...	...	...	...	...
Cu(II) gluconate	...	...	...	...	•	...	...	...
Cu(II) hydroxide	•	...	...	...	ù	...	...	•
Cu(I) iodide	...	...	...	...	...	...	...	•
Cu(II) nitrate	•	...	ù	•	•	•	•	...
Cu(I) oxide	...	...	•	ù	...	•	•	•
Cu(II) oxide	•	ù	•	•	•	•	•	•
Cu(I) soaps	•	...	ù	...	...	...	...	•
Cu(II) sulfate	...	...	...	...	...	...	...	...

Compound	Organic reactions	Paper, paper products	Pigments, dyes	Pollution control catalysis	Printing, photocopying	Pyrotechnics, rocketry	Wood preservatives
Cu(I) acetate	•	...	ù	...	•	...	...
Cu(II) acetate	•	...	...	...	...	...	ù
Cu(I) acetylide	•	...	...	...	...	...	...
Cu(II) arsenate	•	...	...	...	...	...	...
Cu(I) bromide	•	...	...	...	ù	...	...
Cu(II) bromide	•	...	...	...	...	...	...
Cu(II) carbonate	•	...	...	...	...	...	...
Cu(I) chloride	•	...	...	...	ù	•	...
Cu(II) chloride	•	ù	•	•	•	•	...
Cu(II) oxychloride	•	...	...	...	...	...	...
Cu(II) chromate	•	...	...	ù	...	•	•
Cu(II) ferrate	...	...	...	...	•	...	...
Cu(II) fluoroborate	...	...	...	...	...	...	•
Cu(II) gluconate	...	...	...	...	•	...	...
Cu(II) hydroxide	•	ù	...	...	...	...	...
Cu(I) iodide	•	...	...	...	ù	...	...
Cu(II) nitrate	•	ù	...	...	...	•	•
Cu(I) oxide	•	...	...	ù	•	•	...
Cu(II) oxide	•	ù	...	•	•	•	•
Cu(I) soaps	•	...	ù	•	•	...	•
Cu(II) sulfate	•	ù	•	•	•	•	•

Source: Ref 2

pools and to treat drinking water to remove color and odor caused by algae. As discussed below, in some countries copper compounds are used to control water-borne diseases, such as schistosomiasis (bilharzia) in humans and liver fluke in animals.

Copper Molluscides

Helminthic disease, or helminthiasis, is a generic name for illnesses caused by worm parasites. Japanese researchers reported the first use of Bordeaux mixture-type copper-based molluscides to combat human schistosomiasis (a parasitic disease transmitted by snails) in 1913 and 1915. Copper sulfate has since been extensively applied in Africa, mainly in Sudan, Egypt, and Brazil. Other copper-base molluscides include copper pentachlorophenate; copper carbonate; copper tartrate; copper arsenite-acetate (Paris green); copper (I) oxide; copper (I) chloride; copper (II) acetylacetonate; copper dimethyldithio-carbamate, and copper resinate (Ref 127). Less than 1 ppm of copper in water is capable of controlling the snails which transmit such diseases.

Copper in the Chemicals Industry

Copper and inorganic copper compounds are important catalysts in the synthesis of a variety of organic chemicals, including polymers. Organo-copper compounds have similar uses, for example, copper acetylide as a catalyst in vinylation and ethynylation reactions of acetylene. Copper-chromium oxide is a well-known hydrogenation catalyst, and it has been recommended for the reduction of aldehydes and ketones to alcohols, of esters to alcohols, and of amides to amines. Copper itself can be used as a hydrogenation catalyst, for example in the reduction of nitro compounds to their corresponding amines, and in some cases of oximes and nitriles to amines, and aldehydes and ketones to alcohols. At elevated temperatures, copper catalyzes the dehydration of primary and secondary alcohols to aldehydes and ketones. A new field has opened for copper compounds in the second half of the 20th century through the discovery of the advantages of organometallic copper compounds. They are replacing Grignard (magnesium organometallic compounds) in organic synthesis for polymer production. Also, copper compounds are used in oxidation and other catalysts.

Copper phthalocyanine and related compounds are important pigments used in inks, paints, and dyes. Copper oxide is used in glazes and so-called cuprous glasses, including the familiar ruby glass.

Table 5.20 lists industrial uses for a number of copper compounds (Ref 141, 142).

REFERENCES

1. W.C. Butterman, Copper, *Mineral Commodity Profiles 1983*, U.S. Bureau of Mines, Washington, D.C., 1984
2. *The Economics of Copper 1990*, 5th ed., Roskill Information Services, Ltd., London, Aug 1990
3. H. Gadberry, "Technical Requirements and Market Potential for Copper Compounds in Commodity Applications," INCRA Report No. 378, New York, NY, Sept 1987
4. "Tejados de Cobre y sus Accesorios," brochure, Centro Espanol de Informacion del Cobre, Madrid
5. "Le Cuivre et ses Alliages en Couverture, Bardage et Evacuation," brochure, Centre d'Information des MJtaux Non Ferreux, Bruxelles
6. Architectural Applications, *Copper/Brass/Bronze Design Handbook*, CDA Pub. 405/7R, Copper Development Association Inc., New York, NY
7. "Copper...the building systems metal," CDA Pub. 401/1, Copper Development Association Inc., Greenwich, CT, 1991
8. "Scheda Tecnica Capitolato," brochure, TEGOSTIL, Zanoletti Metalli, Milan, Italy
9. "Canadian Copper," No. 116, Canadian Copper and Brass Development Association, Don Mills, Ontario, 1989
10. "Canadian Copper," No. 122, Canadian Copper and Brass Development Association, Don Mills, Ontario, 1990
11. "Copper in Architecture," brochure, Copper Development Association Inc., New York, NY, 1996
12. V. Cassidy, Copper Stages Comeback in High Rise Roofing, *Mod. Met.*, Aug 1984
13. "Copper Roofing Systems," CDA Pub. 403/1, Copper Development Association Inc., Greenwich, CT, 1981
14. "Canadian Copper," No. 112, Canadian Copper and Brass Development Association, Don Mills, Ontario, 1988
15. Copper Rides Crest of U.S.A. Restoration Wave, *Copper Topics*, No. 62, Copper Development Association Inc., New York, NY
16. *Snips*, Vol 55 (No. 2), April 1986
17. "Canadian Copper," No. 104, Canadian Copper and Brass Development Association, Don Mills, Ontario, 1986
18. Copper-Theme Curtain Wall Tower is Centerpiece of St. Petersburg, Florida, *Copper Topics*, No. 71, Copper Development Association Inc., New York, NY
19. *Snips*, Vol 43 (No. 4), March 1984
20. "Canadian Copper," No. 115, Canadian Copper and Brass Development Association, Don Mills, Ontario, 1988
21. D. Rodriguez, *Investigacion Sobre Cubiertas de Cobre y sus Accesorios*, Informe final convenio P.V.C.-PROCOBRE, Escuela de Arquitectura, Pontificia Universidad Católica de Chile, April 1991

22. Use of Power Roofing Equipment for Copper Increasing, Say Manufacturers, *Copper Topics*, No. 70, Copper Development Association Inc., New York, NY

23. M. Haselbach, *Sonderdruck: Oberflächenverhalten von Kupferbauteilen an der Atmosphäre*, IKZ 4/1973, Deutsches-Kupfer Institut, reprinted in 1982 and 1986

24. "Chemische Färbungen von Kupfer und Kupferlegierungen," brochure, Deutsches-Kupfer Institut, Berlin, Nov 1974

25. H. Oguchi, T. Ichihashi, and S. Niyama, "Chemical Coloring of Copper Roofs by Spraying," Report No. 1, Japan Copper Development Association

26. "Comment Colorer le Cuivre at ses Alliages," CICLA Pub. Nos. 110, 111, Centre d'Information Cuivre Laitons Alliages, Bruxells

27. "Chemisches Färben von Kupfer und Kupferlegierungen," DKI Information Pub. Order No. 16, Deutsches-Kupfer Institut, Berlin

28. Sheet Copper Applications, *Copper/Brass/Bronze Design Handbook*, CDA Pub. 401/0, Copper Development Association Inc., Greenwich, CT, 1990

29. D.C. Hemming, *The Evaluation of Pilot Scale Operation to Produce Basic Copper Sulfate Patina on Copper by Accelerated Process*, Project 141, International Copper Research Association Inc., New York, NY, 1969

30. Copper Flashing, Gutters, Downspouts: Usage Up Along with Copper Roofing, *Copper Topics*, No. 64, Copper Development Association Inc., New York, NY, 1988

31. "Creative Design in Architecture, Copper/Brass/ Bronze: Built-in Gutters," CDA Pub. 401/9, Copper Development Association Inc., Greenwich, CT, 1989

32. Creative Design in Architecture, Building Expansion Joints, *Copper/Brass/Bronze Design Handbook*, CDA Pub. 408/70, Copper Development Association Inc., New York, NY

33. M. Earley, R. Murray, and J. Caloggero, *The National Electrical Code 1990 Handbook*, 5th ed., National Fire Protection Association, Quincy, MA, 1990

34. W.T. Black, The Outlook for Copper Wire and Cable Markets in the 1990s, paper presented at *Wire and Cable Focus Symposium* (Cherry Hill, NJ), 22 Sept 1992

35. A. Brieva and L. Bastias, Normativa General de Instalaciones Públicas y Domiciliarias, Agua Potable, Alcantarillado, Electricidad, Gas, Pavimentos, *Editorial Jurídica de Chile*, Paper No. 47, Oct 1978

36. Catalog of Eletrical Conductors, No. 751, MADECO Corporation, Santiago, Chile, 1990

37. *Copper Building Wire Product Handbook*, CDA Pub. 601/0, Copper Development Association Inc., New York, NY

38. "Expand Your Business the Smart Way," brochure, Smart House L.P., Upper Marlboro, MD, 1991

39. "Technical Guide to Smart House," brochure, Smart House L.P., Upper Marlboro, MD, 1990

40. "Smart-Redi Hits the Market," Update No. 28, Smart House L.P., Upper Marlboro, MD, July/Aug 1991

41. "Smart House for Electrical Code Officials," brochure, Smart House L.P., Upper Marlboro, MD, July 1991

42. "Smart House Object-Oriented Applications," brochure, Smart House L.P., Upper Marlboro, MD, Sept 1990

43. "Smart-Redi Hits the Market," Update No. 28, Smart House L.P., Upper Marlboro, MD, July/Aug 1991

44. W.B. Hampshire, Tin and Tin Alloys, *Properties and Selection: Nonferrous Alloys and Special-Purpose Materials*, Vol 2, *ASM Handbook*, ASM International, Materials Park, OH, 1990, p 517–526

45. Copper Tube for Plumbing, Heating, Air Conditioning, and Refrigeration, *Copper/Brass/Bronze Product Handbook*, CDA Pub. 404/0R, Copper Development Association Inc. Greenwich, CT, 1990

46. Service Experience with Copper Plumbing Tube, *Mater. Prot. Perform.*, Vol 11 (No. 2), 1972, p 48–53

47. "Copper in Domestic Heating Systems," CDA TN39, Copper Development Association (UK), Potters Bar, Herts, June 1988

48. "Canadian Copper," No. 117, Canadian Copper and Brass Development Association, Don Mills, Ontario, 1989

49. Corrosion of Copper and Copper Alloys, *Properties and Selection: Nonferrous Alloys and Special-Purpose Materials*, Vol 2, *ASM Handbook*, ASM International, Materials Park, OH, 1990

50. Solar Energy Systems, *Copper/Brass/Bronze Design Handbook*, Copper Development Association Inc., Greenwich, CT

51. "Chauffage Central avec Canalisations en Cuivre," brochure, CICLA, Centre d'Information Cuivre Laitons Alliages, Brussels, 1970

52. "Copper Alloys in Refrigeration," CDA TN14, Copper Development Association (UK), Potters Bar, Herts, Nov 1972

53. Copper for Air Conditioning and Refrigeration Applications, *Copper/Brass/Bronze Product Handbook*, CDA Pub. 123/4, Copper Development Association Inc., Greenwich, CT, 1984

54. Using Waste Heat from Air Conditioners to Heat Water in Restaurants, *Copper/Brass/Bronze Product Handbook*, CDA Pub. 402/6A, Copper Development Association Inc., New York, NY

55. Copper for Hot and Cold Potable Water Systems, *Heat./Piping/Air Cond.*, May 1978

56. "Rame Notizie 5," brochure, Vol 2 (No. 5), Instituto Italiano del Rame, Milan, 1989 (Also UNI-CIG 8723 and UNI-CIG 9034 for 5 bar following the UNI 6507/86 standard)

57. Superintendencia de Servicios Electricos y de Gas (SEGTEL), Circular No. 2677 GD-558 GI-70, Norma NSEG 28, Ministerio del Interior, Repdblica de Chile, p 82

58. Copper Natural Gas Systems, *Canadian Copper*, No. 14E, Canadian Copper and Brass Development Association, Don Mills, Ontario, 1991

59. "The Romance of Brass," CDA Pub. 501/0, Copper Development Association Inc., New York, NY

60. M. Pohl, "Erneuerung haustechnischer Anlagen in Altbauten," DKI Informations Pub. Order No. 171, Deutsches-Kupfer Institut, Berlin, 1984
61. "Gazzetta Ufficiale della Republica Italiana," No. 128 NE140, Instituto Italiano del Rame, Milan, April 1987
62. "Fire Sprinkler Systems: A Design Guide," CDA Pub. 408/1R, Copper Development Association Inc., Greenwich, CT, 1991
63. R. Gewain, Fire Sprinklers and Plastic Pipe, *Southern Building,* Vol 403 (No. 6A), May/June 1986
64. A. Woollaston and O. von Franqué, The Role of Copper in the Building and Construction Industry, *Proc. Copper '90 Symposium* (Västerås, Sweden), The Bern Press, Dorset, 1–3 Oct 1990
65. "Fire Sprinkler Case Studies: Pelican Bay Apartments," CDA Pub. 410/5, Copper Development Association Inc., Greenwich, CT, 1985
66. "Turbine-Generator Reference Lists," No. B78, NEI Parsons Ltd., Heaton Works, Newcastle-upon-Tyne, UK, 1988
67. "High Conductivity, Outokumpu Oxygen Free High Purity Copper for All Common Electrical Uses of Regular Grade UNS C10200 OF," brochure, Outokumpu Copper Oy, Helsinki
68. "Alambres y Cables Eléctricos," brochure, Cobre Cerrillos S.A.(COCESA), Santiago, Chile
69. "Hobart Welding Guide," No. EW-385, Hobart Brothers Company, Troy, OH, 1980
70. *NEC Standards for Welding Equipment,* Section 630
71. R. Bates and Y. Junel, Lithium-Cupric Oxide Cells, *Handbook of Batteries and Fuel Cells,* D. Linden, Ed., McGraw-Hill, New York, NY, 1984
72. "Product Range, Technology, Discharge Curves, Lithium 1.5 Volt-DC Series Technical Information," brochure, SAFT (UK), Ltd., London
73. G.E. Lagos, Lithium Materials for High Power Batteries, paper presented at *Chilean-EEC Workshop on Materials Science* (Pucón, Chile), March 1992
74. "Copper in Electrical Contacts," CDA TN23, Copper Development Association (UK), Potters Bar, Herts, July 1977
75. A. Knowlton, *Standard Handbook for Electrical Engineers,* McGraw-Hill, New York, NY, 1949
76. Y.S. Shen, P. Lattari, and H. Wiegand, Electrical Contact Materials, *Properties and Selection: Nonferrous Alloys and Special-Purpose Materials,* Vol 2, *ASM Handbook,* ASM International, Materials Park, OH, 1990, p 840–868
77. "Bänder und Drähte aus Kupferweskstoffen für Bauelemente der Elektrotechnik und der Elektronik," DKI Information Pub. Order No. 1020, Deutsches-Kupfer Institut, Berlin
78. J. Crane and A. Khan, New Connector Materials for the Automobile, *Proc. Third Japan International SAMPE Symposium,* 7–10 Dec 1993
79. E. Schaefer, "Technische Anforderung und Marktaussichten," brochure, Deutshes Kupfer Institut, Berlin, Oct 1988
80. "CDA Market Study: The Use of Copper and Copper Alloys in Electronic Connections, 1980–1985," brochure, Copper Development Association Inc., Greenwich, CT, Aug 1983
81. D. Tyler and A. Khan, High Conductivity Copper Alloys: Tailoring Performance to Application, *Proc. EHC '93, Copper in Technology and Energy Efficiency* (Birmingham, AL), 7–10 Dec 1993, Copper Development Association (UK), Potters Bar, Herts, 1993
82. W. Duckworth and B. Sherwin, Fiber is Fascinating but Copper Can Still Compete, *Telephony,* Nov 1991
83. D. McCarty and G. Hooper, Telcos are Spending Over $3 Billion a Year on Copper Installation and Replacement, *Outside Plant,* Vol 9 (No. 10), 1991, p 31–46
84. C. Maneesin, Manufacturing Foamskin Insulation Jelly Filled Telephone Cable Employing Polypropylene Skin, *Wire J. Int.,* Dec 1991, p 70–71
85. S. Fleming and M.B. McLaughlin, ADSL: The On-Ramp to the Information Highway, *Telephony,* 12 July 1993, p 20–25
86. W.D. Waring and T.R. Hsing, Fiber Upgrade Strategies Using High Bit Rate Copper Technologies for Video Delivery, *J. Lightwave Tech.,* Vol 10 (No.11), 1992
87. J. Saxtan, The Quiet Evolution of Copper Cable Keeps Up with Today's Needs, *Outside Plant,* Vol 9 (No. 10), 1991, p 31–46
88. J. Saxtan, The Gloves Are Off as Copper and Fiber Meet with a "Handshake," *Outside Plant,* Vol 9 (No. 10), 1991, p 31–46
89. S. Nuckel, Copper Industry Stronger Than Many Might Expect, *Outside Plant,* Vol 9 (No. 10), 1991, p 31–46
90. W.T. Black, Copper's Evolving Role in Telecommunications, presented at *Wire & Cable Focus 94* (Philadelphia, PA), 21 Sept 1994
91. A Renaissance for Copper, *Outokumpu Copper News,* Outokumpu Copper Oy, Helsinki, Vol 27 (No. 2), 1990
92. Superconductor Industry, News Item, Fall 1990, citing *Japan Times,* 11 July 1990
93. "Light Wall Copper-Nickel Condenser Tube," Copper/Brass/Bronze technical report, Copper Development Association Inc., New York, NY, Aug 1973
94. C. Depommier, *Statistical Data Relative to Stress Corrosion Cracking of Copper Tubes in Buildings,* Universite Libre de Bruxelles, Dec 1981
95. R.A. Goyer, Toxicity of Metals, *Properties and Selection: Nonferrous Alloys and Special-Purpose Materials,* Vol 2, *ASM Handbook,* ASM International, Materials Park, OH, 1990, p 1251–1252
96. "Materials and Finishes Guide: Food Service Equipment," brochure, National Sanitation Foundation, Ann Arbor, MI, Oct 1985
97. H.C. Rippel, *Cast Bronze Bearing Design Manual,* 2nd ed., Cast Bronze Bearing Institute Inc., Cleveland, OH, Sept 1979. Reprinted by Copper Development Association Inc., New York, NY, 1993

98. "Selecting Bronze Bearing Materials," CDA Pub. 702/6, Copper Development Association Inc., New York, NY

99. H.C. Rippel and S. Heller, *User's Manual, Cast Bronze Hydrodynamic Sleeve Bearing Performance Tables*, Cast Bronze Bearing Institute Inc., Cleveland, OH, 1970

100. J.L. Tevaarwerk and W.A. Glaeser, *Handbook for the Use of Computerized Journal Bearing Design Programs HYDRO and BOUND*, Copper Development Association Inc., New York, NY, 2 Sept 1993

101. H.C. Rippel, *User's Manual, Cast Bronze Hydrostatic Bearing Design Manual*, 2nd ed., Cast Bronze Bearing Institute Inc., Cleveland, OH, 1975

102. "Usage of Copper and Copper Alloys in the USA," Copper Development Association Inc., Greenwich, CT, 1991

103. "Minimum Tin Content Solders For Auto Radiators," Copper/Brass/Bronze technical report, Copper Development Association Inc., Greenwich, CT, Aug 1983

104. R. Webb and W.F. Yu, "Stress Distribution and Stress Reduction in Copper/Brass Radiators," SAE paper No. 870183, Society of Automotive Engineers, Warrendale, PA, Feb 1987

105. M. Forbes, Tin and Its Uses, *The Manufacture of Vehicle and Industrial Radiators, Parts I, II, and III*, International Tin Research Institute, Greenford, Middlesex, 1965

106. "Methods for Reducing the Cost of Manufacturing Automotive Radiators from Copper-Based Material," Final Annual Report, INCRA Project No. 219, International Copper Research Association, Inc., New York, NY, April 1977

107. R. Webb, T. Burkett, and H. Foust, "Assessment of Automotive Radiator Technology and the Future of Copper/Brass Radiators," SAE paper No. 850043, Society of Automotive Engineers, Warrendale, PA, March 1985

108. "Zinc-Base Solders for Copper and Brass Automotive Radiators," Copper/Brass/Bronze technical report, CDA Pub. 801/3R, Copper Development Association Inc., Greenwich, CT

109. G.G. Page, Some Operating Conditions Influencing Attack of Non-Ferrous Tubular Condenser and Heat-Exchange Equipment, *Proc. First Int. Congress on Metallic Corrosion* (London), 10–15 April 1961, Butterworth, London, 1962

110. "2nd Symposium Kupfer-Werkstoffe: Eigenschaften, Verarbeitung, Anwendung," brochure, Deutsches-Kupfer Institut, Berlin, Oct 1988

111. P. Kille, Einsatz von Kupfer-Werkstoffen in der Automobilelektrik Am Beispiel des Golf, *2. Symposium Kupfer-werkstoffe*, Deutsches-Kupfer Institut, Berlin, Oct 1988

112. *Canadian Copper*, No. 129, Canadian Copper and Brass Development Association, Dons Mills, Ontario, 1992

113. H. Meigh, Designing Aluminum Bronze Castings, Technical file reprints No. 116, *Engineering*, Aug 1983

114. "Aluminum Bronze, Essential for Industry," CDA Pub. 86, Copper Development Association (UK), Potters Bar, Herts

115. T. Cullen, X-Ray Spectrometric Analysis of Alloyed Copper, *Development in Applied Spectroscopy*, Vol 3, 1964, p 97–103

116. Copper-Nickel Hulled Boats Around the World, from New York to Genoa, are Free of Fouling, *Copper Topics*, No. 63, Copper Development Association Inc., Greenwich, CT

117. "Copper-Nickel Alloys," CDA TN30, Copper Development Association (UK), Potters Bar, Herts, Sept 1982

118. A Return to Copper Hulls?, *Marine Eng. Log*, June 1982

119. A. Tuthill, "Guidelines for the Use of Copper Alloys in Seawater, CDA Pub. 88/8, Copper Development Association Inc., Greenwich, CT, Feb 1988

120. L. Sandor, Copper-Nickel for Ship Hull Construction—Welding and Economics, *Weld. J.*, Dec 1982

121. "Copper-Nickel Cladding for Offshore Structures," CDA Pub. TN37, Copper Development Association (UK), Potters Bar, Herts

122. "Materials for Seawater Pipeline Systems," CDA Pub. TN38, Copper Development Association (UK), Potters Bar, Herts

123. H. Pircher and B. Ruhnland, "Use of Copper-Nickel Cladding on Ship and Boat Hulls," CDA Pub. TN36, Potters Bar, Herts, 1985

124. Copper-Clad Steel Hulls Results Good, *Copper Topics*, CDA Pub. 703/6, Copper Development Association Inc., Greenwich, CT, May 1986

125. *Copper-Nickel Sheathing Study—Phase II; Two-Year Service Performance of Test Panels on the ARCO TEXAS*, final report, U.S. Department of Transportation, Maritime Administration, Research & Development Contract MA-81-SAC-10074, 31 May 1984

126. H.H.A. Dollwet and J.R.C. Sorenson, Historic Uses of Copper Compounds in Medicine, *Tr. Elem. Med.*, Vol 2 (No. 2), 1985, p 80–87

127. V.M. Shorrocks and B.J. Alloway, "Copper in Plant, Animal and Human Nutrition," CDA Pub. TN35, Copper Development Association (UK), Potters Bar, Herts, 1985

128. J.O. Nriagu, Ed., *Copper in the Environment. Part 1: Ecological Cycling; Part 2: Health Effects*, Wiley-Interscience, 1979

129. *The Economics of Copper 1992*, 6th ed., Roskill Information Services Ltd., June 1992, p 111–114

130. E.J. Underwood, *Trace Elements in Human and Animal Nutrition*, 3rd ed., Academic Press, New York, NY, 1977

131. R. Braude, *J. Agric. Sci.*, Vol 35, 1945, p 1163

132. B.C. Cooke, Recent Advances in Animal Nutrition, *Copper in the Environment, Part 1: Ecological Cycling; Part 2: Health Effects*, Wiley-Interscience, 1979, p 59–62

133. N.F. Childers, Nutrition of Fruit Crops, Temperate, Sub-tropical, Tropical, *Horticultural Publications*, Rutgers State University, New Brunswick, NJ, 1966

134. A. Scaife and M. Turner, Diagnosis of Mineral Disorders in Plants, *Vegetables,* Vol 2, HMSO, London, 1983

135. H.B. Sprague, *Hunger Signs in Crops*, David McKay Company, New York, NY, 1949

136. T. Wallace, *The Diagnosis of Mineral Deficiencies in Plants by Visual Symptoms*, HMSO, London, 1961

137. B.J. Alloway, The Bioavailability of Copper in the Terrestrial Environment, *Proc. Copper '90, Refining, Fabrication, Markets* (Vaesteras, Sweden), The Bern Press, Dorset, 1–3 Oct 1990

138. F.F. Heyroth and J. Cholak, Copper in Biology, *Copper*, Rheinhold, New York, NY, 1954, p 853–863

139. F. van Assche and H. Clijsters, A Biological Test System for the Evaluation of the Phytotoxicity of Metal-Contaminated Soils, *Environmental Pollution*, Vol 66, 1990, p 157–172

140. D.C. Martens, *The Evaluation of Copper in Pig Manure*, Project No. 292, International Copper Research Association Inc., New York, NY, 1989

141. J. Faust and R. Fröböse, Organocopper Compounds, *Gmelin Handbook of Inorganic Chemistry,* Part 2, 8th ed., Springer-Verlag, New York, NY, 1983

142. R.A. Benkeser and H. Gilman, Copper in Organic Chemistry, *Copper,* New York, NY, 1954, p 834–852

6

Copper in the Environment

Introduction

Copper is a naturally occurring element with an average concentration of about 50 ppm in the earth's crust. It is present as an essential element in all higher plants and animals and is one of the few chemical elements that occurs mineralogically in the elemental form as metallic copper. It occurs in at least 160 minerals as oxides and as sulfides in the cuprous, +1, state and in the cupric, +2, state. It is a strongly sulfur-seeking (chalcophile) element and tends to be concentrated in sulfide deposits. In igneous rocks, it occurs mainly as finely divided sulfides. It is present in waters with a range of 1 to 15 ppb in seawater and 0.4 to 150 ppb in fresh water with an average of about 3 and 5 ppb, respectively (Ref 1, 2).

All soils contain copper where it is an essential element for plant life and agricultural production; however, in some cases, the total content may be so low that there is insufficient copper available for healthy crop or cattle growth. In such cases, copper must be added as a soil supplement if the land is to be used for agricultural purposes (Ref 3, 4).

This chapter discusses the environmental impact of copper on plant and animal living cycles, including humans.

Copper Essentiality

Copper serves as a micronutrient in plants and animals, including humans, and thus is essential to most life forms. A typical dose-response curve for micronutrients, such as copper, is shown in Fig. 6.1. There are three zones of criticality—deficiency, adequacy, and toxicity. Another way of looking at the body's physiological response to metals is shown in Fig. 6.2. Here, metals that are essential to metabolism exhibit a range of benefit

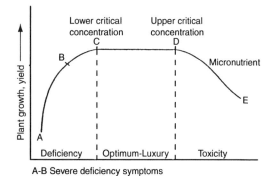

A-B Severe deficiency symptoms
B-C Mild deficiency

Fig. 6.1 A typical dose-response curve for micronutrients such as copper in humans, animals, and plants (Ref 4)

over the range of concentration. Metals that are not essential will cause a deleterious response as concentration increases without a range of benefit. The concentrations for these zones will differ depending on the life form and micronutrient. The deficiency side of the curve can be divided into severe effects, which are manifested by distinct stress symptoms at the lowest concentrations, ranging to moderate to subclinical symptoms at the higher concentrations. Likewise, in the toxicity concentrations, the subclinical toxic symptoms may begin to occur at the downward slope of the graph with severe symptoms at the higher concentrations. Both low and high concentrations of copper can lead to death in many species (Ref 3–6).

Copper in the Human Body

Because copper is essential to the metabolic functions of humans, it normally occurs in all bodily organs, tissues, and fluids. Its concentration varies depending upon a number of factors that will be discussed.

In Organs and Tissues. The total amount of copper in humans, based on individual tissue analyses, is estimated to range from 50 to 120 mg for adults and 14 mg for a full-term newborn infant. This level is equivalent to between 0.71 and 1.71 mg/kg for an adult and approximately 4 mg/kg for an infant. Copper is generally bound to proteins or to other organic compounds. Because ionic copper has the ability to rapidly combine with a variety of organic compounds, copper generally does not appear in the human body as free ions but rather as metallo-organic compounds. Table 6.1 shows typical wet-copper concentrations in several organs for adults and infants. The liver and the brain contain together one-third of the total copper of the body. Most other organs contain 15% of the liver contents (around 1 mg/kg), except the heart, lungs, kidneys, stomach, small intestine,

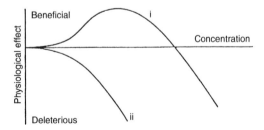

Fig. 6.2 Comparision of the physiological response to the uptake of: (i) an essential metal and (ii) an exogenous nonessential metal (Ref 7)

pancreas, and bones, which contain higher levels. There is a higher concentration of copper in the livers of newborn mammals, including humans. The high concentration of copper in the fetal liver is considered remarkable. Not only is there a massive buildup of liver copper in the normal child during the last three months of pregnancy, but the effect lasts for several months following birth, after which time the liver copper will have normally reverted to adult levels. This circumstance is discussed in more detail in the following (Ref 4, 7–11).

In Vital Fluids. All fluids of the body contain one or more copper complexes. Very high concentrations are found in the bile and, to a lesser extent, in the pancreatic juice. Copper concentrations in cerebrospinal fluid is 1.1 to 2.0 µg/mL. Saliva contains 0.02 to 0.22 µg/mL. The bile provides the major excretory route for copper, and its copper content has been measured to range from 0.24 to 5.4 µg/mL. The mechanism of copper transference from liver cells into bile is estimated to occur via carrier molecules, such as glutathione (Ref 7, 9, 12, 13).

Total blood copper levels in healthy humans normally range from 0.7 to 1.5 µg/mL. It has been found that variations in copper level in blood serum and plasma are indicative of the health of an individual. Copper contents in plasma are slightly higher for women than for men (1 to 1.3 µg/mL and 0.9 to 1.2 µg/mL, respectively). The major portion of copper in whole blood is divided among five separate fractions: erythrocyte superoxide dismutase; an unidentified copper complex in the erythrocytes; plasma albumin; plasma ceruloplasmin, and plasma amino acids. Each of these fractions has a unique role in copper transport and utilization in the human body. The most important of these is ceruloplasmin (or ferroxidase) because it contains 90 to 95% of the copper in the blood. Nine percent of the remaining 10% is bound at a specific site of plasma albumin and amino acids. Two to four percent (others say 1%) is bound to amino acids, especially histidine. Some of these are chelated by low molar mass serum compounds. The two major complexes are copper-albumin, transcuprein, and a ternary copper-albumin-histidine complex (Ref 14,15).

Ceruloplasmin is a multifunctional protein involved not only in the mobilization of plasma iron but also copper transport and regulation of biogenic amines. It has a molecular weight of 134,000, and its molecule contains 6 to 6.6 atoms of copper. Ceruloplasmin forms part of the blood plasma. This protein is believed also to serve as ferroxidase facilitating the incorporation of ferric iron into transferrin. Copper is an important adjunct to iron in the formation of hemoglobin. The concentration of ceruloplasmin—generally between 20 and 40 mg/dL of plasma or serum in the normal adult—rises, occasionally reaching 100 mg/dL, during the later stages of pregnancy when estrogens are administered and in a number of pathologic states that include infection. Plasma copper concentrations may increase 200 to 300% above normal during the duration of some diseases. Values can also fluctuate with age, sex, exercise, health, and circadian condition. There are also hormonal influences. During pregnancy, the level of copper in the blood of the mother almost doubles just before delivery. Values return to normal one to two months after childbirth. Women on oral contraceptives have been shown to have an increased serum copper concentration (up to twice normal) and to accumulate more total body copper. This increase is a common side effect of the estrogen hormone and explains the mechanism of copper increase during pregnancy (Ref 4,12,16,17).

Erythrocyte and plasma copper concentrations are approximately equal: 89 ± 11.4 µg/100 mL, packed erythrocytes, 109 ± 17 µg/100 mL, plasma. Normally about 90 to 95% of the copper in the plasma is bound to ceruloplasmin, a plasma protein in humans. In other animal species, the ceruloplasmin fraction may be smaller. Some are chelated by low molar mass serum compounds. Copper found in erythrocytes is either associated with superoxide dismutase (SOD) (Cu/Zn SOD, 90% or 60% in humans) or with a complex mixture of amino acids. There are species' differences in ceruloplasmin blood levels. The avian has little or none, and mammalian levels follow the order: pig > man > rat > sheep > cattle > dog (Ref 3,17,18).

Copper is mainly lost from the body in the feces by means of the bile system. However, considerable amounts of copper may be found in sweat, and some studies have indicated that this could account for as much as 45% of the total dietary intake of copper. On the other hand, urine contains very low levels of copper, and it has been estimated that a negligible amount of copper intake is lost in

Table 6.1   Copper in organs and tissues of adults and infants

| Organs or fluids | Adults | | Infants, typical values |
	Mean value	Minimum and maximum single values	
Adrenal gland	2.1 ± 0.62 mg/kg wet	1.4–4.4 mg/kg wet	...
Blood common	1.0 mg/L	0.6–1.3 mg/L	...
	1.1 ± 0.24 mg/kg wet	0.79–1.7 mg/kg wet	
Bones	Not calculated	1–4 mg/kg	...
Rib	0.52 ± 0.21 mg/kg wet	0.23–0.79 mg/kg wet	...
Cerebellum	6.2 ± 1.2 mg/kg wet	2.7–8.0 mg/kg wet	...
Cerebrum	4.8 mg/kg	...	3.7 mg/kg
	6.3 mg/kg wet	...	...
	5.1 ± 1.4 mg/kg wet	2.9–8.4 mg/kg wet	...
Fat	Not calculated	0.24– 0.28 mg/kg wet	...
Heart	3.0 mg/kg	...	2.6 mg/kg
	3.3 ± 0.67 mg/kg wet	2.2–4.6 mg/kg wet	...
Intestine	Not calculated	1.1–4.3 mg/kg	...
Small	2.1 ± 0.48 mg/kg wet	1.3–3.2 mg/kg wet	...
Large	1.7 ± 0.40 mg/kg wet	1.0–2.6 mg/kg wet	...
Kidney	2.9 mg/kg wet	(a)	3.1 mg/kg
	2.0 mg/kg wet	...	...
	2.6 ± 0.38 mg/kg wet	1.8–3.4 mg/kg wet	...
Liver	7.4 mg/kg	(a)	19.0 mg/kg
	5.1 mg/kg wet	3–9.5 mg/kg	...
	9.9 ± 5.5 mg/kg wet	2.1–23 mg/kg wet	...
Lung	3.1 mg/kg	1.1–4.8 mg/kg	...
	1.3 ± 0.24 mg/kg wet	0.81–1.9 mg/kg wet	...
Muscles	2.2 mg/kg	0.7–3.9 mg/kg	...
	0.92 ± 0.29 mg/kg wet	0.41–1.7 mg/kg wet	...
Pancreas	1.5 ± 0.31 mg/kg wet	0.83–2.1 mg/kg wet	...
Skin	0.71 ± 0.29 mg/kg wet	0.27–1.5 mg/kg wet	...
Spleen	1.2 ± 0.23 mg/kg wet	0.83–1.7 mg/kg wet	...
Stomach	Not calculated	2.4–3.7 mg/kg	...
Testicles	0.94 ± 0.17 mg/kg	0.61–1.2 mg/kg	...
Ovary	0.97 ± 0.28 mg/kg wet	0.41–1.3 mg/kg wet	...
Trachea	0.85 ± 0.21 mg/kg wet	0.37–1.2 mg/kg wet	...

(a) Not reported; all values refer to net weight. Source: Ref 4, 10, 11

urine. Levels of copper in urine from women in nonpolluted farming regions in Japan, measured in 1987 to 1988, averaged 36.9 µg/L. It has been reported that healthy adults have a copper excretion through the urine of less than 100 µg/L and that the usual value is around 20 µg/L. Menses also constitute an excretion vehicle for copper (Ref 4, 5, 12, 14).

Metabolism of Copper

Biological Functions. Radioactive tracer studies have shown that copper is mainly absorbed in the acid medium of the upper gastrointestinal tract in the stomach and small intestine. Most of the copper absorbed from the gastrointestinal tract binds to plasma albumin after 1 to 3 h of ingestion. A small fraction (1%) is chelated to amino acids, especially histidine; some of the copper histidine complex also is reported to bind specifically to albumin. After 48 h of having injected "labeled" ^{67}Cu, the largest fraction of copper was found in brain and kidneys as metallothionein. In a group of other nine organs, the fraction of copper in thioneine was equal to that in CuZn-SOD. Metallothionein appears as a major copper transport vehicle in the form of Cu(I)-thiolates in copper homeostatic processes. Not all the copper in the diet is absorbed however, and estimates vary from 25% to more than 60% depending on a number of factors. Copper taken orally is absorbed in the gastrointestinal tract and is preferentially captured by the liver. The liver plays a central role in copper metabolism. All body tissues require copper for normal metabolism, but some have greater metabolic needs than others. Their content of copper reflects this, such as the brain and the heart (Ref 3, 4).

Hepatic copper homeostasis comprises four functions: distribution through blood serum in the form of ceruloplasmin; incorporation in the liver in essential enzymes like cytochrome C and SOD; cytoplasmic storage incorporated in low M_r proteins, metallothioneins; or secretion into bile for later excretion via feces (Ref 8).

Rheumatoid arthritis, osteoarthritis, and other chronic degenerative diseases have been an area of active research interest since hypercupremia (associated with ceruloplasmin increase) was first observed in rheumatoid arthritis in 1938. In 1951, it was discovered that copper complexes can be successful in treating arthritis. The anti-inflammatory reactivity of copper(I)-thioneine has been evaluated as an effective new treatment concept based upon the decontamination properties of its OH

radical and single oxygen radical O^{-2}. Actually, it had been found by 1994 that a series of copper coordination compounds catalyze the dismutation of the superoxide radical O^{-2} *in vivo*. Suggested equations are (Ref 4,18–21):

$$Cu^{2+} + O^{-2} \rightarrow Cu^+ + O_2$$

$$Cu^+ + O_2 + 2H^+ \rightarrow Cu^{2+} + H_2O_2$$

Anti-inflammatory, antirheumatic, and immunosuppressive effects caused by glucocorticoids are attributed to an increased synthesis of thioneine with concomitant inhibition of phagocytes. Thus, scavenging of activated oxygen species combined with a catalytic dismutation of superoxide, provides support to the wound healing process (Ref 22).

Copper-thioneine topically applied to inflamed areas is reported to accelerate maturation of procollagen to collagen and increase angiogenesis (Ref 22).

Anticarcinogenic reactivity of copper-dischiff bases with superoxide dismutase-like activity has been investigated. They are efficient antitumor agents in the Walker 256 carcinosarcoma and in the human K 562 erythroleukemia cell line. Compounds tested were $CuPu(Py)_2$ and $CuPu(Im)_2$ or {[N,N'-bis(2-pyridylmethylene-1, 4-butanediamine](N, N', N'', N''')}-copper(II), and {[1, 8-di(2-imidazolyl)-2, 7-diazaoctadiene-1,7]-(N, N', N'', N''')}-copper (II) (Ref 23).

Copper is also required for bone formation, connective tissue development, myelinization of the spinal cord, keratinization, and tissue pigmentation (Ref 24).

The presence of competing metals, such as zinc or cadmium, or of copper-complexing agents, such as ascorbic acid, in the diet can have a marked effect on the absorption of copper. Large doses of iron may also reduce copper absorption. Ascorbic acid not only chelates copper but enhances iron absorption. This exchange has been demonstrated in children suffering from severe protein malnutrition in Perú. The presence of sulfides also inhibits absorption. The insolubility of cupric sulfide renders copper unavailable for the formation of the copper complexes required for transporting the metalloelement across the intestinal mucosa (Ref 4, 5, 15).

Evidence associating lipid metabolism to copper has come from the identification of chemicals that can: (a) increase plasma cholesterol and at the same time inhibit copper absorption and retention, such as ascorbic acid, cadmium, fructose, glucose, histidine, sucrose, and zinc; or (b) decrease plasma cholesterol and simultaneously improve copper absorption and retention, including calcium, clofibrate, and sodium phytate (Ref 25).

The copper taken in with food and drinking water, after absorption, enters the blood where it binds to serum albumin. It is carried in this form by the blood to the liver where it is complexed either with ceruloplasmin or to metallothionein. Ceruloplasmin provides the transport mechanism for copper through the blood stream. Metallothionein is a sulfur-base protein that forms mercaptides with copper, zinc, and cadmium and provides a storage medium for these elements. It is active in the liver, the intestinal mucosa, and elsewhere. About 5% of the liver copper is normally associated with metallothionein and the greater part of the remainder is used for the production of ceruloplasmin.

Excess copper is ultimately excreted through the bile in the feces in the form of complexes with both high- and low-molecular weight proteins. There is still some dispute as to whether or not copper can be reabsorbed in the intestine from bile complexes.

Pregnancy and the Neonate. Copper plays a major part in the rate of fetal growth and in early postnatal development. During pregnancy, the ceruloplasmin content of the mother's plasma increases 2 to 3 times normal with a consequential reduction of the copper level of the liver. Following delivery, this level returns to normal within a month. Conversely, the copper level of the fetus' liver is built up during the last trimester to the point that at birth it is 6 to 18 times that of an adult's. Simultaneously, the newborn's ceruloplasmin level is the lowest at the time of birth than it will ever be again. The unusually high concentrations of copper in liver and other tissues of the neonate appear to represent a reserve for synthesis of ceruloplasmin and other copper-containing proteins to meet metabolic needs immediately following birth. This ensures adequate supplies for growth in the first few months of living when the diet is restricted to milk, which is low in copper. Thus, it is possible that the concentration of copper in the maternal diet may be involved in determining the rate of growth of the fetus and its postnatal development (Ref 26, 27).

Dietary Requirements. Because copper is needed for normal metabolism and prevention of disease, care should be taken to assure that dietary intake provides required amounts of copper. Copper is normally obtained from food and water intake and, when necessary, by dietary supplements.

As copper is present in minute quantities in most food and water, nutritional needs of humans are generally believed to be adequately met through proper diet. It is common practice to add copper to infant formula and animal feeds, particularly swine, as a dietary supplement. It is contained in vitamin and mineral preparations, particularly in formulations for the elderly. Copper gluconate, copper sulfate, and cuprous iodide are approved by the U.S. Food and Drug Administration for use as nutrients in conventional foods, infant formula, special dietary foods, and vitamin supplements. Copper gluconate and copper sulfate are used as nutrients in milk products, processed fruit juices, soft candy, snack foods, beverages, chewing gum, and baby and infant formula. Cuprous iodide is used as a source of iodine in iodized table salt.

The human adult body contains 50 to 120 mg of copper (between 1.4 and 2.1 mg of copper per kilogram of body weight) and a full-term newborn infant of normal gestation will contain about 14 mg of copper. The infant body contains up to three times the adult's amount per unit weight, consistent with the fact that infant metabolic needs are much greater than those of adults. Daily intake for a normal adult should be 1.5 to 3 mg and for children, 0.7 to 2.0 mg, which are the estimated safe and adequate intake values established by the Food and Nutrition Board of the U.S. National Academy of Sciences. It has been found that many modern human diets do not supply adequate amounts of copper and other metalloelements. This is because of low-copper diets and the fact that the body is only capable of absorbing 35 to 55% of the copper in the diet. Surveys conducted in the 1970s showed that customary adult diets in the United States for 84% of the population contained only 0.82 mg Cu/day; in New Zealand, 79% of the population's diet contained less than 2 mg Cu/day. Swedish studies showed a daily intake by their citizens of 1.6 mg Cu/day; in Denmark, the value was 1.7 mg Cu/day; in the United Kingdom, the mean value found was 1.0 mg Cu/day. A 1989 study in Eastern Germany showed intakes of 0.55 to 0.77 mg/day for women and 0.75 to 0.95 mg/day for men. A 1987 to 1988 study in the United States confirmed that mean intakes of copper failed to reach even the bottom of the range for both adults and for children. The World Health Organization (WHO) has established the safe range of mean copper intakes for adults to be up to 12 mg/day (170 µg/kg/day) and for infants to be up to 150 µg/kg/day (Ref 28–34)

The WHO recommends that infants receive 80 µg of Cu/kg/day and older children 40 µg. For the premature infant, however, 100 to 200 µg/kg/day is recommended. The U.S. Food and Drug Administration's minimum specification for copper in infant formulas is 60 µg/100 kcal and 100 µg/100 kcal for premature infants. The recommended maximum is 200 µg/100 kcal, although formulas containing as much as 300 µg/100 kcal have been fed to premature infants for as long as a month without adverse effects. Most prepared infant formulas contain copper as a dietary supplement at the level of 40 to 90 µg/100 mL. The Committee on Nutrition of the American Academy of Pediatrics recommends 60 µg/100 mL of copper in infant formulas (Ref 35, 36).

Copper in Food. Table 6.2 gives the typical copper content of a number of human and animal foods. Based on average food and water concentrations, most drinking water contributes a small proportion of the daily copper intake. Nevertheless, the U.S. National Academy of Sciences suggests that the extra copper contributed by water is a dietary safety factor that should be maintained if feasible. They point out that the potential for toxicity from the levels of copper in drinking water is extremely low (Ref 5, 37, 38).

Human milk contains 0.1 to 0.7 µg/mL of copper depending upon the stage of lactation, with the higher value in the first week. Human colostrum contains 0.2 to 1.2 µg/mL, and cow's milk contains 0.02 to 0.2 µg/mL. Thus, cow's milk is more copper deficient than human milk. In addition to the small amount of copper in milk, it must be considered that not all the copper contained in milk is bioavailable. This effect is more pronounced in newborns than in adults (Ref 14, 39).

The degree of absorption of copper from milk varies. Neonates are able to assimilate as much as 60% of the copper contained in human milk but only 30% of the copper contained in cow's milk. One explanation for this is that the copper in human milk is concentrated in the whey fraction and that in cow's milk is concentrated in the casein fraction. Because the digestibility of cow's milk casein may be low in infants, the bioavailability of copper bound to cow's milk casein may also be low. In addition, there are other factors that restrict copper absorption. It has been shown that lactose inhibits the absorption of copper from milk. Also, infant cereals fed concurrently with milk have been shown to significantly reduce the bioavailability of copper from human milk; however, the same effect has not been found with cow's milk (Ref 40–43).

Copper in Drinking Water. Copper is naturally present in most bodies of water, and, thus, it is present in drinking water. Copper can appear in drinking water from three sources: (1) in the natural water through natural contamination from

Table 6.2 Copper contents in foodstuffs

Foodstuff	Edible portion, mg Cu/100 g
Vegetables	
Whole grained wheat	0.47 – to 0.821
Quaker oats	0.23 – 0.74
Cornflakes	0.023 – 0.20
Rice	0.031 – 0.36
Rye flour	0.07 – 0.80
Wheat flour	0.04 – 0.70
Wheat grit	0.46 – 1.99
Whole-grained wheat bread	0.063 – 1.45
Rye bread	0.04 – 0.68
Wheat bread (white)	0.019 – 0.34
Paste ware	0.02 – 0.46
Potatoes (raw)	0.04 – 0.272
Potatoes (skinless, cooked)	0.11 – 0.15
Sugar	0.01 – 0.24
Cocoa	1.94 – 5.0
Peanuts	0.27 – 1.0
Hazelnuts	0.21 – 1.35
Almonds	0.14 – 1.411
Coconuts	0.019 – 7.0
Beans (white)	0.044 – 1.5
Beans (green)	0.03 – 1.36
Peas (green)	0.045 – 0.80
Carrots	0.011 – 0.508
Celery bulbs	0.01 – 0.15
Beets	0.015 – 0.26
Onions	0.01 – 0.546
Cabbage	0.04 – 0.350
Parsley	0.02 – 0.53
Spinach	0.035 – 1.87
Lettuce	0.01 – 0.546
Tomatoes	0.04 – 0.350
Grapefruit	0.02 – 0.53
Oranges	0.004 – 0.13
Apples	0.014 – 0.14
Cherries	0.07 – 0.14
Strawberries	0.02 – 0.440
Raspberries	0.024 – 0.31
Pineapples	0.020 – 0.08
Bananas	0.066 – 0.21
Coffee (brewed)	0.004 – 0.09
Tea (brewed)	0.007 – 0.031
Tea leaves	1.59 – 4.8
Meats, poultry, seafood, dairy products	
Beef	0.012 – 0.68
Veal	0.04 – 0.25
Pork	0.011 – 0.52
Poultry meat	0.011 – 0.50
Heart	0.29 – 0.34
Liver	0.32 – 10.8
Kidneys	0.042 – 0.53
Seawater fish	0.011 – 0.55
Fresh water fish	0.012 – 0.34
Shellfish	0.048 – 13.705
Eggs (entire)	0.03 – 0.23
Human milk	0.02 – 0.07
Cow's milk	0.005 – 0.07
Cheese	0.010 – 1.5
Butter	0.006 – 0.392
Margarine	0.001 – 0.060

Source: Ref 5

rocks or soil flooded as the water collected in bigger streams, (2) as a result of water treatment practice using copper sulfate as an algicide, and (3) as a corrosion product from copper and copper alloys used in the plumbing system. Byproducts of the corrosion of copper and other metals by aggressive (low pH, low hardness) waters is the most prevalent of the three sources.

The question of the allowable limit of copper in drinking water is an ongoing issue. In the United States under the Safe Drinking Water Act, Lead and Copper Rule, established in December, 1992, the maximum allowable copper content of drinking was established as 1300 µg/L. Under a secondary standard, not enforceable by the federal government, the recommended standard was set at 1000 µg/L. These levels apply to samples of the first liter drawn from the kitchen or bathroom tap of water that has stood quiescent in the plumbing system for at least six hours (Ref 44, 45).

Historically, the level of 1000 µg/L has been based on taste and is a recommended level in most jurisdictions. In 1984, the World Health Organization (WHO) recommended that water retained 12 h or more in copper tubes should not contain more than 3000 µg/L copper. The reason for this limit was also based on taste. In 1992, the WHO issued a revised Provisional Guideline for Drinking Water, which suggested a level of 2000 µg/L for health effects and 1000 µg/L for comfort effects (Ref 46, 47).

Germany and most of Europe has a limit of a copper content of 3000 µg/L for drinking water after a period of 12 h stagnation. In this case, 3000 µg/L has been considered to be appropriate even for newborn children (Ref 48).

Deficiency of copper may cause effects ranging from lack of growth, loss of hair and skin pigmentation, to hyperlipidemia and damage to the blood vessels, and may even lead to sudden death by heart attack. Severe copper deficiency is associated with cardiovascular abnormalities in several animal species. In humans, the effects of severe copper deficiency are present in those with Menkes Syndrome, an x-linked genetic disorder associated with a widespread defect of copper transport. This disease is discussed in the section "Diseases of Copper Metabolism" in this chapter (Ref 49, 50).

Copper deficiency can decrease the percentage of the total plasma cholesterol that is bound to high-density lipoprotein and increase the percentage bound to low-density lipoprotein. The implications of these results are that the risk of arteriosclerosis in which fatty nodules collect on the inner walls of the arteries, is higher when the high-density lipoprotein cholesterol level is low and low-density lipoprotein cholesterol is high (Ref 51).

Full-term infants are able to withstand the stresses of a mildly copper deficient diet for several months after birth. In contrast, premature infants with reduced storage of copper in the liver are much more likely to develop copper-deficiency pathologies. Chronic diarrhea can be a predisposing cause of copper deficiency, as was the case in the study of a group of malnourished young children in Perú. Copper supplements modulated but did not correct the severity of the deficiency. Copper deficiency in the neonate has also been correlated with sudden infant death syndrome and with Schwachman syndrome (Ref 52–55).

Growing children have a high requirement for copper that cannot usually be met by cow's milk. Young children who become copper deficient due to the exclusive consumption of cow's milk, or of a copper-free diet, are reported to develop hypochromic microcytic anemia, hypoferremia, and hypoproteinemia. It is thought that the disturbances to blood formation are mainly due to defects in iron transport in the gastrointestinal tract induced by copper deficiency. Important reversal of the symptoms is reported following oral administration of copper.

Copper in Medicine

Copper was used in medicine as a curative aid for a number of ailments thousands of years before the recognition of copper's role in the human body. The first record of its use can be found in the Smith Papyrus, an ancient Egyptian medical text written between 2600 and 2200 B.C. (Ref 56). This text records the use of copper to treat infected chest wounds and to sterilize drinking water. An excellent treatise on the subject of the historic uses of copper in medicine from ancient times to the present is given by Dollwet and Sorenson (Ref 57).

Most historical medicinal uses of copper were based on its early recognized bactericidal and fungicidal properties; however, when a number of ailments reportedly cured by the use of copper are viewed in terms of modern medical knowledge, another property of copper—its anti-inflammatory capability—becomes evident. Among these afflictions were headaches, arthritis, stomach ulcers, and "tremors" (epilepsy and St. Vitus Dance). Another medical application, used until the last quarter of the century, is its use as an emetic. Copper sulfate ingested in a dosage of about 300 mg is an emetic agent (approximately 300 mg/L Cu if administered in 250 mL of water). The death of a patient who was given a normal dose but who had previously had a surgically altered gastrointestinal track, which enhanced the absorption of the drug, showed the sensitivity of this procedure and caused its use to fall out of favor. In Nigeria, copper's emetic effect is used in religous rites "to purge one's problems and impurities" in the form of a spiritual water or "green water." One sample of green water was found to contain 13,390 mg/L Cu (Ref 58–60).

Some of the disease states studied in animals and humans that are influenced by copper include: wound healing, convulsions, some forms of cancer, diabetes, radiation damage to tissue, and microbial infections. Recent evidence suggests that copper also plays an important role in the immune system. Impairment of immune function may be highly correlated with an increasing incidence of infection and higher mortality rates observed in copper-deficient animals. Copper has also been shown to be effective against dental plaque buildup when used in mouthwash (Ref 61–63).

Today, in spite of the buildup of a significant body of literature providing scientific evidence of the medicinal role of copper in the human body, its use as a pharmaceutical is not supported by the general medical profession. Although the role of copper in the fighting of disease is not disputed, the general impression is that the body is capable of providing the correct amount of copper from dietary intake. Nonetheless, copper is added to infant formulas and some vitamin preparations, particularly those for geriatric use, as part of the mineral constituents to supplement those received in normal food and water intake.

Copper Toxicity

Despite the practical benefits of copper to man, over the ages there have been repeated discussions as to whether copper is a poison or a medicine. In *Historia Naturalis*, published in A.D. 77, Pliny documented that copper is more of a medicine than a poison. This debate has periodically surfaced over the intervening years. In 1900, Lewin said, with reference to its toxicological significance, that no other substance had produced so numerous contradictory opinions, experiences, and experimental results as copper. Unfortunately, in spite of all the knowledge gained since the turn of the century about copper metabolism and ho-

meostasis in humans, this uncertainty continues (Ref 64, 65).

The regular daily intake of 2 to 5 mg Cu in adults is considered normal and, thus, free from harmful effects. Higher dosages, of 32 mg, can have an astringent effect and cause nausea; 80 to 132 mg may lead to vomiting (copper's emetic effect). Such doses, leading to acute toxicity, have been reported in instances where water for tea was taken from a badly corroded water heater, cocktails were stored in a corroded cocktail shaker, or copper tubing was used in a carbonated soft drink vending machine. Copper sulfate, in gram quantities, can be lethal and has been used as a vehicle for murder and suicide in India. A single dose of 10 to 20 g of copper sulfate can cause death in humans. Also, copper sulfate, when used as a bacteriocidal agent in the treatment of burns, can be lethal if the burned area is sufficiently large for a large amount of copper to be absorbed. The only confirmed manifestation of chronic copper toxicity in humans is in the case of Wilson's disease, a genetic abnormality that leads to a reduced ability to remove copper from the body, as is discussed in the following (Ref 7, 66–72).

It is especially dangerous to introduce copper or copper compounds into the human body parenterally or by inhaling such that it is deposited in lung alveolae. In spite of this, no poisoning of workers in copper mines, smelters, or refineries has been demonstrated. A 1967 study for the FAO/WHO of Chilean copper miners, all of whom had worked in the mines, smelters, or refineries for at least 20 years, showed no indication of copper toxicosis. Also, in spite of the wide use of copper and copper alloys in metal manufacturing, there is little evidence of occupational health problems among copper metal workers. Inhalation of copper powder or copper-containing fume may cause "metal fever," which disappears after one or two days. Symptoms are reported to be headache, sweating, nausea, and exhaustion. Workman spraying vineyards with Bordeaux mixture ($CuSO_4$ and lime) tend to develop what has been called "vineyard sprayers lung," which, in some instances, has been reported to be fatal. These workers developed interstitial pulmonary lesions and scar tissue containing deposits of copper (Ref 73–76).

Copper Pathologies

Lipid Metabolism and Cardiovascular Disease. Lipid metabolism is related with cardiovascular disease because it may result in hypercholesterolemia. The first data to link the metabolism of copper with that of cholesterol were obtained in 1970. In the ensuing period, numerous studies have shown the relationship between copper deficiency and elevated cholesterol. Lipid abnormalities have been associated with copper in several species, including humans (Ref 77–79).

It has been noted that humans with ischemic heart disease exhibit several characteristics that are also found in animals suffering from copper deficiency: abnormal electrocardiograms, decreased myocardial copper, glucose intolerance, hypercholesterolemia, hyperuricemia, necrosis of myocardial cell, and sudden death (Ref 49).

Diseases of Copper Metabolism. There are two rare inherited diseases which are distortions of normal copper metabolism, Wilson's disease and Menkes syndrome. These lead to deep alterations in metabolism and ultimately can cause death (Ref 80).

Wilson's disease affects 27 per 1,000,000 (approximately 1 in 30,000) people. It was first described by Samuel A.K. Wilson in 1912 and named *progressive lenticular degeneration*. It was proven to be of genetic origin in 1948 and to actually be a family of diseases in 1993. The basic characteristic of Wilson's disease is a massive accumulation of copper in the liver and brain due to the inability to transport copper out of the affected tissues via the blood protein, ceruloplasmin, and the normal bile excretion mechanism. At an advanced stage of the disease, the accumulation of the copper leads to disorders of the nervous system (tremors), abnormal behavior patterns, and to pathological lesions especially in the liver. In extreme cases, light brown circles, referred to as Kayser-Fleischer rings, are seen surrounding the iris. These are caused by the deposition of copper salts in the cornea (Ref 80–83).

Patients suffering from Wilson's disease are treated by chelation therapy with such drugs as penicillamine, 2, 3-dimercaptopropanol (British Anti-Lewisite, BAL), or trimethylene tetramine (TETRA) that modulate copper distribution and remove it through the urinary tract. Zinc, being an antagonist of copper in the body, has also been proven to be an effective therapeutic agent for Wilson's disease. Although some physicians recommend a low copper diet for Wilson's disease patients, a person cannot eat little enough copper to avoid the progression of the disease. Other physicians, who do not restrict the diet of their patients, feel that control of the disease by chelation therapy is adequate (Ref 79).

Menkes syndrome was described for the first time in 1962. It is an x-linked genetic disease of males. This is a copper-deficiency disease. Patients show abnormally low amounts of copper in liver and serum. Symptoms appear before the third month of life and the illness usually terminates the life of the child before the fifth or sixth year. The disease is characterized by rapidly progressive cerebral degeneration, bone lesions resembling those seen in scurvy, and elongation and tortuosity of the cerebral arteries. Death generally occurs by the age of three years. The terminal event in Menkes patients is severe, progressive neurodegeneration and frequently rupture of major arteries or arterial thrombosis (Ref 50, 51).

No cure or treatment has proven to be effective and attempts to control the course of the disease with copper therapy has been unsuccessful. Orally administered copper to Menkes patients is very poorly absorbed, with the copper tending to accumulate within the mucosal cells. Intravenous copper injections have been proven only partially successful; however, continuing research appears promising (Ref 84).

The effects of copper deficiency in the nutrition of healthy individuals' and Menkes syndrome patients' symptomatology have been compared. Both subject groups showed similar but not identical connective-tissue pathologies and imperfections of the arterial walls. However, it has been found that the intracellular metabolism of copper in a Menkes syndrome-affected cell is quite different from that of normal cells. The main difference in pathology between Menkes syndrome and nutritional copper deficiency relates to the simultaneous iron metabolism. Although anemia is often observed in nutritional copper-deficiency in many species, it does not occur in the genetic copper-deficiency disease (Ref 49).

In addition to these pathologies, mention should be made of another disease that has been alleged, but never proven, to be caused by chronic copper overdose. This disease is a liver cirrhosis affecting infants or small children that is still under further analysis at publication. This is commonly known as *Indian childhood cirrhosis* (ICC) due to the fact that it was first observed in India and the majority of diagnosed cases have been in India. It is now more widely known as *idiopathic copper toxicosis* (ICT), and cases have been observed in a number of other countries, including the United States, Germany, Italy, Australia, Malaysia, Mexico, and Kuwait (Ref 85–95).

Copper was first associated with ICT by the excessively high copper content of the liver in its victims. Copper levels in ICT livers have been observed as high as those in the livers of victims of Wilson's disease; however, this level is achieved much more rapidly in ICT.

ICT has been labeled "the disease of theories." One theory has it that, because copper and brass cookware is used extensively in India, copper overload is caused by the ingestion of milk, containing excess copper resulting from its preparation and storage in contact with this cookware. However, it is reported that the disease also occurs in households that use steel or aluminum cookware. Further, there have been no cases reported in Korea, another society that extensively uses copper and brass cooking ware. In Germany, the presence of the disease has been related to high-copper levels in drinking water. On the other hand, in the United States, three towns have been identified with drinking water copper levels between 8.5 and 8.8 mg/L. No deaths from cirrhosis or any form of liver disease were reported for children of those towns between the years that the data are available, 1969 and 1991 (Ref 38–40, 96–99).

A recent retrospective study of 135 ICT-type cases that had occurred in the Tyrol, in the western part of Austria, between 1900 and 1974 indicated that all of the victims were genetically related—all bearing one of five family names. During this time, it was also customary to use copper cooking utensils to prepare baby food. The conclusion reached is that intake of elevated concentrations of copper can trigger ICT in those infants who are already genetically predisposed to the disease (Ref 100).

Because a familial relationship seems to be present in most cases, the probability of a genetic defect has also been theorized to be the primary cause of the disease. ICT generally affects children up to their fourth year of life. If caught early enough, it seems to be treatable by chelation therapy; however, too often the damage has progressed too far for chelation therapy to be effective (Ref 101).

A number of hepatic diseases from extraneous causes can result in a high-copper content in the liver and, thus, give a false indication of copper toxicity. Because the normal route of copper excretion is via the bile, several conditions that interfere with biliary excretion causes copper to accumulate in the hepatocytes of the liver. However, under these conditions, copper is not deposited in other tissues as occurs in Wilson's disease. Primary biliary cirrhosis is one such disease. In advanced cases, the concentration of copper in the liver can be in excess of that found in Wilson's

disease. Prolonged extrahepatic biliary obstruction, primary sclerosing cholangitis and the various forms of familial cholestasis of infancy comprise a number of syndromes, some of which are probably genetically determined, such as Byler's disease, whereas other prolonged cholestasis following biliary atresia, may be acquired (Ref 102).

Copper in the Ecosystem

In examining copper's effects on the ecosystem, it is useful to begin with copper's usage in agriculture where it serves as both an essential element and as a toxin for the control of fungus and disease in both plants and animals. Table 6.5 gives a listing of copper compounds commonly used in agriculture for correction of copper deficiency. Copper "needles" are also implanted subcutaneously in animals to overcome copper deficiency. Copper also plays a prominent role in water treatment, where it serves as an algicide to remove color and taste from potable water. These subjects are discussed in Chapter 5, "Applications," and are mentioned here because some of copper's positive values are based on what otherwise would be considered as negative values when the ecosystem is concerned.

Copper in Soils

All soils contain copper; it is an essential element for plant life and agricultural production. In this relationship, copper appears as a micronutrient. The same role is assumed by the entire first-row transition elements of the periodic system: V, Cr, Mn, Fe, Co, Ni, Cu, and Zn (Ref 103). The most abundant among these elements in humans (in terms of weight percentages) are Fe, Zn, and Cu. The same order obtains with decreasing contents in the body tissues. The amounts are 5 to 7 g Fe, 1 g Zn, and 100 mg Cu.

Many factors are responsible for variations in both total and available copper contents of soils. The copper content of a soil is the result of input of metal from three different sources: (1) minerals in the soil parent material (weathered rock, decayed vegetation); (2) anthropogenic inputs onto land, such as chemical additives, fungicides for special culture broth, such as wine grapes, sewage sludge, and manures; and (3) deposition from the atmosphere, such as volcanic ash, power-plant fly ash, mining dusts, and municipal dusts (Ref 104).

Within the soil profile, copper is concentrated in the surface horizons as a result of cycling through vegetation, atmospheric deposition, and adsorption by the soil organics (Ref 103).

Copper in Water and Sediments

Copper is a naturally occurring element, and therefore, it is present in most bodies of water from natural sources as well as from man-made (anthropogenic) sources. In the United States, for example, samples of 1577 sources of raw, untreated surface waters showed a minimum copper content of 1 μg/L to a maximum of 280 μg/L with a mean of 15 μg/L. In fresh water, it has been shown that the bioavailability of copper is a function of the pH and the hardness (carbonate concentration) of the water. When present, humic materials bind more than 90% of the dissolved copper (Ref 105).

As shown in Table 6.4, the copper content of oceans can vary from less than 1 μg/L to 3000 μg/L. In the open ocean, the copper content ranges from 0.1 μg/L to 3.9 μg/L for an average of 0.8 μg/L of seawater. Copper increases in ecosystems may result from natural processes, such from Eolian input, and geothermal sources or anthropogenic sources, such as from waste discharge, rain water runoff, air-borne dusts, and marine antifouling paints. Of these, rain water runoff is probably the most important, providing localized input into coastal waters from both natural and man-made sources (Ref 105–107).

The geochemistry of dissolved copper in seawater is distinguished by the fact that it shows depletion near the surface and progressive enrichment at lower depths. This progressive depletion of copper towards the surface is thought to be biological in origin. Copper is required for normal growth functions of marine organisms, such as plankton, which take up copper. The biogenic organic debris, which mostly originates in the planktonic zone, is biodegraded as the particles sink to the bottom. The release of copper from this debris results in the enrichment of copper at depth (Ref 108).

As in fresh water, the copper in seawater can exist in several chemical and physical forms including insoluble particulates, soluble inorganic and organic complexes, and various oxidation states of copper ions. The distribution of these forms determines the bioavailability and, thus, the potential for toxicity to various biological species. The distribution of copper between ionic and complexed forms (inorganic and organic) in seawater is complicated. The fraction of organically bound dissolved copper can follow a wide distribution ranging from 3 to 99.9% of the total dissolved copper (Ref 109).

One of the reasons for this distribution is the seasonal aspect of the planktonic biomass. The period of high-copper complexing capacity can be expected to occur during or just after a period of high phytoplanktonic bloom. Other factors affecting the complexing of copper in water are the amount and nature of humic substance present and the presence of other metallic ions, which may be more or less attracted to the same humic material and, thus, displace copper from the complex (Ref 110).

The redox conditions of the water will determine the distribution of Cu^{+2} and Cu^{+1}. Both oxidation states form complexes with chloride ion.

Table 6.3 Copper compounds commonly used for treating copper deficiency in crops

Compound	Formula	Copper, %	Forms
Copper sulfate	$CuSO_4 \cdot 5H_2O$	25	Crystals
Bordeaux mixture (copper sulfate + calcium hydroxide)	$CuSO_4 \cdot 3Cu(OH)_2 + 3CaSO_4$	12.75	Powder suspension
Copper oxychloride	$3Cu(OH)_2CuCl_2$	52	Powder
Copper oxychloride	$3Cu(OH)_2CuCl_2$	25	Liquid(a)
Copper hydroxide	$Cu(OH)_2$	25	Liquid(a)
Copper hydroxide	$Cu(OH)_2$	50	Powder
Cuprous oxide	Cu_2O	89	Powder
Cupric chloride	$CuCl_2$	17	Powder
Copper slags	CuO, Cu silicates	1.5–5.0	Powder
Copper frits	Cu silicates	40–50	Powder(a)
Copper ethylene diamine	$Na_2CuEDTA$	14	Powder(a)
Tetra acetic acid	$Na_2CuEDTA$	7.5	Liquid(a)
Copper hydroxyethylene diamine	NaCuHEEDTA1	14	Powder(a)
Triacetic acid	...	...	...
Copper lignosulphonate	...	4	Liquid(a)
Copper phenolic acid	...	5	Liquid(a)
Copper polyflavanoid	...	6	Liquid(a)
Copper methoxypropane	...	5–6	Liquid(a)

(a) Formulated compound. Source: Ref 3

Cu^{+2} also forms complexes with carbonate ion and Cu^{+1} forms complexes with sulfides. In the low-oxygen condition of the marine environment at depth, sulfide complexes followed by the chloride complexes are the main chemical forms of copper.

The Biogeochemical Cycle and Bioaccumulation. Processes occurring during and after the entry of copper into an aqueous environment can cause a change in metal species in solution with a resultant change in the biological availability and levels of accumulation of the metal. The impact of copper on biota of water systems depends on the oxidation state of the copper and on the fate and distribution of the copper in the water and in the sediment. The copper may be present in both soluble and particulate forms. Copper in soluble forms may remain as such or it may associate with particles in suspension or in the sediment. Pro-

Table 6.4 Copper concentrations in oceans, marine and estuarine waters: levels in unfiltered and "filtered" water

Location	(Average)/Range	Location	(Average)/Range
North Sea and Baltic Sea Region		**Orient and Indian Ocean Region**	
Baltic Sea		Mekong River Coast, South China Sea	D 1.2 – 17.1 µg/L
Northern Baltic Sea		Saigon River Station	D 1.4, 9.6 µg/L
ASV-labile	D(189, 361) ng/L	Qiantang-jiang Estuary, East China Sea	T 17.20, 26.00 µg/L
ASV-total	D(379, 537) ng/L	Erhjen Chi (River) Estuary, Taiwan	
Gotland Deep	D(0.05, 0.72) µg/L	River water	? 56.61 – 793.50 ppb
Klaypeda Inlet	D 2.4, 9.2 µg/L	Mariculture water	? 5.53 – 86.88 ppb
Vassorfärden Bay, Finland	?(69) µg/L	Takasaki Seto, Japan	
Kirsiu Marios Lagoon, Lithuania,	0.004 – 0.016 mg/L	Surface seawater	? 0.39 – 0.99 µg/L
North Sea		Bottom seawater	? 0.42 – 0.70 µg/L
Framvaren Fjord, Norway	T 0.13 5.00 nM	Uranouchi Bay, Japan	
Scheldt Estuary, Belgium	D 0.84 – 2.6 µg/L	Surface seawater	? 0.38 µg/L
Firth of Forth, Scotland	D(1.05, 2.59) µg/L	Bottom seawater	? 0.87 µg/L
		Boso Peninsula	
North Atlantic Ocean Region		Surface seawater	? 0.20, 0.34 µg/L
Lavos Region, Portugal	D ND, 10.7 µ/L	Bottom seawater	? 1.24 µg/L
Sargasso Sea		Antarctic Ocean-Indian Section	D 0.28, 4.05 nmol/kg
Deep sea	T 0.79, 2.2 nM	India	
Surface transects	T.09, 3.3 nM	Arabian Sea	D (5.73, 8.00) µg/L
United States of America		Purna River Estuary	D (3.8, 7.6) µg/L
Pettaquamscutt Estuary, RI	D 0.13 – 0.53 µ/L	Lakshadweep Lagoon	
Mississippi River Delta, LA	D (18.3, 23.8) nmol/kg	Inside lagoon	D 0.69, 4.68 µg/L
Vero Beach, FL	D (100) ng/L	Outside lagoon	D 0.96, 3.27 µg/L
North Cotentin, France	D (0.13 – 0.80) µg/L	Mindhola River Estuary	D 2.7 – 15.9 µg/L
	T (0.25 – 1.20) µg/L		
		North Pacific Region	
South Atlantic Region		USSR, Gulf of Peter the Great	? (0.70, 1.90) µg/L
Argentina		Northeast Pacific-Deep sea	T 1.6, 3.9 nM
Blanca Bay	D (1.7 – 3.3) µg/L	United States of America	
	T (1.7 – 3.3) µg/L	Santa Monica Basin, CA	T 1.4, 3.0 nM
Embudo Channel	D (6.80) µg/L	Puget Sound, WA—Upper	
Bermejo Channel	D (5.30) µg/L	50 µM of the sea surface	
Falsa Bay	D (4.70) µg/L	Elliott Bay	D 3.38 – 259.34 µg/L
Verde Bay	D (4.20) µg/L		T 9.75 – 3215.34 µg/L
South Atlantic	D 0.8, 2.3 µg/L	Commencement Bay	D 2.91 – 45.72 µg/L
Antarctic Ocean	D 0.7, 2.2 µg/L		T 3.85 – 631.72 µg/L
	? (8.88) µg/L	Central Sound	D 4.16, 4.68 µg/L
	D (0.52) µg/L		T 15.78, 37.37 µg/L
South Shetland Islands, Antarctic	? 2.3 µg/dm³	Sequim Bay	D 0.97 – 5.18 µg/L
			T 0.97 – 12.71 µg/L
Mediterranean Region		Port Angeles	D 11.65 µg/L
Black Sea	T 0.8, 6.92 nM		T 15.76 µg/L
Dnieper-Bug Lagoon, USSR	D 40.6 to 95.2%		
Ligurian Sea, Italy		**South Pacific Region**	
Surface layer (0 to 50 m)	D(3.57 – 6.6) µg/L	Australia	
Deep layer (200 to 2000 m)	D(0.721 – 72) µg/L	Great Barrier Reef Lagoon	D(0.47) µg/L
Boka Kotorska Bay, Yugoslavia		Shark Bay Area	$D 0.13 – 2.92 ppb
Kotor (10 m)	? 28.8 nmol/m³	South Neptune Island Station	T 0.21 – 2.2 µg/L
Herzeg Novi (16 m)	? 21.9 nmol/m³	Gulf St. Vincent	
Malta Coast	T(0.58) µg/L	Offshore	T < 0.5 – 1.8 µg/L
Alexandria Coast, Egypt	D (2.7, 4.2) µg/L	Inshore	Tz 0.5 – 13 µg/L

Note: "Filtered" means any mechanism used to reduce the concentration of particulate matter. T = total (unfiltered water) copper. D = "dissolved" copper. ? = no indication given, assumed to be total (unfiltered water) copper. $ = values present in reference. Source: Ref 106

cesses that control copper reactions with particles include sorption, chelation, coprecipitation, and biological concentration (Ref 108, 111).

The biological response to copper is a function of the amount of cupric ion (Cu^{+2}) present. This is determined by the oxidation state of the system and the degree of complexation of the copper with other inorganic or organic species. Biological effects, as well as copper levels, in organisms are a function of the uptake of the metal and its interaction with the systems' metabolic processes. As the copper levels in organisms depend, at the end, on its availability in the environment, this must form part of any consideration of the biological importance of copper. In this connection, it is important to appreciate that the bioavailability of copper in the environment can be quite different from the amount measured as total copper (Ref 6).

The uptake of copper by aquatic organisms is a function of the concentration of copper in the aquatic environment and the ability of the individual species to regulate its metal content. The "aquatic environment" necessarily has different connotations depending on the nature of the organism and where it receives its nourishment. Copper concentration in the water, for example, is important for phytoplankton and seaweeds; however, copper concentration in sediment is important to sea grasses. Likewise, observations of a number of marine mammals in German coastal waters indicate that copper and zinc have vital functions in the metabolism of higher developed animals, and there is evidence that they can regulate the metal levels in their organs (Ref 112, 113).

Ecobalance

With increasing concerns for material production and usage from an environmental point of view, the concept of ecobalance is being adopted in the choice of materials to be used in applications. Life-cycle assessment is one of the tools used to examine the environmental cradle-to-grave consequences of making and using products or providing services. Whereas material choices and purchasing decisions have been customarily based on suitability for the application, price and delivery, under the ecobalance concept, new elements, generally expressed in terms of monetary cost, have been added such as land usage, energy consumption (crude oil, coal, and natural gas), air pollution, health factors, and waste disposal. Thus far, society has treated many of these extra material costs in an external manner, as for example,

payment for refuse disposal. However, under the ecobalance concept, more and more attention is being given to the concept of internalizing these costs in the cost of the product at the time of purchase. The health and environmental impacts of each product manufactured, and the processes used in their manufacture and ultimate disposal, are taken into consideration by the purchaser and by the use of a life-cycle assessment for each product under consideration. Thus far, there are only two measurable factors in a life cycle assessment—the energy consumed in a material or product's production and usage and the degree of recycling practiced (Ref 114).

Energy Intensity

Of the commonly used metals, copper has one of the lowest energy intensities for production. Although the absolute quantity of energy required for production varies depending on ore grade and other factors, a value for newly mined and refined copper derived from a 0.5% ore grade, based on values reported in the literature, largely based on 1970s data, is approximately 30 MWh/t. A comparable value for aluminum is 75 MWh/t and for steel 8 MWh/t. The majority of the energy required to produce copper is consumed in the mining, milling, and concentrating the ore. Table 6.5 shows the breakdown by process as of 1980. The fabrication of copper and copper-alloy shapes requires an additional 2 to 5 MWh/t with thin sheet and wire at the upper end of this range (Ref 116–122).

From the data in Table 6.5, it is apparent that most current publications are based on data collected prior to 1980—the date of the data collected for this table. It should be mentioned that the copper industry achieved significant energy reductions in its mining and smelting operations during

Table 6.5 Energy requirements for copper production

Operation	Energy, MWh/t	Percent of total
Open pit mining	6.6	19 – 25
Milling	14.1	40 – 52
Smelting	2.0 – 7.5	8 – 21
Converting	0.3 – 2.1	1 – 6
Gas cleaning	2.1 – 2.7	8
Electrorefining	1.8 – 2.1	6 – 7
Total	**27 – 35**	**100**

Note: Significant changes have been made, particularly in mining and smelting, since this information was first collected. However, the industry has thus far not been successful in agreeing on what might be considered typical numbers. In the meantime, these are the only referenceable numbers that have any credulence. Source: Ref 122

the 1980s, which, while documented in company reports, is not documented in published information. Therefore, the numbers reported should be considered as representing the upper end of that which is in actual practice at this publication date. More recent unpublished estimates for primary refined copper indicate that an average value of 22 to 27 MWh/t is more representative of 1990s practice.

The energy intensity for the recyling of copper varies by the purity of the scrap. Clean scrap, that requires only remelting, requires only approximately 1 MWh/t. Scrap that requires elecrolytic refining requires approximately 6 MWh/t and that which must be purified by resmelting requires approximately 14 MWh/t (Ref 121).

Because many applications of copper, particularly alloys, utilize scrap rather than virgin metal, the energy intensity of that metal is a function of the proportion of scrap used. For example, in a copper and brass automotive radiator, typically using 40% scrap in its makeup, mainly as brass in the tubes and header plates, the energy intensity is 20 MWh/t, not the 30 MWh/t of newly produced copper.

Recycling

Recycling is environmentally beneficial in that it reduces the amount of contaminents placed in the natural environment by man's usage. It also provides benefits such as energy savings, reduced emissions associated with some savings, and reduced volume of wastes. Recycling also ensures continuity for the world's metallic resources. Figure 6.3 shows the percentage of copper consumed derived from the direct use of copper scrap for the Western World between 1970 and 1990. Europe and the United States are the major users of recycled copper scrap. In addition to copper scrap consumed in these countries, there is an additional fraction of scrap that is shipped from OECD countries to non-OECD countries for their reprocessing and consumption. This amount rose in value from US$100 million in 1984 to US$350 million in 1993. In the period 1953 to 1993, an average of 40% of the refined copper consumed was derived from recycled material in the western world. In the United States in 1994, 1.7 million tons of copper was recovered as scrap. This amount comprised 41% of the total national copper consumption. Of this, 11% was added to smelters, 3% was added to refineries, and the remainder was used directly in the makeup of alloys. Consequently, secondary copper plays a major role in the global copper market (Ref 123–126).

Approximately 94% of all copper produced is in metallic form and, therefore, is recyclable. The copper that is not recycled is combined as chemicals or as powders that are used in dissipative applications such as agricultural and water treatment chemicals and as paint pigments. An analysis of the consumption curves shown in Fig. 6.4 shows that it requires an average of 35 years for the metal recovered as scrap to be equivalent to the total amount of metal consumed in any given year. Because most uses of copper, for example, wire and cable and plumbing tube, are used in capital goods

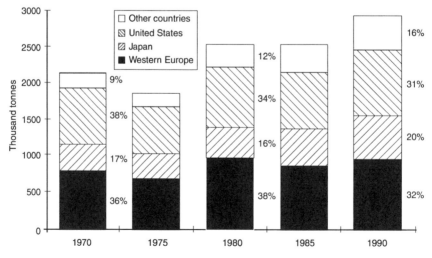

Fig. 6.3 Direct use of copper scrap in Western Europe, Japan, the United States, and other countries, 1970 to 1990 (excludes Eastern Europe, Russia, and China) (Ref 123)

that see a service life of 50 or more years, the 35-year figure reflects the fact that some of the scrap recovered is *prompt* scrap; for example, that recovered from manufacturing operations. Most copper consumed is not available for recycling until the building or equipment in which it is used has served its useful function. Economics, (for example, market prices) may also discourage recovery. This happens, for instance, in the case of buried cables, where the cable may not be excavated after it has served its useful life unless it is cost-effective to do so (Ref 123).

As indicated in Fig. 6.4, there is a great deal of variability in the usage of copper scrap in any given year. This is primarily due to the fact that the scrap collection and recycling business is a commodity business and it is very sensitive to market price. Often, scrap dealers hold back their supply of scrap hoping for a higher price. Figure 6.5 shows the correlation between copper scrap recycling as a function of the market price for copper in the United States. The price of copper scrap is a function of the price of refined copper and varies according to the degree of purity of the scrap type. Because of its value, copper is one of the most recycled of all of the metals in common use. The Institute of Scrap Recycling Industries, in the United States, publishes a yearly scrap specification circular for use in trade. This circular lists 53 classes of copper and copper-alloy scrap in commerce. The current market price of many of

these classifications are regularly published in trade magazines (Ref 126–129).

In the United States, of the 40% or more of refined copper consumption derived from scrap each year, approximately one-third comes from used objects (old scrap), and two-thirds from manufacturing wastes (new scrap). Both old scrap and new scrap are recognized commodities whose prices are publicly quoted weekly. It is rare that copper in any form, including turnings and powders, appears in landfills or other terminal disposal points because of its high intrinsic value.

Copper scrap re-enters the production process in any of several ways depending on its quality (Ref 121):

- *No. 1 scrap:* Scrap that is of cathode quality and which requires only melting and casting is handled in the same manner as cathode copper. This may be manufacturing wastes of rod or bare wire. Energy requirement for recycle is approximately 1.2 MWh/t.
- *No. 2 scrap:* Unalloyed scrap, such as wire and plumbing tube and copper that is contaminated with other metals (such as might have been used in plating, soldering, or brazing), may be introduced into the primary smelter. Here it is refined as part of the primary production and re-enters the market in the form of high-grade cathode. Energy requirement for recycle is approximately 5.6 MWh/t.

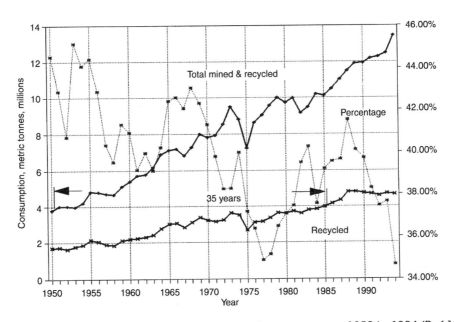

Fig. 6.4 World consumption of mined copper and copper scrap, 1950 to 1994 (Ref 124)

- *No. 3 scrap:* Low grade scrap of variable composition (10 to 88% copper) is smelted in various smelting furnaces and then fire- and electro-refined to high-grade cathode. This may be done by primary smelters or special smelters that specialize in the recovery of copper and other metals from scrap. Energy requirement for recycle is approximately 14 MWh/t.

- *No. 4 scrap:* Alloyed scrap, which consists mainly of brasses, bronzes, and cupronickels, is reintroduced into brass mills for the makeup of new alloys. Such batches of scrap are carefully sorted according to composition. Some compositional adjustments are done by air oxidation to remove aluminum, chromium, silicon, iron, and sometimes tin in the form of slag. However, care must be taken so as to avoid oxidizing valuable desirable metals such as zinc in brasses and tin in bronzes. Scrap, which is high in lead, is used in the manufacture of leaded-brass alloys that are valuable for their free-machining characteristics. The commercial use of such alloys provides a safe repository for lead from solders that would otherwise reappear in the environment. Energy requirement for recycle is approximately 2.3 MWh/t.

In the United States in 1994, 27% of the copper scrap consumed was processed by smelting and refining, 8% by melting and refining, and the remainder by direct use in alloys (Ref 128).

REFERENCES

1. E.J. Underwood, *Trace Elements in Human and Animal Nutrition,* 3rd ed., Academic Press, New York, NY, 1977, p 101–115
2. R.L. Parker, Composition of the Earths Crust, *Data of Geochemistry,* 6th ed., M. Fleischer, Ed., U.S. Geological Survey paper 440-D, 1967
3. V.M. Shorrocks and B.J. Alloway, "Copper in Plant, Animal and Human Nutrition," CDA Pub. TN35, Copper Development Association (UK), Potters Bar, Herts, 1985
4. V.M. Shorrocks, "Copper and Human Health—a Review," CDA Pub. TN34, Copper Development Association (UK), Potters Bar, Herts
5. G.E. Cartwright and M.M. Winthrobe, Copper Metabolism in Normal Subjects, *Am. J. Clin. Nutr.,* Vol 14, 1994, p 224–231
6. E. Nieboer and W.E. Sanford, *Essential, Toxic and Therapeutic Functions of Metals (Including Determinants of Reactivity),* Elsevier Science, New York, NY, 1985
7. H.J. Holtmeier, M. Kuhn, and I. Merz, On the Medical and Nutritional-Physiological Significance of Copper as a Trace Element for Humans (Über Die Medizinische und Ernährungs-Physiologische Bedeutung von Kupfer Als Spurenelement für den Menschen), *Metall.,* Vol 32, 1978
8. R. Uauy, personal communication, Institute for Nutrition and Food Technology, University of Chile, Santiago, Chile, Aug 1993

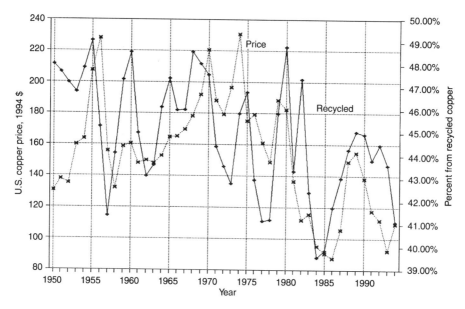

Fig. 6.5 Correlation between amount of copper recycled with the price of copper in the United States (Ref 128, 129)

9. M.J. Ettinger, in *Copper Metabolism and Diseases of Copper Metabolism in Copper Proteins and Copper Enzymes,* René Lontie, Ed., CRC Press, Boca Raton, FL, 1984

10. "Copper-Food-Health" ("Kupfer-Lebensmittel Gesundheit"), DKI Information Pub. Order No. 1.019, Deutsches Kupfer-Institut, Berlin

11. K. Sumino, K. Hayakawa, T. Shibata, and S. Kitamura, Heavy Metals in Normal Japanese Tissues, *Arch. Environ. Health,* Vol 30 (No. 10), 1975

12. L. Friberg, G.F. Nordberg, and V.B. Vouk, Ed., Chapter 10, *Specific Metals,* Vol II, *Handbook on the Toxicology of Metals,* 2nd ed., Elsevier Science, Amsterdam, 1986

13. G.W. Evans and W.T. Johnson, Copper Homeostasis, *Inflammatory Diseases and Copper,* J.R.J. Sorenson, Ed., Humana Press, Clifton, NJ, 1982, p 3–15

14. C.A. Owen, Biological Aspects of Copper, *Copper in Medicine and Biology Series,* Noyes Publications, Park Ridge, NJ, 1982, p 7–8

15. G.W. Evans, Copper Homeostasis in the Mammalian System, *Physio. Rev.,* Vol 53, 1973, p 535–570

16. L. Ryden and I. Bjork, Reinvestigation of Some Physico-Chemical and Chemical Properties of Human Ceruloplasmim, *Biochem.,* Vol 15, 1976, p 3411–3417

17. B.L. O'Dell and B.J. Campbell, Trace Elements and Metabolic Function, in *Comprehensive Biochemistry,* Vol 21, M. Florkin and E.H. Stotz, Ed., Elsevier, New York, NY, 1971, p 179–182, 191–203

18. U. Weser, C. Steinkühler, and G. Rotilio, Copper Complexes and Free Radicals, Chapter 3, Section C in *Handbook on Metal-Ligand Interactions in Biological Fluids 1,* G. Berthon, Ed., Marcel Dekker, New York, NY, 1995

19. O. Iwami, T. Watanabe, H. Nakatsuka, and M. Ikeda, Levels of Copper in the Urine of Women from Non-polluted Farming Regions of Japan, *The Science of the Total Environment,* Vol 108, 1991, p 191–203

20. K. Felix, H. Jürgen, and U. Weser, Cu(I)-Thionein Release from Copper-Loaded Yeast Cells, *Biol. Met.,* Vol 2, 1989, p 50–54

21. J.R.J. Sorensen, Copper Complexes Offer a Physiological Approach to Treatment of Chronic Diseases, *Progress in Medicinal Chemistry,* Vol 26, G.P. Ellis and G.B. West, Ed., Elsevier Science, New York, NY, 1989, p 39–568

22. R. Miesel, H.J. Hartmann, and U. Weser, Anti-Inflammatory Reactivity of Copper(I)-Thionein, *Inflammation,* Vol 14 (No. 5), 1990, p 471–483

23. R. Miesel and U. Weser, Anticarcinogenic Reactivity of Copper-Dischiffbases with Superoxide Dismutase like Activity, *Free Rad. Comms.,* Vol 11 (No. 1–3), 1990, p 39–51

24. *Federal Health Notice (Bundesgesundheitsblatt),* Vol 34 (No. 7), 1991

25. L.M. Klevay, Cholesterol and Cuprotropic Chemicals, *Trace Elements in Man and Animals—TEMA 5, Proc. of the Fifth Int. Symp. on Trace Elements in Man and Animals,* Commonwealth Agricultural Bureaux, Farnham Royal, Slough, UK, 1985, p 180–183

26. F.H. Morriss, Tracer Minerals, *Seminars in Perinatology,* Vol 3, 1979, p 367–379

27. P.A. Pleban, B.S. Numerof, and F.H. Wirth, Trace Element Metabolism in the Fetus and Neonate, *Clinics in Endocrinology and Metabolism,* Vol 14 (No. 3), 1985, p 545–566

28. E.R. Frisell, *Human Biochemistry,* McMillan, New York, NY, 1982

29. *National Academy of Sciences Food and Nutrition Board,* National Academy Press, Washington, D.C., 1980

30. J.R. Turnlund, W.R. Keyes, H.L. Anderson, and L.L. Acord, Copper absorption and retention in young men at three levels of dietary copper by use of the stable isotope 65Cu, *Am. J. Clin. Nutr.,* Vol 49, 1989, p 870–878

31. L.M. Klevay, An Appraisal of Current Human Copper Nurtiture, *Inflamatory Diseases and Copper,* J.R.J. Sorenson, Ed., Humana Press, Clifton, NJ, 1982

32. U. Krause, M. Anke, and W. Arnhold, Copper Intake of Humans in the GRD, *6th Int. Trace Element Symp.,* Vol 2, Universitätsbibliothek Hannover und Technische Informationsbibliothek, Leipzig, 1989

33. H.S. Wright, H.A. Guthrie, M.Q. Wang, and V. Bernardo, The 1987–1988 Nationwide Food Consumption Survey: An Update on the Nutrient Intake of Respondents, *Nutrition Today,* 1991, Vol 26 (No. 3), p 21–27

34. Nutrition (draft), *Copper,* World Health Organization, 11 Nov 1991

35. M.S. Brown, Physiologic Anemia of Infancy: Nutritional Factors and Abnormal States, in *Developmental and Neonatal Hematolgy,* J.A. Stockman and C. Pochedly, Ed., Raven Press, New York, NY, 1988, p 275–295

36. K.M. Hambridge and N.F. Krebs, Upper Limits of Zinc, Copper and Manganese in Infant Formulas, *J. Nutr.,* Vol 119, 1989, p 1861–1864

37. E.B. Grunau, The Significance of Copper in Human Nutrition (Die Bedeutung des Kupfers in der menschlichen Ernährung), *Metall.,* Vol 24 (No. 2), 1970, p 191–196

38. "Drinking Water and Health," The National Research Council, National Academy of Science, Washington, D.C., 1980, p 312–315

39. C. Glazier and B. Lonnerdal, Mineral Distribution in Human Milk and Cow's Milk after In-Vitro Proteolysis, *FASEB J.,* Vol 2, 1988, p 207

40. B. Lonnerdal, J.G. Bell, and C.L. Keen, Copper Absorption from Human Milk and Infant Formulas: Effects of Copper and Zinc Concentrations, in *Trace Elements in Man and Animals—TEMA 5, Proc. of the Fifth Int. Symp. on Trace Elements in Man and Animals,* Commonwealth Agricultural Bureaux, Farnham Royal, Slough, UK, 1985, p 312–315

41. S.J. Fomon, *Infant Nutrition,* W.B. Saunders Co., Philadelphia, PA, 1974, p 370

42. J.J. Strain, Milk Consumption, Lactose and Copper in the Aetiology of Ischaemic Heart Disease, *Medical Hypotheses,* Vol 25, 1988, p 99–101

43. J.G. Bell, C.L. Keen, and B. Lonnerdal, Effect of Infant Cereals on Zinc and Copper Absorption during Weaning, *Am. J. Dis. Child.,* Vol 141, 1987, p 1128–1132

44. H.H. Dieter, E. Meyer, and R. Möller, Kupfer-Vorkommen, Bedeutung und Nachweis, in *Drinking Water Regulations (Die Trinkwasserverordnung),* 3rd ed., K. Aurand et al., Ed., Erich Schmidt Verlag, Berlin, 1991, p 472–491

45. Drinking Water Regulations: Maximum contamination level goals and national primary drinking water regulations for lead and copper, *Fed. Regist.,* Vol 53 (No. 160), 1988, p 31516–31578

46. *Guidelines for Drinking Water Quality, Recommendations,* Vol 1 and 2, World Health Organization, Geneva, Switzerland, 1984

47. *Guidelines for Drinking Water Quality, Recommendations,* Vol 1, 2nd ed., World Health Organization, Geneva, Switzerland, 1993

48. I.M. Sayre, International Standards for Drinking Water, *J. AWWA,* Vol 80 (No. 1), 1980, p 53–60

49. L.M. Klevay, Ischemic Heart Disease: Toward a United Theory, in *Role of Copper in Lipid Metabolism,* K.Y. Lei and T.P. Carr, Ed., CRC Press, Boca Raton, FL, 1990, p 233–267

50. D.M. Danks, Disorder of copper transport, *The Metabolic Basis of Inherited Disease,* C. Scriver, A.L. Beaudet, W.S. Sly, and D. Valle, Ed., McGraw-Hill, New York, NY, 1989, p 1411–1431

51. D.M. Danks, Copper Deficiency in Humans, in *Biological Roles of Copper, Ciba Foundation Symposium,* Vol 79, Exerpta Medica, Amsterdam, Netherlands, 1980, p 209–221

52. M.S. Brown, *Physiologic Anemia of Infancy in Nutritional Factors and Abnormal States in Developmental and Neonatal Hematology,* J.A. Stockman III and C. Pochedly, Ed., Raven Press, New York, 1980

53. A. Cordano, J.M. Baertl, and G.C. Graham, Copper Deficiency in Infancy, *Pediatr.,* Vol 34, 1934, p 324–336

54. G.M. Reid, Sudden Infant Death Syndrome: Congenital Copper Deficiency, *Medical Hypotheses,* Vol 23, 1987, p 167–175

55. C.R. Paterson and K.G. Wormsley, Hypothesis: Schachman's Syndrome of Exocrine Pancreatic Insufficiency may be Caused by Neonatal Copper Deficiency, *Ann. Nutr. Metab.,* Vol 32, 1988, p 127–132

56. J.H. Breasted, *The Edwin Smith Papyrus,* The University of Chicago Press, Chicago, IL, 1930

57. H.H.A. Dollwet and J.R.J. Sorenson, Historic uses of copper compounds in medicine, *Trace Elements in Medicine,* Vol 2, 1985, p 80–87

58. Gastrointestinal Drugs, *Remington's Pharmaceutical Sciences,* Mack Publishing, 1970, p 806–807

59. R.S. Stein, D. Jenkins, and M.E. Korns, Death after Use of Cupric Sulfate as Emetic, *JAMA,* Vol 235 (No. 8), 1976, p 801

60. E. Sontz and J. Schwieger, The "green water" syndrome: copper-induced hemolysis and subsequent acute renal failure as consequence of a religious ritual, *Am. J. Med.,* Vol 98, 1995, p 311–315

61. J.R.J. Sorenson, Copper Complexes: A Physiologic Approach to Treatment of Chronic Diseases, *Comp. Therapy,* Vol 11 (No. 4), 1985, p 49–69

62. J.R. Stabel and J.W. Spears, Effect of Copper on Immune Function and Disease Resistance in Advances in Experimental Medicine and Biology 258, *Copper Bioavailability and Metabolism,* Plenum Press, New York, NY, 1989

63. S.M. Waler and G. Rolla, Comparison between plaque inhibiting effect of chlorheridine and aqueous solutions of copper and silver ions, *Scand. J. Dent. Res.,* Vol 90, 1982, p 131–133

64. *The Elder Pliny's Chapters on Chemical Subjects,* Part II, Translated by K.G. Bailey, Ed., 1932, p 25–55

65. L. Lewin, Investigations on Copper Workers, *Deutsch. Med. Wochnschr.,* Vol 26, 1900, p 689–694

66. P.O. Nicholas, Food Poisoning due to Copper in the Morning Tea, *The Lancet,* Vol 10, 1968, p 40–42

67. A.B. Semple, W.H. Parry, and D.E. Philips, Acute Copper Poisoning—An Outbreak Traced to Contaminated Water from a Corroded Geyser, *The Lancet,* Vol 2, 1960, p 700–701

68. J. Wyllie, Copper Poisoning at a Cocktail Party, *Am. J. Publ. Health,* Vol 47, 1957, p 617

69. J.H. LeVan and E.L. Perry, "Copper Poisoning on Shipboard," Pub. Health Report 76, U.S. Public Health Service, 1961, p 334

70. J.D. McDonald and R. Cook, Copper Poisoning Associated with a Dispensing Machine, *Enviro. Health,* Vol 82, 1974, p 153

71. A.K.R. Chowdbury, S. Ghosh, and D. Pal, Acute Copper Sulfate Poisoning, *J. Indian Med. Assn.,* Vol 36, 1961, p 330–336

72. I.H. Scheinberg, Copper Toxicity, in *Encyclopedia of Occupational Health,* Vol 1, International Labor Organization, Geneva, 1983, p 587

73. I.H. Scheinberg, personal communication, 1990, based on unpublished study for the FAO/WHO, 1967

74. W.H. Lyle, J.E. Payton, and M. Hui, Haemodialysis and Copper Fever, *The Lancet,* Vol 1, 19 June 1976, p 1324–1325

75. T.G. Villar, Vineyard Sprayer's Lung: Clinical Aspects, *Am. Rev. Respir. Dis.,* Vol 110, 1974, p 545–555

76. J.C. Pimentel and A.P. Menezes, Vineyard Sprayers' Lung: A New Occupational Disease, *Thorax,* Vol 24, 1969, p 678–688

77. L.M. Klevay, Hypercholesterolemia in Rats Produced by an Increase in the Ratio of Zinc to Copper Ingested, *Am. J. Clin. Nutr.,* Vol 26, 1973, p 1060

78. L.M. Klevay, L. Inman, L.K. Johnson, M. Lawler, J.R. Mahalko, D.B. Milne, H.C. Lukaski, W. Bolonchuk, and H.H. Sandstead, Increased Cholesteron in Plasma in a Young Man during experimental Copper Depletion, *Metabolism,* Vol 33, 1984, p 1112

79. S. Reiser, A. Powell, C.-Y. Yang, and J.J. Canary, Effect of Copper Intake on Blood Cholesterol and its Lipoprotein Distribution in Men, *Nutr. Rep. Int.*, 1987, Vol 36, p 551

80. I.H. Scheinberg and I. Sternlieb, *Wilson's Disease,* Vol XXIII, *Major Problems in Internal Medicine,* L.H. Smith, Ed., W.B. Saunders Co., Philadelphia, PA, 1984, p 16

81. S.A.K. Wilson, Progressive lenticular degeneration: a familial nervous disease associated with cirrhosis of the liver, *Brain,* Vol 34, 1912, p 295–507

82. J.N. Cumings, The copper and iron content of the brain and liver in normal and in hepatolenticular degeneration, *Brain,* Vol 71, 1948, p 410–415

83. R.E. Tanzi et al., The Wilson's disease gene is a copper transporting ATPase with homology to the Menkes disease gene, *Nature Genetics,* Vol 5, 1993, p 345–350

84. B. Sarjar, K. Lingertat-Walsh, and J.T.R. Clarke, Copper-histidine therapy for Menkes disease, *J. Pediatr.,* Vol 123, 1993, p 828–830

85. M.S. Tanner, Indian childhood cirrhosis, *Recent Advances in Paediatrics,* Vol 8, S.R. Meadow, Ed., Churchill Livingston, Edinburgh, Scotland, 1986, p 103–120

86. M. Adamson et al., Indian Childhood Cirrhosis in an American Child, *Gastroenterology,* Vol 102 (No. 5), 1992, p 1771–1777

87. P. Schramel, J. Müller, U. Meyer, M. Wei, and R.J. Eife, Nutritional Copper Intoxication in Three German Infants with Severe Liver Cell Damage (Features of Indian Childhood Cirrhosis), *J. Trace Elem. Electrolytes Health Dis. 2,* 1988, p 85–89

88. J. Müller, M. Wei, U. Weser, P. Schramel, B. Wiebecke, B. Belohradsky, and G. Hübner, Fatal copper storage disease of the liver in a German infant resembling Indian childhood cirrhosis, *Virchows Arch. A,* Vol 411, 1987, p 379–385

89. J. Müller, U. Meyer, B. Wiebecke, and G. Hübner, Copper storage disease of the liver and chronic dietary copper intoxication in two further German infants mimicking Indian childhood cirrhosis, *Path. Res. Pract.,* Vol 183, 1988, p 39–45

90. G. Maggiore et al., Idiopathic hepatic copper toxicosis in a child, *J. Pediatr. Gastro. Nutr.,* Vol 6, 1987, p 980–983

91. J. Walker-Smith and J. Blomfield, Wilson's disease or chronic copper poisoning?, *Archives of Disease in Childhood,* Vol 48, 1973, p 476–479

92. C.T. Lim, Wilson's disease in a 2-year old child, *J. Sing. Paed. Soc.,* Vol 81, 1979, p 99–103

93. Valencia, M.P. Gaboa, and J. Medina, Copper overload and cirrhosis in four Mexican children, *Lab. Invest.,* Vol 68 (No. 1), 1993

94. I.A. Aljajeh et al., Indian childhood cirrhosis-like disease in an Arab child: A brief report, *Virchows Arch.,* Vol 424, 1994, p 225–227

95. A.P. Narang, H. Mfsood, and G.M. Jan, Copper and Indian Childhood Cirrhosis (ICC), *Trace Elements in Medicine,* Vol 6 (No. 4), 1989, p 183

96. M.S. Tanner, A.H. Kantarjian, S.A. Bhave, and A.N. Pandit, Early Introduction of Copper Contaminated Milk Feeds as a Possible Cause of Indian Childhood Cirrhosis, *The Lancet,* Vol 2, 1983, p 992–995

97. B. Bhandari and S. Sharda, Indian Childhood Cirrhosis: A Search for Cause and Remedy, *Indian J. Pediatr.,* Vol 51, 1984, p 135–138

98. R.J. Eife, M. Müller-Hocker, M. Kellner, S. Arleth, A. Schmolz, M. Weiss, Ch. Bender-Gotze, P. Schramel, and H. Holtmann, Copper Water Tube as Basis for Immune Deficiency and Early Lethal Liver Cirrhosis (of the Indian Childhood Cirrhosis) (Kupferwasserleitungen als Ursache für Immundefizienz und Fruhkindiche Letale Leberzirrhose (vom Typ der Indian Childhood Cirrhosis)), *Padiat. Prax.,* Vol 36, 1987/88, p 69–76

99. I.H. Scheinberg and I. Sternlieb, Is Non-Indian Childhood Cirrhosis Caused by Excess Dietary Copper?, *The Lancet,* 1994, Vol 344, p 1002–1004

100. H.F. Müller, H. Freichtinger, H. Berger, and W. Müller, Non-Indian childhood cirrhosis type Tyrol— A review of cases from 1900–1995, *7th Int. Symp. on Wilson's/Menkes' Disease* (Vienna, Austria), 25–27 Aug 1995

101. B. Bhandari and A.N. Pandit, D-penicillamine in the therapy of Indian childhood cirrhosis, *Indian J. Pediatr.,* Vol 54, 1987, p 587–590

102. J.M. Walshe, Copper: Its Role in the Pathogenesis of Liver Disease, *Seminars in Liver Disease,* Vol 4 (No. 3), 1984, p 252–263

103. J.O. Nruigu, The global copper cycle, Part I, Ecological Cycling, *Copper in the Environment,* J.O. Nruigu, Ed., Wiley-Interscience, New York, NY, 1979

104. D.E. Baker, Copper, in *Heavy Metals in Soils,* B.J. Alloway, Ed., John Wiley and Sons, New York, NY, 1990, p 151–174

105. D.J.H. Phillips, The use of biological indicator organisms to monitor trace metal pollution in marine and estuarine environments—A review, *Environ. Pollut.,* Vol 13, 1977, p 281–318

106. A.G. Lewis, The Biological Importance of Copper—A Literature Review, Final Report INCRA Project 223, International Copper Association, Ltd., New York, NY, June 1990

107. J.-M. Martin and H.L. Windom, Present and future roles of ocean margins in regulating marine biogeochemical cycles of trace elements, *Ocean Margin Processes in Globlal Change, Dahlem Workshop,* R.C. Mantoura, J.-M. Martin, and R. Wollast, Ed., John Wiley and Sons, Chichester, NY, 1991

108. E.A. Boyle, Copper in Natural Waters, Ecological Cycling, Part I, *Copper in the Environment,* J.O. Nruigu, Ed., Wiley-Interscience, New York, NY, 1979, p 77–88

109. P.J. Wangersky, Biological Control of Trace Metal Residence Time and Speciation: A Review and Synthesis, *Mar. Chem.,* Vol 18, 1986, p 269–297

110. Z. Kazarac, M. Plavsic, B. Cosovic, and D. Vilicic, Interaction of Cadmium and Copper with W-Surface-Active Organic Matter and Complexing Agents by

Marine Phytoplankton, *Mar. Chem.,* Vol 26, 1989, p 313–330

111. F.L. Harrison, Effects of physicochemical form on copper availability to aquatic organisms, Aquatic Toxicology and Hazard Assessment, Seventh Symposium, ASTM special technical pub. 854, R.D. Cardwell, R. Perdy, and R.C. Bahner, Ed., American Society of Testing and Materials; Philadelphia, PA, 1985

112. D.C. Burrel and J.R. Schubel, Seagrass Ecosystems: A Scientific Perspective, Int. Workshop (Leiden, Netherlands), Oct 1973, C. McRoy, P. Helfferich, and C. Helfferich, Ed., New York, NY, 1977, p 195–232

113. G.W. Bryan, in *Marine Pollution,* R. Johnson, Ed., Academic Press, London, 1976, p 185–302

114. F. Consoli et al., Ed., *Guidelines for Life-Cycle Assessment: A "Code of Practice,"* Society of Environmenatal Toxicology and Chemistry, Pensacola, FL, 1993

115. H.H. Kellogg, Energy for Metal Production in the 21st Century in Productivity and Technology in the Metallurgical Industries, Vol 145, The Metallurgical Society, Warrendale, PA, 1989

116. K. Yoshiki-Gravelsins, J. Toguri, and R. Choo, Energy Consumption, *JOM,* Vol 46 (No. 5), 1994, p 15–20

117. S. Young and W. Vanderburg, Applying Environmental Life Cycle Analysis to Materials, *JOM,* Vol 46 (No. 4), 1994, p 22–27

118. L.O. Gullman, Energibehov vid framställning av aluminium och alternativa material, Technical report 823, Gränges Essem, 1974

119. D. Forrest and J. Szekely, Global Warming and the Primary Metals Industry, *JOM,* Vol 43 (No. 12), 1991, p 23–30

120. E. Arpaci and T. Vendura, Recycling von Kupferwerkstoffen, *Metall.,* Vol 47 (No. 4), 1993, p 340–345

121. C.L. Kusik and C.B. Kenahan, Energy Use Patterns for Metal Recycling, *Metal Resources and Energy,* Butterworth, Boston, MA, 1983, p 138

122. C.H. Pitt and M.E. Wadsworth, "An Assessment of Energy Requirements in Proven and New Copper Processes," Contract EM-78-S-07-1743, U.S. Department of Energy, University of Utah, Salt Lake City, UT, 31 Dec 1980

123. *Metal Statistics (Metallstatistik) (1984–1994),* 82nd ed., Metallgesellschaft AG, Frankfurt am Main, 1995

124. *Metal Statistics (Metallstatistik) 1993,* Metallgesellschaft AG, Frankfurt am Main, 1994

125. U. Hoffmann, "A Statistical Review of Metal Scrap and Residues," unpublished study of Environmental Issues Section, Commodities Division, United Nations Conference on Trade and Development, presented at the Global Workshop on the Applicability of Decision 2 of the Basel Convention (Dakar, Senegal), 15–17 March 1995

126. J.J. Jolly, J.F. Papp, and P. Plunkert, "Recycling-Nonferrous Metals," U.S. Department of the Interior, Bureau of Mines, 1991

127. "Guidelines for Nonferrous Scrap," Scrap specifications circular NF-88, Institute of Scrap Recycling Industries, 1988

128. "Copper Supply & Consumption—Annual Data, 1995," brochure, Copper Development Association Inc., New York, NY, 1995

129. American Bureau of Metal Statistics, Howell, NJ

Appendix I

World Copper Development and Information Centers

Africa
Copper Development Association (PTY) Ltd.
53 Randell Road
P.O. Box 14785
Wadeville, Germiston, 1422
South Africa
Tel: +27-11-824-3916
Fax: +27-11-824-3120
E-mail: copdevsa@icon.co.za

Asia
Copper Development Center - South East Asia
87 Beach Road
04-02 Chye Sing Building 189695
Singapore
Tel: +65-334-3828
Fax: +65-334-6221
E-mail: cdcsea@signet.com.sg

International Copper Association, Ltd.
Beijing Office
Canway Building, Room 1016
66 Nan Li Shi Lu
Beijing 100045
People's Republic of China
Tel: +86-10-6804-2450
Fax: +86-10-6802-0990
E-mail: hankqiu@public3.bta.net.cn

International Copper Association, Ltd.
Indian Office
892 Vishwa-Kutir, Ground Floor
Shankar Ghanekar Marg
Dadar (West) Mumbai 400 028
India
Tel: +91-22-432-9396

Fax: +91-22-432-9411
E-mail: icagmpmu@bom3.vsnl.net.in

Indian Copper Development Centre
27-B, Camac Street
Calcutta 700016
India
Tel: +91-33-247-5724
Fax: +91-33-247-2763
E-mail: icdc.india@smb.sprintrpg.ems.vsnl.net.in

Japan Copper Development Association
Konwa Building, 12-22, 1-Chome
Tsukiji, Chuo-ku
Tokyo 104
Japan
Tel: +81-3-3542-6631
Fax: +81-3-3542-6599

Australia
Copper Development Association of Australia, Ltd.
Westfield Towers, Level 7, Suite 1
100 William Street
Sydney, NSW 2011
Australia
Tel: +61-2-9380-2000
Fax: +61-2-9380-2666

Europe
Copper Benelux
Avenue de Tervurenlaan 168, b10
Brussels 1150
Belgium
Tel: +32-2-777-7090
Fax: +32-2-777-0099

Copper Development Association
Verulam Industrial Estate
224 London Road
St. Albans, Hertz AL1 1AQ
England
Tel: +44-1727-731205
Fax: +44-1727-731216

Centre d'Information du Cuivre Laitons
  et Alliages
30, Avenue de Messine
Paris 75008
France
Tel: +33-1-4225-2567
Fax: +33-1-4953-0382
E-mail: centre@cuivre.org

Deutsches Kupfer-Institut
Am Bonneshof 5
Dusseldorf 40474
Germany
Tel: +49-211-479-6300
Fax: +49-211-479-6310
E-mail: dki@dki.d.eunet.de

Hellenic Copper Development Institute
74 L. Riankour St.
Athens 151 23
Greece
Tel: +30-1-640-4407
Fax: +30-1-640-4463
E-mail: copper@netor.gr

Magyar Rezpiaci Kozpont, Kft
P.O. Box 62
Oktober 6u.7 (328.szoba)
Budapest 1241
Hungary
Tel: +36-1-266-6793
Fax: +36-1-266-6794
E-mail: hcpc.bp@ind.eunet.hu

Instituto Italiano del Rame
Via Corradino D'Ascanio 1
Milano 20142
Italy
Tel: +39-2-8930-1330
Fax: +39-2-8930-1513
E-mail: iir-servizi@wirenet.it

Polish Copper Promotion Centre
Pl. 1 Maja 1-2
Wroclaw 50-136
Poland

Tel: +48-71-781-2502
Fax: +48-71-781-2504
E-mail: copperpl@wroclaw.top.pl

Scandinavian Copper Development Association
Vasteras S-721-88
Sweden
Tel: +46-21-198-620
Fax: +46-21-198-035

Centro Español de Informacion del Cobre
Princessa, 79
Madrid 28008
Spain
Tel: +34-1-544-8451
Fax: +34-1-544-8884

Latin America
Argentina & Uruguay:

ICA Argentina
Reconquista 559
Piso 6B
Buenos Aires 1003
Argentina
Tel: +54-11-4314-1159
Fax: +54-11-4314-2322
E-mail: icaba@www.ofiservices.com

Procobre—Instituto Brasileiro Do Cobre
Avenida Brigadeiro Faria Lima 2128
Sao Paulo SP 01451-903
Brazil
Tel: +55-11-816-6383
Fax: +55-11-816-6383
E-mail: procobre@br.homeshopping.com.br

Procobre—Chile
Santo Domingo 551—2 Piso, Oficina 201
Santiago
Chile
Tel: +56-2-639-2874
Fax: +56-2-638-1200
E-mail: hsierra@reuna.cl

Procobre—Mexico
Av. Sonora No. 166, Piso 1
Col. Hipodromo
Mexico D.F. 06100
Mexico
Tel: +52-5-211-1201
Fax: +52-5-286 7723
E-mail: 74052.3111@compuserve.com

Procobre—Peru
Calle Francisco Grana No. 671
Magdalena
Lima 17
Peru,
Tel: +51-1-261-4067
Fax: + 51-1-460-1616
E-mail: rrproc@amauta.rcp.net.pe

North America
Canadian Copper and Brass Development
 Association
49 The Conway West, Suite 415

North York, ON M3C 3M9
Canada
Tel: +1-416-391-5599
Fax: +1-416-391-3823
E-mail: coppercanada@onramp.ca

Copper Development Association Inc.
260 Madison Avenue, 16th Floor
New York, NY 10016
USA
Tel: +1-212-251-7212
Fax: +1-212-251-7234
E-mail: rpayne@cda.copper.org

Appendix II

International Editorial Review Committee

(Affiliation at time of review)

Russell J. Taylor
CSIRO
Australia

Jean Farge
Noranda Minerals, Inc.
Canada

Alan G. Lewis
University of British Columbia
Canada

Ricardo Uauy
INTA-University of Chile
Chile

Ilpo Koppinen
Outokumpu, Oy
Finland

W. Durrschnabel
Weiland-Werke, AG
Germany

Otto von Franque
Deutsches Kupfer Institut
Germany

Shiro Sato
Sumotomo, Ltd.
Japan

Brett Wells
Industrial Research, Ltd.
New Zealand

Brian Alloway
University of London
United Kingdom

J. Hector Campbell
Consultant
United Kingdom

Peter Gilbert
Consultant
United Kingdom

E. Henry Chia
American Fine Wire Corp.
United States

Terry F. Bower
T. F. Bower & Assoc.
United States

Michael King
ASARCO, Inc.
United States

Michael Myers
Handy & Harmon
United States

Dale Peters
Copper Development Association Inc.
United States

Horace Pops
Essex Group, Inc.
United States

Thomas D. Schlabach
AT&T Bell Laboratories
United States

John R. J. Sorenson
Univ. of Arkansas for Medical Science
United States

Derek Tyler
Olin Corporation
United States

Joseph Winter
Olin Corporation
United States

Subject Index

F

G

Alloy Index